Advances in Experimental Medicine and Biology

More information about this series at http://www.springer.com/series/5584

Jen-Chywan Wang • Charles Harris
Editors

Glucocorticoid Signaling

From Molecules to Mice to Man

Springer

Editors
Jen-Chywan Wang, PhD
Department of Nutritional
Science & Toxicology
University of California Berkeley
Berkeley, CA, USA

Charles Harris, MD, PhD
Division of Endocrinology
Metabolism & Lipid Research
Department of Internal Medicine
Washington University School of Medicine
St. Louis, MO, USA

ISSN 0065-2598 ISSN 2214-8019 (electronic)
Advances in Experimental Medicine and Biology
ISBN 978-1-4939-2894-1 ISBN 978-1-4939-2895-8 (eBook)
DOI 10.1007/978-1-4939-2895-8

Library of Congress Control Number: 2015945236

Springer New York Heidelberg Dordrecht London

Printed on acid-free paper

Springer Science+Business Media LLC New York is part of Springer Science+Business Media (www.springer.com)

Acknowledgements

JCW would like to thank his parents and family for support. He would also like to thank his graduate and postdoctoral mentors, Daryl Granner and Keith Yamamoto, his former and current lab members, and many collaborators and friends for their immense help during his research career.

CH would like to thank his parents, Stewart and Helen; his wife, Audrey; and his son, Benjamin for their support. He would like to thank countless mentors who helped form his perspective as a physician scientist. Specifically, he would like to thank basic science mentors, Diane Lipscombe, Eugene M Johnson Jr., and Robert V. Farese Jr., and clinical mentor, James Blake Tyrrell, who introduced him to the clinical care of patients with the fascinating Cushing syndrome.

Contents

Contributors

Margarita Arango-Lievano, Ph.D. Inserm U1191, CNRS UMR5203, Institute for Functional Genomics, Montpellier, France

Sue C. Bodine, Ph.D. Department of Neurobiology, Physiology and Behavior, University of California, Davis, Davis, CA, USA

Department of Physiology and Membrane Biology, University of California, Davis, Davis, CA, USA

Andrew C.B. Cato, Ph.D. Institute of Toxicology and Genetics, Karlsruhe Institute of Technology, Karlsruhe, Baden Württemberg, Germany

Tzu-Chieh Chen, M.S. Department of Nutritional Sciences and Toxicology, University of California Berkeley, Berkeley, CA, USA

Baruch Frenkel, D.M.D., Ph.D. Department of Orthopaedic Surgery, Keck School of Medicine, Institute for Genetic Medicine, University of Southern California, Los Angeles, CA, USA

Department of Biochemistry and Molecular Biology, Keck School of Medicine, Institute for Genetic Medicine, University of Southern California, Los Angeles, CA, USA

Aaron R. Friedman, B.A., Ph.D. Department of Integrative Biology, University of California, Berkeley, Berkeley, CA, USA

J. David Furlow, Ph.D. Department of Neurobiology, Physiology and Behavior, University of California, Davis, Davis, CA, USA

Anna C. Geraghty, B.A. Department of Integrative Biology, University of California, Berkeley, CA, USA

Anthony N. Gerber, M.D., Ph.D. Department of Medicine, National Jewish Health, University of Colorado, Denver, Denver, CO, USA

Julie E. Goodwin, M.D. Department of Pediatrics, Yale University School of Medicine, New Haven, CT, USA

Daryl K. Granner, M.D. Department of Molecular Physiology and Biophysics, University of Iowa Carver College of Medicine, Iowa City, IA, USA

Department Internal Medicine, Fraternal Order of Eagles Diabetes Research Center, University of Iowa Carver College of Medicine, Iowa City, IA, USA

Roldan M. de Guia, Ph.D. Department of Molecular Metabolic Control, German Cancer Research Center, Center for Molecular Biology and University Hospital Heidelberg, Heidelberg, Germany

Charles Harris, M.D., Ph.D. Division of Endocrinology, Metabolism and Lipid Research, Department of Internal Medicine, Washington University School of Medicine, St. Louis, MO, USA

Stephan Herzig, Ph.D. Department of Molecular Metabolic Control, German Cancer Research Center, Center for Molecular Biology and University Hospital Heidelberg, Heidelberg, Germany

Freddy Jeanneteau, Ph.D. Inserm U1191, CNRS UMR5203, Institute for Functional Genomics, Montpellier, France

Daniela Kaufer, Ph.D. Department of Integrative Biology, University of California, Berkeley, CA, USA

Taiyi Kuo, Ph.D. Department of Nutritional Sciences and Toxicology, University of California Berkeley, Berkeley, CA, USA

W. Marcus Lambert, Ph.D. Department of Microbiology, New York University School of Medicine, New York, NY, USA

Allison McQueen, B.S. Department of Nutritional Sciences and Toxicology, University of California Berkeley, Berkeley, CA, USA

Sebastiaan H. Meijsing, Ph.D. Department of Computational Molecular Biology, Max Planck Institute for Molecular Biology, Berlin, Germany

Emmanuel Oppong, Ph.D. Institute of Toxicology and Genetics, Karlsruhe Institute of Technology, Karlsruhe, Baden Württemberg, Germany

Miles A. Pufall, M.S., Ph.D. Department of Biochemistry, Carver College of Medicine, Holden Comprehensive Cancer Center, Iowa City, IA, USA

Samuel A. Sakhai, Ph.D. Department of Psychology, University of California, Berkeley, CA, USA

Shawn N. Shirazi, B.A. Department of Integrative Biology, University of California, Berkeley, CA, USA

Jeremy W. Tomlinson, FRCP, Ph.D. Oxford Centre for Diabetes, Endocrinology & Metabolism, University of Oxford, Radcliffe Department of Medicine, Oxford, UK

Jan Tuckermann, Ph.D. Institute for Comparative Molecular Endocrinology, University of Ulm, Ulm, Germany

Jen-Chywan Wang, Ph.D. Department of Nutritional Sciences and Toxicology, University of California Berkeley, Berkeley, CA, USA

Wendy White, M.D. Department of Diabetes and Endocrinology, Eisenhower Medical Center, Rancho Mirage, CA, USA

Conor Woods, M.B. B.Ch., M.R.C.P. Department of Diabetes and Endocrinology, St. Vincent's University Hospital, Dublin, Leisnter, Ireland

Keith R. Yamamoto, Ph.D. Cellular & Molecular Pharmacology, University of California, San Francisco, San Francisco, CA, USA

Part I
Introductory Materials

Chapter 1
Regulatory Actions of Glucocorticoid Hormones: From Organisms to Mechanisms

Daryl K. Granner, Jen-Chywan Wang, and Keith R. Yamamoto

Abstract The history of glucocorticoid hormone research is an excellent example of "bedside to bench" investigation. It started with two very insightful clinical observations. Thomas Addison described the syndrome of what came to be known as adrenal hormone insufficiency and Harvey Cushing the syndrome of glucocorticoid hormone excess. These dramatic and life-threatening conditions spawned 150 years of active research that has involved many disciplines; indeed some of the fundamental observations of molecular biology are the result of this work. We have a fundamental knowledge of how glucocorticoids regulate gene transcription, their major effect. The challenge facing current and future investigators is to discern how to use this information to make these powerful therapeutic agents safer and more effective.

Dedication We dedicate this chapter to Gordon M. Tomkins. Gordon was a visionary who, after direct exposure to the Paris bacterial genetics group in the early 1960s, quite clearly foresaw the field of mammalian gene regulation. He was one of the founders of the discipline now known as Molecular Endocrinology. Most importantly, as regards the topic of this book, his scientific passion was glucocorticoid action. A generation of young scientists was fortunate to spend time in his laboratory; many others were influenced by his writings, entertaining lectures and the informal talks he gave during his many visits to universities and research institutes. Gordon was a direct mentor to two of us, D.K.G. and K.R.Y., and a second generation mentor to J.-C.W.

D.K. Granner, M.D. (✉)
Department of Molecular Physiology and Biophysics, University of Iowa Carver College of Medicine, 169 Newton Road, 4312 PBDB, Iowa City, IA 52242, USA

Department Internal Medicine, Fraternal Order of Eagles Diabetes Research Center, University of Iowa Carver College of Medicine, 169 Newton Road, 4312 PBDB, Iowa City, IA 52242, USA

Vanderbilt University School of Medicine, Nashville, TN, USA
e-mail: daryl-granner@uiowa.edu

J.-C. Wang, Ph.D.
Department of Nutritional Sciences and Toxicology, University of California Berkeley, 119 Morgan Hall, Berkeley, CA 94720-3104, USA
e-mail: walwang@berkeley.edu

K.R. Yamamoto, Ph.D.
Cellular & Molecular Pharmacology, University of California, San Francisco, 600 16th Street, Genentech Hall S572D, Campus Box 2280, San Francisco, CA 94143-2280, USA
e-mail: yamamoto@ucsf.edu

J.-C. Wang, C. Harris (eds.), *Glucocorticoid Signaling*, Advances in Experimental Medicine and Biology 872, DOI 10.1007/978-1-4939-2895-8_1

Keywords Glucocorticoids • Gene expression • Nuclear receptors • Transcription factors • DNA elements

The survival of multicellular organisms requires rapid and efficient adaptation to an ever-changing external environment. Many mechanisms have evolved to ensure the effective coupling of external cues to internally generated signals that affect

Table 1.1 The pleiotropic effects of glucocorticoid hormones

Effects on intermediary metabolism
1. Increase glucose production by: (a) increasing the delivery of amino acids and glycerol (the gluconeogenic substrates) from peripheral tissues; (b) increasing the rate of gluconeogenesis by increasing the amount and activity of several key enzymes; and (c) "permitting" other metabolic reactions to operate at maximal rates
2. Inhibiting the uptake of glucose by tissues excepting the nervous system, heart and red blood cells
3. Increase hepatic glycogen deposition by promoting the activation of glycogen synthase
4. Promote lipolysis, but can cause lipogenesis in some sites (face and trunk) especially at higher than physiologic levels
5. Promote protein metabolism. This is an anabolic effect, particularly in liver, at physiologic levels. Can be catabolic in certain conditions as a means of supplying amino acids for gluconeogenesis
Effects on host mechanisms
1. Suppress the immune response. These hormones cause a species- and cell type-specific lysis of lymphocytes
2. Suppress the inflammatory response by: (a) decreasing the number of circulating leukocytes and the migration of tissue leukocytes; (b) inhibiting fibroblast proliferation; (c) inducing lipocortins, which by inhibiting phospholipase A2, blunt the production of the potent anti-inflammatory prostaglandins and leukotrienes; and (d) inhibit the action of NF-kB by increasing synthesis of the inhibitor IkB; GR tethering to p65 subunit of NF-kB, competing with Pol II CTD kinase P-TEFb association with p65
Effects on development/differentiation
1. Development of the lung, including the production of surfactants and an inducible sodium channel
2. Development of neural crest-derived chromaffin cells (catecholamine production) in the adrenal medulla. The delivery of high concentrations of glucocorticoids to the medulla through the intra-adrenal portal system allows for the induction of phenylethanolamine-N-methyltransferase, which catalyzes the conversion of norepinephrine to epinephrine
Other effects
1. Necessary with catecholamines for maintenance of normal blood pressure and cardiac output
2. Required for maintenance of normal water and electrolyte balance, perhaps by restraining ADH release (H_2O) and by increasing angiotensinogen ($Na+$). These effects contribute to the effect on blood pressure
3. Necessary, with the hormones of the adrenal medulla, allowing the organism to respond to stress

complex processes such as the response to food deprivation, exercise, stress, trauma, and infection. A constant supply of energy is of central importance in all of these functions. Glucocorticoid hormones are named for the central role they play in glucose homeostasis, which is an important source of energy for all tissues; in particular, the brain depends almost entirely on glucose metabolism.

Glucocorticoids are also of interest because of the direct and indirect effects they have on a large number of apparently diverse physiologic and biochemical processes (Table 1.1). And, they play an important therapeutic role as life-saving replacement treatment in adrenal insufficiency (Addison's disease; Chap. 4), as a key component in the therapy of certain malignancies (Chap. 14), as an immunosuppressant in transplantation and autoimmune diseases and as an anti-inflammatory agent (Chap. 9). It is amazing that this class of hormones, which are small (cortisol is 362 Da), relatively simple derivatives of cholesterol, can accomplish so much. The current understanding of how these hormones work started first with a description of the adrenal glands, then with a remarkable clinical observation that led to more than 150 years of research that has employed, and also helped formulate, many of the basic principles of physiology, biochemistry and molecular biology. This introductory chapter will trace the discoveries that have led to our current understanding of glucocorticoid action, and will attempt to set the stage for the succeeding chapters that delve more deeply into the different processes affected by these interesting hormones.

The Importance of the Adrenal Glands Is Established

The story begins in 1563 when Eustachius described two small organs, located near and just above (ad-) the kidneys (renal) in humans. Adrenal glands, as they were subsequently named, are found across vertebrates, but their role in biology remained unknown until the mid-1800s when Kolliker placed the adrenals among the group of ductless glands that communicate only with the blood system. Adrenals were shown to consist of two discrete areas: a firm outer layer, or cortex, and a soft, spongy inner layer, the medulla. The function of each of these areas was unknown, and the concept of hormones, molecules synthesized in one organ and transported through the vascular system to one or more distant target organs, was not formulated until the studies of the control of secretin secretion were reported by Bayliss and Starling in 1901–1902 [1, 2].

Thomas Addison made a brilliant clinical-pathologic observation in 1855 that really launched this field of research. He described a syndrome that included intense skin pigmentation, weakness, feeble pulse, and general debility with a fatal outcome in a group of 11 patients, all of whom had small, diseased or absent adrenal glands. The title of his monograph, published posthumously, was "On the Constitutional and Local Effects of Disease of the Suprarenal Capsule" [3]. Curiously, also in 1855, Claude Bernard first described his studies on the glycogenic function of the liver whereby this organ "prepares sugar at the expense of the elements of the blood passing through it" [4], a concept of central importance once the full manifestations of adrenal insufficiency were known. Bernard also first wrote about the importance of maintaining a "stable internal environment", which led to the concept of "homeostasis", a term first used by Cannon in 1926 [5]. Although presented in the same year, the observations by Addison and Bernard were not connected for about 75 years.

Brown-Séquard showed that adrenalectomy resulted in the death of experimental animals [6], as was observed in humans affected with what had become known as Addison's disease. Attempts to treat persons (or experimental animals subjected to adrenalectomy) commenced in the latter part of the nineteenth century, even though the active agent(s) in the adrenals was unknown. Aqueous extracts of the entire adrenal often had effects on heart rate and blood pressure, but did not resolve the life-threatening symptoms of Addison's disease. With clarification of the medullary source of adrenaline (epinephrine) and its subsequent purification and synthesis, the separate role of this glandular structure, and its role in the sympathetic nervous system, became apparent. The search for the critical adrenal cortical factor "cortin" became the focal point of interest during this time and continued during the early part of the twentieth century. Cortin was suspected of being a hormone, a concept which Starling had by then defined [7], but the structure of cortin was a complete mystery.

The Active Adrenal Cortical Hormones Are Identified and Synthesized

The identification of the active hormone proved to be an arduous task. A major breakthrough came when organic solvents were used in place of water to make adrenal cortical extracts; the subsequent demonstration of the lipophilic nature of steroids explained this important discovery. These extracts led to the survival of adrenalectomized animals and in the improvement of patients with Addison's disease [8, 9]. The race to discovery continued, and in the 1920s and 1930s many groups, most notably those of Kendall and Reichstein, developed techniques for the crystallization of adrenal corticosteroids and the subsequent synthesis of many of these molecules [10, 11]. The significant difficulties encountered in this work were understood much later when it was realized that there are dozens of steroids in the adrenal cortex, most of which are intermediates in the synthesis of the active hormones from cholesterol. The problem was made even more difficult by the fact that many of these molecules co-purified and co-crystallized. To complicate matters even further, very small molecular changes had large effects on activity, thus the chemical synthesis had to be very precise [10, 11]. These obstacles were eventually overcome. Cortisone, synthesized by Sarett in 1947 [12], was the first glucocorticoid to be extensively used clinically.

The Metabolic and Therapeutic Effects of Glucocorticoids Are First Explored

By the early part of the twentieth century, when various clinical parameters could be reliably quantitated, the syndrome of Addison's disease was further expanded to include metabolic and renal components. The inability to maintain glucose homeostasis, coupled with extreme insulin sensitivity (hypoglycemia), was a serious problem in patients with Addison's disease. In time this was attributed to a reduced ability of the liver to convert amino acids or glycerol into glucose (impaired gluconeogenesis) or to convert glycogen into glucose, a validation of the early ideas formulated by Claude Bernard. These observations also led to studies of the hormonal regulation of these processes, as is discussed in detail below. The renal manifestations include excessive retention of potassium and diuresis associated with excessive loss of sodium in the urine, which contributes to the severe hypotension noted in these patients. The eventual availability of molecules of known structure led to the categorization of adrenal corticosteroids into glucocorticoids and mineralocorticoids, according to their predominant, but not exclusive, action. Hydrocortisone (cortisol) and corticosterone are the major glucocorticoids in humans and rodents, respectively; aldosterone is the major mineralocorticoid. Persons with primary adrenal insufficiency are now usually treated with both a glucocorticoid and a mineralocorticoid.

The production of adrenal androgens (mostly androstenedione) was defined later, based in part on the serendipitous synthesis of steroids with androgenic activity in the course of efforts to make glucocorticoids, and on subsequent clinical observations, which helped explain why some persons with adrenal hyperplasia develop masculinization. The structures of the primary adrenal hormones are shown in Fig. 1.1.

A second clinical observation played a major role in advancing this research field. Harvey Cushing, in 1912, described a syndrome in which an adenoma of the anterior pituitary gland caused hypertrophy of the adrenal glands and a characteristic set of clinical signs and symptoms [13]. The manifestations of Cushing's disease were, in many ways, the opposite of those seen in Addison's disease: hypertension, fluid retention, weight gain, obesity with ectopic fat deposition, hyperglycemia with insulin resistance, masculinization, and thin friable skin with bruising, among others. Now known to be due to excessive, unsuppressed release of adrenocorticotropic hormone (ACTH) from a tumor of the basophilic cells in the anterior pituitary (or corticotrophin releasing hormone (CRH) from the hypothalamus), the condition is mimicked when excessive amounts of exogenous glucocorticoids are administered therapeutically, or when primary adrenal tumors overproduce the hormones, so-called Cushing's syndrome. Philip Hench, a colleague of Kendall at the Mayo Clinic, first used a glucocorticoid (cortisone) to treat persons with rheumatoid arthritis [14]. This treatment had remarkable beneficial effects and led to its subsequent use as an anti-inflammatory/immunosuppressant agent. Unfortunately, clinical

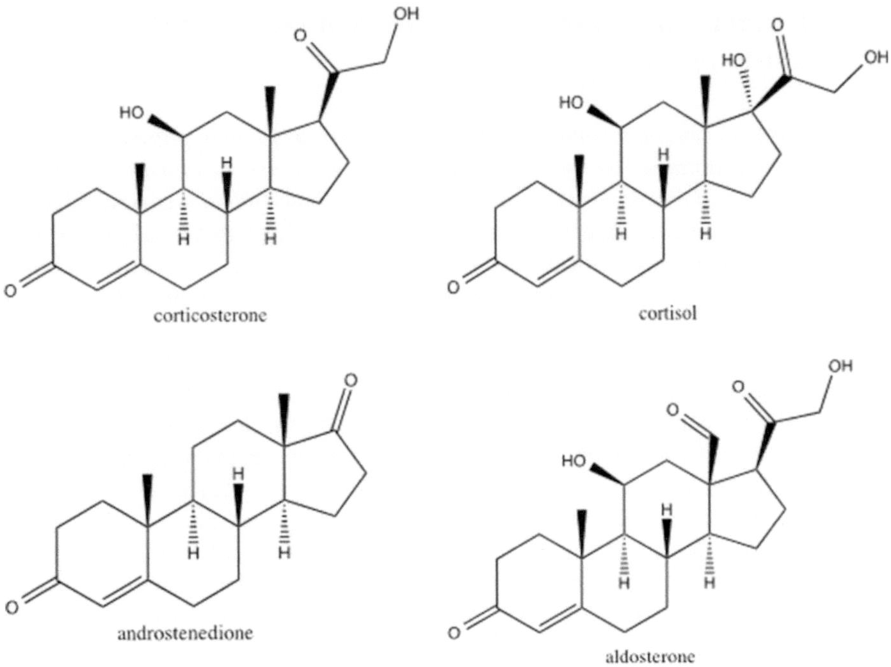

Fig. 1.1 Basic structures and trivial names of major hormones of the adrenal cortex

remission requires long-term use at high doses, and this often leads to the serious complication of Cushing's syndrome with its attendant, devastating complications.

These studies, in collection, are an excellent example of how early endocrine research evolved. The general sequence of discovery was: (1) ablate a gland of interest; (2) observe and quantitate the physiologic and biochemical events that ensue; (3) isolate, purify and synthesize the putative hormone; and (4) prove the role of the latter by replacing the pure hormone and restoring normal homeostasis. The next challenge was to elucidate how glucocorticoids accomplish all these physiologic and pathophysiologic events.

Once the physiologic effects of glucocorticoid deficiency or excess were defined, and pure molecules were readily available, the question became "what exactly do these hormones do and how do they do it?" Emphasis was placed early on the regulation of glucose metabolism because of the notable effects glucocorticoids appeared to have on this process and because of a considerable body of relevant knowledge which had been developing contemporaneously. Important concepts such as: (a) enzymes have a unique structure, (b) precise metabolic pathways exist and they are coordinated, (c) proteins turnover independent of cell replication, and (d) enzyme adaptation (induction and repression) were all applied to the study of glucocorticoid hormone action.

Enzymes Are Defined and the Metabolic Pathways Are Elucidated

The concept that the conversion of foodstuffs into cellular constituents, or into energy, involved an orderly progression of discrete biochemical reactions, each catalyzed by an enzyme, first began attracting attention as early as 1752 when de Réaumur showed that gastric secretions could digest meat [15]. Others demonstrated that saliva could convert starch to sugar, and in 1833 diastase (amylases) was described [16]. This work is associated with the subsequent convention of adding "-ase" to the name of an enzyme, although the word "enzyme" was apparently not used until 1877 [17]. The nature and function of enzymes was unknown. The prevailing theory, based on the fermentation of sugar into alcohol, was that the process required a living cell. When Buchner showed, in 1907, that yeast extracts accomplish the same purpose, the cell-free action of enzymes was established [18].

Investigators had shown that enzymatic activity was associated with proteins, but had not proven that a protein, per se, was capable of this action. In 1926 Sumner purified and crystallized the protein urease; he repeated this with catalase in 1937 [19]. Northrup and Stanley, who studied pepsin, trypsin and chymotrypsin, among several other proteins, presented further proof that enzymes are proteins [20, 21]. And importantly, as Northrop stated, "the enzymatic activity is a property of the protein itself and not due to a non-protein impurity".

Studies by hundreds of investigators, which started even before the exact nature of enzymes was established, led to the construction of the "metabolic chart". The chart presents a picture (although details are still being added) of the complex, interacting metabolic events that occur within a cell. This is truly one of the great scientific accomplishments of the twentieth century, especially when one considers that a great many of the enzymes on this chart were purified using virtually none of the overexpression, chromatographic and affinity techniques available today; very tedious, nonspecific techniques (e.g., salt fractionation, alcohol and acetone fractionation) were among those commonly employed.

Knowledge of the metabolic pathways made it possible to finally understand that the conversion of foodstuffs into cellular constituents or energy involves an orderly progression of discrete biochemical reactions, each catalyzed by an enzyme. This begged the question of whether these processes are regulated, particularly in view of observations such as those that suggested glucocorticoids might play a role in one or more of these processes and thereby account for some of the manifestations of Addison's disease. Subsequent experiments in this area focused on the coordination and regulation of these complex pathways. A number of investigators formulated the hypothesis that hormones might provide the means of metabolic coordination [22]. But another important concept had to be developed before this hypothesis could be tested.

The Concept of Differential Turnover of Cellular Constituents Is Established

Until the late 1930s cells and cellular constituents were thought to turn over at the same rate. The stable components of a cell somehow replicated themselves and were equally distributed into the two daughter cells with cell division. This was a conservative mechanism, but it would not allow for adaptation based on changing the amount of a cellular constituent (protein) independent of cell replication. The pioneering work by Schoenheimer, Shemin, Rittenberg, and others, who were among the first to use isotopes to address biologic questions, showed convincingly that various lipids and proteins have turnover rates different from that of the cell itself, and have different turnover rates within a given cell [23–25]. For example, hepatocytes were found to have a small component of proteins (~3 %) with a $t_{1/2}$ of ~140 days (about the $t_{1/2}$ of the cell) and a much larger component consisting of two subclasses with $t_{1/2}$ values of 4.5 and 12 days [24].

The concept that turnover occurs, and is dynamic, was a major advance, as subtle adjustments of an active synthesis/degradation process could allow for flexible, rapid and accurate adaptive responses to the challenges of an acutely changing external and internal environment. This observation offered the possibility that the enzymatic reactions that govern a certain metabolic pathway could be regulated by changes of the amount and/or activity of one or more enzymes. These changes could be facilitated by intercellular signals (e.g., glucocorticoid hormones) that would allow a cell to respond to various metabolic and environmental challenges.

Enzymes Show Adaptive Changes and Glucocorticoids Regulate Gene Expression

The observation that cellular components turnover led directly to the concept that organisms could show adaptive responses to their environment. Remarkable changes in the amount of enzymes in microorganisms had been demonstrated in the 1940s, generally in response to alterations of substrate concentration [26–28]. This phenomenon was demonstrated in mammalian cells when tryptophan was shown to induce a six to eightfold increase in tryptophan oxygenase (TO) [29]. This effect, which appeared to be an example of substrate induction similar to that observed in bacteria, was rapid and self-limited. However, subsequent studies showed that other amino acids and compounds, which were not substrates of TO, also increase activity of the enzyme. All the substances tested appeared to stress the animals, and the response only occurred in those with an intact pituitary-adrenal axis [30]. Selye had proposed that this axis was involved in the stress response [31], which led to the hypothesis that the adrenal cortical hormones were responsible for the induction of

Table 1.2 The timetable of glucocorticoid induction of hepatic enzymes

Function	Enzyme		
	TAT	TO	PEPCK
Increased enzyme activity	1957	1954	1963
Increased enzyme amount	1962	1962	x
Increased enzyme synthesis	1962	1965	1975
Increased mRNA activity	1976	1973	1977
Increased mRNA amount	1983	1982	1983
Increased transcription	1987	1983	1983

TO. This concept was validated shortly thereafter when a purified glucocorticoid (see above) administered to adrenalectomized rats resulted in the induction of TO [32]. By 1956, many examples of changes of enzyme activity in response to adrenalectomy, thyroidectomy, hypophysectomy, and diabetes were known [33]. When glucocorticoids were also found to induce tyrosine aminotransferase (TAT) [34], the era of research on the hormonal regulation of enzyme induction by these hormones was well underway. The timetable of progress is shown in Table 1.2.

Knox and Mahler observed that the changes in TO activity could result from a change in the amount of enzyme rather than to a change in the catalytic activity of the protein—"the production of a potential increase in metabolism by increasing the amount of enzyme, but without affecting the catalytic activity of a given amount of enzyme may therefore be a general means of metabolic regulation" [29]. This statement seems obvious today but it was presented when much of the effort to determine the mechanism of action of steroid hormones was confined to cell free systems, since a prevailing idea was that they acted to alter catalytic activity by serving as energy transducers, enzyme cofactors or allosteric regulators.

Proof that increased activity of an enzyme was due to an increased amount of the protein required a purified protein, which was used to produce a specific antibody that could then be used to selectively immunoprecipitate the radioactively labeled protein. Such evidence was obtained for TAT [35], and for TO [36]. The concept of turnover implied that an increased amount of protein could result from an increased rate of synthesis, from a decreased rate of degradation, or from some combination of these processes. The theoretical basis for such experiments was defined by Schimke [37] who then showed that tryptophan slowed hepatic TO degradation while hydrocortisone enhanced TO synthesis [38]. By contrast, rat liver TAT synthesis was enhanced by hydrocortisone without an effect on degradation [39].

Tryptophan and tyrosine are not significant gluconeogenic substrates, so the regulation of TO and TAT served mostly as model systems for studies of enzyme regulation. By contrast, phosphoenolpyruvate carboxykinase (PEPCK), which catalyzes the conversion of oxaloacetate (from pyruvate) to phosphoenolpyruvate, is a major gluconeogenic enzyme. Thus, the demonstration of the induction of PEPCK by glucocorticoids was especially significant [40].

Glucocorticoids Regulate the Transcription of Specific Genes

An enormous conceptual advance occurred as a result of the studies of the regulation of the E. coli lac operon by Jacob and Monod [41]. Two major concepts arose from these, and subsequent, experiments. The first was that genes consist of structural and regulatory components. The second was the "information flow" hypothesis, which states that genes direct the synthesis of a messenger RNA (mRNA), which then directs the synthesis of the corresponding protein. The studies of the *lac* operon gave immediate direction to studies of regulation of gene expression in eukaryotic cells, even though it would take many years to develop the techniques necessary for performing these investigations.

Gordon Tomkins was one of the first persons to propose that the approach Jacob and Monod used to study gene regulation in prokaryotes might be applied to the analysis of the hormonal regulation of enzyme synthesis in cultured mammalian cells. This idea was not enthusiastically accepted at first, to say the least, but the demonstration that glucocorticoids induce TAT in cultured H4IIE and HTC hepatoma cells was a game-changer [42, 43]. A subsequent study showed that the basic observations of the induction of TAT in liver were replicated in HTC cells [44], and it soon became clear that the ability to precisely control the hormonal environment, select for mutants, synchronize cells, adapt them to growth as single cells in suspension, etc. offered a system amenable to the molecular biology studies that were to follow.

The conceptual framework used to analyze how glucocorticoids affect enzyme synthesis was applied to the analysis of the role of mRNA in this process. Changes of the rate of synthesis of a specific enzyme could result from changes of the translational activity of a fixed amount of mRNA, or a changed amount of mRNA from either an alteration of mRNA stability or of its rate of synthesis (transcription). Unfortunately, many of the glucocorticoid-regulated enzymes exist in very small amount; TO, TAT and PEPCK are each present at ≤ 1 % of cytosolic protein. Later studies showed, as expected, that the corresponding basal levels of the mRNAs for these enzymes comprise less than 0.1 % of total poly A+ RNA in hepatocytes [45]. The basal rate of PEPCK gene transcription is 0.01 % of the total [46]. Because procedures had to be developed to account for this lack of abundance (there were no commercially available reagents or kits, DNA had to be sequenced by manual procedures, none of the genes had been isolated or characterized, etc.), it took many years before specific assays of mRNA activity, amount, or transcription, measured by various cell-free translation systems or by hybridization to specific cDNA probes, were established. In the meantime, results obtained from experiments using various inhibitors of RNA synthesis were used to make inferences about the mediating role of mRNA. Since many of these compounds, most notably actinomycin D, inhibited glucocorticoid induction of TAT [43, 47], TO [47], and PEPCK [40], it was assumed that ongoing mRNA synthesis was necessary for the response [45].

The assumption that hormones regulate mRNA synthesis was directly tested in more than a decade of research starting in the early 1970s. mRNA activity was assessed by translating total nuclear poly A+ RNA in a wheat germ or reticulocyte lysate translation system. The amount of radiolabel incorporated into specific protein (again detected by immunoprecipitation) was compared to that in the total protein synthesized. The amount of mRNA was assessed by hybridization to specific cDNA probes once those were available. Transcription assays, much easier to perform in cultured cells, used longer cDNA probes to detect the amount of radioisotope incorporated into a specific mRNA. The progression illustrated in Table 1.2 shows that the glucocorticoid-induced increase of enzyme activity is accomplished through an enhanced rate of transcription of the TO [48], TAT [49] and PEPCK [46] genes, measured using an elongation assay. It is noteworthy that 20 years elapsed between first concept and the final accomplishment, which underscores the difficulties encountered in performing these experiments.

Glucocorticoids often do not act in isolation on important metabolic processes. An example is hepatic gluconeogenesis. This process is increased by glucocorticoids and glucagon (cAMP) and decreased by insulin. It thus is of interest to note that each of these hormones affect PEPCK activity in parallel with their effect on gluconeogenesis. The changes in PEPCK activity caused by the hormones are due to changes of specific mRNA amount, which are, in turn, directly proportional to the rate of transcription of the PEPCK gene [46, 50]. The basal rate of transcription of the PEPCK gene is ~100 ppm of total RNA synthesized; glucocorticoids and cAMP each increase this rate several fold and their effects are additive. Insulin inhibits basal and induced transcription, and the insulin effect is dominant [46]. All of these actions could be studied in the H4IIE rat hepatoma cell line, thus, as is discussed below, a system existed for analyzing how several hormones interact at the level of a single gene to regulate an important metabolic function.

A Specific Receptor Mediates the Action of Glucocorticoids

While studies of the effect of glucocorticoids on gene expression were progressing, several investigators were establishing the physiologic and biochemical parameters of what became known as the glucocorticoid hormone signal transduction pathway. The following is a brief summary of these important observations. Cortisol, synthesized in the fasciculata and reticularis zones of the adrenal cortex, is secreted directly into plasma. Cortisol exists in two forms in plasma: (1) bound to transcortin (corticosteroid-binding globulin; CBG) and (2) as free, unbound, cortisol. The latter, which is a small percentage of the total circulating hormone (<10 %), is the biologically active form. Free cortisol readily crosses the plasma membrane, where it initiates action by binding, with high affinity, to a specific glucocorticoid receptor (GR).

The concept of specific receptors for steroid hormones began with the work of Talwar et al., who showed that estradiol binds with high affinity to a uterine cytosolic substance [51], *later shown to be a protein.* Definitive evidence for a GR was presented a few years later, as summarized in comprehensive reviews [52, 53]. Early biochemical studies, performed before the purified GR was available, revealed several key points: (1) GR is located in the cytosol in the absence of ligand. (2) The ligand·GR interaction is of high affinity and is rapidly reversible. (3) The binding of ligands to GR correlates well with the biologic activity of the ligand, and the biologic effect disappears quickly following removal of the ligand. Cortisol, corticosterone and aldosterone all bind to the GR with high affinity, but in humans the dominant glucocorticoid is cortisol because of its much greater plasma concentration. (4) The absence of GR in a cell results in a loss of biologic activity of the ligand. For example, lymphocytes that lack GR are resistant to the cell-killing effects of glucocorticoids. (5) An activation process results in the transfer of GR into the nucleus of target cells. GR is associated with one or more chaperones (i.e., hsp90) in the absence of ligand. This large, multimeric complex dissociates upon ligand binding, and the ligand·GR complex can then translocate into the nucleus. (6) The DNA component of chromatin binds GR. This observation is based on direct binding studies using DNA, and on the observation that DNase treatment of chromatin reduces GR binding. Actually, GR binding to DNA exceeds that to chromatin, which was early evidence that chromatin can occlude transcription factor binding to DNA. (7) All tissues known to respond to glucocorticoids exhibit this pattern of GR behavior. Extension of these studies depended on the availability of purified GR, and the subsequent isolation of a cDNA specific for the GR.

The Glucocorticoid Receptor Is a Transcriptional Regulatory Factor

The GR is not an abundant protein, is relatively unstable, and forms complexes with several other proteins, so its purification is difficult. In the late 1970s Gustafsson and colleagues reported substantial success in purifying the GR [54, 55]. A single polypeptide of ~90 kDa, with hormone binding properties virtually identical to those of crude cytosol preparations, was obtained. The sequence of GR, deduced from the open reading frames of cDNAs from human [56], rat [57] and mouse [58], show that this molecule has been highly conserved across evolution.

Evidence accumulated in the 1970s showed that glucocorticoid receptors bind to DNA [52, 59, 60], but selective binding, in a region likely to affect gene transcription, was not possible until the early 1980s when purified GR became available. Several studies suggested that the possible association of binding and function might be established by analyzing the glucocorticoid-enhanced production of mammary tumor virus (MTV) [61], an effect due to enhanced transcription of chromosomally-integrated MTV proviral DNA [62, 63]. The virus contains all the information required for glucocorticoid action in the MTV long-terminal repeat (LTR) segment [64–66], but the regions of the LTR involved, and the specifics of how the hormone accomplishes this induction, had not been established.

The specific binding of purified glucocorticoid receptor to DNA was demonstrated by Payvar and colleagues in 1981 [67]. A number of GR binding sequences (GBSs) in MTV DNA were identified in this and subsequent investigations [67–71]. The identification of specific receptor-DNA interactions *in vitro* allowed investigators to next test whether they were sufficient to regulate transcription and whether they were independent from the DNA region in the LTR known to be required for transcription initiation.

The strategy employed to demonstrate that ligand-receptor binding to specific DNA segments could affect the transcription of a specific gene involved the construction of a reporter gene, based in this case on the thymidine kinase (*tk*) gene, which has a promoter that is not responsive to glucocorticoids. A 340 bp sequence of MTV LTR DNA previously shown to contain several sites that bind the ligand-GR complex [67–71], and known to be lacking in transcription initiation sequences, was inserted into the reporter gene in various positions and orientations upstream from the *tk* promoter. The addition of dexamethasone (a potent, synthetic glucocorticoid) to these fusion genes led to robust induction of *tk* transcription from the endogenous *tk* promoter [72]. This study led to several important conclusions: (1) the MTV LTR contains a "glucocorticoid response element", or GRE, the first response element (*i.e.*, a genomic segment that confers a particular transcriptional regulatory effect *in vivo*) and the prototype for all hormone response elements (HREs); (2) the GR is a transcriptional regulatory factor, the first such factor identified that is encoded in a eukaryotic genome; (3) the GR combines the two functions of a receptor, ligand binding and signal transduction, in a single molecule; (4) the location and orientation of the GRE relative to the transcription initiation site is quite flexible, revealing a general functional explanation for the phenomenon of transcriptional "enhancement", which had been described in the SV40 tumor virus; and (5) the transcription enhancing function afforded by the GR-GRE interaction is separable from the process of transcription initiation [72, 73].

The Glucocorticoid Receptor Is Modular, Containing Discrete Functional Domains

The successful isolation of a cDNA molecule for GR was a landmark achievement [74]. Biochemical, immunologic and genetic studies had suggested that the GR has at least three functional domains: DNA-binding (DBD), ligand binding (LBD) and an N-terminal modulator (NTD) (Fig. 1.2 and [55, 59, 75, 76]). The availability of a GR cDNA allowed investigators to express and test different portions of the molecule for functional activity, to perform domain swap experiments within the same molecule, and to ligate various regions of GR to completely unrelated molecules to test for transference of function. Selected studies, summarized below, led to the conclusion that GR, the founding member of the nuclear receptor superfamily, is indeed composed of several functional domains, or modules (Fig. 1.2) and these modules are independent from one another.

The N-terminal region of the GR varies in length and amino acid sequence from other members of the nuclear receptor superfamily [77]. It contains antigenic sites and a transcription activation domain (AF1), deletion of which results in a partial reduction of activation by GR. AF1 is highly acidic, and it contains many phosphorylation sites which can affect the basic functions of the receptor i.e., ligand,

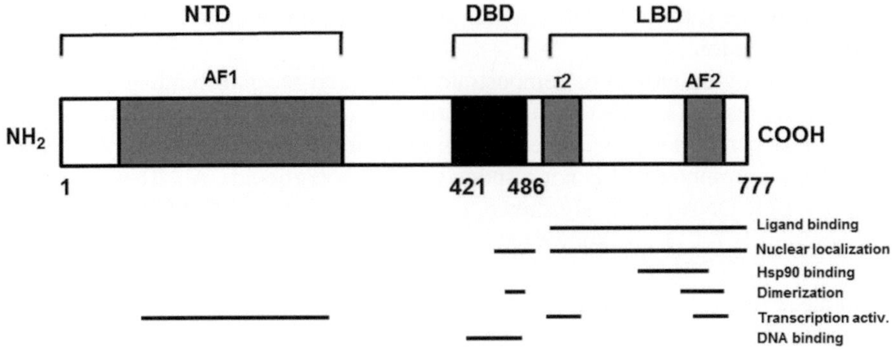

Fig. 1.2 Schematic diagram of the structure of the human glucocorticoid receptor. The amino- and carboxy-termini are illustrated as positions 1 and 777, respectively. The major domain structures: amino-terminal (NTD), DNA-binding (DBD) and ligand binding (LBD) are demarcated by *brackets* at the top. AF1, τ2 and AF2 represent the transactivation domains. Regions corresponding to specialized functions are shown as *horizontal lines* at the bottom

DNA and co-regulator binding [78–80]. This domain is particularly important for assembly of the co-regulators involved in the assembly of an active transcription apparatus. Additional domains functional in transcription activation are located in the LBD (AF2 and tau 2) and in the dimerization region of the DNA binding domain. Like AF1, AF2 and tau 2 are highly acidic, but they appear to be structurally unrelated to AF1. The three domains function from different positions relative to promoters, and when fused to heterologous DBDs [79, 81].

The DNA-binding domain (DBD) is a highly conserved, cysteine-rich, ~75 amino acid segment located in the middle of the molecule (Fig. 1.2). Analysis of the cDNA-deduced structure of the DBD revealed an arrangement of eight cysteine residues (perfectly conserved in all members of the receptor superfamily) that form two zinc finger structures, each of which coordinates the association of a zinc atom to four cysteine residues [82]. The first zinc finger is primarily involved in DNA binding, whereas the second finger stabilizes this binding. GBS recognition is a function of the first, or N-terminal, zinc finger domain. A three amino acid segment subdomain, the P-box, is critical to binding specificity for all nuclear receptors. In GR this sequence is Gly-Ser-Val, whereas in the estrogen receptor it is Glu-Gly-Ala [83].

Modularity of receptor function was demonstrated in domain swap experiments. When the DBD of the GR was placed into the corresponding region of the estrogen receptor, this chimeric receptor converted a gene that was normally glucocorticoid-responsive into an estrogen responsive one [84]. In another experiment, the GR DBD was replaced by the DBD of the bacterial transcription repressor Lex A, a helix-turn-helix transcription factor. This chimeric molecule activated transcription from a Lex promoter-operator construct in response to dexamethasone [85].

The GR binds to GREs as a dimer. The DBD of each receptor monomer contains a five amino acid segment, the D box, which is located between the two most N-terminal cysteine residues at the base of the second finger. The D box is involved in mediating the dimerization and cooperative binding of two receptor molecules to a GRE. Mutations of the D-box compromise the cooperativity of occupying the two half-sites of the GBS *in vitro*, and reduce GR activity *in vivo*. NMR and X-ray crystallographic studies show that receptors interact with the GRE as a dimer in a head-to-head arrangement where each arm of the palindromic DNA element contacts a single receptor molecule [82, 86].

The C-terminal ligand-binding domain (*LBD*) (Fig. 1.2) is a ~260 amino acid segment that displays greater sequence divergence that the DBD among nuclear receptors. Several widely separated and discontinuous amino acids form the hormone-binding surfaces of this domain [81, 87]. The LBD has other functions, including hormone-dependent nuclear localization (the NL1 and NL2 subdomains), hsp90 binding and transcription activation (AF2 and tau2). Starting with the binding of hormone to the LBD, a sequence of events links these functions. First, hsp90 appears to dissociate from the LBD, resulting in a conformational change of the receptor. This has two consequences. The nuclear localization signals are functional, so GR translocates stably to the nucleus where it can now function as a transcriptional regulatory factor [88].

Interestingly, inactivation of functional domains by the hsp90-associated unliganded LBD is maintained when the LBD is repositioned to the N-terminus of GR, and even when the LBD is fused to unrelated proteins, such as the viral E1a protein [88]. The mechanism of this global inactivation is unknown. As expected, however, GR lacking the LBD, and the inhibitory action of the bound hsp90, activates transcription constitutively [79, 85].

These representative experiments provide direct evidence of the modular nature of the GR, and show that each of the domains can function independently. Importantly, however, they do not imply that the domains do not interact functionally in the intact GR. Indeed, as described below, there is abundant evidence of extensive allosteric signaling between domains.

The GBS Is Sufficient for GR-Regulated Transcription Activation in a Reporter Context

The availability of purified receptors and the promoter regions of several hormone responsive genes led to tests of the relationship between in vitro glucocorticoid binding sequence (GBS) activity and in vivo GR-mediated transcriptional regulatory activity. In these experiments, purified GR was bound to cloned target gene promoter regions, and the bound sequences determined by DNase I footprinting [68, 89]. To discover segments of DNA sufficient to confer hormonal regulation, various promoter-proximal fragments were ligated into expression vectors bearing basal promoter elements (*e.g.*, from the thymidine kinase (*tk*) gene) driving a reporter gene (e.g. chloramphenicol acetyltransferase (CAT) or luciferase). The chimeric construct was transfected into a recipient cell, hormone was added and expression of the reporter gene quantitated. Such experiments demonstrated that the canonical idealized GBS represents a large family of 15 base pair elements related to the sequence GGTACAnnnTGTTCT [77], and inferred that the GBS is sufficient for regulatory activity [83, 89, 90]. Importantly, however, as described below, GBSs are not essential for GR regulation, and *bona fide* GREs associated with chromosomal GR responsive genes are substantially more complex than GBSs.

Context-Specific Glucocorticoid Regulation Is Specified by Complex Arrays of DNA Elements and Binding Proteins

The notion that natural genes, be they chromosomal or viral, would be regulated by glucocorticoids solely by GR binding to GBSs (termed "simple GREs" in early work) was soon challenged by experimental observations: two new classes of GREs were described [91]. *Composite GREs* are ~0.5–2 kb compound elements [92] composed of one or more GBSs together with binding sites for one, or more typically, multiple nonreceptor transcriptional regulatory factors, producing functional cross-talk between the different classes of factors that affects the regulatory outcome. *Tethering GREs* are also compound elements encompassing binding sequences for nonreceptor regulators, but lack GBSs; GR associates at tethering GREs through protein·protein interactions with one or more of the bound nonreceptor factor instead of directly with DNA.

The first composite element was described at the proliferin gene, wherein a GBS and an AP-1 binding site are contiguous (Fig. 1.3). The proteins *c-fos* and *c-jun*, two prominent members of the phorbol ester-activated AP-1 transcription factor family, activate proliferin expression *via* either *c-jun* homodimers or *c-fos·c-jun* heterodimers. GR regulates proliferin expression through its GBS, but only in the presence of AP-1, further activating the homodimeric species and repressing the heterodimer [93]. The ratio of *c-jun* to *c-fos* in the cell is therefore the determinant of whether GR mediates a positive or negative transcription response.

GR represses transcription at tethering GREs proximal to the collagenase gene, where GR associates through protein·protein interactions with a specifically bound c-Jun subunit of AP-1 [94, 95], and to the IL-8 gene, where GR binds to the p65 subunit of a NFκB factor bound at a κB binding sequence [96]. Hence, tethering elements can be viewed as a subset of composite elements in which both GR and nonreceptor regulators occupy a specific genomic site, but that GRE activity in these tethering contexts proceeds in the absence of GBS elements.

Involvement of nonreceptor factors in GR-mediated regulation was also suggested for MTV, where binding sites for nuclear factor-1 (NF-1) and octamer transcription factor-1 (Oct-1), appear to increase glucocorticoid-induced transcription [97]. The binding sites for these factors lie between the proximal GRE region and the TATA-box (Fig. 1.3). Binding of NF-1 *in vivo* occurs only in the presence of hormone, but there is no evidence of a direct GR·NF-1 interaction. In this case, a GR-induced alteration in chromatin structure, resulting in loss of a nucleosome near

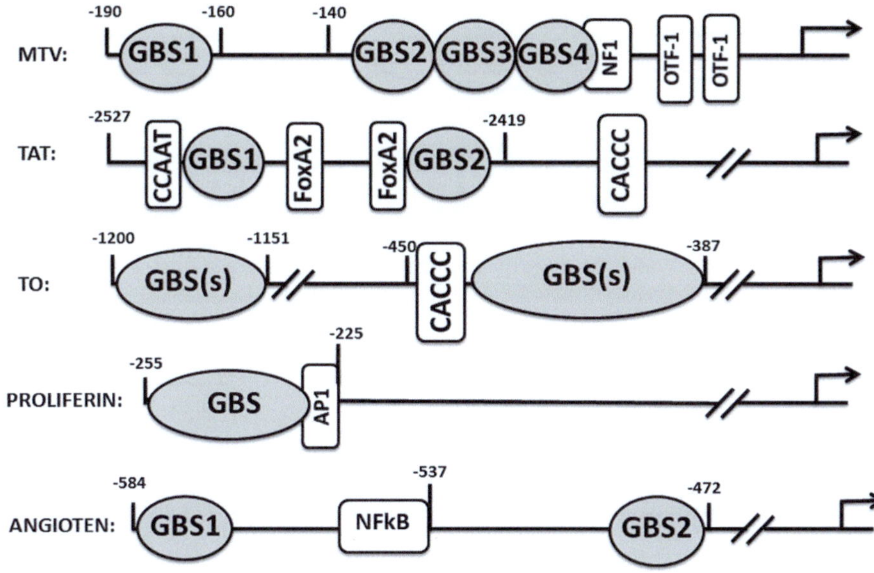

Fig. 1.3 Presumptive glucocorticoid binding sequences (GBSs) proximal to several glucocorticoid-regulated genes. The mammary tumor virus (MTV), tyrosine aminotransferase (TAT), tryptophan oxygenase (TO), proliferin, and angiotensinogen (angioten) promoter-proximal regions are illustrated. Glucocorticoid receptor binding sequences (GBSs) each of which binds a GR dimer *in vitro*, are shown as *shaded ovals* or *circles*. Binding sequences for various nonreceptor transcriptional regulatory factors are shown as *open forms*. The *numbers* indicate the approximate positions of the elements in relation to the transcription initiation site (*arrow*)

the transcription initiation site may "open" the chromatin structure, facilitating binding of general transcription factors, which are required for transcription initiation [98, 99]. Mutation of the NF-1 binding site reduced modestly the glucocorticoid response, whereas mutation of the two Oct-1 binding sites located between the NF-1 site and the TATA box (Fig. 1.3) resulted in a markedly reduced response of the reporter gene to glucocorticoids [100]; studies suggesting cooperative DNA binding between GR and Oct-1 on the MTV promoter imply a direct protein·protein interaction between these factors.

As more candidate GREs close to target genes have been analyzed, it became apparent that all contain binding sequences for various non-receptor regulatory factors, either together with or in the absence of GBSs. As illustrated in Figs. 1.3 and 1.4, this includes candidate GREs at the TAT, TO and PEPCK genes, the first proteins known to be induced by glucocorticoids, as well as other GC-regulated genes such as proliferin and angiotensinogen. Deletion or mutation of these accessory DNA elements, with the associated loss of binding of the cognate transcription factor, blunts or abolishes the response of these promoters to glucocorticoids in reporter assays (for review see [101]). Also important were the precise sequences of the GBSs, as were their spacing, multiplicity and location. It may be significant in this

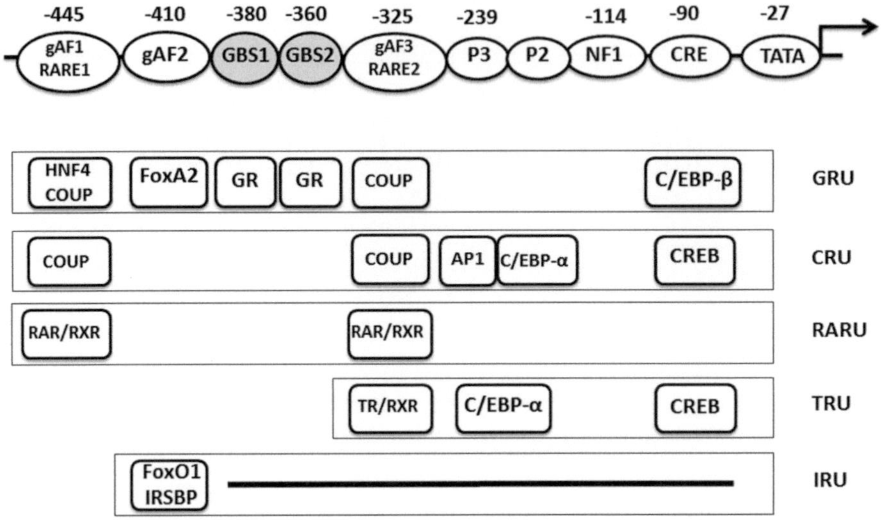

Fig. 1.4 Promoter-proximal region of the PEPCK gene. Binding sequences for various regulatory and general transcription factors are shown across the *top line*. The *numbers* above these represent the approximate center of each element with respect to the transcription initiation site (demarcated by the *arrow*). The five *boxes* below represent hormone-specific response units, and the regulatory factors thought to be bound, for glucocorticoids (GRU), cyclic AMP (CRU), retinoic acid (RARU), thyroid hormone (TRU) and insulin (IRU) that mediate each hormone's regulation of PEPCK transcription. Insofar as the CRU, RARU, TRU and IRU impact the actions of the GRU under physiological conditions, all of these response units interact functionally, cooperating or competing, to comprise the PEPCK gene hormone response domain. Details are described in the text

regard that the receptors for glucocorticoids, mineralocorticoids, progestins and androgens have very similar DBDs, and all can regulate transcription through an idealized GBS, despite their very different physiologic actions *in vivo* (for review see [101]; and see below).

It is now apparent that GREs (and by extension, all genomic transcriptional response elements) are nucleation centers for the dynamic assembly and disassembly of multifactor regulatory complexes containing context-specific combinations of >100 different genome-associated regulatory and coregulatory proteins associated through protein·DNA and protein · protein interactions [102]. It is the combinatorial assembly of these complexes, displaying both precision in a given context and plasticity to shift readily to different assembly instructions in a different context, that enables GCs and GR to regulate arrays of distinct gene transcription networks with exquisite gene, cell and physiologic specificity. These three crucial contexts are conveyed to GR by a combination of cellular signals (GR interactions with hormone or ligand, DNA sequence, and other regulatory proteins), which are received, interpreted and integrated as allosteric alterations in GR conformation [86, 90]. In turn, these conformations define functional GR surfaces that serve as "assembly instructions" for multi-component transcriptional regulatory complexes comprised of unique combinations of regulatory factors and co-regulators [103].

The PEPCK GRE Provides Finely Tuned, Versatile Regulation of Transcription

A key element of regulatory crosstalk is the integration of information from multiple signaling pathways, which occurs at response elements and provides precise control of remarkably complex biological processes involving gene networks in multiple organs and tissues. For example, metabolic processes such as gluconeogenesis involve interacting regulatory mechanisms that provide additive, synergistic, positive, negative and dominant control. To illustrate this, we focus here on the integration of multiple hormone signals in particular cultured cell contexts, at the level of gene transcription, that govern expression of hepatic phosphoenolpyruvate carboxykinase (PEPCK), which catalyzes a critical, first step in gluconeogenesis. The amount and activity of the cytosolic form this enzyme (PEPCK-C) is determined by several hormones through their effects on transcription of the PEPCK gene (*Pck-1*) [46]. Glucagon (acting through cyclic AMP), retinoic acid,[1] thyroid hormone and glucocorticoids stimulate transcription, whereas insulin and glucose exert dominant, inhibitory effects [46, 50].

In the contexts examined, several DNA elements with associated regulatory factors have been implicated in the glucocorticoid response of the PEPCK gene. A promoter-proximal segment, denoted here as a glucocorticoid response unit (GRU), contains two adjacent GBSs (GBS1 and GBS2) located between positions −349 to −395 relative to the transcription initiation site (Fig. 1.4). GBS1 and GBS2 can each bind GR dimers *in vitro*, albeit with 30-fold lower affinity than that of the idealized GBS, probably because they resemble the latter at only 7/12 and 6/12 positions, respectively [105]. Unlike the TAT GRE, neither of the PEPCK GBSs, by themselves, confers a glucocorticoid response in transient reporter assays [105]. Rather, glucocorticoid regulation in this context requires flanking DNA bearing binding sequences for non-receptor regulators collectively termed glucocorticoid accessory factor elements (gAF)[2]: gAF1 binds HNF-4 α and COUP-TF [106], both members of the nuclear receptor family; gAF2 binds FoxA2 (HNF-3 β) [107, 108]; and gAF3 binds COUP-TF (Fig. 1.4) [109]. Deletion of either GBS1 or GBS2 compromises the glucocorticoid response; the effect of GBS1 is greater than that of GBS2, but both are required for maximal activity in the contexts tested [105].

[1] This effect helps explain an observation made more than 50 years ago. Wolf et al. showed that, in vitamin A deficient rats, hepatic cholesterol and fatty acid synthesis, the citric acid cycle, glycogen metabolism and glycolysis were all normal. Gluconeogenesis, however, was markedly impaired [104].

[2] The GC accessory factor elements in the PEPCK gene promoter were originally referred to as AF1-3. As the designation of the transactivation domains in nuclear receptors became known as AF1 and AF2, the DNA elements in the PEPCK gene were designated gAF1-3.

Deletion of either gAF1/gAF3 (gAF1 > gAF3) or the gAF2 element reduces the glucocorticoid response by ~50 %; deletion of both abolishes the response. Chromatin immunoprecipitation (ChIP) assays revealed that HNF-4 α, COUP-TF and FoxA2 occupy the GRU in the absence of dexamethasone; as expected GR does not [110]. Those gAF proteins increase the affinity of GR for GBS1 and GBS2, as demonstrated by fluorescence anisotropy [111], thereby nucleating assembly of a transcriptional regulatory complex that includes co-regulators SRC-1, CBP/p300, PGC-1, FoxO1, FoxO3, which appear to assemble through protein · protein interactions [112–114]. ChIP assays confirmed recruitment of those factors, as well as polymerase II, to the PEPCK GRU following addition of dexamethasone [110].

Interestingly, gAF1 and gAF3 also contain binding sequences for heterodimers of retinoic acid receptor (RAR) and retinoid X receptor (RXR), and in the absence of dexamethasone but in the presence of retinoic acid, activate PEPCK gene transcription (Fig. 1.4) [115]. Thus, gAF1 and gAF3, in one context GRU components, serve in another as a retinoic acid response unit (RARU).

Similarly, the GRU/RARU, in the absence of their cognate signals, serves as a cyclic AMP response unit (CRU) to enhance PEPCK transcription upon glucagon stimulation (Fig. 1.4); in that context, the CRU includes some DNA elements and corresponding regulatory factors that overlap with those in the GRU and RARU, as well as others that are distinct [116, 117].

Finally, the dominant effect of insulin is mediated, in part, through the multifunctional gAF2 element [118]; an epigenetic effect involving insulin-induced demethylation of arginine-17 on histone H3 may also be operative [110]. FoxO1 is clearly involved in this insulin effect, as it rapidly leaves the IRU (see below) and exits the nucleus in response to insulin [119]. The ChIP assay was used to demonstrate how quickly this dominant inhibition happens. H4IIE cells[3] were treated with dexamethasone and the maximally active transcription complex was allowed to assemble. Within 3 min following the addition of insulin, p300, FoxO1and FoxO3 are removed from the IRU and by 10 min most components of the assembly are at, or below, the basal level, including polymerase II [110].

This complex system, comprised of overlapping but distinct composite response units and associated regulatory factors, underscores the importance of finely tuned homeostatic regulation of gluconeogenesis. A mutation of one DNA element or accessory factor blunts, but does not abolish, the effect of a given stimulatory hormone. Indeed, the complete loss of the response to one of these hormones (or the absence of the hormone itself) blunts, but does not abolish, the positive regulation of gluconeogenesis. By contrast, insulin stands alone as the hormonal inhibitor of gluconeogenesis. Under physiological conditions, the actions of the five hormones (and other cellular signals) are integrated in a highly context-dependent manner, so with respect to glucocorticoid-mediated regulation, all of the response units and their regulatory factors operate in aggregate, cooperating or competing, as the PEPCK GRE.

---

[3] All the experiments described in this section were performed using this cell line, which was derived from a rat hepatoma [42].

Multiple Context Effects Determine GRE Activities and Mechanisms

Molecular analyses showing that glucocorticoid-responsive genes are governed by composite GREs, bolstered by the dissection of multi-hormonal regulation of PEPCK expression, demonstrate that the remarkable specificity of GR-mediated transcriptional regulation emerges from the integration of multiple context effects that converge on GR in a given setting.

The consequences of this context dependence are striking: some specific GR binding regions (GBRs; genomic segments occupied by GR in certain contexts *in vivo*) are functional GREs in some cell types but not in others; a given GRE-bound regulatory complex may activate transcription in one setting and repress in another; a GC ligand may be a strong agonist in one cell or gene context, a weak agonist in another, and an antagonist in a third. It is this extreme context dependence that enables GR to orchestrate different transcription networks in every cell and tissue type, transducing a simple molecular signal into an array of distinct physiological outcomes (Table 1.1).

The remarkable effects of context, imposed as noted above by the integrated actions of multiple classes of signaling inputs and mediated by allosteric transitions in GR structure, result in the assembly of regulatory complexes with different compositions, structures and mechanistic actions on transcription. The production and accumulation of mRNA is itself an exceedingly complicated process, comprised of coupled ordered reactions—initiation, elongation, splicing, cleavage, polyadenylation, termination, nuclear export, degradation—that are themselves complex and each a potential point of positive or negative regulation. Quantitative methods for distinguishing these processes have been devised [120, 121] but typically only mRNA accumulation, the endpoint of all of these steps, is monitored by methods such as qPCR or RNA-seq. To date, GR has been implicated in regulating initiation [122], elongation [96] and stabilization [123], but other steps have not been ruled out.

Application of novel survey methods together with a better understanding of the mechanistic effects of coregulatory factors resident in GR-containing regulatory complexes [124, 125] will provide the full picture of the varied strategies by which GR modulates gene transcription. In framework, it is apparent that the coregulators can catalyze chromatin remodeling, chemically modify histones and other factors, and recruit or occlude general transcription factors; each of these activities could potentially alter one or more steps in mRNA production or accumulation.

The powerful and varied actions of GR on specific gene transcription, together with the knowledge that eukaryotic transcriptional regulatory complexes, especially in higher metazoans such as *Drosophila* or mammals, can operate from very long range, raise the challenging question of the determinants of which GRE(s) will confer regulation on which target gene. Chromatin immunoprecipitation (ChIP) methods for determination of the genomic sites of occupancy by GR *in vivo*, assessed initially on selected genomic regions (ChIP-qPCR) and subsequently across the entire genome (ChIP-seq) have large numbers of GR binding regions (GBRs), many of which are specific to particular contexts, such as cell type or developmental stage. As the sensitivity, range and discrimination of the methods have increased, the number of reported GBRs has increased, from hundreds to thousands to tens of thousands [126–132].

Which of these GBRs are functional GREs, and which GC-responsive target gene is controlled by which GRE? It is thought that not every GBR is a functional GRE, at least in the restricted contexts examined, as some can be deleted without apparent effect, and many are located many megabases from GC-responsive genes. Lacking unequivocal methods for assessing activity in native chromosomal environments (rather than reporters or transgenes), researchers have resorted to proximity, "assigning" GRE activity to one or more GBRs that neighbor GC-responsive genes. Even with this proviso, most assigned GREs are >10 kb from their presumptive target promoters [127]. Whether GBRs that appear to lack function in one context may function in another context is an intriguing untested possibility.

To date, only a single GBR, which resides some 25 kb downstream of the transcription start site for the Per2 circadian rhythm regulatory gene, has been unequivocally assigned to its target gene [133]. This was discovered only because a deletion constructed for another purpose fortuitously removed the GBS and the mouse that ensued had lost GR regulation of Per2 expression [133]. Fortunately, powerful and facile gene editing methods such as CRISPR [134, 135] now make possible targeted changes in genome sequence that will enable GRE activities on specific target genes to be determined with certainty.

The ability to analyze GC regulated genes across the entire genome, in all organs and tissues, enables rather direct access to many interesting and important questions. For example, can relationships be discerned between regulatory complex structure and mechanism and the physiologic processes that they control, e.g., complexes involved in GC-mediated anti-inflammation or immunosuppression. If this were the case, it might then be possible to design small molecules that target those genes without affecting the different assemblies involved in metabolism, growth and development, and thus avoid or minimize the devastating side effects that complicate, and limit, current glucocorticoid therapy. This might not have to be an "all or none" phenomenon. As informed by the regulation of the PEPCK gene, where a complex array of hormone signals, accessory proteins and co-regulators appears to allow for a linear degree of gene expression from 0 to 100 %, subtle modifications of the expression of one class of genes, with full expression of another, may be sufficient to control unwanted side effects.

Epilogue

Here we have provided an historical perspective on the progression of research on the actions of glucocorticoids, and how early attempts to treat the consequences of adrenal insufficiency eventually evolved to become key drivers of the conceptual and experimental understanding of metazoan transcriptional regulation. The outstanding contributors to this volume build on our overview to describe in detail many of the physiologic and pathophysiologic effects of these hormones, and the molecular mechanisms involved in these processes.

Future work promises to reveal still more regulatory processes, perhaps demonstrating roles for intranuclear position, chromosome topology, disassembly of regulatory complexes, or hormone transport across plasma membranes. Other studies will illuminate the relationships between such regulatory mechanisms and the physiologic outcomes they specify, relationships currently shrouded in the complexity of combinatorial processes yet to be deeply understood. And with those advances could come the capacity for the prediction and design of mechanisms, and of hormone-like ligands that trigger or inhibit them, This, in turn, will lead to a more selective, effective and safer class of therapeutics.

Acknowledgements The authors wish to thank Ms. Stacie Vik and Allison McQueen for their assistance in preparing the manuscript and Michael Stallcup for his incisive comments and encouragement during its preparation. We thank the many people who, over the years, performed experiments in our laboratories. All of their contributions could not be discussed, but all are a part of this story. We also are grateful to the legion of colleagues who, over years of their own work, and in formal and informal discussions, made our participation in this field so interesting.

References

1. Bayliss WM, Starling FH. The movements and innervation of the small intestine. J Physiol. 1901;26:125–38.
2. Bayliss WM, Starling FH. The mechanism of pancreatic secretion. J Physiol. 1902;28: 325–53.
3. Addison T. On the constitutional and local effects of disease of the suprarenal capsule. In: A collection of the published writings of the late Thomas Addison MD. London: New Sydenham Society; 1868.
4. Bernard C. Remarques sur le sécrétion du sucre dans la foie, faites à l' occasion de la communication de M Lehman. CR Acad Sci Paris. 1855;40:589–92.
5. Cannon WB. The wisdom of the body. New York: WW Norton; 1932.
6. Henderson J. Ernest starling and "hormones": an historical commentary. J Endocrinol. 2005;184:5–10.
7. Starling FH. Croonian lecture: On the chemical correlation of the functions of the body. Lancet. 1905;2:339–41.
8. Swingle WW, Pfiffner JJ. The revival of comatose adrendectomized cats with an extract of the suprarenal cortex. Science. 1930;72:75–6.
9. Hartman FA, Brownell KA. The hormone of the adrenal cortex. Science. 1930;72:76.
10. Kendall EC. The development of cortisone as a therapeutic agent. In: Nobel lectures in physiology or medicine: 1942–1962. Amsterdam: Elsevier; 1964. p. 270–288.
11. Reichstein T. Chemistry of the adrenal cortex hormones. In: Nobel lectures in physiology or medicine: 1942–1962. Amsterdam: Elsevier; 1964. p. 291–308.
12. Sarett LH. A new method for the preparation of 17α-hydroxy-20-ketopregnanes. J Am Chem Soc. 1948;70:1454–8.
13. Cushing H. The basophil adenomas of the pituitary body and their clinical manifestations (pituitary basophilism). Bull Johns Hopkins Hosp. 1932;50(4):137–95.

14. Hench PS. The reversibility of certain rheumatic and non-rheumatic conditions by the use of the cortisone or of the pituitary adrenocorticotropic hormone. In: Nobel lectures in physiology or medicine 1942–1962. Amsterdam: Elsevier; 1964. p. 311–43.
15. de Réaumur RAF. Observations sur la digestion des oiseaux. Hist Acad Roy Sci. 1752;266:461.
16. Payen A, Persoz J-F. Memoire sur la diastase, les principaux produits de ses réactions, et leurs applications aux arts industriels. Ann Chim Phys. 1833;53:73–92.
17. Kühne W. On the behavior of various organized and so-called unformed ferments. Verb. d. Nat. Med. Ver. Heidelberg. 1877;1:190–8.
18. Buchner E. Cell-free fermentation. In: Nobel lectures in chemistry 1901–1921. Amsterdam: Elsevier; 1966. p. 103–20.
19. Sumner JB. The chemical nature of enzymes. In: Nobel lectures in chemistry 1901–1921. Amsterdam: Elsevier; 1964. p. 114–21.
20. Northrop JH. The chemical nature of enzymes. In: Nobel lectures in chemistry 1942–1962. Amsterdam: Elsevier; 1964. p. 124–34.
21. Stanley WM. The chemical nature of enzymes. In: Nobel lectures in chemistry 1942–1962. Amsterdam: Elsevier; 1964. p. 138–57.
22. Levine R. Insulin action: 1948–80. Diabetes Care. 1981;4(1):38–44.
23. Schoenheimer R. The dynamic state of body constituents. Cambridge: Harvard University Press; 1942.
24. Shemin D, Rittenberg D. Life span of human red blood cell. J Biol Chem. 1946; 166:627–36.
25. Ballou JE, Thompson RC. Studies of metabolic turnover with tritium as a tracer: V. The predominantly non-dynamic state of body constituents in the rat. J Biol Chem. 1956; 223:795–809.
26. Dubos RJ. The adaptive production of enzymes by bacteria. Bacteriol Rev. 1940;4(1):1–16.
27. Monod J. The phenomenon of enzymatic adaptation and its bearings on problems of genetics and cellular differentiation. Growth. 1947;11:223–89.
28. Spiegelman S. Nuclear and cytoplasmic factors controlling enzymatic constitution. Cold Spring Harbor Symp Quant Biol. 1946;11:256–77.
29. Knox WE, Mehler AH. Adaptive increase of tryptophan peroxidase-oxidase system of liver. Science. 1951;113:237–8.
30. Knox WE. Two mechanisms which increase in vivo liver tryptophan peroxidase activity: specific enzyme adaptation and stimulation of the pituitary-adrenal system. Br J Exp Pathol. 1951;32(5):462–9.
31. Selye H. The general adaptation syndrome and the diseases of adaptation. J Clin Endocrinol Metab. 1946;6:117–230.
32. Thompson JF, Mikuta ET. Effect of total body x-irradiation on tryptophan peroxidase activity of rat liver. Proc Soc Exp Biol Med. 1954;85(1):29–32.
33. Knox WE, Auerbach VH, Lin ECC. Enzymatic and metabolic adaptation in animals. Physiol Rev. 1956;36:164–254.
34. Lin ECC, Knox WE. Adaption of the rat liver-tyrosine-alpha-ketoglutarate transaminase. Biochim Biophys Acta. 1957;26:85–8.
35. Kenney FT. Induction of tyrosine-α-ketoglutarate transaminase in rat liver. II: enzyme purification and preparation of antitransaminase. J Biol Chem. 1962;237(5):1605–9.
36. Feigelson P, Greengard O. Immunochemical evidence for increased titers of liver tryptophan pyrrolase during substrate and hormonal enzyme induction. J Biol Chem. 1962;237:3714–7.
37. Schimke RT. Control of enzyme levels in mammalian issues. Adv Enzymol Relat Areas Mol Biol. 1973;37:135–87.
38. Schimke RT, Sweeney EW, Berlin CM. The roles of synthesis and degradation in the control of rat liver tryptophan pyrrolase. J Biol Chem. 1965;240:322–31.
39. Kenney FT. Induction of tyrosine-α-ketoglutarate transaminase in rat liver: IV. Evidence for an increase in the rate of enzyme synthesis. J Biol Chem. 1962;237:3495–8.
40. Shrago E, Lardy HA, Nordlie RC, Foster DO. Metabolic and hormonal control of phosphoenolpyruvate carboxykinase and malic enzyme in rat liver. J Biol Chem. 1963;238:3188–92.

41. Jacob F, Monod J. Genetic regulatory mechanisms in the synthesis of protein. J Mol Biol. 1961;3:318–56.
42. Pitot H, Peraino C, Morse P, Potter V. Hepatomas in tissue culture compared with adapting liver in vivo. Natl Cancer Inst Monogr. 1964;13:229–45.
43. Thompson EB, Tomkins GM, Curran JF. Induction of tyrosine alpha-ketoglutarate transaminase by steroid hormones in a newly established tissue culture cell line. Proc Natl Acad Sci U S A. 1966;56:296–303.
44. Granner DK, Hayashi S-I, Thompson EB, Tomkins GM. Stimulation of tyrosine aminotransferase synthesis by dexamethasone phosphate in cell culture. J Mol Biol. 1968; 35:291–301.
45. Granner DK, Beale EG. Regulation of the synthesis of tyrosine aminotransferase and phosphoenolpyruvate carboxykinase by glucocorticoid hormones. In: Biochemical actions of hormones, vol. 12. New York: Academic; 1985. p. 89–138.
46. Sasaki K, Cripe T, Koch S, et al. Multihormonal regulation of PEPCK gene transcription: the dominant role of insulin. J Biol Chem. 1984;259:15242–51.
47. Greengard O, Acs G. The effect of actinomycin on the substrate and hormonal induction of liver enzymes. Biochim Biophys Acta. 1962;61:652–3.
48. Danesch U, Hashimoto S, Renkawitz R, Schütz G. Transcriptional regulation of the tryptophan oxygenase gene in rat liver by glucocorticoids. J Biol Chem. 1983;258(8):4750–3.
49. Schmid E, Schmid W, Jantzen M, Mayer D, Jastorff B, Schütz G. Transcription activation of the tyrosine aminotransferase gene by glucocorticoids and cAMP in primary hepatocytes. Eur J Biochem. 1987;165(3):499–506.
50. Lucas PC, O'Brien RM, Mitchell JA, et al. A retinoic acid response element is part of a pleiotropic domain in the phosphoenolpyruvate carboxykinase gene. Proc Natl Acad Sci U S A. 1991;88:2184–8.
51. Talwar GP, Segal SJ, Evans A, Davison OW. The binding of estradiol in the uterus: a mechanism for depression of RNA synthesis. Proc Natl Acad Sci U S A. 1964;52:1059–66.
52. Rousseau GG, Baxter JD. Glucocorticoid receptors. In: Glucocorticoid hormone action. Heidelberg: Springer; 1979. p. 49–77.
53. Baxter JD. Glucocorticoid hormone action. Pharmacol Ther B. 1976;2(3):605–59.
54. Wrange O, Carlstedt-Duke J, Gustafsson JÅ. Purification of the glucocorticoid receptor from rat liver cytosol. J Biol Chem. 1979;254:9284–90.
55. Wrange O, Okret S, Radojćić M, Carlstedt-Duke J, Gustafsson J. Characterization of the purified activated glucocorticoid receptor from rat liver cytosol. J Biol Chem. 1984; 259:4534–41.
56. Hollenberg SM, Moldenhauer C, Ong ES, et al. Primary structure and expression of a functional human glucocorticoid receptor cDNA. Nature. 1985;318:635–41.
57. Miesfeld R, Rusconi S, Godowski PJ, et al. Genetic complementation of a glucocorticoid receptor deficiency by expression of cloned receptor cDNA. Cell. 1986;46(3):389–99.
58. Danielson M, Northrop JP, Ringold GM. The mouse glucocorticoid receptor: mapping of functional domains by cloning, sequencing and expression of wild-type and mutant receptor proteins. EMBO J. 1986;5(10):2513–22.
59. Yamamoto KR, Gehring U, Stampfer MR, Sibley CH. Genetic approaches to steroid hormone action. Recent Prog Horm Res. 1976;32:3–32.
60. Yamamoto KR. Steriod hormone receptors: interaction with deoxyribonucleic acid and transcription factors. Annu Rev Genet. 1985;19:209–52.
61. Ringold GM, Yamamoto KR, Tomkins GM, Bishop M, Varmus HE. Dexamethasone-mediated induction of mouse mammary tumor virus RNA: a system for studying glucocorticoid action. Cell. 1975;6(3):299–305.
62. Ringold GM, Yamamoto KR, Bishop JM, Varmus HE. Glucocorticoid-stimulated accumulation of mouse mammary tumor virus RNA: increased rate of synthesis of viral RNA. Proc Natl Acad Sci U S A. 1977;74(7):2879–83.
63. Ucker DS, Ross SR, Yamamoto KR. Mammary tumor virus DNA contains sequences required for its hormone-regulated transcription. Cell. 1981;27:257–66.

64. Huang AL, Ostrowski MC, Berard D, Hager GL. Glucocorticoid regulation of the Ha-Mu SV p21 gene conferred by sequences from mouse mammary tumor virus. Cell. 1981;27:245–55.
65. Yamamoto KR, Chandler VL, Ross SR, Ucker DS, Ring JC, Feinstein SC. Integration and activity of mammary tumor virus genes: regulation by hormone receptors and chromosomal position. Cold Spring Harb Symp Quant Biol. 1981;45:687–97.
66. Groner B, Kennedy N, Rahmsdorf U, Herrlich P, Van Ooyen A, Hynes NE. Introduction of a proviral MMTV gene and a chimeric MMTV-tk gene into L cells results in their glucorticoid responsive expression. In: Hormones and cell regulation. Amsterdam: Elsevier Biomedical Press; 1982. p. 217–28.
67. Payvar R, Wrange O, Carlstedt-Duke J, Okret S, Gustafsson JǺ, Yamamoto KR. Purified glucocorticoid receptors bind selectively in vitro to a cloned DNA fragment whose transcription is regulated by glucocorticoids in vivo. Proc Natl Acad Sci U S A. 1981;78:6628–32.
68. Payvar F, Firestone GL, Ross S, et al. Multiple specific binding sites for purified glucocorticoid receptors on mammary tumor virus DNA. J Cell Biochem. 1982;19:241–7.
69. Govindan MV, Spiess E, Majors J. Purified glucocorticoid receptor-hormone complex from rat liver cytosol binds specifically to cloned mouse mammary tumor virus long terminal repeats in vitro. Proc Natl Acad Sci U S A. 1982;79(17):5157–61.
70. Geisse S, Scheidereit C, Westphal HM, Hynes NE, Groner B, Beato M. Glucocorticoid receptors recognize DNA sequences in and around murine mammary tumour virus DNA. EMBO J. 1982;1(12):1613–9.
71. Pfahl M. Specific binding of the glucocorticoid-receptor complex to the mouse mammary tumor proviral promoter region. Cell. 1982;31:475–82.
72. Chandler VL, Maler BA, Yamamoto KR. DNA sequences bound specifically by glucocorticoid receptor in vitro render a heterologous promoter hormone responsive in vivo. Cell. 1983;33(2):489–99.
73. Yamamoto KR. Transcriptional enhancement by specific regulatory protein: DNA complexes. In: Ginsberg HS, Vogel HJ, editors. Transfer and expression of eukaryotic genes. New York: Academic; 1983. p. 79–92.
74. Miesfeld R, Okret S, Wikstrom AC, Wrange O, Gustafsson JA, Yamamoto KR. Characterization of a steroid hormone receptor gene and mRNA in wild-type and mutant cells. Nature. 1984;312(5996):779–81.
75. Carlstedt-Duke J, Okret S, Wrange O, Gustafsson JǺ. Immunochemical analysis of the glucocorticoid receptor: identification of a third domain separate from the steroid-binding and DNA-binding domains. Proc Natl Acad Sci U S A. 1982;79:4260–4.
76. Westphal HM, Moldenhauer G, Beato M. Monoclonal antibodies to the rat liver glucocorticoid receptor. EMBO J. 1982;1:1467–71.
77. Evans RM. The steroid and thyroid hormone receptor superfamily. Science. 1988;240:889–95.
78. Ismaili N, Garabedian MJ. Modulation of glucocorticoid receptor function via phosphorylation. Ann N Y Acad Sci. 2004;1024:86–101.
79. Hollenberg SM, Evans RM. Multiple and cooperative trans-activation domains of the human glucocorticoid receptor. Cell. 1988;55:899–906.
80. Galliher-Beckley AJ, Cidlowski JA. Emerging roles of glucocorticoid receptor phosphorylation in modulating glucocorticoid hormone action in health and disease. IUBMB Life. 2009;61(10):979–86.
81. Carson-Jurica MA, Schrader WT, O'Malley BW. Steroid receptor family: structure and functions. Endocr Rev. 1990;11(2):201–20.
82. Luisi BF, Xu WX, Otwinowski Z, Freedman LP, Yamamoto KR, Sigler PB. Crystallographic analysis of the interaction of the glucocorticoid receptor with DNA. Nature. 1991; 352:497–505.
83. Mader S, Kumar V, de Verneuil H, Chambon P. Three amino acids of the oestrogen receptor are essential to its ability to distinguish an oestrogen from a glucocorticoid-responsive element. Nature. 1989;339:271–4.
84. Green S, Chambon P. Oestradiol induction of a glucocorticoid-responsive gene by a chimaeric receptor. Nature. 1987;325(6099):75–8.

85. Godowski PJ, Picard D, Yamamoto KR. Signal transduction and transcriptional regulation by glucocorticoid receptor-LexA fusion proteins. Science. 1988;241(4867):812–6.
86. Watson LC, Kuchenbecker KM, Schiller BJ, Gross JD, Pufall MA, Yamamoto KR. The glucocorticoid receptor dimer interface allosterically transmits sequence-specific DNA signals. Nat Struct Mol Biol. 2013;20(7):876–83.
87. Danielson M. Structure and function of the glucocorticoid receptor. In: Parker MG, editor. Nuclear hormone receptors: molecular mechanisms, cellular functions, clinical abnormalities. London: Academic; 1991. p. 39–78.
88. Picard D, Salser SJ, Yamamoto KR. A movable and regulable inactivation function within the steroid binding domain of the glucocorticoid receptor. Cell. 1988;54(7):1073–80.
89. LaBaer J, Yamamoto KR. Analysis of the DNA-binding affinity, sequence specificity and context dependence of the glucocorticoid receptor zinc finger region. J Mol Biol. 1994;239:664–88.
90. Meijsing SH, Pufall MA, So AY, Bates DL, Chen L, Yamamoto KR. DNA binding site sequence directs glucocorticoid receptor structure and activity. Science. 2009; 324(5925):407–10.
91. Miner JN, Yamamoto KR. Regulatory crosstalk at composite response elements. Trends Biochem Sci. 1991;16:423–6.
92. So AY, Cooper SB, Feldman BJ, Manuchehri M, Yamamoto KR. Conservation analysis predicts in vivo occupancy of glucocorticoid receptor-binding sequences at glucocorticoid-induced genes. Proc Natl Acad Sci U S A. 2008;105(15):5745–9.
93. Diamond MI, Miner JN, Yoshinaga SK, Yamamoto KR. Transcription factor interactions: selectors of positive or negative regulation from a single DNA element. Science. 1990;249(4974):1266–72.
94. Jonat C, Rahmsdorf HJ, Park KK, et al. Antitumor promotion and antiinflammation: down-modulation of AP-1 (Fos/Jun) activity by glucocorticoid hormone. Cell. 1990;62(6): 1189–204.
95. Miner JN, Diamond MI, Yamamoto KR. Joints in the regulatory lattice: composite regulation by steroid receptor-AP1 complexes. Cell Growth Differ. 1991;2(10):525–30.
96. Luecke HF, Yamamoto KR. The glucocorticoid receptor blocks P-TEFb recruitment by NFkappaB to effect promoter-specific transcriptional repression. Genes Dev. 2005; 19(9):1116–27.
97. Truss M, Beato M. Steroid hormone receptors: interaction with deoxyribonucleic acid and transcription factors. Endocr Rev. 1993;14(4):459–79.
98. Zaret KS, Yamamoto KR. Reversible and persistent changes in chromatin structure accompany activation of a glucocorticoid-dependent enhancer element. Cell. 1984;38(1):29–38.
99. Richard-Foy H, Hager GL. Sequence-specific positioning of nucleosomes over the steroid-inducible MMTV promoter. EMBO J. 1987;6(8):2321–8.
100. Brüggemeier U, Kalff M, Franke S, Scheidereit C, Beato M. Ubiquitous transcription factor OTF-1 mediates induction of the MMTV promoter through synergistic interaction with hormone receptors. Cell. 1991;64(3):565–72.
101. Lucas PC, Granner DK. Hormone response domains in gene transcription. Annu Rev Biochem. 1992;61:1131–73.
102. Yamamoto KR, Darimont BD, Wagner RL, Iniguez-Liuhi JA. Building transcriptional regulatory complexes: signals and surfaces. Cold Spring Harb Symp Quant Biol. 1998;63: 587–98.
103. Chodankar R, Wu DY, Schiller BJ, Yamamoto KR, Stallcup MR. Hic-5 is a transcription coregulator that acts before and/or after glucocorticoid receptor genome occupancy in a gene-selective manner. Proc Natl Acad Sci U S A. 2014;111(11):4007–12.
104. Wolf G, Lane MD, Johnson BC. Studies on the function of vitamin A in metabolism. J Biol Chem. 1957;225:995–1008.
105. Imai E, Stromstedt PE, Quinn PG, Carlstedt-Duke J, Gustafsson JÅ, Granner DK. Characterization of a complex glucocorticoid response unit in the phosphoenolpyruvate carboxykinase gene. Mol Cell Biol. 1990;10(9):4712–9.

106. Hall RK, Sladek FM, Granner DK. The orphan receptors COUP-TF and HNF-4 serve as accessory factors required for induction of phosphoenolpyruvate carboxykinase gene transcription by glucocorticoids. Proc Natl Acad Sci U S A. 1995;92(2):412–6.
107. Wang JC, Stromstedt PE, O'Brien PM, Granner DK. Hepatic nuclear factor 3 is an accessory factor required for the stimulation of phosphoenolpyruvate carboxykinase gene transcription by glucocorticoids. Mol Endocrinol. 1996;10:794–800.
108. Wang JC, Stafford JM, Scott DK, Sutherland C, Granner DK. The molecular physiology of hepatic nuclear factor 3 (HNF3) in the regulation of gluconeogenesis. J Biol Chem. 2000; 275:14717–21.
109. Scott DK, Mitchell JA, Granner DK. The orphan receptor COUP-TF binds to a third glucocorticoid accessory factor element within the phosphoenolpyruvate carboxykinase gene promoter. J Biol Chem. 1996;271:31909–14.
110. Hall RK, Wang XL, George L, Koch SR, Granner DK. Insulin represses phosphoenolpyruvate carboxykinase gene transcription by causing the rapid disruption of an active transcription complex: a potential epigenetic effect. Mol Endocrinol. 2007;21(2):550–63.
111. Stafford JM, Wilkinson JC, Beechem JM, Granner DK. Accessory factors facilitate the binding of glucocorticoid receptor to the PEPCK gene promoter. J Biol Chem. 2001; 276(43):39885–91.
112. Wang JC, Stafford JM, Granner DK. SRC-1 and GRIP1 coactivate transcription with hepatocyte nuclear factor 4. J Biol Chem. 1998;273:30847–50.
113. Wang JC, Stromstedt PE, Sugiyama T, Granner DK. The phosphoenolpyruvate carboxykinase gene glucocorticoid response unit: identification of the functional domains of accessory factors HNF3β and HNF4 and the necessity of proper alignment of their cognate binding sites. Mol Endocrinol. 1999;13:604–18.
114. Yoon JC, Puigserver P, Chen G, et al. Control of hepatic gluconeogenesis through the transcriptional coactivator PGC-1. Nature. 2001;413:131–8.
115. Lucas PC, Forman BM, Samuels HH, Granner DK. Specificity of a retinoic acid response element in the phosphoenolpyruvate carboxykinase gene promotor: consequences of both retinoic acid and thyroid hormone receptor binding. Mol Cell Biol. 1991;11(10):5164–70.
116. Short JM, Wynshaw-Boris A, Short HP, Hanson RW. Characterization of the phosphoenolpyruvate carboxykinase (GTP) promoter-regulatory region. J Biol Chem. 1986;261:9721–6.
117. Bokar JA, Roesler WJ, Vandenbark GR, Kaetzel DM, Hanson RW, Nilson JH. Characterization of the cAMP responsive elements from the genes for the alpha-subunit of glycoprotein hormones and phosphoenolpyruvate carboxykinase (GTP). Conserved features of nuclear protein binding between tissues and species. J Biol Chem. 1988;263(36):19740–7.
118. O'Brien RM, Bonovich MT, Forest CD, Granner DK. Signal transduction convergence: phorbol esters and insulin inhibit phosphoenolpyruvate carboxykinase gene transcription through the same 10-base-pair sequence. Proc Natl Acad Sci U S A. 1991;88(15):6580–4.
119. Matsumoto M, Pocai A, Rossetti L, DePinho RA, Accili D. Impaired regulation of hepatic glucose production in mice lacking the forkhead transcription factor Foxo1 in liver. Cell Metab. 2007;6:208–16.
120. Churchman LS, Weissman JS. Nascent transcript sequencing visualizes transcription at nucleotide resolution. Nature. 2011;469(7330):368–73.
121. Kwak H, Fuda NJ, Core LJ, Lis JT. Precise maps of RNA polymerase reveal how promoters direct initiation and pausing. Science. 2013;339(6122):950–3.
122. Ucker DS, Yamamoto KR. Early events in the stimulation of mammary tumor virus RNA synthesis by glucocorticoids. Novel assays of transcription rates. J Biol Chem. 1984; 259(12):7416–20.
123. Peterson DD, Koch SR, Granner DK. The 3′ noncoding region of phosphoenolpyruvate carboxykinase mRNA contains a glucocorticoid-responsive mRNA-stabilizing element. Proc Natl Acad Sci U S A. 1989;86:7800–4.

124. Bittencourt D, Wu D-Y, Jeong KW, et al. G9a functions as a molecular scaffold for assembly of transcriptional coactivators on a subset of glucocorticoid receptor target genes. Proc Natl Acad Sci U S A. 2012;109(48):19673–8.
125. Engel KB, Yamamoto KR. The glucocorticoid receptor and the coregulator Brm selectively modulate each other's occupancy and activity in a gene-specific manner. Mol Cell Biol. 2011;31(16):3267–76.
126. Yu C-Y, Mayba O, Lee JV, et al. Genome-wide analysis of glucocorticoid receptor binding regions in adipocytes reveal gene network involved in triglyceride homeostasis. PLoS One. 2010;5:1–13.
127. So AY-L, Chaivorapol C, Bolton E, et al. Determinants of cell-and gene-specific transcriptional regulation by the glucocorticold receptor. PLoS Genet. 2007;94:1–16.
128. Kuo T, Lew M, Mayba O, et al. Genome-wide analysis of glucocorticoid receptor-binding sites in myotubes identifies gene networks modulating insulin signaling. Proc Natl Acad Sci U S A. 2012;109(28):11160–5.
129. Grontved L, John S, Baek S, Liu Y, et al. C/EBP maintains chromatin accessibility in liver and facilitates glucocorticoid receptor recruitment to steroid response elements. EMBO J. 2013;32(11):1568–83.
130. John S, Sabo PJ, Thurman RE, et al. Chromatin accessibility pre-determines glucocorticoid receptor binding patterns. Nat Genet. 2011;43(3):264–8.
131. Polman JA, Welten JE, Bosch DS, et al. A genome-wide signature of glucocorticoid receptor binding in neuronal PC12 cells. BMC Neurosci. 2012;13:118.
132. Schiller BJ, Chodankar R, Watson LC, Stallcup MR, Yamamoto KR. Glucocorticoid receptor binds half sites as a monomer and regulates specific target genes. Genome Biol. 2014;15(7):418.
133. So AY, Bernal TU, Pillsbury ML, Yamamoto KR, Feldman BJ. Glucocorticoid regulation of the circadian clock modulates glucose homeostasis. Proc Natl Acad Sci U S A. 2009; 106(41):17582–7.
134. Haurwitz RE, Jinek M, Wiedenheft B, Zhou K, Doudna JA. Sequence- and structure-specific RNA processing by a CRISPR endonuclease. Science. 2010;329(5997):1355–8.
135. Hochstrasser ML, Doudna JA. Cutting it close: CRISPR-associated endoribonuclease structure and function. Trends Biochem Sci. 2015;40(1):58–66.

Chapter 2
Molecular Biology of Glucocorticoid Signaling

Margarita Arango-Lievano, W. Marcus Lambert, and Freddy Jeanneteau

Abstract Well-defined as signaling hormones for the programming of cell type-specific and context-dependent gene expression signatures, glucocorticoids control experience-driven allostasis. One unifying model is that glucocorticoids help maintaining the integrity and plasticity of cellular networks in changing environments through the mobilization of cellular energy stores, profiling of gene expression, and changes in the electrical and morphological properties of cells. The nucleus is their primary site of action, yet recent discoveries point to additional gene transcription-independent functions at the plasma membrane of neuronal synapses. Glucocorticoids are secreted factors that reflect intrinsically the changes coming from the external world, temporally and regionally, during development and adulthood. In this review, we will enumerate the properties and signaling attributes of glucocorticoids and their receptors that characterize them as allostatic modulators. The molecular mechanisms used to support their role at the synapse will be highlighted.

Keywords Allostasis • Transcription • MAPK • Phosphorylation • Context-dependent signaling

Introduction

Glucocorticoid hormones are chemical messengers that signal via alterations of gene expression in cells expressing cognate receptors. Nuclear receptors are an evolutionarily conserved class of transcription factors that regulate gene expression in a cell type-specific and context-dependent manner. Importantly, ligand-activated

M. Arango-Lievano, Ph.D. • F. Jeanneteau, Ph.D. (✉)
Inserm U1191, CNRS UMR5203, Institute for Functional Genomics,
141 rue de la Cardonille, Montpellier Cedex 05 34094, France
e-mail: margarita.arango@igf.cnrs.fr; freddy.jeanneteau@igf.cnrs.fr

W.M. Lambert, Ph.D.
Department of Microbiology, New York University School of Medicine,
550 First Avenue, New York, NY 10016, USA
e-mail: wil2009@med.cornell.edu

J.-C. Wang, C. Harris (eds.), *Glucocorticoid Signaling*, Advances in Experimental
Medicine and Biology 872, DOI 10.1007/978-1-4939-2895-8_2

nuclear receptors are able to integrate multiple signals giving rise to distinct patterns of gene expression, and rapid non-genomic signaling outcomes yet to be discovered. Converging signaling pathways act in part by altering receptor phosphorylation that fosters the recruitment of interacting co-regulatory molecules in the mitochondria, cytoplasm, synapse and nucleus. This chapter explores how glucocorticoid-mediated allostasis employs rapid non-genomic and slow genomic mechanisms together with ongoing neuronal activity, thus providing a molecular framework for understanding normal and pathological glucocorticoid functions. In particular, the mechanisms underlying the slow appearance of glucocorticoid resistance in numerous human disease-states are not understood although several molecular correlates have been identified. One hypothesis that shows great promise is that the survival/ growth MAPK pathway modulates glucocorticoid-mediated allostasis, in part, through the phosphorylation of the glucocorticoid receptors. This chapter synthesizes our current understanding of glucocorticoid signaling with an emphasis on cellular networks of the brain.

Ligands

Sources

Glucocorticoids are cholesterol-derived steroid hormones that prepare the organism to adapt to changing environments during development and adulthood. Metabolism, immunity, cognition, circadian learning and allostatic response (defined as the physiological processes engaged to return to homeostasis [1]) to physical or psychological threats are well-characterized physiological responses that engage glucocorticoid hormone signaling. Cortisol, the major endogenous glucocorticoid in humans (corticosterone in rodents), is secreted by the adrenal cortex where biosynthetic enzymes are produced [2]. The classical view is that cortisol reaching the brain comes from the adrenal gland. Yet, the expression of some of these enzymes like 11βhydroxysteroid dehydrogenase (11βHSD1) exists in other tissues like the brain, which supports the possibility of local production sites for glucocorticoids and related hormones called neurosteroids [3]. Therefore, adrenal glands are the major source of glucocorticoids but the presence of biosynthetic or degradation enzymes modulate the availability of these hormones at target organs. Future investigations shall clarify the impact of the presence of 11βHSD1 in the brain. Interestingly, it is the brain that processes sensory information and activates the hypothalamic-pituitary-adrenal (HPA) axis that triggers the secretion of cortisol. Hypothalamic paraventricular neurons secrete corticotropin releasing hormone (CRH) and arginine-vasopressin (AVP) into the portal vessels to reach the anterior lobe of the pituitary gland and stimulate adrenocorticotropic hormone (ACTH) secretion. Circulating ACTH binds to cognate receptor in the cortex of the adrenal gland, which stimulates the (i) biosynthesis and (ii) release of cortisol in the bloodstream. Through the vasculature, crossing the blood–brain-barrier, cortisol can readily access every organ to prepare a coordinated cellular allostatic response.

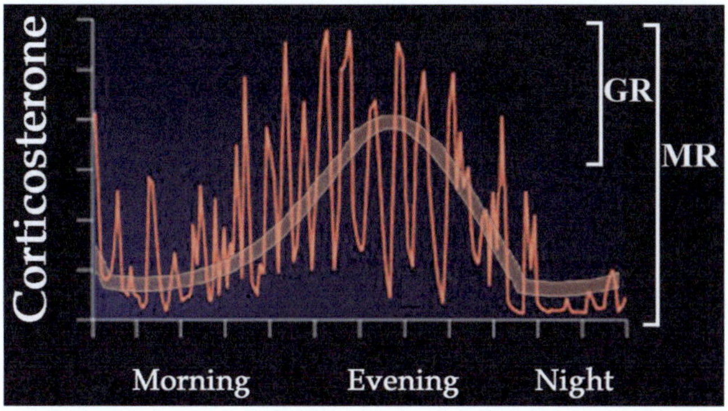

Fig. 2.1 Typical superimposed ultradian (*fine line*) and circadian (*thick line*) circulating corticosterone plasma level. Due to a very high affinity for corticosterone, MR are constantly activated contrary to GR only activated at secretion peaks (adapted from [6])

Thanks to robust anti-inflammatory and immunosuppressive effects, glucocorticoids are a mainstay of treatment for numerous inflammatory and immune diseases, and against trauma despite a wide range of side effects in multiple organs (brain, bone, liver, lung, eye, muscle). Side effects of glucocorticoid therapy that involve the brain include manifestations of emotional liability, psychosis, gain or loss of appetite, insomnia, and memory impairments.

Secretion Modes

Pulsatile

Glucocorticoids are secreted in synchrony with circadian rhythms and in response to stress. At rest, the HPA axis displays a circadian pattern of activity, which produces a cortisol secretion peak during the active period of the day and a trough during the inactive phase (Fig. 2.1). A recent study demonstrates that glucocorticoid peaks and troughs are critical determinants of circadian learning [4]. That is, mice learning at glucocorticoid peaks (evening) acquire a learned motor coordination task better than mice trained at glucocorticoid trough (morning). If learning is sensitive to glucocorticoid circadian rhythms, recall of the learned motor task is not sensitive to glucocorticoid peaks and troughs. This study illustrates how naturally occurring glucocorticoid oscillations impacts distinct phases required for behavioral adaptation to novelty. Circadian activation of the HPA axis is controlled by the suprachiasmatic nucleus, which functions as a light–dark oscillator with direct output projections to the hypothalamic PVN. Thus, neurons of the suprachiasmatic nucleus order CRH-producing neurons to activate the HPA axis, at the time period that precedes the awakening and active phase [5–7].

Superimposed to the circadian rhythms are the ultradian oscillations. The frequency of ultradian pulsatory release of glucocorticoid is much more rapid than that of circadian rhythms. An ultradian oscillation usually resolves within a 2 h period [8–10]. The coincidence of ultradian and circadian peaks and troughs produces the highest amplitudes of glucocorticoid oscillations [6] (Fig. 2.1). Counter-intuitively, glucocorticoid secretion at ultradian trough during the circadian peak may be lower than that resulting from ultradian peak at circadian trough. One question that arises from this notion is whether learning at the ascending ultradian peak is more effective than learning at the ultradian trough. According to proposed mathematical models, this ultradian oscillation is the consequence of the slightly delayed feedforward control of pituitary ACTH on glucocorticoid release and the feedback control of adrenal glucocorticoids on ACTH release that depends on an underlying CRH drive but does not necessitate CRH pulsatility [11]. This model contrast with the best described endocrine system controlled by ultradian oscillations in which the oscillating secretion of gonadotropin-releasing hormone from the hypothalamus directly controls pulses of LH and FSH in the pituitary [12, 13]. Disruption of circadian glucocorticoid rhythms in animal models is sufficient to impair learning and memory and to produce symptoms of depression [4, 14]. Consistently, perturbation of glucocorticoid oscillations have been associated with numerous diseases of the central nervous system [6].

In Response to Stress

Circulating levels of glucocorticoids not only vary as a result of intrinsic rhythms, they are also secreted in response to stress. Polysynaptic limbic circuits converge on the hypothalamic paraventricular nucleus (PVN) to activate CRH-producing neurons [15]. These input circuits involve multiple brain regions including the hippocampus, prefrontal cortex, amygdala, bed nucleus of the stria terminalis (BNST), and locus coeruleus (LC). The intricacy of these circuits allow for a diversity of stressor specific responses like physical stress or emotionally arousing experiences, leading to the activation of the HPA axis. Deactivation of the HPA axis is critical to prevent escalation of glucocorticoid levels beyond physiological range. A remarkable feature of this pathway is the ability of glucocorticoid to exert feedback inhibition on CRH, AVP, ACTH and glucocorticoid levels, resulting in the deactivation of the HPA axis. The dexamethasone-suppression test allows for the determination of the strength of the negative feedback in humans and animal models. About 50 % of depressed individuals in multiple cohorts present with diminished negative feedback [16]. Such signs of glucocorticoid resistance gave rise to the glucocorticoid hypothesis of depression [17]. Therefore, maintenance of the HPA axis involves homeostatic equilibrium between the activation pathway and inhibitory feedback.

Availability

Bioavailability of glucocorticoids to target tissues is not only regulated by its secretion patterns but also by carrier proteins in the blood. Corticosteroid binding globulin (CBG) a high affinity plasma protein and albumin bind up to 95 % of plasma glucocorticoids. Only the remaining unbound fraction is free to diffuse across plasma membranes and bind receptors, be degraded by the liver or processed by converting enzymes [18]. CBG is saturated at glucocorticoid concentrations corresponding to the peak concentration of each pulse (400–500 nmol cortisol for humans), permitting an increase of free cortisol only at these time points [6, 19]. CBG-glucocorticoid coupling regulation, allows for spatial and temporal modulation of glucocorticoid availability. Minute temperature rise leads to a reversible decrease in CBG affinity for glucocorticoids, permitting in case of fever or inflammation an increment of glucocorticoid availability [19]. Inflammation sites are rich in proteinases, like elastase that cleaves CBG such that steroid binding is irreversibly lost [20, 21]. Glucocorticoid availability in target tissues is also regulated by locally expressed 11βHSD enzymes (see Chap. 16 for more discussion on this topic). These exist in two isoforms that convert corticosteroid precursors into cortisol and vice-versa [22]. Type 2 (11βHSD-2) catalyzes inactivation by converting cortisol in inactive cortisone. Type 1 (11βHSD-1) catalyzes activation by converting inactive keto-forms like cortisone in active cortisol, locally within the cells. 11βHSD-1 is abundant in the liver, adipose tissue and brain, where it plays a role in ageing related cognitive decline, as demonstrated by stable learning ability of 11βHSD-1 KO mice [23, 24]. Finally, to access target tissues like the brain, glucocorticoids need to cross the blood–brain barrier, a specialized layer of endothelial cells. Despite their lipophilic nature permitting diffusion through plasma membranes, transporters pump glucocorticoids out of the intracellular space. Among these transporters, multidrug resistance-p-glycoproteins 1a and b (Mdr1) are expressed by endothelial cells of the blood–brain barrier and cortical and hippocampal neurons [25]. Interestingly, *Mdr1a* expression in neurons is upregulated following seizures [25, 26].

Receptors

Early postnatal development is characterized by a glucocorticoid hypo-responsive period during which the organism is highly vulnerable to extreme environmental changes. Extensive research using early life stress paradigms indicates that glucocorticoid signaling is not fully mature during embryogenesis and the first weeks of life. In consequence, pups are unable to cope with extreme stress and easily develop defects that persist during adulthood. For example, early life stress (or glucocorticoid treatment during the same period) impairs neuronal growth and differentiation that could result in neural circuit wiring defects responsible for the development of cognitive and neuropsychiatric illnesses during adulthood [27]. It is believed that such critical period of glucocorticoid responsiveness depends on receptor expression patterning that is temporally and spatially regulated.

To date, glucocorticoids have been shown to signal trough two nuclear receptors, the mineralocorticoid receptor (MR), and the glucocorticoid receptor (GR), denominations that correspond to their peripheral activity, as MR regulates electrolyte balance, and GR is involved in gluconeogenesis. Both of these receptors belong to the nuclear hormone receptor family. Nuclear hormone receptors are comprised of

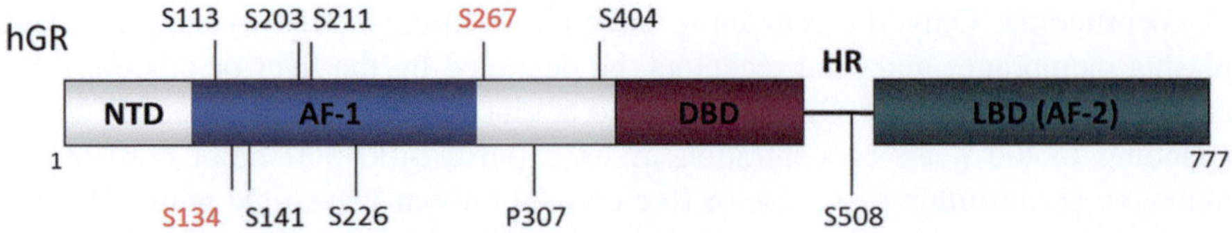

Fig. 2.2 A schematic diagram of the functional domains and best characterized phosphorylation sites of human GR. Sites in *red* are BDNF-dependent sites. *NTD* amino terminal domain, *AF-1* activation function-1, *DBD* DNA-binding domain, *HR* hinge region, *LBD* ligand-binding domain, *AF-2* activation function-2, *S* serine, and *P* proline

multiple, independent functional domains [28]. These include (i) a variable amino-terminal domain (NTD or A/B region), containing an activation function (AF)-1 with ligand-independent transactivation activity; (ii) a conserved DNA binding domain (DBD or C region) involved in the recognition of specific DNA ligand sequences; (iii) a variable hinge region (D) connecting the DBD to the ligand binding domain (LBD); (iv) and a conserved LBD (E) containing an additional region for transactivation activity named AF-2. Some receptors contain an additional highly variable carboxyl-terminal region of unknown function (Fig. 2.2). Of these functional domains, the NTD is the most variable among nuclear receptors in terms of length and sequence similarities and is the major target for ligand-dependent phosphorylation at multiple serine residues [29, 30].

The Cloned Receptor Subtypes

Mineralocorticoid Receptor

MR is encoded by the *NR3C2* gene. The human gene is composed of ten exons and eight introns. The first two exons, 1β and 1α undergo alternative splicing, giving rise to two different mRNA forms [31]. This alternative transcription is under control of two distinct promoters whose activities are tissue specific and regulated during development. The alternative spliced exons encode the 5′ untranslated region of the transcript and translation starts at exon 2. Both transcript variants result in the same protein of 984 amino acids [31, 32]. In the adult brain, MR is mostly expressed in the dorsolateral septum and hippocampus. The highest levels of MR receptors are detected in the pyramidal layer of the CA1 and CA2 region of the hippocampus and the granular layer of the dentate gyrus, with lower expression on the pyramidal layer of the CA3 region of the hippocampus [33, 34]. Outside the brain, the main ligand for MR is aldosterone, a corticoid hormone also secreted by the adrenal gland at a concentration 100-fold less than glucocorticoids. Having a similar affinity for glucocorticoids and mineralocorticoids, the coexpression of MR with 11βHSD-2 in peripheral organs like the kidney provides deterministic signaling properties. Indeed, the degradation of intra-kidney cortisol by 11βHSD-2 allows MR to bind to the less abundant aldosterone [22]. Because 11βHSD-2 is mostly absent from the adult brain, the preferred MR ligand within the brain is cortisol in humans and corticosterone in rodents. Corticosterone binding affinity to MR K_d of 0.1–0.3 nM is very high compared to that of GR K_d of 2–5 nM [35, 36]. One consequence of this very high affinity is the relatively high occupation of MR with endogenous corticosterone that can reach 80 % at a low corticosterone circadian trough level. This translates into a constant activation of MR signaling regardless of circadian and ultradian rhythms (Fig. 2.1). What could be the function of a hippocampal glucocorticoid receptor that would be constantly activated? The hippocampus plays an important role in the HPA axis regulation as demonstrated by lesion studies [37]. Using different pharmacological antagonists, it was demonstrated that blocking MR receptor activates the HPA axis, suggesting a role of MR on the tonic inhibition of the HPA axis [36, 38, 39]. Additionally, dentate granular neurons death resulting from adrenalectomy can be rescued by hormonal replacement with aldosterone, a specific MR receptor agonist [40]. Aldosterone treatment also enhances the proliferation and survival of newly-born granule cells of naïve mice [41]. Given that GR activation inhibits granule cell proliferation in the hippocampus, it is believed that a balance of GR/MR activation along the glucocorticoid circadian and ultradian oscillations controls the electrical activity and number of newly-born neurons in the adult hippocampus.

Glucocorticoid Receptor

GR is encoded by the *NR3C1* gene, comprising 9 exons and 11 introns, and the protein is coded from exons 2 to 9. Exons 1 and 9 undergo alternative splicing. Alternative splicing of exon 9 gives rise to two isoforms: the most prevalent GRα and, and GRβ form that has a shorter C terminus transactivation region [42, 43]. GRα is the majoritarian form and classically shuttles between the cytoplasm and nucleus depending on signaling. GRβ resides permanently in the nucleus and acts as a dominant negative inhibitor of the GRα isoform, but can also directly regulate genes that are not regulated by the α isoform [43, 44]. Additional isoforms of GR are generated by alternative translation via different initiation sites giving rise to eight variants with truncated N termini. The resulting isoforms all bind glucocorticoids with similar affinities but have different transcriptional activity depending on the presence or not of the AF1 domain [43]. The expression of GR is ubiquitous and several isoforms can be co-expressed in many tissues adding to the complexity of GR signaling. Consequently, GR signaling depends on tissue-specific isoforms, expression level and glucocorticoid availability (time at exposure, duration and dose). Given that mice lacking NR3C1 die of lung maturation defects [45, 46], further tissue specific gene inactivation studies provided valuable information regarding the role of limbic circuits in the control of the HPA axis and allostatic responses to drugs of abuse and fear [47–49].

In stark contrast with the MR, it is unlikely that GR is activated at circadian and ultradian glucocorticoid troughs due to its low affinity for endogenous corticoids. This is one important feature of glucocorticoid signaling that alternates between cycles of MR followed by a MR/GR co-activation both of which are critical for cellular networks of the limbic system known to co-express both receptor subtypes (Fig. 2.1). For instance, glucocorticoid circadian peaks are as important as glucocorticoid troughs for learning and memory. If training at glucocorticoid peaks facilitates the acquisition of a motor coordination task, the glucocorticoid trough is paramount for its retention. In vivo imaging studies indicate that glucocorticoid peaks via GR enhance the formation of dendritic spines, the post-synaptic entity of excitatory synapses whereas the following troughs via MR permits the elimination of pre-existing old spines [4]. Overall, the coincidence of glucocorticoid naturally occurring circadian oscillations with procedural learning may favor the patterning of dendritic spines needed to adjust neural circuit wiring with behavioral demands.

Sub-cellular Distribution

The classic view is that both MR and GR reside in the cytoplasm of cells and shifts to their nucleus upon ligand binding for transcriptional regulation. Yet, unliganded GR can be found in the nucleus unbound to chromatin and yet, readily accessible to cortisol [50–52]. Sub-cellular distribution is actually controlled by the rates of import and export through the nuclear pores, the directionality being determined by glucocorticoid binding and recruitment of specific co-factors like importins, HSP90, FKBP52 and FKBP51. For instance, cytoplasmic location is favored in the absence of ligand, when GR is bound to a HSP90-FKBP51-based chaperone complex. This chaperone complex stabilizes the receptor, represses its regulatory activities, and favors a conformation that facilitates ligand binding [53]. Upon glucocorticoid binding, the receptor undergoes a conformational change and is released from the cytoplasmic chaperone-complex. Structural rearrangements elicited upon ligand binding notably expose nuclear localization signal sequences facilitating the binding to import proteins and active transport through the nuclear pores. Once in the nucleus, the interaction of MR and GR with DNA ligands and transcription factors specify the transcriptional targets involved in the glucocorticoid genomic signaling. The nucleus is not the only organelle where DNA ligands for the GR have been characterized [54]. Such discovery opens new avenue for understanding cell autonomous metabolic effects of glucocorticoids through the mitochondria [55] (Fig. 2.3). Since then, mitochondrial GR has been purified in brain extracts from mice subjected to chronic stress [56]. The functional role of mitochondrial GR is still in its infancy but experiments that forced GR localization to the mitochondria suggests it is highly toxic [57]. Beyond the dogmatic genomic effects, which are slow at onset (minutes to hours from the transcription to the bioactivity of regulated genes) exists rapid non-genomic effects that could rely upon the subcellular distribution of the receptors. One example is illustrated by the effects of glucocorticoids on neuronal excitability that cannot be accounted for genomic effects [58, 59], because they are not only faster than expected from a transcriptional response, but also resistant to protein synthesis inhibitors [58, 60].

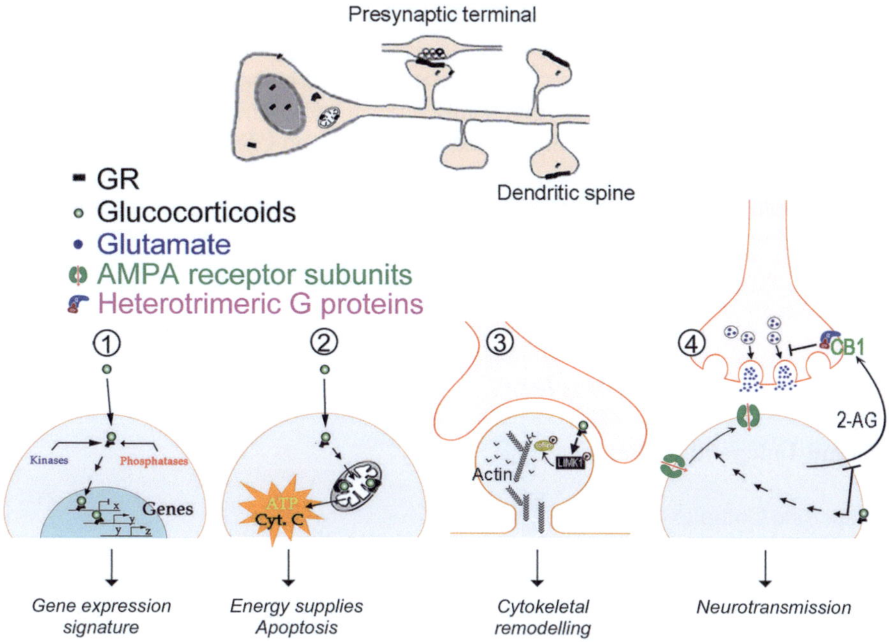

Fig. 2.3 Subcellular distribution of GR associates with distinct signaling outcomes. Neuronal GR is abundant in the nucleus, somatic cytoplasm and less abundant but detectable in the mitochondria and synaptic terminals like post-synaptic dendritic spines. (*1*) Ligand activated nuclear GR binds to DNA to induce or repress gene transcription, which target Y may also function as transcription factor to trigger a second wave of glucocorticoid response involving the gene Z. GR phosphorylation is tightly control by kinases and phosphatases that regulates the binding of biased signaling co-factors. (*2*) Ligand activated GR can translocate to the mitochondria to regulate the production of ATP and the release of cytochrome C known to endanger the survival prognosis of targeted cells. (*3*) Ligand-activated GR signaling from a membrane origin rapidly activates the LIMK1-Cofilin pathway that impinges on the turnover of the actin cytoskeleton. As a result, glucocorticoids rapidly enhance the formation of post-synaptic dendritic spines. (*4*) The postsynaptic membrane bound GR first facilitates neurotransmission by enhancing the transport of AMPA receptor subunits to the active zone of the synapse, and diminish neurotransmission through a pre-synaptic mechanism that requires the release of a retrograde messenger (endocannabinoids) that employs a G-protein coupled receptor (CB1) to suppress the release of the excitatory neurotransmitter glutamate

Other Receptors

Rapid glucocorticoid effects on neuronal excitability can be reproduced with a functional synthetic glucocorticoid conjugated with BSA that cannot penetrate the plasma membrane nor access the cytoplasmic and nuclear receptors. Such glucocorticoid signaling of membrane-bound origin has been replicated in numerous paradigms and frequently associated with a MR type of pharmacology because it can be abolished by a MR antagonist and does not occur in mice lacking MR [61].

Yet, in some tissues, the membrane-bound glucocorticoid receptor is also sensitive to GR antagonists and GR knockdown at least on dendritic spine formation [4]. Localization of MR and GR in close proximity to the plasma membrane notably at the synapse was evidenced by biochemical purification and electron microscopy [60, 62–65]. The existence of membrane-bound MR and GR has been functionally proved in other models [66, 67] (Fig. 2.3). Despite the evidence, the pharmacology of membrane-bound cortisol receptors supports the possibility that several trans-membrane proteins coupled to small heterotrimeric G proteins and sensitive to MR and or GR could signal rapidly.

Associated Protein Complexes

Ligand-Independent Partners

Chaperone Complex

The unliganded MR and GR are permanently scanning the cellular environment thanks to a large complex of chaperone proteins that control the architecture of the glucocorticoid ligand binding domain. Numerous GR and MR binding partners have been characterized. Here, we present only a few examples. Invariably, HSP90 serves as scaffold for the 'foldosome' complex (HSP70, HSP40 and Hop) that molds the ligand-binding pocket of GR, in an ATP-dependent manner [53, 68, 69]. Binding of p23 by this heterocomplex, further stabilizes the conformational change [69]. The release of Hop from the complex leads to the dynamic and competitive addition of other co-chaperones like the immunophilins FKBP51, FKBP52 and cyp 40 and the immunophilin-like phosphatase PP5 that mediate changes in receptor mobility, ligand affinity and signaling capacity [70, 71]. For example, FKBP52 facilitates GR signaling activity by bringing together GR and motor proteins onto the microtubules for active nuclear transport [70, 72]. In contrast, binding of FKBP51 to the chaperone supercomplex attenuates glucocorticoid signaling by facilitating the nuclear export of GR. The robust induction of FKBP51 expression by glucocorticoid, unlike FKBP52, suggests that glucocorticoid signaling feedback can also be cell autonomous [73].

Ligand Dependent Partners

Numerous proteins have been functionally associated with the liganded MR and GR. By no means is it the goal of this chapter to enumerate them. Only a small selection of relevant partners will be emphasized, as it is dependent on post-translational modifications and subcellular distribution.

Post-translational Modifications

With low basal phosphorylation at rest, MR and GR get hyperphosphorylated upon ligand binding offering binding sites for the 14-3-3 family of proteins. Despite multiple phosphorylated serine or threonine residues identified in the N-terminal domain of GR [74], only a select number of sites have been functionally characterized. For instance, glucocorticoid-dependent phosphorylation at conserved serines 203 (S203), 211 (S211), 226 (S226), and 404 (S404) in the human GR numbering scheme impacts on transcriptional capabilities at specific target gene. Three of these sites demonstrate some interdependency upon one another [75]. All four sites exhibit low basal phosphorylation in the absence of cortisol as long as the protein phosphatase PP5 remains associated with GR and the HSP90 chaperone complex [76]. S203 phosphorylation, however, is characteristically higher without hormone compared to S211 and S226 sites. Upon glucocorticoid binding, the S203 phosphorylated form of the receptor appears perinuclear, suggesting that GR phosphorylated at S203 does not participate in DNA-bound GR transcriptional regulation [75]. Consistent with this idea, S203 phosphorylated GR does not occupy select GREs whereas S211 and S226 phospho-isoforms can bind to DNA ligands in the genome [77]. Classically, phosphorylation of GR at S211 serves as surrogate marker of ligand-activated GR because basal phosphorylation is null in absence of glucocorticoids and ligand-induced S211 phosphorylation correlates with the magnitude of GR transcriptional activity [77]. In mitotic cells, S211 phosphorylation depends on CDK2 whereas CDK5 or ERK may phosphorylate GR in postmitotic neurons [78, 79]. In contrast, phosphorylation at S226 by JNK serve as docking site for specific cofactors that increases nuclear export and reduces GR transcriptional activation [77, 80, 81]. S404, a substrate of GSK3β reduces GR transactivation by increasing the turnover of the liganded GR and by hindering GR-mediated NF-κB repression [82].

Other post-translational modifications are dependent on the phosphorylation code of GR. For example, JNK-dependent phosphorylation of S246 enhances sumoylation at K293 and K277, which are implicated in the control of GR transactivation at multiple GREs [83, 84]. Lastly, ligand-mediated GR degradation by the proteasome requires ubiquitination of K419, which is also dependent upon phosphorylation, as a phospho-defective GR mutant is resistant to ligand-induced degradation [84, 85]. Functionally, GR phosphorylation at rat serine 232 (analogous to S211 in humans), responds to neurotoxic insults to induce specifically the ectopic expression of *hdac2* that correlates with the occurrence of cognitive defects [86]. GR phosphorylation at serine 232 also increases in the rodent brain after exposure to glucocorticoids or stress [87, 88]. The MR is also phosphorylated at rest and upon ligand binding, yet the functional characterization of individual sites is far less advanced [89]. Although the kinases have more or less been characterized for most sites, it is unclear how glucocorticoid signaling can activate kinases.

Nuclear Trafficking

It was long believed that nuclear translocation of GR necessitated the release from the chaperone complex [90]. The emerging picture is that the entire complex moves along the microtubule cytoskeleton depending on the recruitment of subtype specific immunophilins, importins and exportins [91, 92]. GR has two nuclear localization signals, NL1 and NL2. NL1 supports rapid and hormone independent translocation [93], whereas NL2 facilitates slower hormone dependent nuclear import [84, 94]. Several importins interact with GR NL regions and mediate its nuclear translocation across the nuclear pore complex. Importins 7 and 8 bind NL1 and NL2, whereas importin α/β bind exclusively to NL1 [94]. Evidence of importins and nuclear pore proteins binding with components of the chaperone complex shed light into the molecular mechanism of nuclear import. Importin β and nuclear pore glycoprotein Nup62 bind GR and HSP90, p23, and FKBP52 to accelerate nuclear import rate [92, 95]. Nuclear export is dependent on exportin/CRM1 as pharmacological blockade abolishes GR translocation to the cytosol [52]. Nuclear calcium levels also impacts upon GR nuclear residency via a direct interaction between the NES sequences and the calcium-sensing protein calreticulin and other related co-factors such as SRC-1 and 14-3-3σ [79, 84, 96–98].

DNA Binding

Despite a high degree of homology between MR and GR, common protein complexes, and similar post-translational modifications such as ubiquitination, sumoylation and phosphorylation, MR and GR affect distinct gene targets [99]. Yet, GR and MR can form heterodimers [100, 101], which may modulate transcription of a few target genes in a unique way that differ from either GR or MR homodimers. One described example is the serotonin receptor 5HT1A gene which transcription is repressed by GR and MR alone but GR/MR heterodimers exert an even stronger inhibition [102].

Zinc fingers of the DNA binding domain mediate of GR-DNA interaction that usually requires the receptor to dimerize with itself or other transcription factors. Yet, the GRdim mutant that cannot homodimerize is still capable of robust transcriptional regulation at multiple but not all targets-genes [43]. The DNA sequence also acts as an allosteric ligand that influences GR structure and activity [99]. Classically, GREs have been grouped into three classes: (i) simple GREs;(ii) composite GREs; (iii) and tethering elements [103]. Each class requires GR binding in different conformations and orientations as follows. Simple GREs are most often inverted, repeat hexameric sequences separated by three nucleotides, supporting the homodimerization model of GR. A few negative GREs (nGREs) that mediate GR-dependent repression have been well characterized [104]. The inverted repeat IR nGREs found in numerous glucocorticoid-repressed genes [105, 106] support a model whereby direct binding of GR as monomer is not always required but may instead involve the recruitment of specific co-factors like nuclear receptor corepressors 1 and 2 (N-CoR) and (SMRT) [105, 107]. Composite GREs contain non-GR binding sequences surrounding a core GR-binding site important for gene-specific synergistic or antagonistic regulation. GR may interact physically and/or functionally with the transcription factors associated with those motifs. An increasing number of mechanistic and genome-wide studies suggest that most functional transactivating GREs are composite elements, composed of binding sites for GR as well as additional DNA-binding regulatory factors that act in synergy with GR in the particular context of each GRE [108, 109]. Such composite elements might explain some of the cell-type specific regulation by GR and is discussed in additional detail in the following Chapter.

Transcription Cofactors

The typical inverted U-shape responses to increasing dose of glucocorticoids could be in part explained by "squelching" or titration of a limiting co-factor required for full response by the activated GR [110–112]. At high concentrations of activated GR, a factor becomes limiting for full GR activity and as such, transcriptional activation is reduced. Thus, the squelching model emphasizes the availability of transcription co-factors. In the case of tethering GREs, it is the DNA bound co-factor of MR and GR that specifies the glucocorticoid-targets genes. Such factors are AP-1, CREB, NF-κB, or STAT5, C/EBP, SP-1, Egr-1 and others less characterized. Most tethering GREs do not contain canonical sequence motifs for GR, but rather contain motifs for the interacting transcription factor that may also mediate induction of target genes in the absence of glucocorticoids. For example, tethering of GR to AP-1 or NF-κB alters the assembly of coactivator complexes and recruits the corepressor glucocorticoid receptor interacting protein 1 (GRIP1) [113, 114]. In the case of direct GR-DNA binding, the transcriptional output depends on recruitment of a distinct class of co-factors that include P300/CBP, HDAC2, p160, MED1,14, and SWI/SNF complex. These factors are thought to affect chromatin structure or the stability of transcriptional machinery at the transcription start site. Coactivator complexes are assembled through interactions with GR's AF-1 or more commonly the AF-2 domain [115]. The p160 family of coactivators include SRC-1, SRC-2/GRIP1/transcriptional intermediary factor 2 (TIF2), and SRC-3 (also known as pCIP/ACTR/AIB1/RAC3). P160 proteins often increase GR transcriptional activity, but GRIP1 was shown to act both as a coactivator and corepressor via intrinsic activation and repression domains [113, 116]. GRIP1 specifically utilizes its repression domain at AP-1 and NF-κB tethering GREs, whereas SRC-1 and SRC-2 lack this domain and fail to assist repression at AP-1 [82]. Many p160 proteins facilitate transcriptional activation by recruiting p300/CBP. P300/CBP contains potent acetyltransferase activity that target histones and other proteins. GR has also been shown to interact with a number of histone deacetylases (HDACs) including: HDAC 1 and 6 as coactivators [117–119] and HDAC2 as both a coactivator and corepressor, and its recruitment depend on S211 phosphorylation [82, 86, 119, 120]. Through its AF1 and AF2 domains GR interacts with MED14 and MED1 two components of the mediator complex [121]. The mediator complex is a large multiunit complex that bridges transcription factors and RNA pol II, affecting transcription initiation [122] and elongation [123]. The assembly of select mediator subunits alters promoter responsiveness to various transcription factors, so distinct subunits and confirmations are important for GR transcriptional activity. In cell-based reporter assays, overexpression of MED14 enhances GR transcriptional activity, while MED1 only enhanced GR activity in the context of MED14 [124].

Signaling

A large number of well-documented rapid and slow cellular responses to glucocorticoids contribute to glucocorticoid-mediated allostasis. It is remarkable that slow effects can either oppose or reinforce the rapid effects. In this section, we will describe a few examples illustrating how pertinent the spatial and temporal resolution of glucocorticoid signaling is to adjust neuronal networks to changing environments.

Minutes

Rapid glucocorticoid signaling has been particularly studied in the context of neuronal excitability within the range of seconds to minutes. For instance, the excitability of dorsal hippocampal CA1 neurons is sensitive to glucocorticoids within minutes of exposure as demonstrated by an increase in the frequency of spontaneous excitatory neurotransmission (Fig. 2.3). Pharmacological and genetic proof of concept studies revealed the contribution of the membrane-bound MR [125, 126]. Pharmacological characterization of downstream signaling revealed the requirement of the ERK1/2 pathway [126]. An increase in frequency of spontaneous excitatory neurotransmission usually results from an increased quantal release of the neurotransmitter glutamate, and extracellular glutamate levels are augmented following acute glucocorticoid treatment. As much as glucocorticoids affect presynaptic neurotransmitters release, rapid glucocorticoid effects on the post-synaptic neuronal terminal were also characterized. Indeed, glucocorticoids through the membrane-bound MR rapidly (few minutes) increase the mobility and dwell time at the post-synaptic density of glutamate receptors GluR2-AMPAR [127] (Fig. 2.3). Such effects are usually interpreted as enhancing neurotransmission. Similar observations were reported in the basoloateral amygdala (BLA) that yet, expresses much lower MR levels than the hippocampus [66]. Interestingly, the rapid increase of neuronal excitability by the membrane-bound MR is context dependent as it varied with a history of glucocorticoid exposure or stress [66]. That is, a second pulse stimulation of brain slice preparations with corticosterone after washout of a first pulse decreased the excitability of BLA neurons that a single pulse could increase [21, 66]. This suggests that time at exposure with glucocorticoids also determines the signaling outcome. This picture is not the rule because glucocorticoids cause a rapid (few minutes) suppression of excitatory synaptic inputs in hypothalamic

CRH- and AVP-releasing neurons of the paraventricular nucleus (PVN) [58, 128]. Physiologically, this effect contributes to one of the rapid components of the negative feedback that deactivates the HPA axis. In this paradigm, a membrane-bound GR pathway was highlighted using a synthetic membrane impermeant GR agonist that curiously cannot be abolished by a specific GR antagonist, suggesting the involvement of a membrane bound GR with non-canonical pharmacological features [58]. Suppression of PVN neuronal excitation employs endocannabinoids released from the post-synaptic membrane and signaling through the CB1 receptor at the pre-synaptic terminals to suppress glutamate release [58, 129, 130]. Given this signaling pathway involves small heterotrimeric G protein, this finding reconciles previous data indicating that glucocorticoid signaling of membrane origin is sensitive to antagonists of MR, GR and G proteins [58, 129]. Finally, glucocorticoids enhance dendritic spine formation in the living cortex within minutes of exposure via GR-mediated activation of a LIMK-cofilin pathway that impinges on the dynamics of the post-synaptic actin cytoskeleton [4] (Fig. 2.3).

Hours

While the rapid glucocorticoid signaling serves immediate purposes like the deactivation of the HPA axis, slow glucocorticoid signaling is classically viewed as an adaptive response of cellular networks to changing environments like the suppression of CRH transcription to maintain HPA axis homeostasis [131]. Both the rapid and slow signaling components often studied separately shall be considered as integrated response overtime. For instance, it is intriguing that only slow effects of GR and MR phosphorylation on transcriptional activity have been studied given that phosphorylation occurs within minutes of stimulation with cortisol. In the hippocampus, GR mediated genomic effects (transcription and RNA decay) follow the rapid increase of hippocampal excitability mediated by MR in order to reduce neuronal excitability by changing the expression levels of signaling molecules and ion channels like the L type calcium channel [21, 132]. Such a short-term increase of hippocampal neurons excitability is viewed as means to consolidate stress-related memory whereas the following decrease in excitability is viewed as means to protect hippocampal cell networks from noise information [21]. This picture is not the rule because glucocorticoids increase excitability in the BLA that persists after drug washout and is sensitive to protein synthesis inhibitors, GR inhibitors, and absent in GR knockout mice, pointing towards a genomic GR-dependent mechanism [66, 133].

Cell-type specific GR-dependent genomic effects are well-illustrated by the transcription of *crh*, the major molecular trigger of the HPA axis that is increased in the amygdala but decreased in the hypothalamic PVN as a function of elevated levels of glucocorticoids [134]. In the cortex, learning-dependent weaving of neural networks relies on the formation and elimination of synaptic connectivity as a function of glucocorticoid levels. At glucocorticoid circadian trough when only MR is activated, elimination of dendritic spines compensate for the increased GR-dependent spine formation that occurred during the glucocorticoid circadian peak [4]. This result is interpreted as if learning associated new spines are offset by the elimination of pre-existing old spines within distinct temporal domains to shape neural circuits as a function of novelty.

Context-Dependent Glucocorticoid Signaling

Glucocorticoid-mediated allostasis employs genomic and non-genomic mechanisms together with ongoing, experience-driven neural activity mediated by excitatory amino acids neurotransmitters, neurotrophic factors such as BDNF, neuropeptides such as CRH and cell adhesion molecules (NCAM) [135, 136]. Therefore, it is particularly interesting to discuss how glucocorticoid activity superimposes with other signaling pathways. For instance, the enhancement of emotionally arousing memories by glucocorticoids requires co-incident norepinephrine signaling in the amygdala [137]. Norepinephrine is capable of potentiating ligand-induced GR transcriptional capabilities and DNA binding in a PI3K-dependent manner [138]. Adrenergic receptor agonists, generally used in the treatment of asthma, synergistically enhance GR-dependent transcription in a cAMP/PKA-dependent manner. Thus, maximal GR transcriptional capabilities are achieved approximately at tenfold lower glucocorticoid concentrations in the presence of cAMP-elevating drugs [139]. Another set of factors are proinflammatory cytokines, such as interleukin IL1, IL-2, IL-6, TNF-a, and interferon (IFN)-a, signaling pathways that alter GR signaling and neuroendocrine function [140, 141]. Co-incident growth factor signaling may also intersect with GR actions. For instance, bFGF and IGF1 can enhance ligand-induced GR transcriptional capabilities in a PI3K-dependent manner [138]. The neurotrophin BDNF through its receptor TrkB can also specify the transcription of select GR-mediated genes [142] that may be relevant for glucocorticoid-induced memory consolidation of fear and inhibitory avoidance [143]. Mechanistically, BDNF signaling results in phosphorylation of rat GR at serines 134 and 267, which fosters cofactor recruitment to promote a novel gene expression signature. Thus, BDNF utilizes GR as a transcription factor to alter glucocorticoid-regulated transcription in neurons [142] (Fig. 2.4). Similarly, GR phosphorylation at rat serine 232 responds to neurotoxic insults to induce specifically the ectopic expression of *hdac2* that correlates with the occurrence of cognitive defects [86]. Pharmacological blockade of HDAC2 activity resolved cognitive impairments produced by the prolonged activation of GR in a model of Alzheimer's disease featuring chronic high glucocorticoid levels known to worsen neuropathological features [86, 144, 145]. In contrast, the expression of HDAC2 is downregulated in glucocorticoid-resistant cases of severe asthma and pharmacological activation of HDAC2 with theophiline and antioxidants present therapeutic value [146, 147]. Glucocorticoid activity may also diverge as a function of estrogen signaling, which increases the expression of the GR phosphatase-PP5 [148]. The opposing effect of cortisol on *crh* expression in the hypothalamus and the amygdala is another good example of

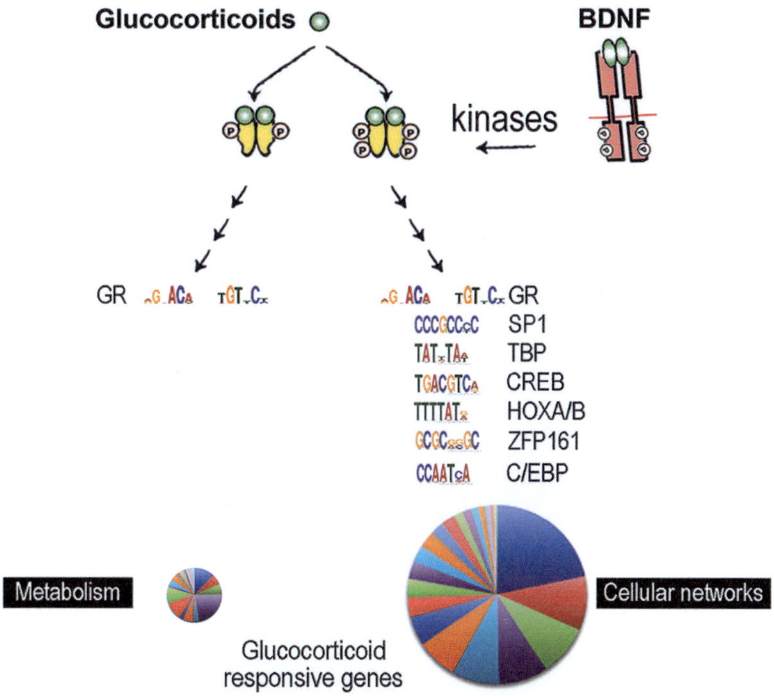

Fig. 2.4 Glucocorticoid transcriptional effects diverge as a function of BDNF signaling. On top of glucocorticoid-mediated GR phosphorylation exists a parallel converging pathway that allows for site-specific GR phosphorylation by the neurotrophic factor BDNF and its receptor TrkB. The resulting hyperphosphorylated GR can activate or repress new select target genes enriched with the indicated transcription factors binding sites, as well as potentiate the expression of GR-sensitive genes. Thus, BDNF-induced GR phosphorylation rewrites the GR transcriptome toward a cellular network signature by fostering the recruitment of phospho-specific cofactors

context-dependent glucocorticoid signaling [134]. It is the CREB co-activator CRTC2 that determines the ability of neurotrophins and glucocorticoids to activate or suppress hypothalamic CRH expression, respectively [112, 149] (Fig. 2.5).

Conclusions

One paramount feature of the body's allostasis resides in the flexibility of the naturally occurring glucocorticoid rhythms and signaling to changing environments. For instance, glucocorticoid-mediated allostasis is critical for behavioral adaptation to novelty, stress coping, learning and memory. Disruption of glucocorticoid circadian and ultradian rhythms is a hallmark of numerous diseases notably neuropsychiatric. In contrast, administration of glucocorticoids, which are mainstay of treatment for

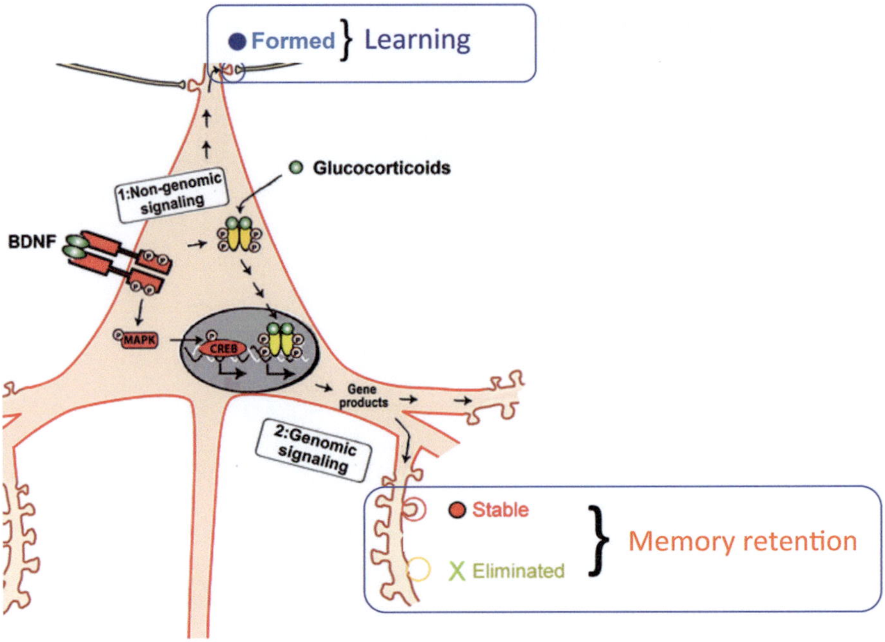

Fig. 2.5 Glucocorticoids employ GR-dependent mechanisms to enhance the formation and the elimination of post-synaptic dendritic spines within distinct temporal domains. Rapid transcription-independent GR signaling facilitates the formation of dendritic spines associated with learning capabilities whereas the slow genomic GR/MR signaling accounts for the stabilization as well as the elimination of pre-existing old spines, a process that is critical for memory retention. The patterning of dendritic spines through the processes of formation and elimination is critical to adjust neural connectivity networks in changing environments

numerous disorders, can produce side effects related to brain functions, like psychosis, depression and memory loss. One hypothesis to account for these effects is that glucocorticoid resistance whether innate or acquired increases the vulnerability to neurotoxic insults that slowly contribute to the development of numerous disorders of the nervous and immune systems. Two examples: glucocorticoids do not cause but worsen the neuropathological feature of Alzheimer's disease and glucocorticoid resistance that interferes with inflammation, increases the sensitivity to develop a common cold. Several studies suggest that the growth/survival MAPK pathway regulates glucocorticoid signaling because diseases featuring glucocorticoid resistance also exhibit disrupted MAPK activity like asthma and depression. Evolution may have selected the MAPK pathway to cope with the allostatic overload of stress and form new memories. One putative mechanism that shows great promises is that MAPKs modulate glucocorticoid activities in part through the phosphorylation of the glucocorticoid receptors. Indeed, MAPK-mediated modulation of GR function appears to be a central player in the development of glucocorticoid resistance. Future functional characterization of the GR and MR phosphorylation codes could help comprehend how glucocorticoid actions can be changed from harmful to protective.

Acknowledgement We are thankful to Michael Garabedian (New York University) for support and Inserm's AVENIR funding program.

References

1. McEwen BS, Wingfield JC. The concept of allostasis in biology and biomedicine. Horm Behav. 2003;43(1):2–15.
2. de Kloet ER, Vreugdenhil E, Oitzl MS, Joëls M. Brain corticosteroid receptor balance in health and disease. Endocr Rev. 1998;19(3):269–301.
3. Compagnone NA, Mellon SH. Neurosteroids: biosynthesis and function of these novel neuromodulators. Front Neuroendocrinol. 2000;21(1):1–56.
4. Liston C, Cichon JM, Jeanneteau F, Jia Z, Chao MV, Gan WB. Circadian glucocorticoid oscillations promote learning-dependent synapse formation and maintenance. Nat Neurosci. 2013;16(6):698–705.
5. Keller-Wood ME, Dallman MF. Corticosteroid inhibition of ACTH secretion. Endocr Rev. 1984;5(1):1–24.
6. Lightman SL, Conway-Campbell BL. The crucial role of pulsatile activity of the HPA axis for continuous dynamic equilibration. Nat Rev Neurosci. 2010;11(10):710–8.
7. Kalsbeek A, Buijs RM. Output pathways of the mammalian suprachiasmatic nucleus: coding circadian time by transmitter selection and specific targeting. Cell Tissue Res. 2002;309(1):109–18.
8. de Kloet ER, Sarabdjitsingh RA. Everything has rhythm: focus on glucocorticoid pulsatility. Endocrinology. 2008;149(7):3241–3.
9. Cook CJ. Measuring of extracellular cortisol and corticotropin-releasing hormone in the amygdala using immunosensor coupled microdialysis. J Neurosci Methods. 2001;110(1–2): 95–101.
10. Jasper MS, Engeland WC. Synchronous ultradian rhythms in adrenocortical secretion detected by microdialysis in awake rats. Am J Physiol. 1991;261(5 Pt 2):R1257–68.
11. Walker JJ, Terry JR, Lightman SL. Origin of ultradian pulsatility in the hypothalamic-pituitary-adrenal axis. Proc Biol Sci. 2010;277(1688):1627–33.
12. Papavasiliou SS, Zmeili S, Khoury S, Landefeld TD, Chin WW, Marshall JC. Gonadotropin-releasing hormone differentially regulates expression of the genes for luteinizing hormone alpha and beta subunits in male rats. Proc Natl Acad Sci U S A. 1986;83(11):4026–9.
13. Wildt L, Hausler A, Marshall G, et al. Frequency and amplitude of gonadotropin-releasing hormone stimulation and gonadotropin secretion in the rhesus monkey. Endocrinology. 1981;109(2):376–85.
14. Gregus A, Wintink AJ, Davis AC, Kalynchuk LE. Effect of repeated corticosterone injections and restraint stress on anxiety and depression-like behavior in male rats. Behav Brain Res. 2005;156(1):105–14.
15. Ulrich-Lai YM, Herman JP. Neural regulation of endocrine and autonomic stress responses. Nat Rev Neurosci. 2009;10(6):397–409.
16. Holsboer F, Ising M. Stress hormone regulation: biological role and translation into therapy. Annu Rev Psychol. 2010;61:81–109.
17. Pariante CM, Miller AH. Glucocorticoid receptors in major depression: relevance to pathophysiology and treatment. Biol Psychiatry. 2001;49(5):391–404.
18. Rosner W. The functions of corticosteroid-binding globulin and sex hormone-binding globulin: recent advances. Endocr Rev. 1990;11(1):80–91.
19. Cameron A, Henley D, Carrell R, Zhou A, Clarke A, Lightman S. Temperature-responsive release of cortisol from its binding globulin: a protein thermocouple. J Clin Endocrinol Metab. 2010;95(10):4689–95.
20. Klieber MA, Underhill C, Hammond GL, Muller YA. Corticosteroid-binding globulin, a structural basis for steroid transport and proteinase-triggered release. J Biol Chem. 2007;282(40):29594–603.

21. Joels M, Sarabdjitsingh RA, Karst H. Unraveling the time domains of corticosteroid hormone influences on brain activity: rapid, slow, and chronic modes. Pharmacol Rev. 2012; 64(4):901–38.
22. Wyrwoll CS, Holmes MC, Seckl JR. 11beta-hydroxysteroid dehydrogenases and the brain: from zero to hero, a decade of progress. Front Neuroendocrinol. 2011;32(3):265–86.
23. Yau JL, Noble J, Kenyon CJ, et al. Lack of tissue glucocorticoid reactivation in 11beta-hydroxysteroid dehydrogenase type 1 knockout mice ameliorates age-related learning impairments. Proc Natl Acad Sci U S A. 2001;98(8):4716–21.
24. Seckl JR, Walker BR. 11beta-hydroxysteroid dehydrogenase type 1 as a modulator of glucocorticoid action: from metabolism to memory. Trends Endocrinol Metab. 2004;15(9): 418–24.
25. Pariante CM. The role of multi-drug resistance p-glycoprotein in glucocorticoid function: studies in animals and relevance in humans. Eur J Pharmacol. 2008;583(2–3):263–71.
26. Volk HA, Burkhardt K, Potschka H, Chen J, Becker A, Loscher W. Neuronal expression of the drug efflux transporter P-glycoprotein in the rat hippocampus after limbic seizures. Neuroscience. 2004;123(3):751–9.
27. Akers KG, Nakazawa M, Romeo RD, Connor JA, McEwen BS, Tang AC. Early life modulators and predictors of adult synaptic plasticity. Eur J Neurosci. 2006;24(2):547–54.
28. Kumar R, Thompson EB. The structure of the nuclear hormone receptors. Steroids. 1999;64(5):310–9.
29. Bodwell JE, Webster JC, Jewell CM, Cidlowski JA, Hu JM, Munck A. Glucocorticoid receptor phosphorylation: overview, function and cell cycle-dependence. J Steroid Biochem Mol Biol. 1998;65(1–6):91–9.
30. Ismaili N, Garabedian MJ. Modulation of glucocorticoid receptor function via phosphorylation. Ann N Y Acad Sci. 2004;1024:86–101.
31. Zennaro MC, Keightley MC, Kotelevtsev Y, Conway GS, Soubrier F, Fuller PJ. Human mineralocorticoid receptor genomic structure and identification of expressed isoforms. J Biol Chem. 1995;270(36):21016–20.
32. Martinerie L, Munier M, Le Menuet D, Meduri G, Viengchareun S, Lombes M. The mineralocorticoid signaling pathway throughout development: expression, regulation and pathophysiological implications. Biochimie. 2013;95(2):148–57.
33. Van Eekelen JA, Jiang W, De Kloet ER, Bohn MC. Distribution of the mineralocorticoid and the glucocorticoid receptor mRNAs in the rat hippocampus. J Neurosci Res. 1988; 21(1):88–94.
34. Gesing A, Bilang-Bleuel A, Droste SK, Linthorst AC, Holsboer F, Reul JM. Psychological stress increases hippocampal mineralocorticoid receptor levels: involvement of corticotropin-releasing hormone. J Neurosci. 2001;21(13):4822–9.
35. Reul JM, de Kloet ER. Two receptor systems for corticosterone in rat brain: microdistribution and differential occupation. Endocrinology. 1985;117(6):2505–11.
36. Reul JM, Gesing A, Droste S, et al. The brain mineralocorticoid receptor: greedy for ligand, mysterious in function. Eur J Pharmacol. 2000;405(1-3):235–49.
37. Jacobson L, Sapolsky R. The role of the hippocampus in feedback regulation of the hypothalamic-pituitary-adrenocortical axis. Endocr Rev. 1991;12(2):118–34.
38. Ratka A, Sutanto W, Bloemers M, de Kloet ER. On the role of brain mineralocorticoid (type I) and glucocorticoid (type II) receptors in neuroendocrine regulation. Neuroendocrinology. 1989;50(2):117–23.
39. Cole MA, Kalman BA, Pace TW, Topczewski F, Lowrey MJ, Spencer RL. Selective blockade of the mineralocorticoid receptor impairs hypothalamic-pituitary-adrenal axis expression of habituation. J Neuroendocrinol. 2000;12(10):1034–42.
40. Woolley CS, Gould E, Sakai RR, Spencer RL, McEwen BS. Effects of aldosterone or RU28362 treatment on adrenalectomy-induced cell death in the dentate gyrus of the adult rat. Brain Res. 1991;554(1–2):312–5.
41. Fischer AK, von Rosenstiel P, Fuchs E, Goula D, Almeida OF, Czeh B. The prototypic mineralocorticoid receptor agonist aldosterone influences neurogenesis in the dentate gyrus of the adrenalectomized rat. Brain Res. 2002;947(2):290–3.

42. Duma D, Jewell CM, Cidlowski JA. Multiple glucocorticoid receptor isoforms and mechanisms of post-translational modification. J Steroid Biochem Mol Biol. 2006;102(1–5):11–21.
43. Kadmiel M, Cidlowski JA. Glucocorticoid receptor signaling in health and disease. Trends Pharmacol Sci. 2013;34(9):518–30.
44. Lewis-Tuffin LJ, Jewell CM, Bienstock RJ, Collins JB, Cidlowski JA. Human glucocorticoid receptor beta binds RU-486 and is transcriptionally active. Mol Cell Biol. 2007;27(6): 2266–82.
45. Kellendonk C, Eiden S, Kretz O, et al. Inactivation of the GR in the nervous system affects energy accumulation. Endocrinology. 2002;143(6):2333–40.
46. Tronche F, Kellendonk C, Kretz O, et al. Disruption of the glucocorticoid receptor gene in the nervous system results in reduced anxiety. Nat Genet. 1999;23(1):99–103.
47. Howell MP, Muglia LJ. Effects of genetically altered brain glucocorticoid receptor action on behavior and adrenal axis regulation in mice. Front Neuroendocrinol. 2006;27(3):275–84.
48. Kolber BJ, Roberts MS, Howell MP, Wozniak DF, Sands MS, Muglia LJ. Central amygdala glucocorticoid receptor action promotes fear-associated CRH activation and conditioning. Proc Natl Acad Sci U S A. 2008;105(33):12004–9.
49. Schmidt MV, Sterlemann V, Wagner K, et al. Postnatal glucocorticoid excess due to pituitary glucocorticoid receptor deficiency: differential short- and long-term consequences. Endocrinology. 2009;150(6):2709–16.
50. Nishi M, Ogawa H, Ito T, Matsuda KI, Kawata M. Dynamic changes in subcellular localization of mineralocorticoid receptor in living cells: in comparison with glucocorticoid receptor using dual-color labeling with green fluorescent protein spectral variants. Mol Endocrinol. 2001;15(7):1077–92.
51. Piwien Pilipuk G, Vinson GP, Sanchez CG, Galigniana MD. Evidence for NL1-independent nuclear translocation of the mineralocorticoid receptor. Biochemistry. 2007;46(5): 1389–97.
52. Hache RJ, Tse R, Reich T, Savory JG, Lefebvre YA. Nucleocytoplasmic trafficking of steroid-free glucocorticoid receptor. J Biol Chem. 1999;274(3):1432–9.
53. Dittmar KD, Banach M, Galigniana MD, Pratt WB. The role of DnaJ-like proteins in glucocorticoid receptor.hsp90 heterocomplex assembly by the reconstituted hsp90.p60.hsp70 foldosome complex. J Biol Chem. 1998;273(13):7358–66.
54. Lee SR, Kim HK, Song IS, et al. Glucocorticoids and their receptors: insights into specific roles in mitochondria. Prog Biophys Mol Biol. 2013;112(1–2):44–54.
55. Du J, Wang Y, Hunter R, et al. Dynamic regulation of mitochondrial function by glucocorticoids. Proc Natl Acad Sci U S A. 2009;106(9):3543–8.
56. Adzic M, Lukic I, Mitic M, et al. Brain region- and sex-specific modulation of mitochondrial glucocorticoid receptor phosphorylation in fluoxetine treated stressed rats: effects on energy metabolism. Psychoneuroendocrinology. 2013;38(12):2914–24.
57. Sionov RV, Cohen O, Kfir S, Zilberman Y, Yefenof E. Role of mitochondrial glucocorticoid receptor in glucocorticoid-induced apoptosis. J Exp Med. 2006;203(1):189–201.
58. Di S, Malcher-Lopes R, Halmos KC, Tasker JG. Nongenomic glucocorticoid inhibition via endocannabinoid release in the hypothalamus: a fast feedback mechanism. J Neurosci. 2003;23(12):4850–7.
59. Di S, Maxson MM, Franco A, Tasker JG. Glucocorticoids regulate glutamate and GABA synapse-specific retrograde transmission via divergent nongenomic signaling pathways. J Neurosci. 2009;29(2):393–401.
60. Groeneweg FL, Karst H, de Kloet ER, Joels M. Mineralocorticoid and glucocorticoid receptors at the neuronal membrane, regulators of nongenomic corticosteroid signalling. Mol Cell Endocrinol. 2012;350(2):299–309.
61. Karst H, Berger S, Turiault M, Tronche F, Schutz G, Joels M. Mineralocorticoid receptors are indispensable for nongenomic modulation of hippocampal glutamate transmission by corticosterone. Proc Natl Acad Sci U S A. 2005;102(52):19204–7.
62. Qiu S, Champagne DL, Peters M, et al. Loss of limbic system-associated membrane protein leads to reduced hippocampal mineralocorticoid receptor expression, impaired synaptic plasticity, and spatial memory deficit. Biol Psychiatry. 2010;68(2):197–204.

63. Prager EM, Brielmaier J, Bergstrom HC, McGuire J, Johnson LR. Localization of mineralocorticoid receptors at mammalian synapses. PLoS One. 2010;5(12), e14344.
64. Wang CC, Wang SJ. Modulation of presynaptic glucocorticoid receptors on glutamate release from rat hippocampal nerve terminals. Synapse. 2009;63(9):745–51.
65. Johnson LR, Farb C, Morrison JH, McEwen BS, LeDoux JE. Localization of glucocorticoid receptors at postsynaptic membranes in the lateral amygdala. Neuroscience. 2005;136(1): 289–99.
66. Karst H, Berger S, Erdmann G, Schutz G, Joels M. Metaplasticity of amygdalar responses to the stress hormone corticosterone. Proc Natl Acad Sci U S A. 2010;107(32): 14449–54.
67. Evanson NK, Tasker JG, Hill MN, Hillard CJ, Herman JP. Fast feedback inhibition of the HPA axis by glucocorticoids is mediated by endocannabinoid signaling. Endocrinology. 2010;151(10):4811–9.
68. Dittmar KD, Demady DR, Stancato LF, Krishna P, Pratt WB. Folding of the glucocorticoid receptor by the heat shock protein (hsp) 90-based chaperone machinery. J Biol Chem. 1997;272(34):21213–20.
69. Stancato LF, Silverstein AM, Gitler C, Groner B, Pratt WB. Use of the thiol-specific derivatizing agent N-iodoacetyl-3-[125I]iodotyrosine to demonstrate conformational differences between the unbound and hsp90-bound glucocorticoid receptor hormone binding domain. J Biol Chem. 1996;271(15):8831–6.
70. Pratt WB, Silverstein AM, Galigniana MD. A model for the cytoplasmic trafficking of signalling proteins involving the hsp90-binding immunophilins and p50cdc37. Cell Signal. 1999;11(12):839–51.
71. Chen S, Smith DF. Hop as an adaptor in the heat shock protein 70 (Hsp70) and hsp90 chaperone machinery. J Biol Chem. 1998;273(52):35194–200.
72. Pratt WB, Toft DO. Steroid receptor interactions with heat shock protein and immunophilin chaperones. Endocr Rev. 1997;18(3):306–60.
73. Jaaskelainen T, Makkonen H, Palvimo JJ. Steroid up-regulation of FKBP51 and its role in hormone signaling. Curr Opin Pharmacol. 2011;11(4):326–31.
74. Miller ML, Jensen LJ, Diella F, et al. Linear motif atlas for phosphorylation-dependent signaling. Sci Signal. 2008;1(35):ra2.
75. Wang Z, Frederick J, Garabedian MJ. Deciphering the phosphorylation "code" of the glucocorticoid receptor in vivo. J Biol Chem. 2002;277(29):26573–80.
76. Wang Z, Chen W, Kono E, Dang T, Garabedian MJ. Modulation of glucocorticoid receptor phosphorylation and transcriptional activity by a C-terminal-associated protein phosphatase. Mol Endocrinol. 2007;21(3):625–34.
77. Chen W, Dang T, Blind RD, et al. Glucocorticoid receptor phosphorylation differentially affects target gene expression. Mol Endocrinol. 2008;22(8):1754–66.
78. Galliher-Beckley AJ, Cidlowski JA. Emerging roles of glucocorticoid receptor phosphorylation in modulating glucocorticoid hormone action in health and disease. IUBMB Life. 2009;61(10):979–86.
79. Kino T, Ichijo T, Amin ND, et al. Cyclin-dependent kinase 5 differentially regulates the transcriptional activity of the glucocorticoid receptor through phosphorylation: clinical implications for the nervous system response to glucocorticoids and stress. Mol Endocrinol. 2007;21(7):1552–68.
80. Itoh M, Adachi M, Yasui H, Takekawa M, Tanaka H, Imai K. Nuclear export of glucocorticoid receptor is enhanced by c-Jun N-terminal kinase-mediated phosphorylation. Mol Endocrinol. 2002;16(10):2382–92.
81. Avenant C, Kotitschke A, Hapgood JP. Glucocorticoid receptor phosphorylation modulates transcription efficacy through GRIP-1 recruitment. Biochemistry. 2010;49(5):972–85.
82. Galliher-Beckley AJ, Williams JG, Collins JB, Cidlowski JA. Glycogen synthase kinase 3β-mediated serine phosphorylation of the human glucocorticoid receptor redirects gene expression profiles. Mol Cell Biol. 2008;28(24):7309–22.
83. Davies L, Karthikeyan N, Lynch JT, et al. Cross talk of signaling pathways in the regulation of the glucocorticoid receptor function. Mol Endocrinol. 2008;22(6):1331–44.

84. Beck IM, De Bosscher K, Haegeman G. Glucocorticoid receptor mutants: man-made tools for functional research. Trends Endocrinol Metab. 2011;22(8):295–310.
85. Webster JC, Jewell CM, Bodwell JE, Munck A, Sar M, Cidlowski JA. Mouse glucocorticoid receptor phosphorylation status influences multiple functions of the receptor protein. J Biol Chem. 1997;272(14):9287–93.
86. Graff J, Rei D, Guan JS, et al. An epigenetic blockade of cognitive functions in the neurodegenerating brain. Nature. 2012;483(7388):222–6.
87. Adzic M, Djordjevic J, Djordjevic A, et al. Acute or chronic stress induce cell compartment-specific phosphorylation of glucocorticoid receptor and alter its transcriptional activity in Wistar rat brain. J Endocrinol. 2009;202(1):87–97.
88. Jeanneteau F, Garabedian MJ, Chao MV. Activation of Trk neurotrophin receptors by glucocorticoids provides a neuroprotective effect. Proc Natl Acad Sci U S A. 2008;105(12):4862–7.
89. Faresse N, Vitagliano JJ, Staub O. Differential ubiquitylation of the mineralocorticoid receptor is regulated by phosphorylation. FASEB J. 2012;26(10):4373–82.
90. Denis M, Poellinger L, Wikstom AC, Gustafsson JA. Requirement of hormone for thermal conversion of the glucocorticoid receptor to a DNA-binding state. Nature. 1988; 333(6174):686–8.
91. Davies TH, Ning YM, Sanchez ER. A new first step in activation of steroid receptors: hormone-induced switching of FKBP51 and FKBP52 immunophilins. J Biol Chem. 2002;277(7):4597–600.
92. Vandevyver S, Dejager L, Libert C. On the trail of the glucocorticoid receptor: into the nucleus and back. Traffic. 2012;13(3):364–74.
93. Savory JG, Hsu B, Laquian IR, et al. Discrimination between NL1- and NL2-mediated nuclear localization of the glucocorticoid receptor. Mol Cell Biol. 1999;19(2):1025–37.
94. Freedman ND, Yamamoto KR. Importin 7 and importin alpha/importin beta are nuclear import receptors for the glucocorticoid receptor. Mol Biol Cell. 2004;15(5):2276–86.
95. Echeverria PC, Mazaira G, Erlejman A, Gomez-Sanchez C, Piwien Pilipuk G, Galigniana MD. Nuclear import of the glucocorticoid receptor-hsp90 complex through the nuclear pore complex is mediated by its interaction with Nup62 and importin beta. Mol Cell Biol. 2009;29(17):4788–97.
96. Olkku A, Mahonen A. Calreticulin mediated glucocorticoid receptor export is involved in beta-catenin translocation and Wnt signalling inhibition in human osteoblastic cells. Bone. 2009;44(4):555–65.
97. Holaska JM, Black BE, Rastinejad F, Paschal BM. Ca2+-dependent nuclear export mediated by calreticulin. Mol Cell Biol. 2002;22(17):6286–97.
98. Amazit L, Alj Y, Tyagi RK, et al. Subcellular localization and mechanisms of nucleocytoplasmic trafficking of steroid receptor coactivator-1. J Biol Chem. 2003;278(34):32195–203.
99. Datson NA, van der Perk J, de Kloet ER, Vreugdenhil E. Identification of corticosteroid-responsive genes in rat hippocampus using serial analysis of gene expression. Eur J Neurosci. 2001;14(4):675–89.
100. Liu W, Wang J, Sauter NK, Pearce D. Steroid receptor heterodimerization demonstrated in vitro and in vivo. Proc Natl Acad Sci U S A. 1995;92(26):12480–4.
101. Nishi M, Tanaka M, Matsuda K, Sunaguchi M, Kawata M. Visualization of glucocorticoid receptor and mineralocorticoid receptor interactions in living cells with GFP-based fluorescence resonance energy transfer. J Neurosci. 2004;24(21):4918–27.
102. Ou XM, Storring JM, Kushwaha N, Albert PR. Heterodimerization of mineralocorticoid and glucocorticoid receptors at a novel negative response element of the 5-HT1A receptor gene. J Biol Chem. 2001;276(17):14299–307.
103. Newton R, Holden NS. Separating transrepression and transactivation: a distressing divorce for the glucocorticoid receptor? Mol Pharmacol. 2007;72(4):799–809.
104. Sakai DD, Helms S, Carlstedt-Duke J, Gustafsson JA, Rottman FM, Yamamoto KR. Hormone-mediated repression: a negative glucocorticoid response element from the bovine prolactin gene. Genes Dev. 1988;2(9):1144–54.
105. Surjit M, Ganti Krishna P, Mukherji A, et al. Widespread negative response elements mediate direct repression by agonist-liganded glucocorticoid receptor. Cell. 2011;145(2):224–41.

106. Reddy TE, Pauli F, Sprouse RO, et al. Genomic determination of the glucocorticoid response reveals unexpected mechanisms of gene regulation. Genome Res. 2009;19(12):2163–71.
107. Hudson WH, Youn C, Ortlund EA. The structural basis of direct glucocorticoid-mediated transrepression. Nat Struct Mol Biol. 2013;20(1):53–8.
108. So AY-L, Chaivorapol C, Bolton EC, Li H, Yamamoto KR. Determinants of cell- and gene-specific transcriptional regulation by the glucocorticoid receptor. PLoS Genet. 2007;3(6), e94.
109. Reddy TE, Gertz J, Crawford GE, Garabedian MJ, Myers RM. The hypersensitive glucocorticoid response specifically regulates period 1 and expression of circadian genes. Mol Cell Biol. 2012;32(18):3756–67.
110. Wright AP, Gustafsson JA. Mechanism of synergistic transcriptional transactivation by the human glucocorticoid receptor. Proc Natl Acad Sci U S A. 1991;88(19):8283–7.
111. Lefstin JA, Thomas JR, Yamamoto KR. Influence of a steroid receptor DNA-binding domain on transcriptional regulatory functions. Genes Dev. 1994;8(23):2842–56.
112. Jeanneteau FD, Lambert WM, Ismaili N, et al. BDNF and glucocorticoids regulate corticotrophin-releasing hormone (CRH) homeostasis in the hypothalamus. Proc Natl Acad Sci U S A. 2012;109(4):1305–10.
113. Rogatsky I, Luecke HF, Leitman DC, Yamamoto KR. Alternate surfaces of transcriptional coregulator GRIP1 function in different glucocorticoid receptor activation and repression contexts. Proc Natl Acad Sci. 2002;99(26):16701–6.
114. Glass CK, Saijo K. Nuclear receptor transrepression pathways that regulate inflammation in macrophages and T cells. Nat Rev Immunol. 2010;10(5):365–76.
115. McInerney EM, Rose DW, Flynn SE, et al. Determinants of coactivator LXXLL motif specificity in nuclear receptor transcriptional activation. Genes Dev. 1998;12(21):3357–68.
116. Chinenov Y, Gupte R, Dobrovolna J, et al. Role of transcriptional coregulator GRIP1 in the anti-inflammatory actions of glucocorticoids. Proc Natl Acad Sci. 2012;109(29):11776–81.
117. Espallergues J, Teegarden SL, Veerakumar A, et al. HDAC6 regulates glucocorticoid receptor signaling in serotonin pathways with critical impact on stress resilience. J Neurosci. 2012;32(13):4400–16.
118. Qiu Y, Zhao Y, Becker M, et al. HDAC1 acetylation is linked to progressive modulation of steroid receptor-induced gene transcription. Mol Cell. 2006;22(5):669–79.
119. Govindan MV. Recruitment of cAMP-response element-binding protein and histone deacetylase has opposite effects on glucocorticoid receptor gene transcription. J Biol Chem. 2010;285(7):4489–510.
120. L-b L, Leung DYM, Martin RJ, Goleva E. Inhibition of histone deacetylase 2 expression by elevated glucocorticoid receptor β in steroid-resistant asthma. Am J Respir Crit Care Med. 2010;182(7):877–83.
121. Hittelman AB, Burakov D, Iniguez-Lluhi JA, Freedman LP, Garabedian MJ. Differential regulation of glucocorticoid receptor transcriptional activation via AF-1-associated proteins. EMBO J. 1999;18(19):5380–8.
122. Malik S, Roeder RG. The metazoan Mediator co-activator complex as an integrative hub for transcriptional regulation. Nat Rev Genet. 2010;11(11):761–72.
123. Takahashi H, Parmely TJ, Sato S, et al. Human mediator subunit MED26 functions as a docking site for transcription elongation factors. Cell. 2011;146(1):92–104.
124. Chen W, Rogatsky I, Garabedian MJ. MED14 and MED1 differentially regulate target-specific gene activation by the glucocorticoid Receptor. Mol Endocrinol. 2006;20(3):560–72.
125. Karst H, Joels M. Corticosterone slowly enhances miniature excitatory postsynaptic current amplitude in mice CA1 hippocampal cells. J Neurophysiol. 2005;94(5):3479–86.
126. Olijslagers JE, de Kloet ER, Elgersma Y, van Woerden GM, Joels M, Karst H. Rapid changes in hippocampal CA1 pyramidal cell function via pre- as well as postsynaptic membrane mineralocorticoid receptors. Eur J Neurosci. 2008;27(10):2542–50.
127. Groc L, Choquet D, Chaouloff F. The stress hormone corticosterone conditions AMPAR surface trafficking and synaptic potentiation. Nat Neurosci. 2008;11(8):868–70.
128. Tasker JG, Herman JP. Mechanisms of rapid glucocorticoid feedback inhibition of the hypothalamic-pituitary-adrenal axis. Stress. 2011;14(4):398–406.

129. Di S, Malcher-Lopes R, Marcheselli VL, Bazan NG, Tasker JG. Rapid glucocorticoid-mediated endocannabinoid release and opposing regulation of glutamate and gamma-aminobutyric acid inputs to hypothalamic magnocellular neurons. Endocrinology. 2005;146(10):4292–301.
130. Malcher-Lopes R, Di S, Marcheselli VS, et al. Opposing crosstalk between leptin and glucocorticoids rapidly modulates synaptic excitation via endocannabinoid release. J Neurosci. 2006;26(24):6643–50.
131. Watts AG. Glucocorticoid regulation of peptide genes in neuroendocrine CRH neurons: a complexity beyond negative feedback. Front Neuroendocrinol. 2005;26(3–4):109–30.
132. Karst H, Karten YJ, Reichardt HM, de Kloet ER, Schutz G, Joels M. Corticosteroid actions in hippocampus require DNA binding of glucocorticoid receptor homodimers. Nat Neurosci. 2000;3(10):977–8.
133. Duvarci S, Pare D. Glucocorticoids enhance the excitability of principal basolateral amygdala neurons. J Neurosci. 2007;27(16):4482–91.
134. Makino S, Gold PW, Schulkin J. Corticosterone effects on corticotropin-releasing hormone mRNA in the central nucleus of the amygdala and the parvocellular region of the paraventricular nucleus of the hypothalamus. Brain Res. 1994;640(1–2):105–12.
135. McEwen BS. The ever-changing brain: cellular and molecular mechanisms for the effects of stressful experiences. Dev Neurobiol. 2011;72(6):878–90.
136. Prager EM, Johnson LR. Stress at the synapse: signal transduction mechanisms of adrenal steroids at neuronal membranes. Sci Signal. 2009;2(86):5.
137. Roozendaal B, Okuda S, Van der Zee EA, McGaugh JL. Glucocorticoid enhancement of memory requires arousal-induced noradrenergic activation in the basolateral amygdala. Proc Natl Acad Sci U S A. 2006;103(17):6741–6.
138. Schmidt P, Holsboer F, Spengler D. Beta(2)-adrenergic receptors potentiate glucocorticoid receptor transactivation via G protein beta gamma-subunits and the phosphoinositide 3-kinase pathway. Mol Endocrinol. 2001;15(4):553–64.
139. Kaur M, Chivers JE, Giembycz MA, Newton R. Long-acting beta2-adrenoceptor agonists synergistically enhance glucocorticoid-dependent transcription in human airway epithelial and smooth muscle cells. Mol Pharmacol. 2008;73(1):203–14.
140. Miller AH, Pariante CM, Pearce BD. Effects of cytokines on glucocorticoid receptor expression and function. Glucocorticoid resistance and relevance to depression. Adv Exp Med Biol. 1999;461:107–16.
141. Raison CL, Capuron L, Miller AH. Cytokines sing the blues: inflammation and the pathogenesis of depression. Trends Immunol. 2006;27(1):24–31.
142. Lambert WM, Xu CF, Neubert TA, Chao MV, Garabedian MJ, Jeanneteau FD. Brain-derived neurotrophic factor signaling rewrites the glucocorticoid transcriptome via glucocorticoid receptor phosphorylation. Mol Cell Biol. 2013;33(18):3700–14.
143. Chen DY, Bambah-Mukku D, Pollonini G, Alberini CM. Glucocorticoid receptors recruit the CaMKIIalpha-BDNF-CREB pathways to mediate memory consolidation. Nat Neurosci. 2012;15(12):1707–14.
144. de Quervain DJ, Poirier R, Wollmer MA, et al. Glucocorticoid-related genetic susceptibility for Alzheimer's disease. Hum Mol Genet. 2004;13(1):47–52.
145. Green KN, Billings LM, Roozendaal B, McGaugh JL, LaFerla FM. Glucocorticoids increase amyloid-beta and tau pathology in a mouse model of Alzheimer's disease. J Neurosci. 2006;26(35):9047–56.
146. Ito K, Ito M, Elliott WM, et al. Decreased histone deacetylase activity in chronic obstructive pulmonary disease. N Engl J Med. 2005;352(19):1967–76.
147. Barnes PJ, Adcock IM. Glucocorticoid resistance in inflammatory diseases. Lancet. 2009;373(9678):1905–17.
148. Zhang Y, Leung DY, Nordeen SK, Goleva E. Estrogen inhibits glucocorticoid action via protein phosphatase 5 (PP5)-mediated glucocorticoid receptor dephosphorylation. J Biol Chem. 2009;284(36):24542–52.
149. Altarejos JY, Montminy M. CREB and the CRTC co-activators: sensors for hormonal and metabolic signals. Nat Rev Mol Cell Biol. 2011;12(3):141–51.

Chapter 3
Mechanisms of Glucocorticoid-Regulated Gene Transcription

Sebastiaan H. Meijsing

Abstract One fascinating aspect of glucocorticoid signaling is their broad range of physiological and pharmacological effects. These effects are at least in part a consequence of transcriptional regulation by the glucocorticoid receptor (GR). Activation of GR by glucocorticoids results in tissue-specific changes in gene expression levels with some genes being activated whereas others are repressed. This raises two questions: First, how does GR regulate different subsets of target genes in different tissues? And second, how can GR both activate and repress the expression of genes?

To answer these questions, this chapter will describe the function of the various "components" and how they cooperate to mediate the transcriptional responses to glucocorticoids. The first "component" is GR itself. The second "component" is the chromatin and its role in specifying where in the genome GR binds. Binding to the genome however is just the first step in regulating the expression of genes and transcriptional regulation by GR depends on the recruitment of coregulator proteins that either directly or indirectly influence the recruitment and or activity of RNA polymerase II. Ultimately, the integration of inputs including GR isoform, DNA sequence, chromatin and cooperation with coregulators determines which genes are regulated and the direction of their regulation.

Keywords Transcription • Coregulators • Chromatin • Cis-regulatory elements • Glucocorticoid receptor

Structure of the Glucocorticoid Receptor

Although glucocorticoids have been used clinically from the 1940s [1], it wasn't until 1984 when the coding sequence for its receptor was initially isolated from rat [2] and soon after its human homolog was cloned [3]. The human gene coding for GR

S.H. Meijsing, Ph.D. (✉)
Department of Computational Molecular Biology, Max Planck Institute for Molecular Biology, Ihnestrasse 63-73, Berlin 14195, Germany
e-mail: meijsing@molgen.mpg.de

J.-C. Wang, C. Harris (eds.), *Glucocorticoid Signaling*, Advances in Experimental Medicine and Biology 872, DOI 10.1007/978-1-4939-2895-8_3

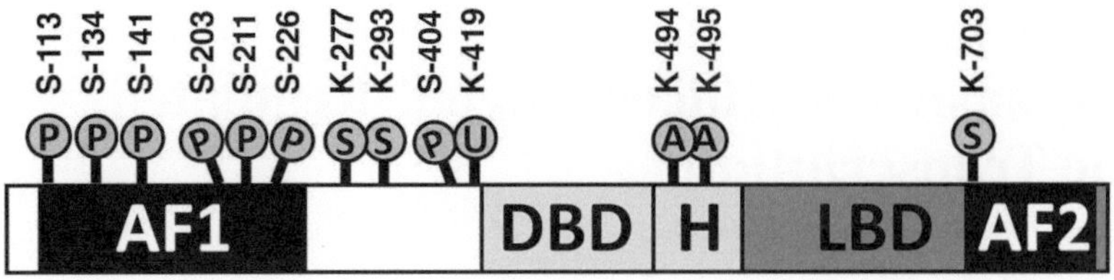

Fig. 3.1 Domain structure of GR and sites of post-translational modifications. Shown are the functional domains of GR: The activation function 1 and -2 (AF1, AF2), the DNA-binding domain (DBD), hinge region (H) and ligand-binding domain (LBD). Also shown are post-translational modifications (phosphorylation (P); acetylation (A); ubiquitination (U); sumoylation (S)) of either Serine (S) or Lysine (K) residues. Amino acid numbering refers to human GR

consists of nine exons and can produce a variety of different gene products through alternative splicing, alternative translational initiation and by post-translational modifications. Here we describe the functional domains of GR and how alternative splicing, translational initiation and post-translational modifications generate receptor isoforms with distinct expression profiles and target genes.

Functional Domains of the Glucocorticoid Receptor

A combination of biochemical (limited proteolysis, [4]) and molecular biological (mutagenesis and domain fusions) approaches have uncovered that the glucocorticoid receptor is a modular protein with several functional domains (Fig. 3.1).

The N-terminal domain of the glucocorticoid receptor contains the activation function 1 domain (AF1, amino acids 77–262, throughout amino acid numbering refers to human GR), which is involved in transcriptional regulation [5]. In contrast to the AF2 domain (see below), the AF1 domain is constitutively active meaning that its activity does not rely on the presence of hormone [5]. GR-dependent transcriptional regulation critically depends on its interaction with several coregulator proteins that either directly or indirectly recruit or influence the activity of RNA polymerase II (the role of coregulators in transcription is described in section "Transcriptional Regulation by GR"). For the AF1 domain these interaction partners include p160, TIF2, DRIP/TRAP and TBP [6]. How these proteins interact with the AF1 domain is largely unknown. No clear interaction domains have been identified in either GR or in the proteins interacting with AF1 and computational predictions and experimental approaches indicate that large parts of the AF1 domain are intrinsically disordered [7]. This may allow the AF1 to adopt different conformations to create interaction surfaces for a variety of coregulators.

DNA binding by GR is mediated by the DNA binding domain (DBD, amino acids 420–480), which is conserved across steroid hormone receptor proteins. GR can bind as a homodimer to DNA sequences consisting of inverted repeats of a loosely defined recognition sequence separated by a three base pair spacer (Fig. 3.2) [8]. The three-dimensional structure shows that the DBD contains several alpha-helices.

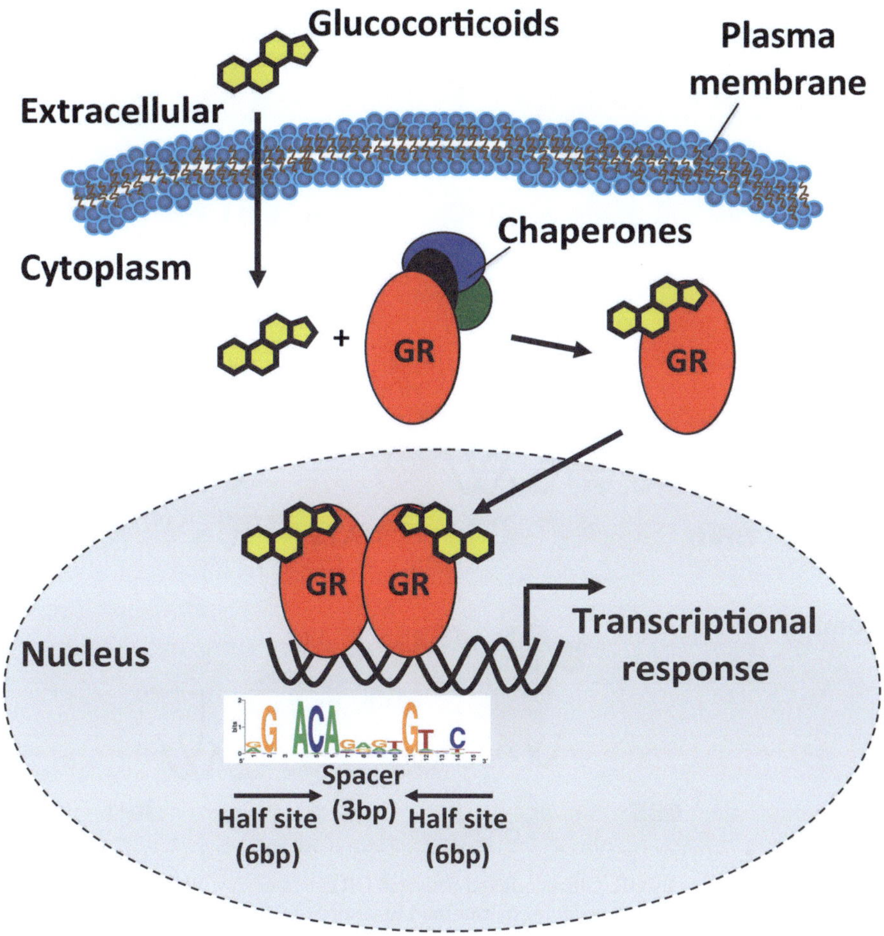

Fig. 3.2 Classical GR signaling pathway. Upon ligand binding, cytoplasmic GR dissociates from chaperone proteins and translocates to the nucleus where it interacts with specific DNA sequences to control the expression of associated target genes

One helix mediates base-specific DNA contacts whereas an alpha-helix at the C-terminus of the DBD makes several non-specific phosphate backbone and minor groove contacts [9, 10]. Two zinc-fingers ascertain proper folding of the DBD to coordinate DNA recognition and dimerization [9, 11, 12]. Like other domains, the DBD interacts with several coregulators including JDP1, JDP2, HMG1, HMG2, GT198 and SET/TAFI-β(beta) [13–16]. In addition to direct DNA binding, GR can also be tethered to the DNA for example via its interaction with activator protein 1 (AP1), NFκ(kappa)B or STAT3 (Fig. 3.3) [17–22]. Interestingly, also here the DBD appears responsible for tethered DNA interactions by directly interacting with the c-Jun/c-Fos or p65 subunits of AP1 and NFκ(kappa)B respectively [17, 21, 23].

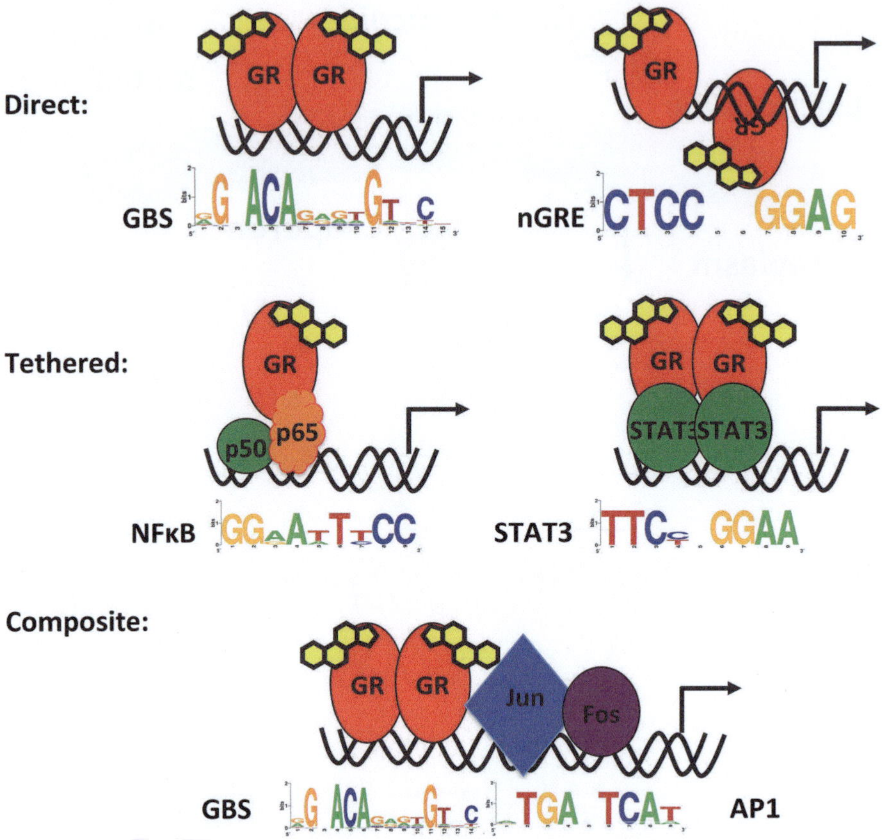

Fig. 3.3 DNA binding by GR. Glucocorticoid-activated GR can interact with DNA either directly (*top*), indirectly via tethering (*middle*) or can bind to composite elements where it engages in cross-talk with neighboring DNA-bound transcriptional regulatory factors (*bottom*)

Arguing for an important role of the DBD in tethered DNA binding, mutations in this domain interfere with GR's function as a transcriptional repressor at sites where it is tethered to the DNA by either AP1 or NFκ(kappa)B [17, 23].

C-terminal to the DBD the hinge region connects the DBD to the ligand binding domain (LBD). The LBD consists of 12 alpha helixes [24] and ligand binding is facilitated by several alpha helixes that together form a hydrophobic pocket [24]. Furthermore, the LBD harbors a second dimerization domain, sequences involved in nuclear translocation upon hormone binding and the activation function 2 (AF2 domain), which mediates the interaction with several coregulators (reviewed in [25]). In the absence of ligand, GR is predominantly cytoplasmic where the interaction of the LBD with chaperone proteins such as hsp90 and p23 keep GR in a hormone-binding competent state (Fig. 3.2) (reviewed in [26]). Ligand binding results in Hsp90 dissociation, nuclear translocation and conformational changes in helix 12

(Fig. 3.2) [26]. These conformational changes facilitate the interaction of the AF2 domain with a variety of coregulators containing LXXLL motifs [24, 26] including p160 coactivator family members SRC1 and GRIP1 [27, 28].

Although the domains of GR can function in isolation, recent studies indicate that the domains of nuclear hormone receptors are both functionally and structurally connected [29–32]. These domain-connections can be rewired depending on the context in which GR is active and accordingly, different combinations of GR domains are required to regulate the expression of individual genes [33].

Creating Functional Diversity: Glucocorticoid Receptor Isoforms and Post-translational Modifications

Although a single gene (*NR3C1*) codes for the glucocorticoid receptor protein, this gene can give rise to several isoforms with unique expression profiles [34, 35]. In addition, post-translational modifications of these isoforms further expand the diversity of responses to glucocorticoids. Besides the predominant GR isoform GRα(alpha), alternative splicing of GR can generate at least four additional isoforms: GRβ(beta), GRγ(gamma), GR-A and GR-P [25]. GRβ(beta) differs from GRα(alpha) in its LBD and is unable to bind hormone [36]. The GRβ(beta) isoform appears to be transcriptionally inactive and can antagonize the activity of GRα(alpha) [37]. Accordingly, increased GRβ(beta) levels have been linked to glucocorticoid resistance in a variety of diseases including asthma, rheumatoid arthritis and acute lymphoblastic leukemia [37]. Use of an alternative splice-donor site generates the GRγ(gamma) isoform, which differs from GRα(alpha) in having a single additional Arginine inserted in the DBD [38]. The Arginine insertion results in gene-specific effects with most genes being unaffected, whereas some genes are regulated more strongly and others more weakly [30]. Consistent with a reduced activity towards certain target genes, GRγ(gamma) has been linked to glucocorticoid resistance in childhood acute lymphoblastic leukemia and small cell lung carcinoma cells [39, 40]. The GR-A and GR-P isoforms lack exons encoding the LBD and consequently lack the ability to bind ligand [41]. Similar to GRβ(beta) and GRγ(gamma), GR-P can antagonize the transcriptional activity of GRα(alpha) and has been linked to glucocorticoid-resistance [42].

Additional GR isoforms are produced as a consequence of alternative translational initiation, which generates GR proteins with shorter N-terminal domains [43]. These translational isoforms differ in their tissue-specific expression and the transcriptional programs they initiate [44]. Another mechanism that can generate functional diversity are post-translational modifications. Such modifications can alter the function of GR and include phosphorylation, acetylation, sumoylation and ubiquitination (Fig. 3.1) [25]. One example of a post-translational modification that influences GR activity is the phosphorylation of Serine residues in the N-terminus of GR. The phosphorylation modulates GR's interaction with coregulators and differentially affects its activity towards individual target genes [45]. Another example is acetylation of GR by CLOCK, a histone acetyltransferase (HAT) involved in circadian rhythm. The CLOCK-dependent acetylation of multiple Lysines in the hinge region of GR interferes with DNA binding resulting in changes in the expression level of a subset of GR target genes [46, 47].

Together, alternative splicing, translational initiation and post-translational modifications generate a variety of GR variants with different target genes. Consequently, differences between cell types and tissues in the expression level of these isoforms and of the enzymes responsible for post-translational modifications likely contribute to the highly tissue-specific effects of glucocorticoids.

Chromosomal Binding of GR

Hormone binding by GR results in nuclear translocation and allows the receptor to bind to specific genomic sequences (Fig. 3.2). The binding of GR to glucocorticoid response elements (GREs) constitutes an essential first step in the regulation of the expression of associated target genes. Here we discuss the contributions of DNA sequence elements and the chromatin landscape in guiding GR to its appropriate genomic destination.

Binding to the Genome: "Classical" GR Binding Sequences

The first described "classical" mode of DNA binding is for liganded GR to associate as a dimer to GR binding sequences (GBSs) [8]. GBSs are typically imperfect palindromic hexameric half-sites separated by a 3bp spacer (Fig. 3.2). Historically, mostly for practical reasons, studies to identify regulatory sequences exploited by GR to regulate target genes were focused on promoter regions and have uncovered numerous promoter-proximal GBSs [48–50]. In support of a role of these GBSs in the regulation of associated target genes, genomic regions that harbor a GBS as well as simply the 15bp GBS are sufficient to facilitate GR-dependent transcriptional activation when localized upstream of heterologous promoters [10, 49]. However, up until recently it was unclear whether promoter-proximal binding by GR is the exception or the rule that governs genomic binding and the control of target gene expression. Technological advances that combine chromatin immunoprecipitation (ChIP) with next generation sequencing (ChIP-seq) allow the unbiased genome-wide identification of GR binding sites [51, 52]. Several ChIP-seq studies have revealed that promoter-proximal binding by GR appears to be the exception and that the majority of GR binding is at promoter-distal locations [53–56]. One representative study showed that for genes that are up regulated in response to glucocorticoid treatment (likely GR target genes), 50 % of the binding sites were located at a distance greater than 10 kb from the transcriptional start-site (TSS) [54]. Even more striking, for down regulated genes the median distance to the TSS was >100 kb [54]. The finding that only the minority of GR binds promoter-proximal is not specific for GR but is also seen for related hormone receptors including ER, PPAR and AR [55, 57, 58]. This suggests that long-range regulation by GR and other hormone receptors might be responsible for the regulation of a large fraction of target genes. In support of this idea, a study using chromatin conformation capture showed that the promoter of the GR-regulated gene *Ciz1* was contacted by a GR binding region located nearly 30 kb away [59].

Bioinformatical analysis of genomic regions bound by GR shows that the canonical 15 bp GBS is highly enriched at such binding sites with one study reporting that 58 % of the bound regions contains a GBS [56]. This underscores the important role of the canonical GBS in guiding GR to specific genomic locations. It does however also hint at the existence of alternative sequences that facilitate GR binding at the remaining 42 % of GR-bound regions.

Binding to the Genome: Other Sequences

Several ChIP-seq studies made the striking observation that only a fraction of all GR binding regions appears to contain the canonical 15 bp GBS [54, 56]. This indicates that GR may be able to bind to very degenerate sequences with the assistance of another transcriptional regulatory factor. Moreover, sequences other than the canonical GBS might be able to recruit GR to specific genomic loci (Fig. 3.3). Such sequences could either bind proteins that tether GR to the DNA or alternatively GR might be able to interact directly with a broader spectrum of DNA sequences. In support of this, studies with the hormone-repressed gene *POMC* uncovered GR-bound sequences that resemble the canonical GBSs somewhat but lack similarity to the consensus motif at key positions [60]. Interestingly, whereas regulation from canonical GBSs is typically associated with activation of transcription, the promoter region of the *POMC* gene mediated transcriptional repression when fused to a luciferase reporter gene and was therefore called negative glucocorticoid response element (nGRE). This repression was lost when the GBS-like sequence was changed to resemble a canonical GBS [60]. In isolation however, this sequence failed to confer repression arguing that its function relies on other functional elements present at the *POMC* promoter [60]. Another class of sequences that has been proposed to directly interact with GR are inverted repeats of CTCC that have a spacing of either 0, 1 or 2 base pairs [61]. These sequences are associated with genes that are repressed by GR. Notably, binding of GR to canonical GBSs strictly requires a 3 bp spacer to position two GR molecules such that they can effectively dimerize [9]. The variable spacing for these nGREs suggests that dimerization might not be required at these nGREs and accordingly structural studies suggest monomeric GR-binding to the half sites (Fig. 3.3) [62].

Together these studies suggest that GR is able to interact directly with a variety of sequence motifs to control the expression of associated target genes.

Binding to the Genome: Tethered Binding

The absence of canonical GBSs in ChIP-seq peaks can also be explained by tethered DNA binding by GR (Fig. 3.3). Tethered GR binding has predominantly been linked to transcriptional repression and has been proposed for several transcriptional regulatory factors including NFκ(kappa)B [21], AP1 [18–20], STAT3 [22, 63] and NGFI-B [64]. For NFκ(kappa)B, the p65 (RelA) subunit physically interacts with GR [21] and recruits GR to NFκ(kappa)B response elements [65]. The ability of GR to repress from NFκ(kappa)B sites can be recapitulated using reporters plasmids simply harboring NFκ(kappa)B sites driving the expression of a luciferase reporter gene arguing that tethered binding to NFκ(kappa)B mediates the repressive effects of GR [66]. Genome-wide studies using ChIP-seq, showed that co-treatment of cells with dexamethasone, a synthetic GR ligand, and with TNFα(alpha) to activate NFκ(kappa)B resulted in GR binding to approximately a thousand additional genomic regions when compared to the binding profile when cells were treated with dexamethasone alone [67]. These additional binding regions are enriched for NFκ(kappa)B binding sites suggesting that tethered binding might occur quite frequently [67]. However, it could also be that part or all of the gained binding is a simple consequence of NFκ(kappa)B-induced changes in chromatin accessibility that makes previously inaccessible GR binding regions available.

AP1 is another factor that physically interacts with GR [18] and has been implicated in tethering GR to DNA [68]. Similar to the observation for NFκ(kappa)B, tethered binding by AP1 is linked to transcriptional repression. This repression can be recapitulated using a luciferase reporter that contains a single copy of the AP1 consensus sequence driving expression of a luciferase reporter gene [69]. Other proteins implicated in tethering GR to the DNA are members of the signal transducer and activator of transcription (STAT) family. GR physically interacts with several STAT proteins including STAT1 [70], STAT3 [63] and STAT5 [71]. Genome-wide profiling of STAT3 and GR binding suggests that GR may be tethered to the DNA by STAT3 at about 300 genomic binding sites and that such binding events are almost exclusively associated with transcriptional repression by GR [22].

Combinatorial Binding by GR and Regulation

Binding sites for GR in the genome are not present in isolation but are surrounded by sequence motifs that can be occupied by other transcriptional regulatory factors (Fig. 3.3). Accordingly, analysis of GR ChIP-seq peaks shows a cell-type specific overrepresentation of various sequence motifs [56]. Recent studies underscore the important role of combinatorial binding in transcriptional regulation by GR and for transcriptional regulatory factors in general [72, 73]. The study by Siersbaek and coworkers analyzed five transcriptional regulatory factors involved in adipogenesis including GR (out of the more than a thousand transcriptional regulatory factors encoded in the human genome) [72]. ChIP-seq of these factors showed combinatorial binding of GR with at least one other factor for >93 % and simultaneous binding of all 5 factors for 25 % of all GR binding events [72]. These "hotspots" of transcriptional regulatory factor binding were also found by the encode consortium that looked at >100 transcriptional regulatory factors [73, 74]. The co-occurrence of a GR binding site with recognition sequences for "partner" transcriptional regulatory factors can give rise to a broad spectrum of signaling cross-talk. A commonly observed type of cross-talk is a synergetic interaction between GR and other transcription factors. For example, knockdown of C/EBPβ(beta) results in a reduction of GR binding at co-occupied sites whereas binding at control sites that are not co-occupied are not affected [72]. The knockdown of factors co-occupying "hotspots" revealed a highly cooperative nature of transcriptional regulatory factor binding at these sites [72]. Synergetic interactions likely reflect at least in part effects of chromatin (chromatin accessibility) where several transcriptional regulatory factors cooperate to keep genomic sites accessible. This might explain the many synergetic interactions with other transcriptional regulatory factors that have been described for GR which include SP1, NF1, STAT3, COUP-TFII and AP1 [75–78].

The cross-talk between GR and other transcriptional regulatory factors at combinatorial binding sites can also be antagonistic. For example, at the osteocalcin promoter, the GR binding site overlaps the TATA box and GR binding thereby antagonizes TFIID binding to the TATA box and transcriptional initiation [79]. Another example of an antagonistic interaction between GR due to overlapping binding sites is found at the prolactin gene where the GR binding sites overlaps site for Oct1 and Pbx1 [80]. GR can also antagonize the activity of other factors via non-overlapping binding sites as was shown for the glutathione S-transferase A2 gene [81]. Here binding of GR to a GBS-like sequence results in the recruitment of the transcriptional co-repressor SMRT to repress C/EBP- and NRF2-mediated activation [81]. For the mouse proliferin gene, depending on the composition of the proteins that bind to the dimeric AP1 binding site, GR can either act antagonistically or synergistically [82].

The complex nature of interactions between GR and other transcriptional regulatory proteins illustrates the complexity of signaling cross-talk occurring at composite elements. This complexity can potentiate the ability of GR to regulate genes in a cell type specific manner and to tailor its activity towards individual genes. Gene-specific effects can for instance be a consequence of differences in the local sequence of the GR binding site. Similarly, the cell-type specific expression and binding of transcriptional regulatory factors that engage in synergistic interactions with GR can explain tissue-specific effects.

DNA Binding: Influence of Chromatin Structure on GR Binding

Another fascinating fact that the genome-wide analysis uncovered is that the genomic binding pattern of GR shows little overlap (<5 % [53]) between cell-types [56], and personal unpublished results). The highly tissue-specific binding by GR

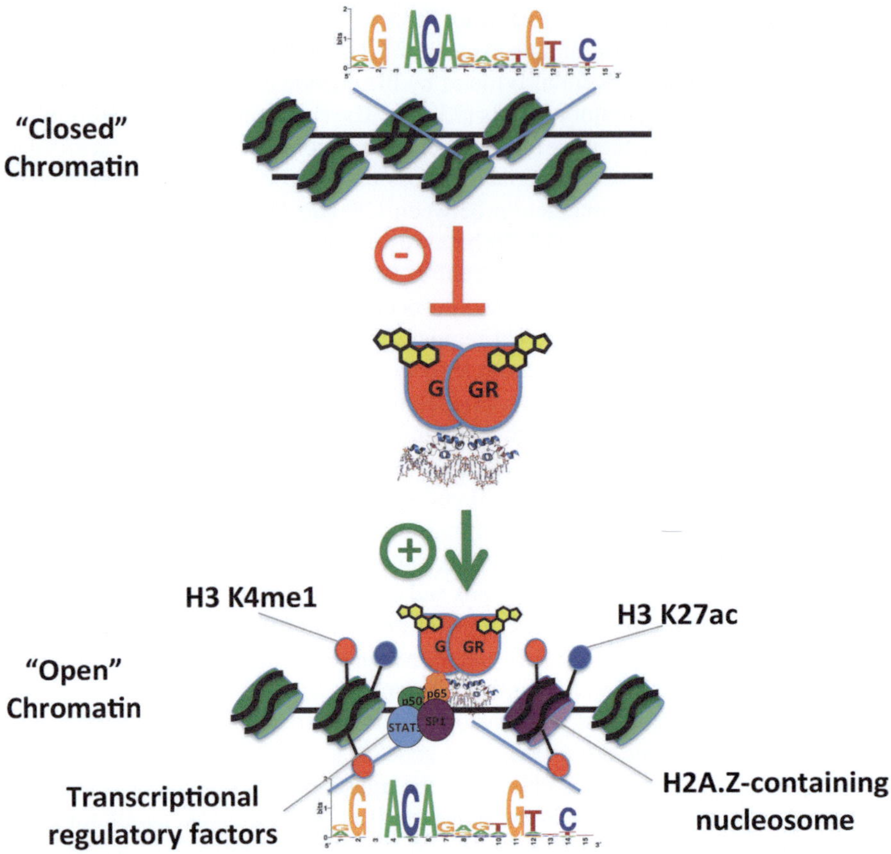

Fig. 3.4 Chromatin features influence DNA binding by GR. Chromatin features either negatively or positively correlating with genomic DNA binding by GR. Negative: closed chromatin (*top*). Positive: closed chromatin, nucleosome free DNA, enrichment for the H2A.Z histone variant, presence of other transcriptional regulatory factors and of histone modifications H3K4me1 and H3K27ac (*bottom*)

indicates that the sequence of GR binding sites alone is insufficient to explain where in the genome GR binds. Thus, DNA sequence specifies where in the genome GR could bind and other inputs including the chromatin landscape are needed to specify where GR actually does bind (Fig. 3.4). One aspect of the chromatin landscape that appears to be a major contributor is chromatin-accessibility as assayed by DNase-I accessibility assays [53, 78]. These studies showed that the majority of GR binding occurs in genomic regions that are DNase-I accessible ("open") prior to hormone treatment [53, 78]. Interestingly, which regions of the genome are actually "open" appears to be highly cell-type specific [53, 73]. So an emerging picture is that cell-type specific "chromatin-accessibility factors", other than GR, specify which regions of the genome are "open" and thereby where in the genome GR can bind and which genes it can regulate. One such "chromatin-accessibility factor" is AP1, whose binding, according to one study, overlaps with >50 % of GR binding sites [78]. Consistent with a role in facilitating GR binding, dominant negative AP1 and depletion of AP1 levels by siRNAs resulted in reduced chromatin accessibility and a loss of GR binding at co-occupied sites [78]. Another factor linking chromatin accessibility and steroid receptor binding is the forkhead box A1 (FoxA1) protein. FoxA1 induces DNase-I hypersensitivity [83, 84] indicative of open chromatin and facilitates GR binding [83]. Conversely, depletion of FoxA1 results in a redistribution of genomic estrogen receptor (ER), androgen receptor (AR) and GR binding [85, 86]. However, for ER depletion of FoxA1 results in an almost complete loss of ER binding [85] whereas for AR and GR FoxA1 depletion results in a redistribution of binding [86] indicating that the role of FoxA1 for GR and AR is more complex than simply facilitating access to the genome.

The observation that transcriptional regulatory factors typically bind together to "hotspots" and mutually stimulate genomic binding [72] suggests that they might cooperatively keep certain regulatory regions open. This can also explain the tissue-specific binding patterns observed for GR due to cell-type-specific expression of these cooperation partners. Notably, not all GR binding occurs at open regions and for a subset of binding events GR appears to act as a "chromatin-accessibility factor" [53, 78]. Analysis of sequence motifs for closed chromatin GR binding sites showed that binding to these sites is mediated by GBSs with high motif scores [87] suggesting that high-affinity binding might be a prerequisite for GR binding at closed chromatin. Interestingly, whereas GR binding sites in "open" chromatin show little overlap between cell types, binding at "closed" sites is often shared between cell types [87] indicating that for these sites GR might not rely on other factors for binding.

Other chromatin features linked to GR binding are the presence of nucleosomes, the post-translational modification state of nucleosomal histones and the presence of histone variants. GR binding sites are enriched for several chromatin features linked to enhancers including monomethylation of histone H3 Lysine 4, acetylation of histone H3 Lysine 27 and enrichment of the histone variant H2A.Z [88, 89] and unpublished data from my group). The enrichment profile of these histone modifications shows a bimodal peak flanking the site of GR-binding, which indicates that GR typically binds to DNA located between two nucleosomes (Fig. 3.4) [88]. However, despite the significant correlation between histone modifications and GR binding, future studies are needed to determine if and how these are causatively connected.

In conclusion, the integration of DNA sequence information, cooperation with other transcriptional regulatory factors and chromatin features appears to determine where in the genome GR binds and ultimately which genes it regulates in a particular cell type.

Transcriptional Regulation by GR

The transcriptional process begins with the recruitment of RNA polymerases to the transcriptional start site (TSS) by the pre-initiation complex (PIC). After recruitment, the RNA polymerases proceed through distinct steps of the transcription cycle: initiation, elongation and termination. RNA polymerases are multi-protein complexes, which change their composition and/or carry different modifications dependent on the step in the transcription cycle. For example, RNA polymerase II is differentially phosphorylated in the C-terminal tail domain (CTD) of its largest subunit dependent on whether it is initiating, elongating or terminating (reviewed in [90]). Gene regulation depends on the action of transcriptional regulatory factors, like GR. GR can exploit a broad spectrum of mechanisms to influence the expression level of genes. Such mechanisms include influencing RNA stability [91–93], sequestering or influencing the activity state of other transcriptional regulatory factors by protein:protein interactions [94, 95] that thus does not require direct interactions of GR with DNA or with the RNA polymerase machinery. Here however, we will focus on transcriptional effects in response to glucocorticoids that involve DNA binding and RNA polymerase II. GR may affect the state of RNA Polymerase II directly (e.g., the phosphorylation state of the CTD or the assembly of the PIC). Alternatively, GR can modulate RNA polymerase II's regulatory role indirectly by recruiting coregulators such as histone modifying enzymes, chromatin remodelers or the mediator complex that bridges the interaction with RNA polymerase II (Fig. 3.5). GR can either increase the transcription rate (hence acting as an activator) or can reduce—or even eliminate—transcription (acting as a repressor). This paragraph presents an overview of different classes of coregulators and their role in mediating the transcriptional effects of GR.

Interaction with Coregulators: Interactions with the Basal Transcriptional Machinery

Perhaps the most straightforward way for GR to influence transcription is by interacting directly or indirectly with components of the basal transcriptional machinery (Fig. 3.5). A direct interaction of the GR's AF1 domain with TBP, which is part of the TFIID component of the pre-initiation complex, suggests that GR can promote transcriptional initiation by recruiting TFIID to promoters of target genes [96]. Furthermore, GR interacts with p300/CBP, which in turn interacts with TFIIB, another component of the pre-initiation complex, and thereby indirectly linking GR to the basal transcriptional machinery [97–99]. The glucocorticoid receptor can also recruit RNA polymerase II via its interaction both physically and functionally with components (MED1 and MED14) of the mediator complex, which interacts with the CTD of RNA polymerase II [100, 101].

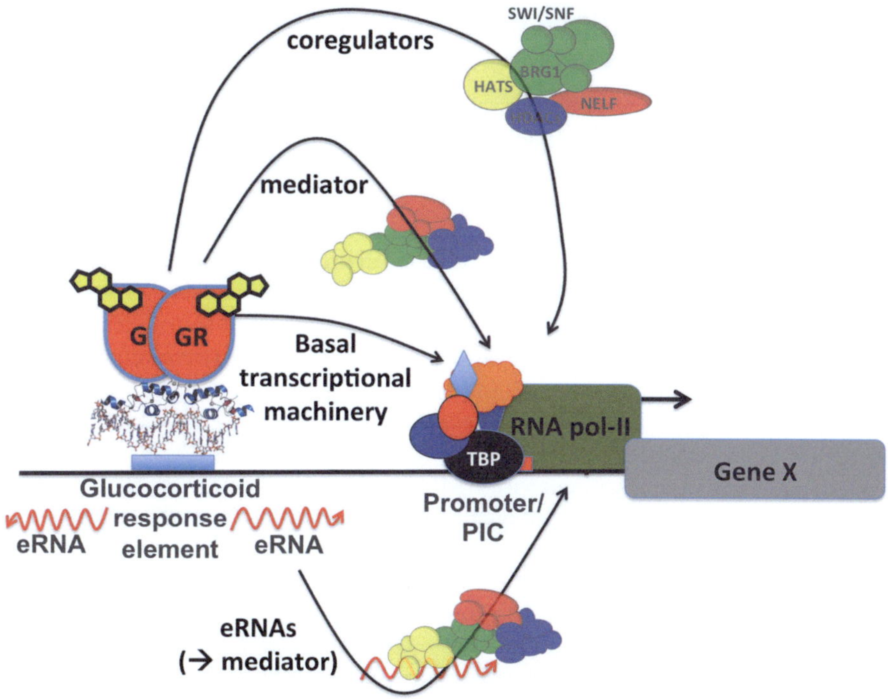

Fig. 3.5 Coregulators and their role in GR-dependent regulation of promoter activity. Overview of interacting coregulators (proteins and RNA) of GR that can either directly or indirectly influence the recruitment or activity of RNA polymerase II and thereby the transcriptional output. *Abbreviations*: *HATS* histone acetyltransferases, *HDACs* histone deacetylases, *NELF* negative elongation factor, *PIC* pre-initiation complex, *TBP* TATA-binding protein, *eRNA* enhancer RNA

Transcriptional control by GR is also exerted at the level of transcriptional elongation, the step in the transcription cycle downstream of transcriptional initiation. For example, GR can interact with proteins that stimulate elongation (elongation factor RNA polymerase (ELL)), resulting in increased levels of transcript [102]. Conversely, at the IL8 gene GR displaces p-TEFb, a factor that stimulates elongation [65]. The displacement prevents the p-TEFb-dependent Serine 2 phosphorylation of the CTD of RNA polymerase II and consequently reduces transcriptional elongation at the IL8 gene [65, 103]. Furthermore, GR interacts with suppressors of elongation (negative elongation factor (NELF), [104]). An elegant study in macrophages shows that GR can either repress the expression of genes at the level of RNA polymerase II recruitment or by recruiting NELF, which results in a pausing of RNA polymerase II [104]. In agreement with a role for NELF in mediating the effects of GR, repression was specifically lost for the elongation-controlled genes in NELF-deficient macrophages [104]. Together these studies indicate that GR directly or indirectly contacts components of the basal RNA polymerase II machinery and thereby can influence gene expression by affecting different stages of the transcription cycle.

Coregulators That Influence Chromatin Structure and Histone Modification States

In eukaryotes DNA accessibility and chromatin structure play an important role in specifying the expression level of genes. Eukaryotic genomes are packaged into chromatin, whose basic repeating unit is the nucleosome [105, 106]. Nucleosomes form by wrapping 147 base pairs of DNA around an octamer of the four core histones (H2A, H2B, H3 and H4) and can be found approximately every 200 base pairs throughout the genome [107]. Their presence affects all DNA-dependent processes, including DNA-repair, DNA replication and transcription. For instance, *in vitro* a chromatinized DNA template prevents RNA polymerase II from initiating transcription [108]. Even before the identification of the coregulators responsible, studies of promoters of hormone-activated genes showed that transcriptional regulation by GR is tightly coupled to chromatin remodeling [109, 110]. Especially studies with the mouse mammary tumor virus (MMTV) have been instrumental in dissecting the steps needed for transcriptional activation (reviewed in [110]). These steps include the recruitment of chromatin modifying enzymes, nucleosome repositioning and changes in sensitivity to nucleases, ultimately resulting in the recruitment of RNA polymerase II. GR-dependent chromatin remodeling is mediated by its interaction with a variety of coregulators that modify chromatin structure and thereby indirectly the recruitment of RNA polymerase II. The first class of GR-interacting chromatin modifiers are members of ATP-dependent chromatin remodelers that can move and remove nucleosomes. Specifically, GR interacts with the ATP-dependent chromatin remodeling complex SWI/SNF [98, 111]. This interaction is mediated by BAF proteins that are part of the SWI/SNF complex [112]. The SWI/SNF complex contains one of two possible core ATP-ase subunits [110] Brm or BRG1 and a physical and functional connection between both ATPase subunits and GR activity has been shown [89, 113, 114]. The interaction with the SWI/SNF complex is essential for GR-dependent transcriptional activation of the MMTV promoter [98]. Here, the SWI/SNF complex repositions nucleosomes to allow other transcriptional regulatory factors and TBP to bind and thereby facilitates the assembly of the pre-initiation complex at the promoter (reviewed in [110]). For endogenous genes, the effects of disrupting Brm or BRG1 activity, by either dominant negative versions of these proteins or by knocking down their expression using siRNA, results in gene-specific effects with some genes being affected whereas others are not [89, 114]. The mechanisms responsible for the facultative requirement for Brm and BRG1 are unknown, but might reflect the fact that for certain genes alternative mechanisms ensure appropriate nucleosome positioning and transcriptional initiation.

The second class of chromatin modifying enzymes that interact with GR are enzymes that post-translationally modify histones. These histone modifications can act as recognition signals for proteins [115]. For example, trimethylated Lysine 4 of histone H3 is recognized by TFIID providing a direct link between histone modifications and the basal transcriptional machinery [116]. Additionally, histone modifications might influence transcription by loosening the chromatin. This occurs when Lysines are acetylated which removes its positive charge thereby reducing the affinity between DNA and histones [117]. One coregulator that acts as a coactivator of GR is the histone acetyltransferase p300 [118]. Conversely, enzymes that remove acetylation marks, histone deacetylases (HDACs), can act as corepressors of GR [119]. Examples of HDACs or complexes containing HDACs that interact with GR are NcoR, SMRT and HDAC2 [61, 119]. Although the activities of HATs and HDACs might be a consequence of "loosening" the chromatin, their role is likely to be more complex. One added level of complexity is that in addition to histones, these enzymes can also modify transcriptional regulatory factors, chaperones like hsp90 and coregulators [47, 120]. For example, acetylation of GR by either CLOCK or GCN5 interferes with GR's ability to interact with DNA [47, 121]. Further illustrating the complexity of the interaction, the GR-interacting coregulator GRIP1 acts as a coactivator for some GR target genes whereas it acts as a corepressor at others [122]. In addition to coregulators that modify the acetylation state of Lysines, GR also interacts with histone modifying enzymes CARM1 and G9a that methylate respectively Arginine or Lysine residues of histones and other proteins [123, 124].

In conclusion, genomic binding by GR coordinates the recruitment of a large variety of coregulator proteins. These coregulators specify the activity as well as the direction of the transcriptional responses to glucocorticoids at individual target genes. However, the underlying mechanisms responsible for the context-specific requirement of coregulators remain largely unknown. One possible explanation could be that the combinatorial binding of GR and another factor creates an interaction surface for coregulators that is not present when these factors bind in isolation. Furthermore, the DNA binding site responsible for GR recruitment appears to play an important role as tethering sites and non-canonical GBSs are typically associated with repression and preferentially recruit corepressors whereas canonical GBSs direct the assembly of regulatory complexes that usually activate transcription. The role of the DNA sequence might in part be explained by the fact that DNA induces sequence-specific conformational changes in the DBD of GR [10, 125]. These conformational changes could be propagated to domains of GR engaged in protein:protein interactions thereby explaining the context specific signaling cross-talk between GR and coregulators.

Concluding Remarks and Future Perspectives

The past decades have generated a wealth of mechanistic insight into how GR orchestrates the transcriptional response of cells and tissues to glucocorticoid hormones. It is becoming increasingly clear that these responses are highly context-specific and that chromatin plays a key role in dictating which transcriptional program is initiation in a particular cell type. In addition to the cell-type-specific effects, GR also appears to have highly gene-specific effects within a cell. This might complicate research as there are perhaps few universally applicable operating principles for GR in transcriptional regulation. It also provides an opportunity to try to activate GR in a targeted way that may selectively affect the expression of a subset of genes and thereby might result in therapeutic usage of glucocorticoids with fewer side effects. One approach in this regard has been to develop synthetic GR ligands with selective activities [126, 127]. Several such ligands do indeed regulate subsets of GR target genes [126, 127]. However, if and to what extend ligands can be identified that display such selectivity towards the therapeutically relevant target genes remains to be seen.

There are still lots of open-ended questions related to transcriptional responses to glucocorticoids that will keep researchers busy for decades to come. For example, ChIP-seq experiments have uncovered thousands of GR binding sites and although regulated genes tend to have more GR binding sites in their vicinity, there are plenty of genes that are not regulated despite having a GR binding site nearby. This raises the question: What distinguishes a productive GR binding site (resulting in the regulation of associated genes) from ones where nearby genes are not regulated? A major complication in answering this question is that binding sites are assigned to a gene based on proximity along the linear DNA chain and not based on established functional connections between binding sites and genes. Some of these binding sites are located 100 s of kb away from the TSS and therefore could just as well be connected to other genes that are perhaps closer when the three dimensional organization of the nucleus is taken into account. Recently developed techniques to edit the genome like zinc finger nucleases (ZFNs), transcription activator-like effector nucleases (TALENs) and the CRISPR/Cas9 RNA-guided system provide the opportunity to disrupt genomic binding sites of GR and thereby to determine functional connections between binding sites and regulated genes [128–130]. Another approach to link binding sites to genes is to systematically determine the physical contacts between GR binding sites and TSSs of genes. Such long-range looping interactions can be identified with the use of chromatin conformation capture (3C)-based techniques and have shown a clear correlation between long-range contacts and transcriptional regulation by transcriptional regulatory factors [131, 132]. A final challenge is to understand how the integration of various inputs warrants that the right genes are expressed at the correct level in response to glucocorticoids. Many of these inputs that modulate the transcriptional responses have been identified including receptor isoform, post-translational modification state, ligand and interaction with other biological macromolecules including proteins and DNA. Likely however, additional inputs exist. For example, the role of the noncoding RNA universe is still largely unexplored and studies with ER have shown that so called enhancer RNAs (eRNAs, see Fig. 3.5) produced at ER binding sites are required for long-range looping and the transcriptional regulation of ER target genes [133]. Ultimately, a detailed knowledge of the signaling inputs and how they are integrated at individual genes will yield a greater understanding of the heterogeneity in GR signaling in health and disease and may one day improve the therapeutic use of glucocorticoids in the clinic.

References

1. Benedek TG. History of the development of corticosteroid therapy. Clin Exp Rheumatol. 2011;29(5 Suppl 68):5–12.
2. Miesfeld R, Okret S, Wikstrom AC, Wrange O, Gustafsson JA, Yamamoto KR. Characterization of a steroid hormone receptor gene and mRNA in wild-type and mutant cells. Nature. 1984;312(5996):779–81.
3. Hollenberg SM, Weinberger C, Ong ES, et al. Primary structure and expression of a functional human glucocorticoid receptor cDNA. Nature. 1985;318(6047):635–41.
4. Carlstedt-Duke J, Stromstedt PE, Wrange O, Bergman T, Gustafsson JA, Jornvall H. Domain structure of the glucocorticoid receptor protein. Proc Natl Acad Sci U S A. 1987; 84(13):4437–40.
5. Kumar R, Thompson EB. Gene regulation by the glucocorticoid receptor: structure:function relationship. J Steroid Biochem Mol Biol. 2005;94(5):383–94.
6. Kumar R, Thompson EB. Transactivation functions of the N-terminal domains of nuclear hormone receptors: protein folding and coactivator interactions. Mol Endocrinol. 2003;17(1):1–10.
7. Kumar R, Thompson EB. Folding of the glucocorticoid receptor N-terminal transactivation function: dynamics and regulation. Mol Cell Endocrinol. 2012;348(2):450–6.
8. Strahle U, Klock G, Schutz G. A DNA sequence of 15 base pairs is sufficient to mediate both glucocorticoid and progesterone induction of gene expression. Proc Natl Acad Sci U S A. 1987;84(22):7871–5.
9. Luisi BF, Xu WX, Otwinowski Z, Freedman LP, Yamamoto KR, Sigler PB. Crystallographic analysis of the interaction of the glucocorticoid receptor with DNA. Nature. 1991; 352(6335):497–505.
10. Meijsing SH, Pufall MA, So AY, Bates DL, Chen L, Yamamoto KR. DNA binding site sequence directs glucocorticoid receptor structure and activity. Science. 2009;324(5925):407–10.
11. Freedman LP, Luisi BF, Korszun ZR, Basavappa R, Sigler PB, Yamamoto KR. The function and structure of the metal coordination sites within the glucocorticoid receptor DNA binding domain. Nature. 1988;334(6182):543–6.
12. Severne Y, Wieland S, Schaffner W, Rusconi S. Metal binding 'finger' structures in the glucocorticoid receptor defined by site-directed mutagenesis. EMBO J. 1988;7(8):2503–8.
13. Hill KK, Roemer SC, Jones DN, Churchill ME, Edwards DP. A progesterone receptor co-activator (JDP2) mediates activity through interaction with residues in the carboxyl-terminal extension of the DNA binding domain. J Biol Chem. 2009;284(36):24415–24.
14. Boonyaratanakornkit V, Melvin V, Prendergast P, et al. High-mobility group chromatin proteins 1 and 2 functionally interact with steroid hormone receptors to enhance their DNA binding in vitro and transcriptional activity in mammalian cells. Mol Cell Biol. 1998;18(8):4471–87.
15. Ichijo T, Chrousos GP, Kino T. Activated glucocorticoid receptor interacts with the INHAT component Set/TAF-Ibeta and releases it from a glucocorticoid-responsive gene promoter, relieving repression: implications for the pathogenesis of glucocorticoid resistance in acute undifferentiated leukemia with Set-Can translocation. Mol Cell Endocrinol. 2008; 283(1-2):19–31.
16. Ko L, Cardona GR, Henrion-Caude A, Chin WW. Identification and characterization of a tissue-specific coactivator, GT198, that interacts with the DNA-binding domains of nuclear receptors. Mol Cell Biol. 2002;22(1):357–69.
17. Tao Y, Williams-Skipp C, Scheinman RI. Mapping of glucocorticoid receptor DNA binding domain surfaces contributing to transrepression of NF-kappa B and induction of apoptosis. J Biol Chem. 2001;276(4):2329–32.
18. Jonat C, Rahmsdorf HJ, Park KK, et al. Antitumor promotion and antiinflammation: down-modulation of AP-1 (Fos/Jun) activity by glucocorticoid hormone. Cell. 1990;62(6): 1189–204.

19. Yang-Yen HF, Chambard JC, Sun YL, et al. Transcriptional interference between c-Jun and the glucocorticoid receptor: mutual inhibition of DNA binding due to direct protein-protein interaction. Cell. 1990;62(6):1205–15.
20. Schule R, Rangarajan P, Kliewer S, et al. Functional antagonism between oncoprotein c-Jun and the glucocorticoid receptor. Cell. 1990;62(6):1217–26.
21. Ray A, Prefontaine KE. Physical association and functional antagonism between the p65 subunit of transcription factor NF-kappa B and the glucocorticoid receptor. Proc Natl Acad Sci U S A. 1994;91(2):752–6.
22. Langlais D, Couture C, Balsalobre A, Drouin J. The Stat3/GR interaction code: predictive value of direct/indirect DNA recruitment for transcription outcome. Mol Cell. 2012; 47(1):38–49.
23. Heck S, Kullmann M, Gast A, et al. A distinct modulating domain in glucocorticoid receptor monomers in the repression of activity of the transcription factor AP-1. EMBO J. 1994;13(17):4087–95.
24. Bledsoe RK, Montana VG, Stanley TB, et al. Crystal structure of the glucocorticoid receptor ligand binding domain reveals a novel mode of receptor dimerization and coactivator recognition. Cell. 2002;110(1):93–105.
25. Oakley RH, Cidlowski JA. The biology of the glucocorticoid receptor: new signaling mechanisms in health and disease. J Allergy Clin Immunol. 2013;132(5):1033–44.
26. Pratt WB, Toft DO. Steroid receptor interactions with heat shock protein and immunophilin chaperones. Endocr Rev. 1997;18(3):306–60.
27. Hong H, Kohli K, Trivedi A, Johnson DL, Stallcup MR. GRIP1, a novel mouse protein that serves as a transcriptional coactivator in yeast for the hormone binding domains of steroid receptors. Proc Natl Acad Sci U S A. 1996;93(10):4948–52.
28. Kucera T, Waltner-Law M, Scott DK, Prasad R, Granner DK. A point mutation of the AF2 transactivation domain of the glucocorticoid receptor disrupts its interaction with steroid receptor coactivator 1. J Biol Chem. 2002;277(29):26098–102.
29. Hall JM, McDonnell DP, Korach KS. Allosteric regulation of estrogen receptor structure, function, and coactivator recruitment by different estrogen response elements. Mol Endocrinol. 2002;16(3):469–86.
30. Thomas-Chollier M, Watson LC, Cooper SB, et al. A naturally occurring insertion of a single amino acid rewires transcriptional regulation by glucocorticoid receptor isoforms. Proc Natl Acad Sci U S A. 2013;110(44):17826–31.
31. Chandra V, Huang P, Hamuro Y, et al. Structure of the intact PPAR-gamma-RXR-nuclear receptor complex on DNA. Nature. 2008;456(7220):350–6.
32. Zhang J, Chalmers MJ, Stayrook KR, et al. DNA binding alters coactivator interaction surfaces of the intact VDR-RXR complex. Nat Struct Mol Biol. 2011;18(5):556–63.
33. Rogatsky I, Wang JC, Derynck MK, et al. Target-specific utilization of transcriptional regulatory surfaces by the glucocorticoid receptor. Proc Natl Acad Sci U S A. 2003;100(24): 13845–50.
34. Nehme A, Lobenhofer EK, Stamer WD, Edelman JL. Glucocorticoids with different chemical structures but similar glucocorticoid receptor potency regulate subsets of common and unique genes in human trabecular meshwork cells. BMC Med Genomics. 2009;2:58.
35. Cao Y, Bender IK, Konstantinidis AK, et al. Glucocorticoid receptor translational isoforms underlie maturational stage-specific glucocorticoid sensitivities of dendritic cells in mice and humans. Blood. 2013;121(9):1553–62.
36. Kino T, Su YA, Chrousos GP. Human glucocorticoid receptor isoform beta: recent understanding of its potential implications in physiology and pathophysiology. Cell Mol Life Sci. 2009;66(21):3435–48.
37. Lewis-Tuffin LJ, Cidlowski JA. The physiology of human glucocorticoid receptor beta (hGRbeta) and glucocorticoid resistance. Ann N Y Acad Sci. 2006;1069:1–9.
38. Rivers C, Levy A, Hancock J, Lightman S, Norman M. Insertion of an amino acid in the DNA-binding domain of the glucocorticoid receptor as a result of alternative splicing. J Clin Endocrinol Metab. 1999;84(11):4283–6.

39. Haarman EG, Kaspers GJ, Pieters R, Rottier MM, Veerman AJ. Glucocorticoid receptor alpha, beta and gamma expression vs in vitro glucocorticod resistance in childhood leukemia. Leukemia. 2004;18(3):530–7.
40. Ray DW, Davis JR, White A, Clark AJ. Glucocorticoid receptor structure and function in glucocorticoid-resistant small cell lung carcinoma cells. Cancer Res. 1996;56(14):3276–80.
41. Sanchez-Vega B, Krett N, Rosen ST, Gandhi V. Glucocorticoid receptor transcriptional isoforms and resistance in multiple myeloma cells. Mol Cancer Ther. 2006;5(12):3062–70.
42. de Lange P, Segeren CM, Koper JW, et al. Expression in hematological malignancies of a glucocorticoid receptor splice variant that augments glucocorticoid receptor-mediated effects in transfected cells. Cancer Res. 2001;61(10):3937–41.
43. Oakley RH, Cidlowski JA. Cellular processing of the glucocorticoid receptor gene and protein: new mechanisms for generating tissue-specific actions of glucocorticoids. J Biol Chem. 2011;286(5):3177–84.
44. Lu NZ, Cidlowski JA. Translational regulatory mechanisms generate N-terminal glucocorticoid receptor isoforms with unique transcriptional target genes. Mol Cell. 2005; 18(3):331–42.
45. Chen W, Dang T, Blind RD, et al. Glucocorticoid receptor phosphorylation differentially affects target gene expression. Mol Endocrinol. 2008;22(8):1754–66.
46. Charmandari E, Chrousos GP, Lambrou GI, et al. Peripheral CLOCK regulates target-tissue glucocorticoid receptor transcriptional activity in a circadian fashion in man. PLoS One. 2011;6(9), e25612.
47. Nader N, Chrousos GP, Kino T. Circadian rhythm transcription factor CLOCK regulates the transcriptional activity of the glucocorticoid receptor by acetylating its hinge region lysine cluster: potential physiological implications. FASEB J. 2009;23(5):1572–83.
48. Karin M, Haslinger A, Holtgreve H, et al. Characterization of DNA sequences through which cadmium and glucocorticoid hormones induce human metallothionein-IIA gene. Nature. 1984;308(5959):513–9.
49. Jantzen HM, Strahle U, Gloss B, et al. Cooperativity of glucocorticoid response elements located far upstream of the tyrosine aminotransferase gene. Cell. 1987;49(1):29–38.
50. Chandler VL, Maler BA, Yamamoto KR. DNA sequences bound specifically by glucocorticoid receptor in vitro render a heterologous promoter hormone responsive in vivo. Cell. 1983;33(2):489–99.
51. Solomon MJ, Larsen PL, Varshavsky A. Mapping protein-DNA interactions in vivo with formaldehyde: evidence that histone H4 is retained on a highly transcribed gene. Cell. 1988;53(6):937–47.
52. Robertson G, Hirst M, Bainbridge M, et al. Genome-wide profiles of STAT1 DNA association using chromatin immunoprecipitation and massively parallel sequencing. Nat Methods. 2007;4(8):651–7.
53. John S, Sabo PJ, Thurman RE, et al. Chromatin accessibility pre-determines glucocorticoid receptor binding patterns. Nat Genet. 2011;43(3):264–8.
54. Reddy TE, Pauli F, Sprouse RO, et al. Genomic determination of the glucocorticoid response reveals unexpected mechanisms of gene regulation. Genome Res. 2009;19(12):2163–71.
55. Yu CY, Mayba O, Lee JV, et al. Genome-wide analysis of glucocorticoid receptor binding regions in adipocytes reveal gene network involved in triglyceride homeostasis. PLoS One. 2010;5(12), e15188.
56. Polman JA, Welten JE, Bosch DS, et al. A genome-wide signature of glucocorticoid receptor binding in neuronal PC12 cells. BMC Neurosci. 2012;13:118.
57. Nielsen R, Pedersen TA, Hagenbeek D, et al. Genome-wide profiling of PPARgamma:RXR and RNA polymerase II occupancy reveals temporal activation of distinct metabolic pathways and changes in RXR dimer composition during adipogenesis. Genes Dev. 2008;22(21):2953–67.
58. Welboren WJ, van Driel MA, Janssen-Megens EM, et al. ChIP-Seq of ERalpha and RNA polymerase II defines genes differentially responding to ligands. EMBO J. 2009;28(10): 1418–28.

59. Hakim O, John S, Ling JQ, Biddie SC, Hoffman AR, Hager GL. Glucocorticoid receptor activation of the Ciz1-Lcn2 locus by long range interactions. J Biol Chem. 2009;284(10): 6048–52.
60. Drouin J, Sun YL, Chamberland M, et al. Novel glucocorticoid receptor complex with DNA element of the hormone-repressed POMC gene. EMBO J. 1993;12(1):145–56.
61. Surjit M, Ganti KP, Mukherji A, et al. Widespread negative response elements mediate direct repression by agonist-liganded glucocorticoid receptor. Cell. 2011;145(2):224–41.
62. Hudson WH, Youn C, Ortlund EA. The structural basis of direct glucocorticoid-mediated transrepression. Nat Struct Mol Biol. 2013;20(1):53–8.
63. Zhang Z, Jones S, Hagood JS, Fuentes NL, Fuller GM. STAT3 acts as a co-activator of glucocorticoid receptor signaling. J Biol Chem. 1997;272(49):30607–10.
64. Philips A, Maira M, Mullick A, et al. Antagonism between Nur77 and glucocorticoid receptor for control of transcription. Mol Cell Biol. 1997;17(10):5952–9.
65. Luecke HF, Yamamoto KR. The glucocorticoid receptor blocks P-TEFb recruitment by NFkappaB to effect promoter-specific transcriptional repression. Genes Dev. 2005; 19(9):1116–27.
66. Bladh LG, Liden J, Dahlman-Wright K, Reimers M, Nilsson S, Okret S. Identification of endogenous glucocorticoid repressed genes differentially regulated by a glucocorticoid receptor mutant able to separate between nuclear factor-kappaB and activator protein-1 repression. Mol Pharmacol. 2005;67(3):815–26.
67. Rao NA, McCalman MT, Moulos P, et al. Coactivation of GR and NFKB alters the repertoire of their binding sites and target genes. Genome Res. 2011;21(9):1404–16.
68. Rogatsky I, Zarember KA, Yamamoto KR. Factor recruitment and TIF2/GRIP1 corepressor activity at a collagenase-3 response element that mediates regulation by phorbol esters and hormones. EMBO J. 2001;20(21):6071–83.
69. Rogatsky I, Waase CL, Garabedian MJ. Phosphorylation and inhibition of rat glucocorticoid receptor transcriptional activation by glycogen synthase kinase-3 (GSK-3). Species-specific differences between human and rat glucocorticoid receptor signaling as revealed through GSK-3 phosphorylation. J Biol Chem. 1998;273(23):14315–21.
70. Aittomaki S, Pesu M, Groner B, Janne OA, Palvimo JJ, Silvennoinen O. Cooperation among Stat1, glucocorticoid receptor, and PU.1 in transcriptional activation of the high-affinity Fc gamma receptor I in monocytes. J Immunol. 2000;164(11):5689–97.
71. Cella N, Groner B, Hynes NE. Characterization of Stat5a and Stat5b homodimers and heterodimers and their association with the glucocortiocoid receptor in mammary cells. Mol Cell Biol. 1998;18(4):1783–92.
72. Siersbaek R, Nielsen R, John S, et al. Extensive chromatin remodelling and establishment of transcription factor 'hotspots' during early adipogenesis. EMBO J. 2011;30(8):1459–72.
73. Thurman RE, Rynes E, Humbert R, et al. The accessible chromatin landscape of the human genome. Nature. 2012;489(7414):75–82.
74. Wang J, Zhuang J, Iyer S, et al. Sequence features and chromatin structure around the genomic regions bound by 119 human transcription factors. Genome Res. 2012;22(9):1798–812.
75. Strahle U, Schmid W, Schutz G. Synergistic action of the glucocorticoid receptor with transcription factors. EMBO J. 1988;7(11):3389–95.
76. Lerner L, Henriksen MA, Zhang X, Darnell Jr JE. STAT3-dependent enhanceosome assembly and disassembly: synergy with GR for full transcriptional increase of the alpha 2-macroglobulin gene. Genes Dev. 2003;17(20):2564–77.
77. De Martino MU, Alesci S, Chrousos GP, Kino T. Interaction of the glucocorticoid receptor and the chicken ovalbumin upstream promoter-transcription factor II (COUP-TFII): implications for the actions of glucocorticoids on glucose, lipoprotein, and xenobiotic metabolism. Ann N Y Acad Sci. 2004;1024:72–85.
78. Biddie SC, John S, Sabo PJ, et al. Transcription factor AP1 potentiates chromatin accessibility and glucocorticoid receptor binding. Mol Cell. 2011;43(1):145–55.
79. Stromstedt PE, Poellinger L, Gustafsson JA, Carlstedt-Duke J. The glucocorticoid receptor binds to a sequence overlapping the TATA box of the human osteocalcin promoter: a potential mechanism for negative regulation. Mol Cell Biol. 1991;11(6):3379–83.

80. Subramaniam N, Cairns W, Okret S. Glucocorticoids repress transcription from a negative glucocorticoid response element recognized by two homeodomain-containing proteins, Pbx and Oct-1. J Biol Chem. 1998;273(36):23567–74.
81. Ki SH, Cho IJ, Choi DW, Kim SG. Glucocorticoid receptor (GR)-associated SMRT binding to C/EBPbeta TAD and Nrf2 Neh4/5: role of SMRT recruited to GR in GSTA2 gene repression. Mol Cell Biol. 2005;25(10):4150–65.
82. Diamond MI, Miner JN, Yoshinaga SK, Yamamoto KR. Transcription factor interactions: selectors of positive or negative regulation from a single DNA element. Science. 1990;249(4974):1266–72.
83. Belikov S, Astrand C, Wrange O. FoxA1 binding directs chromatin structure and the functional response of a glucocorticoid receptor-regulated promoter. Mol Cell Biol. 2009; 29(20):5413–25.
84. Cirillo LA, Lin FR, Cuesta I, Friedman D, Jarnik M, Zaret KS. Opening of compacted chromatin by early developmental transcription factors HNF3 (FoxA) and GATA-4. Mol Cell. 2002;9(2):279–89.
85. Hurtado A, Holmes KA, Ross-Innes CS, Schmidt D, Carroll JS. FOXA1 is a key determinant of estrogen receptor function and endocrine response. Nat Genet. 2011;43(1):27–33.
86. Sahu B, Laakso M, Ovaska K, et al. Dual role of FoxA1 in androgen receptor binding to chromatin, androgen signalling and prostate cancer. EMBO J. 2011;30(19):3962–76.
87. Gertz J, Savic D, Varley KE, et al. Distinct properties of cell-type-specific and shared transcription factor binding sites. Mol Cell. 2013;52(1):25–36.
88. Grontved L, John S, Baek S, et al. C/EBP maintains chromatin accessibility in liver and facilitates glucocorticoid receptor recruitment to steroid response elements. EMBO J. 2013;32(11):1568–83.
89. John S, Sabo PJ, Johnson TA, et al. Interaction of the glucocorticoid receptor with the chromatin landscape. Mol Cell. 2008;29(5):611–24.
90. Buratowski S. Progression through the RNA polymerase II CTD cycle. Mol Cell. 2009;36(4):541–6.
91. Murasawa S, Matsubara H, Kizima K, Maruyama K, Mori Y, Inada M. Glucocorticoids regulate V1a vasopressin receptor expression by increasing mRNA stability in vascular smooth muscle cells. Hypertension. 1995;26(4):665–9.
92. Lasa M, Brook M, Saklatvala J, Clark AR. Dexamethasone destabilizes cyclooxygenase 2 mRNA by inhibiting mitogen-activated protein kinase p38. Mol Cell Biol. 2001;21(3): 771–80.
93. Ing NH. Steroid hormones regulate gene expression posttranscriptionally by altering the stabilities of messenger RNAs. Biol Reprod. 2005;72(6):1290–6.
94. Beermann F, Schmid E, Ganss R, Schutz G, Ruppert S. Molecular characterization of the mouse tyrosinase gene: pigment cell-specific expression in transgenic mice. Pigment Cell Res. 1992;5(5 Pt 2):295–9.
95. Ratman D, Vanden Berghe W, Dejager L, et al. How glucocorticoid receptors modulate the activity of other transcription factors: a scope beyond tethering. Mol Cell Endocrinol. 2013;380(1–2):41–54.
96. Ford J, McEwan IJ, Wright AP, Gustafsson JA. Involvement of the transcription factor IID protein complex in gene activation by the N-terminal transactivation domain of the glucocorticoid receptor in vitro. Mol Endocrinol. 1997;11(10):1467–75.
97. Shiama N. The p300/CBP family: integrating signals with transcription factors and chromatin. Trends Cell Biol. 1997;7(6):230–6.
98. Fryer CJ, Archer TK. Chromatin remodelling by the glucocorticoid receptor requires the BRG1 complex. Nature. 1998;393(6680):88–91.
99. Almlof T, Wallberg AE, Gustafsson JA, Wright AP. Role of important hydrophobic amino acids in the interaction between the glucocorticoid receptor tau 1-core activation domain and target factors. Biochemistry. 1998;37(26):9586–94.
100. Hittelman AB, Burakov D, Iniguez-Lluhi JA, Freedman LP, Garabedian MJ. Differential regulation of glucocorticoid receptor transcriptional activation via AF-1-associated proteins. EMBO J. 1999;18(19):5380–8.

101. Chen W, Roeder RG. The Mediator subunit MED1/TRAP220 is required for optimal glucocorticoid receptor-mediated transcription activation. Nucleic Acids Res. 2007;35(18):6161–9.
102. Pascual-Le Tallec L, Simone F, Viengchareun S, Meduri G, Thirman MJ, Lombes M. The elongation factor ELL (eleven-nineteen lysine-rich leukemia) is a selective coregulator for steroid receptor functions. Mol Endocrinol. 2005;19(5):1158–69.
103. Conaway JW, Shilatifard A, Dvir A, Conaway RC. Control of elongation by RNA polymerase II. Trends Biochem Sci. 2000;25(8):375–80.
104. Gupte R, Muse GW, Chinenov Y, Adelman K, Rogatsky I. Glucocorticoid receptor represses proinflammatory genes at distinct steps of the transcription cycle. Proc Natl Acad Sci U S A. 2013;110(36):14616–21.
105. Germond JE, Hirt B, Oudet P, Gross-Bellark M, Chambon P. Folding of the DNA double helix in chromatin-like structures from simian virus 40. Proc Natl Acad Sci U S A. 1975;72(5):1843–7.
106. Kornberg RD. Chromatin structure: a repeating unit of histones and DNA. Science. 1974;184(4139):868–71.
107. Richmond TJ, Davey CA. The structure of DNA in the nucleosome core. Nature. 2003;423(6936):145–50.
108. Workman JL, Roeder RG. Binding of transcription factor TFIID to the major late promoter during in vitro nucleosome assembly potentiates subsequent initiation by RNA polymerase II. Cell. 1987;51(4):613–22.
109. Hager GL, Archer TK, Fragoso G, et al. Influence of chromatin structure on the binding of transcription factors to DNA. Cold Spring Harb Symp Quant Biol. 1993;58:63–71.
110. King HA, Trotter KW, Archer TK. Chromatin remodeling during glucocorticoid receptor regulated transactivation. Biochim Biophys Acta. 2012;1819(7):716–26.
111. Yoshinaga SK, Peterson CL, Herskowitz I, Yamamoto KR. Roles of SWI1, SWI2, and SWI3 proteins for transcriptional enhancement by steroid receptors. Science. 1992;258(5088):1598–604.
112. Hsiao PW, Fryer CJ, Trotter KW, Wang W, Archer TK. BAF60a mediates critical interactions between nuclear receptors and the BRG1 chromatin-remodeling complex for transactivation. Mol Cell Biol. 2003;23(17):6210–20.
113. Johnson TA, Elbi C, Parekh BS, Hager GL, John S. Chromatin remodeling complexes interact dynamically with a glucocorticoid receptor-regulated promoter. Mol Biol Cell. 2008;19(8):3308–22.
114. Engel KB, Yamamoto KR. The glucocorticoid receptor and the coregulator Brm selectively modulate each other's occupancy and activity in a gene-specific manner. Mol Cell Biol. 2011;31(16):3267–76.
115. Nightingale KP, O'Neill LP, Turner BM. Histone modifications: signalling receptors and potential elements of a heritable epigenetic code. Curr Opin Genet Dev. 2006;16(2):125–36.
116. Vermeulen M, Mulder KW, Denissov S, et al. Selective anchoring of TFIID to nucleosomes by trimethylation of histone H3 lysine 4. Cell. 2007;131(1):58–69.
117. Bode J, Henco K, Wingender E. Modulation of the nucleosome structure by histone acetylation. Eur J Biochem. 1980;110(1):143–52.
118. Glass CK, Rose DW, Rosenfeld MG. Nuclear receptor coactivators. Curr Opin Cell Biol. 1997;9(2):222–32.
119. Ito K, Barnes PJ, Adcock IM. Glucocorticoid receptor recruitment of histone deacetylase 2 inhibits interleukin-1beta-induced histone H4 acetylation on lysines 8 and 12. Mol Cell Biol. 2000;20(18):6891–903.
120. Kovacs JJ, Murphy PJ, Gaillard S, et al. HDAC6 regulates Hsp90 acetylation and chaperone-dependent activation of glucocorticoid receptor. Mol Cell. 2005;18(5):601–7.
121. Zelin E, Zhang Y, Toogun OA, Zhong S, Freeman BC. The p23 molecular chaperone and GCN5 acetylase jointly modulate protein-DNA dynamics and open chromatin status. Mol Cell. 2012;48(3):459–70.

122. Rogatsky I, Luecke HF, Leitman DC, Yamamoto KR. Alternate surfaces of transcriptional coregulator GRIP1 function in different glucocorticoid receptor activation and repression contexts. Proc Natl Acad Sci U S A. 2002;99(26):16701–6.
123. Chen D, Huang SM, Stallcup MR. Synergistic, p160 coactivator-dependent enhancement of estrogen receptor function by CARM1 and p300. J Biol Chem. 2000;275(52):40810–6.
124. Lee DY, Northrop JP, Kuo MH, Stallcup MR. Histone H3 lysine 9 methyltransferase G9a is a transcriptional coactivator for nuclear receptors. J Biol Chem. 2006;281(13):8476–85.
125. Watson LC, Kuchenbecker KM, Schiller BJ, Gross JD, Pufall MA, Yamamoto KR. The glucocorticoid receptor dimer interface allosterically transmits sequence-specific DNA signals. Nat Struct Mol Biol. 2013;20(7):876–83.
126. Wang JC, Shah N, Pantoja C, et al. Novel arylpyrazole compounds selectively modulate glucocorticoid receptor regulatory activity. Genes Dev. 2006;20(6):689–99.
127. De Bosscher K, Vanden Berghe W, Beck IM, et al. A fully dissociated compound of plant origin for inflammatory gene repression. Proc Natl Acad Sci U S A. 2005;102(44): 15827–32.
128. Mali P, Esvelt KM, Church GM. Cas9 as a versatile tool for engineering biology. Nat Methods. 2013;10(10):957–63.
129. Campbell JM, Hartjes KA, Nelson TJ, Xu X, Ekker SC. New and TALENted genome engineering toolbox. Circ Res. 2013;113(5):571–87.
130. Carroll D. Genome engineering with zinc-finger nucleases. Genetics. 2011;188(4):773–82.
131. Dixon JR, Selvaraj S, Yue F, et al. Topological domains in mammalian genomes identified by analysis of chromatin interactions. Nature. 2012;485(7398):376–80.
132. Jin F, Li Y, Dixon JR, et al. A high-resolution map of the three-dimensional chromatin interactome in human cells. Nature. 2013;503(7475):290–4.
133. Li W, Notani D, Ma Q, et al. Functional roles of enhancer RNAs for oestrogen-dependent transcriptional activation. Nature. 2013;498(7455):516–20.

Chapter 4
Clinical Perspective: What Do Addison and Cushing Tell Us About Glucocorticoid Action?

Charles Harris

Abstract This chapter is distinct from the others in its clinical subject matter. I will attempt to outline the major points of interest in glucocorticoids clinically. To aid the illustration in the evaluation of a patient with Cushing disease I have created a case study.

Keywords Glucocorticoids • Cushing syndrome • Cushing disease • Addison • Hypothalamus • Hypothalamic pituitary adrenal axis (HPA axis) • CRH • ACTH • Adrenal gland • Cortisol

The Hypothalamic Pituitary Adrenal Axis

The HPA axis is a classic endocrine feedback system. The hypothalamus produces CRH in response to circadian cues as well as in response to stress from other regions of the brain. CRH binds to CRH receptors found in the pituitary and stimulates ACTH release. ACTH binds to the ACTH receptors (MC2R) in the adrenal gland and stimulates the synthesis and release of cortisol. Cortisol in turn via the glucocorticoid receptor inhibits release of CRH and ACTH. Glucocorticoid excess can be broken down into endogenous excess and iatrogenic due to exogenous glucocorticoid administration for the treatment of any of a number of diseases. Endogenous glucocorticoid excess could theoretically arise from hyperactivity at each of the levels of the HPA axis. Although there are case reports of excess CRH secretion causing Cushing's syndrome, it is a very rare event. The most common cause of endogenous hypercortisolemia is due to pituitary tumors of the ACTH secreting corticotroph lineage. Excess glucocorticoid exposure

C. Harris, M.D., Ph.D. (✉)
Division of Endocrinology, Metabolism and Lipid Research, Department of Internal Medicine, Washington University School of Medicine, 660 S Euclid Avenue, Campus Box 8127, St. Louis, MO 63110, USA
e-mail: caharris@dom.wustl.edu

J.-C. Wang, C. Harris (eds.), *Glucocorticoid Signaling*, Advances in Experimental Medicine and Biology 872, DOI 10.1007/978-1-4939-2895-8_4

results in Cushing syndrome and when this is due to an ACTH producing pituitary tumor it is known as Cushing disease. Less commonly, a tumor of the adrenal gland can become autonomous of regulation by ACTH and secrete excess glucocorticoids. The least common form of endogenous hypercortisolemia is due to ectopic ACTH syndrome, when a non-pituitary tumor secretes ACTH which can stimulate the adrenal gland. Therefore, endogenous glucocorticoid excess can be broken into ACTH-dependent disease (pituitary tumor and ectopic ACTH) and ACTH-independent disease (adrenal tumor). So the overall evaluation of a patient with signs of glucocorticoid excess is to determine if it is exogenous or endogenous, confirm the diagnosis of endogenous hypercortisolism and determine the source.

Approach to the Patient with Potential Glucocorticoid Excess

It has been greater than 100 years since Harvey Cushing published a case report of 23-year old Minnie G who presented with central obesity, elevated systolic blood pressure and laevulose intolerance. Today, the most common cause of glucocorticoid excess unfortunately is iatrogenic use of glucocorticoid medications for the treatment of inflammatory diseases. Endogenous glucocorticoid excess is a rare disease, but must never be forgotten as a cause of Cushing syndrome. The clinician's job has become even more difficult given the increasing prevalence of metabolic syndrome which is a subset of Cushing syndrome. However, patients with metabolic syndrome have normal serum cortisol levels. The approach to diagnose Cushing's syndrome varies by nation, medical center and practitioner, but I will put forth a general approach. First, the diagnosis of endogenous glucocorticoid excess must be made. This can be done by several complementing approaches. A 24 h urine cortisol is helpful. A midnight serum cortisol would be very helpful because the normal circadian pattern of cortisol secretion is lost in endogenous glucocorticoid excess. Therefore, cortisol is very low in normals and does not suppress at night in patients with endogenous hypercortisolism. Because of the practical difficulties in obtaining such testing (patient laboratories are generally not open at that hour, and it is not practical for patients to go to the ER for a blood draw) salivary cortisol levels have been used to obtain similar information. Salivary tests have the advantage of being less invasive, which not only do patients prefer, but also are less likely to induce a stress response as phlebotomy might. The patient collects a saliva sample at midnight in the convenience of their home and returns the sample to the clinical lab for analysis. In addition to a 24 h urine cortisol, a dexamethasone suppression test is helpful in establishing a diagnosis of endogenous glucocorticoid excess. The patient takes a 1 mg dose of the potent glucocorticoid dexamethasone at 11 p.m. at home and then serum cortisol is measured at 8 a.m. the next day. In normal patients the dexamethasone will suppress CRH and ACTH secretion and therefore cortisol secretion to less than 2. If the patient does not perform their portion of the test properly, i.e., they do not take it at the correct time or do not take it at all, the "normal" AM cortisol will appear non-suppressed and the patient will have a false positive result. In these cases, and in a few cases of altered dexamethasone metabolism [1], one can obtain a serum dexamethasone level to determine if the patient has adequate dexamethasone levels to suppress cortisol production, but this is rarely necessary in practice. Once the diagnosis of endogenous glucocorticoid excess is established, one needs to determine the source. An AM measurement of ACTH indicates whether the patient has ACTH dependent disease (ACTH is normal or high) or ACTH independent disease (ACTH is suppressed). If the ACTH is in the low normal range, the clinician will need to consider both ACTH dependent or independent disease. ACTH independent disease is most likely due to adrenal adenomas and a CT scan is indicated to evaluate for this. If there is a single adenoma, unilateral resection should be curative. If there is bilateral disease one must decide whether to resect the dominant (larger) nodule containing adrenal, pursue definitive curative bilateral adrenalectomy, or medical therapy. If removal of the adrenal containing the dominant nodule is not curative, a second surgery to remove the remaining cortical adenoma may be necessary. This is particularly true in cases of macronodular adrenal hyperplasia. Some centers can perform cortical- sparing resection of such tumors, or depending on the severity of Cushing's syndrome, patient may opt for medical therapy. Cortical sparing adrenalectomy has usually been performed in the setting of medullary disease, such as pheochromocytoma, but has been reported for Cushings due to adrenal tumors as well [2].

The advantage of bilateral adrenalectomy is it is a definitive cure, not just for adrenal Cushing's, but all forms of endogenous hypercortisolism. The disadvantage of this procedure is it renders the patient permanently adrenally insufficient and this has been correlated with decreased quality of life [3]. If bilateral adrenalectomy is pursued for Cushing's disease it leaves open the possibility of accelerated growth of a pituitary tumor. This is because the pituitary tumor has been partially suppressed by cortisol and if this is removed, unopposed growth of the tumor, so called Nelson's syndrome can occur [4]. Furthermore, bilateral adrenalectomy in ectopic ACTH does not address the underlying problem of the ectopic tumor. For ACTH dependent disease the clinician must determine if the source of the ACTH is from the pituitary or an ectopic source. Some centers rely on a second dexamethasone suppression test, using higher amounts of dexamethasone administered over 2 days and the percent dexamethasone suppression as an indicator of pituitary vs. ectopic disease. This "hi-dose dexamethasone suppression test" takes advantage of the fact that pituitary tumors can be partially suppressed, while suppression is completely lacking in ectopic ACTH. However, given that the pre-test probability of pituitary disease in ACTH dependent disease is ~90 % and is comparable to the sensitivity of this high-dose dexamethasone suppression test, I do not use it in my practice. The next step is an MRI of the pituitary gland. Because most pituitary adenomas causing Cushing's disease are small, a negative MRI does not exclude the possibility of Cushing's disease. The sensitivity of MRI is ~3–4 mm and remarkably pituitary tumors smaller than this can cause full-blown Cushing's disease. Generally, if there is an obvious pituitary microadenoma greater or equal to 6 mm on the pituitary MRI, transphenoidal surgery is indicated. Because of the possibility of smaller findings being incidental, if no obvious adenoma is seen on MRI, it is recommended to undergo petrosal venous sampling. The principle of this test is that the pituitary is the point source of ACTH secretion. The closer you get to the pituitary, the higher the concentration of ACTH in the blood. During this study a trained interventional radiologist will sample blood from the periphery as well as near the petrosal sinus. For patients with pituitary secretion of ACTH, such a point source will result in a gradient of ACTH between the petrosal sinus and peripheral blood of two or more. A patient with an ectopic source of ACTH should not have higher ACTH levels in the petrosal sinus because the HPA axis would be suppressed due to the ectopic ACTH mediated secretion of cortisol. The sensitivity can be increased by administering agents that enhance ACTH secretion. The most commonly used agent is CRH. Generally a gradient of 2 before stimulation is considered a positive gradient and a gradient of 3 after stimulation is considered positive. Patients with confirmed pituitary disease should undergo transphenoidal surgery by a neurosurgeon with experience in such cases. It should be noted the importance in correctly diagnosing patient with glucocorticoid excess because a normal patient without glucocorticoid excess would also be expected to have a pituitary gradient of ACTH. If the petrosal sampling points to a non-pituitary source (no gradient means ectopic ACTH) the source of ACTH must be determined. Since the most common source is from bronchial carcinoids, a CT scan of the chest is usually the next test performed. If this does not reveal a single obvious suspicious lesion additional imaging modalities can be used to identify the lesion. In addition, there is case report of pulmonary vein sampling in a patient with ectopic ACTH and bilateral pulmonary nodules to determine the culpable lesion [5].

All patients with endogenous glucocorticoid excess will be rendered adrenally insufficient if they are surgically cured regardless of the tumor type. To assess for surgical cure, cortisol measurements in the blood are made in the immediate post-operative period. If the cortisol level is not very low, surgical cure probably has not occurred and the patient may still have glucocorticoid excess or is at high risk for relapse in the near future. In the case of adrenal disease, chronic excess cortisol suppresses the HPA axis. HPA axis suppression also occurs in ectopic ACTH syndrome. HPA axis suppression occurs in Cushing's disease as well as the normal corticotropes are suppressed. Therefore, all patients will require replacement dosing of glucocorticoids. This period of adrenal insufficiency can last upwards of 12 months. Patients should undergo ACTH stimulation testing prior to weaning off replacement glucocorticoids. Patients with Cushing's disease require long-term post-operative follow-up as recurrence can occur, even years after resection. Periodic urine cortisol and pituitary MRI testing is appropriate. In addition, patients should be educated about their symptoms of glucocorticoid excess and instructed to seek care if these symptoms return.

Medical therapy for endogenous Cushing's syndrome is an unmet need. Currently, two medications are approved for Cushing's syndrome. Korlym (mifepristone) is

FDA approved for hyperglycemia associated with Cushing's syndrome. Mifepristone is a glucocorticoid antagonist. Because of its unique mechanism of action, serum and urine cortisol levels increase but the net effect is decreased signaling through the glucocorticoid receptor. Serum and urine cortisol increase because the GR that mediates peripheral effects of glucocorticoids is the same molecule that mediates negative feedback inhibition in the pituitary. Patients with pituitary disease still have some degree of feedback inhibition by glucocorticoids, i.e., secretion is not entirely autonomous, hence the increase in serum and urine cortisol. Increased circulating cortisol can overcome blockade by the renal 11-β-HSD2 enzyme and lead to apparent mineralocorticoid excess. In fact, many patients on mifepristone experience increased blood pressure and hypokalemia. This can be addressed by combining mifepristone with a mineralocorticoid antagonist such as eplerenone. In addition, prolonged blockade of the progesterone receptor chronically has unknown health effects, particularly in women. Currently, mifepristone is approved by the FDA for the treatment of hyperglycemia associated with endogenous glucocorticoid excess, but it likely has beneficial effects on other organs adversely affected by glucocorticoid excess. The other medication approved for endogenous hypercortisolism is pasireotide, but because of this drug's mode of action, it is only effective against Cushing's disease (pituitary disease). Pasireotide is an octreotide analog with high affinity for the somatostatin receptor 5. It only has efficacy in ~25 % of patients at the highest dose and has adverse effects including hyperglycemia [6]. The anti-fungal ketoconazole has been used off-label to treat endogenous hypercortisolism. Since it is an inhibitor of glucocorticoid biosynthesis, it is effective in all subtypes of endogenous hypercortisolism. Ketoconazole is becoming increasingly difficult to procure because it has fallen out of favor as an anti-fungal, it's only approved indication. Ketoconazole has rare adverse effect of hepatitis and is also known to prolong QT interval (increasing the risk of ventricular tachycardia, a potentially life threatening arrhythmia of the heart), especially when combined with other medications. For these reasons, a specific enantiomer is being developed for the treatment of endogenous hypercortisolism (http://clinicaltrials.gov/show/NCT01838551). Other molecules in development are more specific GR antagonists, other inhibitors of glucocorticoid synthesis including LCI-699, an inhibitor of 11-β-hydroxylase [7].

Case Study

A 45 year old female presents with worsening weight gain, diabetes and hypertension over the last 3 years. During that time she has gradually gained 30 lbs and gone from having a normal blood pressure to elevated blood pressure that is poorly controlled on three blood pressure medications. In addition, she stopped menstruating 2 years ago, and notices some disturbing growth of facial hair. She developed excess urination 1 year ago and was diagnosed with diabetes at that time. On physical exam she is afebrile, hypertensive (BP 180/100) with a pulse of 90, she has a round "moon" facies, dorsorcervical fat (buffalo hump), supraclavicular fat, her heart is beating with regular rate and rhythm, and her breath sounds are normal. Her abdomen shows central obesity and her extremities are thin. Her skin is thin when rubbed between the examiners fingers and she has purple stretch marks (striae) on the sides of her abdomen. She is tearful, when describing her recent problems and adds that her moods have been much more labile over the last 2 years. She is alert and oriented to person, place and time. She has mildly decreased strength and has difficulty rising from a chair without using her hands.

Her current medications include: Lisinopril, hydrochlorothiazide, atenolol, metformin, glipizide. She denies taking any glucocorticoid medications.

She has had some blood work done by her primary care provider

HgbA1C 9.0 (normal <6.0)
Cortisol 22 mcg/dL (done at 9 a.m., normal 9–24 mcg/dL)
ACTH 65 pg/mL (done at 9 a.m., normal 6–55 pg/mL)

You tell her, you suspect she has Cushing's syndrome due to endogenous glucocorticoid excess, but that you will need to run some additional tests. You order a test for 24-h urine cortisol. In addition, you order a blood serum cortisol to be done 8–10 a.m. the morning after she takes a 1 mg dose of dexamethasone that you write a prescription for. Importantly, you tell her to do the dexamethasone suppression test after she has completed the 24-h urine collection.

She returns to the clinic 1 week later with the following results

24-h urine cortisol is 302 mcg (normal <50 mcg), creatinine is 900 mg in a volume of 2.1 L.
Cortisol in the AM 10.0 (after having taken 1 mg of dexamethasone the night prior at 11 p.m.).

The case is informative as to the effects of glucocorticoids on the body. One of the most important elements of this patient's history is that she did not take glucocorticoids. This is important because glucocorticoids are given as medications for several common conditions including asthma, emphysema (COPD), inflammatory bowel disease and rheumatoid arthritis. As a practical point, since patients may not be familiar with the term "glucocorticoid" it is useful to list the names of the commonly used glucocorticoids such as prednisone, hydrocortisone, cortef, dexamethasone, etc. Usually Cushing's syndrome is only seen with systemic glucocorticoid exposure, but it is possible to receive enough systemic exposure from injections into the joint, inhaled preparations, or topical preparations [8–10]. There were a large number of cases of Cushing's syndrome in HIV patients taking the inhaled glucocorticoid fluticasone because the commonly used anti-viral medication ritonavir inhibited the body's system of breakdown for the inhaled glucocorticoid, thereby raising levels of the inhaled glucocorticoid in the blood [11]. Another relevant point is that one cannot rule in or rule out Cushing's disease with the serum cortisol levels alone. Cortisol is secreted in a circadian manner with peaks in the early morning and nadirs in the evening and night. Patients with endogenous hypercortisolemia often have normal peak cortisol levels but have failure to reduce levels in the evening. This is why either a 24-h urine collection or a midnight salivary cortisol test is needed. Glucocorticoids are known to increase adiposity, but the exact mechanisms underlying this are unknown. Glucocorticoids are known to increase appetite, and in fact are used off-label for this purpose for such conditions such as cancer cachexia [12]. Megestrol is a commonly used glucocorticoid appetite stimulant in patients with cachexia due to cancer or AIDS. Glucocorticoids are also known to stimulate adipogenesis, the formation of fat cells from fat cell precursors such as adipocytes. In fact much of what we know about adipocyte biology has been learned using a cell line 3T3-L1 that is a fibroblast, but can be differentiated into adipocytes with a chemical cocktail including glucocorticoids. This patient has evidence of lipodystrophy with the buffalo hump, supraclavicular fat, central obesity, and loss of fat from her arms and legs. Although there is a net gain of fat mass the redistribution of fat in Cushing's syndrome is more prominent. This redistribution is due to increased breakdown of fat in some depots and enhanced synthesis in other. All depots undergo futile cycling of fat continuously breaking it down and resynthesizing it, so it is not difficult to see if the net synthesis/breakdown of the various depots was altered, one could get fat redistribution. In fact, it is remarkable that we are able to keep our fat depots in a constant distribution. Similarly, it is known that glucocorticoids raise blood pressure. This is likely due to effects on the endothelium, the kidney, and the heart (see Chap. 13). In addition, some of the effects of glucocorticoids to raise blood pressure are mediated by the mineralocorticoid receptor and some are mediated through the glucocorticoid receptor. Glucocorticoids can bind to the mineralocorticoid receptor, but when glucocorticoids are present at normal levels, they are inactivated locally in the kidney via the 11-b-HSD2 enzyme. However, when endogenous glucocorticoids are found at very high levels, they can overcome this inhibition and are free to bind to the mineralocorticoid receptor which increases blood pressure and lowers serum potassium. However, it is also clear that glucocorticoids raise blood pressure through the glucocorticoid receptor as well because synthetic glucocorticoids that do not bind the mineralocorticoid receptor also raise blood pressure. Usually, the increase in blood pressure is more severe than that seen in the general population ("essential" hypertension). The fact that this patient's blood pressure is very high while on three blood pressure medications is an indicator of the severity of the problem. The loss of menses can also be directly attributed to glucocorticoids as glucocorticoids suppress the hypothalamic-pituitary gonadal axis. Of course, loss of menses could be due to menopause. This patient had loss of menses at age 43, which is several standard deviations away from the population mean of age 51. If she did not have an obvious cause such as her Cushing's syndrome, other hormone testing would be helpful. It is possible she is having early menopause. In that case her FSH and LH would be elevated due to loss of negative feedback from estrogen that is no longer being produced by her ovaries. If her lack of periods (amenorrhea) is due to Cushing's syndrome, we expect her serum FSH and LH to be low. Ockham's razor, that the simplest explanation is usually the correct one, would dictate her amenorrhea is due to Cushing's. Other common hormonal abnormalities that could cause loss of menses include hypothyroidism, hyperthyroidism, and elevated prolactin levels. These scenarios would involve a distinct set of associated signs and symptoms. The growth of facial hair is common in endogenous glucocorticoid excess. In both ACTH-dependent and independent disease the adrenal gland is stimulated. The adrenal gland makes both glucocorticoids and androgens and the latter class of compounds will cause the increased facial hair in women. The excess urination (polyuria) is a direct consequence of diabetes and the diabetes could also be due to glucocorticoid excess. Although the exact mechanisms for this are not known, glucocorticoids are known to stimulate gluconeogenesis by the liver. Increased production of glucose by the liver can raise the serum glucose causing diabetes. In addition, glucocorticoids cause insulin resistance in peripheral tissues such as muscle and adipose tissue. Normally insulin stimulates these tissues to take up glucose from the blood, but in a state of insulin resistance glucose uptake is reduced, also raising serum glucose. Her skin is thin, because glucocorticoids suppress the synthesis of collagens. It is difficult to appreciate skin thickness visually so it is necessary to pick up the skin, usually on the back of the patients hand, between the examiner's thumb and forefinger. Of course, skin thickness decreases with age in the general population, so this must be taken into account. The glucocorticoids also cause the pigmented stretch marks, known as striae. Striae are virtually pathognomonic for Cushing's syndrome; i.e., when they occur, the patient usually has Cushing's syndrome. Her neurological exam was notable for muscle weakness as evidenced by her difficulty rising from a chair without using her arms. Glucocorticoids cause myopathy or muscle wasting. The pattern shows a proximal myopathy, with the larger muscles towards the center of the body being preferentially affected. Because, these are the muscles required to rise from a seat (without using one's hands), this is a good clinical test for proximal myopathy. Again, there are multiple causes of proximal myopathy, but in the setting of this patient it is likely due to glucocorticoid excess. Glucocorticoids induce muscle wasting by transcriptionally activating genes encoding proteases that breakdown muscle protein into amino acids. These amino acids are sent to the liver where they are converted to glucose. Her neurological history and exam are notable for mood changes and labile moods. Glucocorticoids cause pronounced cognitive changes including depression, psychoses and impairments in learning and memory. The patient's further laboratory testing confirms endogenous glucocorticoid excess as evidenced by increased urine cortisol and failure to suppress cortisol secretion in the dexamethasone suppression test. Ingesting 1 mg of dexamethasone at 11 p.m. should suppress serum cortisol levels to less than 2 mcg/dL the next morning (reference range in the morning in the absence of dexamethasone is 9–24 mcg/dL). Since, a diagnosis of endogenous hypercortisolism has been confirmed we can use the serum ACTH to distinguish whether she has ACTH-dependent or -independent disease. Since her ACTH is above 20, she has ACTH dependent disease. We tell her she needs an MRI of her pituitary gland, which is obtained and shows a 4 mm tumor in her pituitary. Because the tumor is quite small and could be an incidental finding in any person "off the street", we tell her we need to confirm that this small tumor is the source of ACTH that is causing her glucocorticoid excess. In fact the incidence of pituitary tumors in autopsy series and MRI or CT scans of normal people ranges from 5 to 20 % and have been given the humorous name pituitary incidentaloma [13, 14]. If her pituitary tumor had been larger (>6 mm), such petrosal sampling would be deemed unnecessary by some and she could be referred directly to a neurosurgeon. She undergoes petrosal sampling and has a petrosal-peripheral gradient of ACTH of 4 before CRH and 6 after CRH. This confirms that the pituitary is the source of excess ACTH driving her adrenal to produce excess cortisol. You tell her she stands a good chance of being cured following surgical resection of the tumor. You send her to a neurosurgeon that has experience operating on these tumors using the transphenoidal approach. The pituitary is in the center of your head, underneath the region of the brain known as the hypothalamus. A craniotomy (cutting through the skull) can be avoided if the surgeon goes through the sphenoid sinus by cutting through the back of the patient's nose. The patient goes for neurosurgery and the surgeon finds the tumor, excises it and sends it to the pathology lab. The next morning the patient's cortisol level is 0.6 and she is started on glucocorticoids: hydrocortisone 20 mg in the AM and 10 mg at 3 p.m. It does seem counter-intuitive to give glucocorticoids to the patient after their body has been plagued by too much glucocorticoids, but it is necessary as the patient will have adrenal insufficiency because the HPA axis has been suppressed. If the patient is not given glucocorticoids, they could have adrenal crisis and die. Patients can be started on a replacement dose of glucocorticoids. However, patients on a replacement dose often have symptoms of glucocorticoid deficiency, so-called "glucocorticoid withdrawal syndrome", because they are accustomed to much higher doses of glucocorticoids. There is likely a reset rheostat of glucocorticoid "tone". The clinician must weigh the need to wean patients to replacement doses of glucocorticoids to minimize excess glucocorticoid exposure with patient discomfort at physiological levels of glucocorticoids. Patients should be able to be weaned to replacement dose of glucocorticoids within a few months of surgery. At this point, one can assess patients periodically with AM cortisol to determine when their HPA axis will become non-suppressed. I typically check an AM cortisol every 3 months, until the AM cortisol is greater or equal than 7 at which point I perform an ACTH stimulation test. If they pass the ACTH stimulation test, I will slowly withdraw the replacement glucocorticoids. Again here one is balancing minimizing the need for glucocorticoids with the inconvenience and cost of performing too many ACTH stimulation tests. Two-weeks after her surgery our patient develops right calf pain and swelling. She undergoes ultrasound of the veins in her leg which show a deep vein thrombosis. She is started on anticoagulant therapy. Our patient was told to hold her dose of hydrocortisone the morning of her blood test and obtain an AM cortisol 3 month post-operatively. It is 4 (low) so she is told to obtain another test 3 months later (6 months post-operatively) and it is now 9 (lower limit of normal). She returns for an ACTH stimulation test and her cortisol is now 21 45 min after receiving an IM injection of 250 μg of cosyntropin (ACTH). Her dose of hydrocortisone is weaned by 2.5 mg every week (alternating reducing the AM and PM dose) so that after another 6 weeks she is off all medication. Her anticoagulant therapy is discontinued. On her next visit she states her menses have returned and a repeat 24-h urine cortisol is in the normal change. She obtains an MRI 6 months post-operatively to obtain a baseline image of the post-surgical changes. Given all these obvious signs and symptoms, one may ask whether there are any negative health consequences of Cushing's syndrome that are less obvious. In fact, one of the gravest health concerns in these patients is silent: osteoporosis. The rates of fracture in patients with Cushing's syndrome are quite high and higher than one would predict based on patients' bone mineral density [15]. This is likely due to poor bone quality, but also other non-bone related factors that contribute to fractures. Glucocorticoids impact bone on many levels. Glucocorticoids cause osteoblast apoptosis, osteoclast activation. In addition glucocorticoids cause hypogonadism in males and females which has a negative impact on bone health. In addition, glucocorticoids cause GH deficiency further impacting bone. In addition, glucocorticoids act as functional antagonists of vitamin D decreasing absorption of calcium. In addition, the glucocorticoid induced muscle weakness contributes to falls, which increases the likelihood of fractures. Bisphosphonates and recombinant PTH (Forteo, Teriparatide) have been shown to treat glucocorticoid induced osteoporosis. Although it is prudent to treat with either an anabolic or anti-resorptive in a patient with active endogenous hypercortisolism to prevent active bone loss, the case for anti-resorptive after a surgical cure is less clear. This is due to the fact that there is spontaneous recovery of bone mass following cure [16], which theoretically could be blunted by an anti-resorptive.

Approach to the Patient with Adrenal Insufficiency

Many of the same principles discussed above pertain to the converse situation of adrenal insufficiency. Adrenal insufficiency can be primary or secondary. In primary adrenal insufficiency, the problem lies in the adrenal gland. In this case there is loss of negative feedback at the level of pituitary ACTH and the circulating level of this hormone is very high. Since ACTH is derived from a larger protein proopiomelanocortin (POMC) which also yields melanin stimulating peptides the result is the dark pigmentation associated with primary adrenal insufficiency. In primary adrenal insufficiency both glucocorticoid and mineralocorticoid secretion is diminished. Patients with primary adrenal insufficiency will need replacement of glucocorticoid and mineralocorticoids, whereas patients with secondary adrenal insufficiency only require glucocorticoids. This is an important distinction as inappropriate use of mineralocorticoid can have negative effects including hypertension, hypokalemia, and heart failure. Conversely, if one were to omit mineralocorticoids in a patient with primary adrenal insufficiency they would develop hyperkalemia. This is due to the major effects of mineralocorticoids to promote sodium absorption and potassium excretion in the kidney. The distinction between primary and secondary adrenal insufficiency is usually obvious based on patient history, physical exam, and laboratory testing.

Because of the circadian fluctuation in cortisol secretion, sampling is best performed in the morning since reference ranges are available. However, the gold standard for diagnosing adrenal insufficiency is the ACTH stimulation test during which ACTH is injected (either intramuscularly or intravenously) and a cortisol level is obtained 30 and or 60 min after ACTH administration. Although the exact diagnostic cut off is arguable most use a stimulated level of greater than 18, but some use up to 20 as a cut off. Several points are worth considering. A controversial issue of glucocorticoid biology in medicine is the diagnosis of adrenal insufficiency in critically ill patients, e.g., patients in intensive care units. An early clinical trial found the incidence of adrenal insufficiency in such patients to be quite high, but was flawed in that a large number of these patients had received etomidate as an anesthetic prior to intubation [17]. Etomidate is a known inhibitor of glucocorticoid synthesis so the adrenal insufficiency in these patients may have been due to etomidate and not critical illness. In addition, serum cortisol levels in critically ill patients may appear low due to low levels of serum proteins that carry glucocorticoids (again due to critical illness) whereas the more important levels of free cortisol are normal [18]. A final misconception of adrenal insufficiency is the requirement that the serum cortisol increase a preset amount (usually set at 7–9) following administration of ACTH in addition to achieving a set level (of 18–20). In my opinion, this is an illogical requirement. If serum cortisol levels are very high, this should be sufficient evidence for adrenal sufficiency. One may not be able to respond to ACTH if endogenous ACTH is at a maximal effect. More recent studies have not demonstrated a benefit of glucocorticoid treatment for critically ill patients with septic shock [19]. Because adrenal insufficiency is often considered in hospitalized patient, glucocorticoids may be given before such a diagnosis is certain. These patients can still undergo an ACTH stimulation test with glucocorticoids "on board" i.e., one can still stimulate through the potential suppression from glucocorticoid therapy. In this case, it may be helpful to switch the patient to a synthetic glucocorticoid such as dexamethasone that does not interfere with the cortisol assay. Another point to consider is that the loss of the adrenal response to ACTH does not occur acutely in secondary adrenal insufficiency, but rather takes approximately 6 weeks to develop. The clinical implication of this is that a patient with secondary adrenal insufficiency may have a falsely normal ACTH stimulation test because the adrenals have not yet lost the ability to respond. Therefore, the clinical scenario is key to diagnosis and clinicians should not hesitate to treat and then perform repeat ACTH stimulation testing when indicated. Another less commonly used test of adrenal function is the metyrapone test. Metyrapone is an inhibitor of the 11-β-hydroxylase, which catalyzes the last step of cortisol synthesis. In a metyrapone test 2–3 g (the exact dose depending on patient weight) of metyrapone are ingested at midnight, and then cortisol and deoxycortisol are measured 8 h later at 8 a.m. when glucocorticoid synthesis is at its peak circadian phase. With inhibition of 11-β-hydroxylase, cortisol levels fall and in an individual with normal adrenal function there should be build-up of the substrate for 11-β-hydroxylase, 11-deoxycortisol. However, if there is low steroid synthesis in the setting of adrenal insufficiency, the 11-deoxycortisol level will not be elevated. One advantage of this test is that it has an "internal control". If the cortisol level is not low, it means the patient did not take the metyrapone properly. So, in a metyrapone test a normal patient has low cortisol levels and elevated deoxycortisol levels and a patient with AI has low cortisol and deoxycortisol after ingesting metyrapone. Specifically, adequate metyrapone exposure is evident with a cortisol <5 mcg/dL and normal adrenal function is defined as deoxycortisol levels >7 mcg/dL. Of note, in the absence of metyrapone, the normal circulating levels of deoxycortisol are <0.1 mcg/dL. Of course deoxycortisol can be elevated not only with metyrapone, but also with genetic mutations in 11-β-hydroxylase, which is a cause of congenital adrenal hyperplasia, beyond the scope of this chapter. It is essential to determine if the patient has primary or secondary adrenal insufficiency. These distinct entities have different causes, treatments and implications for the patient. Patients with primary adrenal insufficiency have a problem at the level of the adrenal glands themselves. This can be caused by an autoimmune process, hemorrhage, infection. Tuberculosis used to be a common cause, but is rare today. The patient with primary adrenal insufficiency will have an elevated ACTH (due to loss of negative feedback of cortisol on POMC transcription in the pituitary). Because POMC pro-protein is cleaved to melanocyte stimulating hormone peptides in addition to ACTH, the patient with primary adrenal insufficiency will be hyperpigmented. This is evident on the skin as well as on mucous membranes. The patient with primary adrenal insufficiency will require replacement of both glucocorticoids and mineralocorticoids. It is not feasible to replace catecholamines produced by the adrenal medulla, and they are also made by the sympathetic nervous system. The adrenal cortex makes androgens, but in males, the adrenal production is much smaller than that produced in the testes. Androgen replacement in women with primary adrenal insufficiency is a controversial issue [20]. Some have postulated low dose androgens should be replaced to ensure normal sexual function and libido in women. However, no androgen formulation exists that can be safely administered to women without the risk of virilization. Some have also proposed patients with primary adrenal insufficiency would benefit from replacement of DHEA-S, but there are not strong data to support this [21]. DHEA has not been shown to have function other than as a precursor to adrenal hormones. Therefore, if the end-product hormone is being replaced, there should be no need to replace DHEAS. Similarly, in the absence of normal adrenal function, the DHEAS will unlikely be metabolized into end product hormones. Therefore, the standard regimen includes a mineralocorticoid and a glucocorticoid. Although cortisol has inherent mineralocorticoid activity, a high dose would be required to overcome HSD2 in the kidney and would result in iatrogenic Cushing syndrome. The dose of mineralocorticoid should be titrated to the patients' blood pressure and serum potassium. There are a large number of glucocorticoid agonists available for replacement. In my practice, I usually use hydrocortisone with a total daily dose of 15 mg that is broken up into 10 mg in the morning and 5 p.m. mid-afternoon. This is done to try to mimic the circadian fluctuation in serum cortisol levels. Other commonly used oral GC agonists include prednisone. A replacement dose of prednisone is 4–5 mg per day as it is ~4 times more potent that hydrocortisone. Prednisone has MR agonist activity. The synthetic GC dexamethasone is 30 times more potent than hydrocortisone so a replacement dose is usually 0.25 mg and dexamethasone has no mineralocorticoid agonism. Dexamethasone is rarely used for adrenal insufficiency because of its higher potency, difficulty in finding an appropriate replacement dose without inducing iatrogenic Cushing's. It does have a role in the non-compliant patient given its very long biological half-life. Patients with adrenal insufficiency should be monitored for adequate replacement as well as iatrogenic Cushing syndrome. Patients with primary adrenal insufficiency usually require 100–200 mcg of a synthetic mineralocorticoid such as fludrocortisone. Interestingly, the most common cause of glucocorticoid deficiency today is glucocorticoids themselves. When glucocorticoids are taken for a length of time, there is suppression of the hypothalamic-pituitary adrenal axis. There are no clear cut rules as to the amount and duration of exposure needed to suppress the HPA axis. Certainly patients taking greater than 15 mg of prednisone for longer than 3 months are at risk and should undergo ACTH stimulation testing prior to discontinuing therapy If patients are taking prednisone, it is best to slowly lower their dose to 5 mg per day, the equivalent to what the body makes on its own. At that point, I find it helpful to switch to hydrocortisone. Given its short half-life and twice a day dosing to more closely mimic the circadian variation seen with cortisol secretion and may contribute to more paid recovery of the HPA axis. Because of the different potency the replacement dose of hydrocortisone is given as 10 mg first thing in the morning and 5 mg in the afternoon. A common error is to prescribe it merely as "bid" which means twice a day. Patients should be instructed to take the medication first thing in the morning and at 3 p.m., not bedtime, to more closely approximate endogenous rhythms. Once the patient is on replacement glucocorticoids one can perform ACTH stimulation testing every 6 months for 2 years or until they pass. Once they pass, and only once they pass, should their dose of glucocorticoid be lowered. Theoretically, once they pass an ACTH stimulation test they could stop glucocorticoids cold turkey, but I usually do this over a 6-week period having the patient lower their dose of hydrocortisone by 2.5 mg per week. If patients do not pass an ACTH stimulation test after 2 years of being on a replacement dose, it is unlikely they will recover HPA function and will likely require lifelong glucocorticoid treatment. I will end this chapter with another case presentation. A 70 year-old male with severe emphysema presents to the emergency room with nausea and vomiting. He has required oral glucocorticoids to control his emphysema and has been on 20 mg of prednisone daily for 3 years. He ran out of prednisone 1 week ago and has had nausea and vomiting for the past 5 days. On physical exam, he appears Cushingoid with moon facies, central obesity, a buffalo hump and pigmented striae. When he comes to the emergency room he has some laboratory testing of blood which shows a serum cortisol level of 1 mcg/dL (taken at 8 a.m.) and his ACTH is undetectable (<5 pg/mL). This patient has both Cushing's syndrome and adrenal insufficiency! He has iatrogenic exogenous Cushing's syndrome from the prednisone, but adrenal insufficiency from suppression of his hypothalamic-pituitary-adrenal axis from said prednisone.

References

1. Cassidy F, Ritchie JC, Verghese K, Carroll BJ. Dexamethasone metabolism in dexamethasone suppression test suppressors and nonsuppressors. Biol Psychiatry. 2000;47:677–80.
2. He HC, et al. Retroperitoneal adrenal-sparing surgery for the treatment of Cushing's syndrome caused by adrenocortical adenoma: 8-year experience with 87 patients. World J Surg. 2012;36:1182–8. doi:10.1007/s00268-012-1509-0.
3. Hawn MT, Cook D, Deveney C, Sheppard BC. Quality of life after laparoscopic bilateral adrenalectomy for Cushing's disease. Surgery 2002;132:1064–8; discussion 1068–9. doi:10.1067/msy.2002.128482.
4. Nelson DH, Meakin JW, Thorn GW. ACTH-producing pituitary tumors following adrenalectomy for Cushing's syndrome. Ann Intern Med. 1960;52:560–9.
5. Vu L, Theodore PR. Localization of a corticotropin-secreting tumor by thoracoscopic pulmonary venous sampling. N Engl J Med. 2005;353:851–2. doi:10.1056/NEJMc050486.
6. Colao A, et al. A 12-month phase 3 study of pasireotide in Cushing's disease. N Engl J Med. 2012;366:914–24. doi:10.1056/NEJMoa1105743.
7. Bertagna X, et al. LCI699, a potent 11beta-hydroxylase inhibitor, normalizes urinary cortisol in patients with Cushing's disease: results from a multicenter, proof-of-concept study. J Clin Endocrinol Metab. 2014;99:1375–83. doi:10.1210/jc.2013-2117.
8. Yombi JC, Maiter D, Belkhir L, Nzeusseu A, Vandercam B. Iatrogenic Cushing's syndrome and secondary adrenal insufficiency after a single intra-articular administration of triamcinolone acetonide in HIV-infected patients treated with ritonavir. Clin Rheumatol. 2008;27 Suppl 2:S79–82. doi:10.1007/s10067-008-1022-x.
9. Kumar S, Singh RJ, Reed AM, Lteif AN. Cushing's syndrome after intra-articular and intradermal administration of triamcinolone acetonide in three pediatric patients. Pediatrics. 2004;113:1820–4.
10. O'Sullivan MM, Rumfeld WR, Jones MK, Williams BD. Cushing's syndrome with suppression of the hypothalamic-pituitary-adrenal axis after intra-articular steroid injections. Ann Rheum Dis. 1985;44:561–3.
11. Bernecker C, West TB, Mansmann W, Scherbaum WA, Willenberg HS. Hypercortisolism caused by ritonavir associated inhibition of CYP 3A4 under inhalative glucocorticoid therapy. 2 case reports and a review of the literature. Exp Clin Endocrinol Diabetes. 2012;120:125–7. doi:10.1055/s-0031-1297993.
12. Yeh SS, Schuster MW. Megestrol acetate in cachexia and anorexia. Int J Nanomedicine. 2006;1:411–6.
13. Freda PU, et al. Pituitary incidentaloma: an endocrine society clinical practice guideline. J Clin Endocrinol Metab. 2011;96:894–904. doi:10.1210/jc.2010-1048.
14. Molitch ME, Russell EJ. The pituitary "incidentaloma". Ann Intern Med. 1990;112:925–31.
15. Van Staa TP, et al. Bone density threshold and other predictors of vertebral fracture in patients receiving oral glucocorticoid therapy. Arthritis Rheum. 2003;48:3224–9. doi:10.1002/art.11283.
16. Randazzo ME, Grossrubatscher E, Dalino Ciaramella P, Vanzulli A, Loli P. Spontaneous recovery of bone mass after cure of endogenous hypercortisolism. Pituitary. 2012;15:193–201. doi:10.1007/s11102-011-0306-3.
17. Annane D, et al. Effect of treatment with low doses of hydrocortisone and fludrocortisone on mortality in patients with septic shock. JAMA. 2002;288:862–71.
18. Hamrahian AH, Oseni TS, Arafah BM. Measurements of serum free cortisol in critically ill patients. N Engl J Med. 2004;350:1629–38. doi:10.1056/NEJMoa020266.
19. Sprung CL, et al. Hydrocortisone therapy for patients with septic shock. N Engl J Med. 2008;358:111–24. doi:10.1056/NEJMoa071366.
20. Arlt W, et al. Dehydroepiandrosterone replacement in women with adrenal insufficiency. N Engl J Med. 1999;341:1013–20. doi:10.1056/NEJM199909303411401.
21. Alkatib AA, et al. A systematic review and meta-analysis of randomized placebo-controlled trials of DHEA treatment effects on quality of life in women with adrenal insufficiency. J Clin Endocrinol Metab. 2009;94:3676–81. doi:10.1210/jc.2009-0672.

Part II
Effects of Glucocorticoids on Metabolism

Chapter 5
Regulation of Glucose Homeostasis by Glucocorticoids

Taiyi Kuo, Allison McQueen, Tzu-Chieh Chen, and Jen-Chywan Wang

Abstract Glucocorticoids are steroid hormones that regulate multiple aspects of glucose homeostasis. Glucocorticoids promote gluconeogenesis in liver, whereas in skeletal muscle and white adipose tissue they decrease glucose uptake and utilization by antagonizing insulin response. Therefore, excess glucocorticoid exposure causes hyperglycemia and insulin resistance. Glucocorticoids also regulate glycogen metabolism. In liver, glucocorticoids increase glycogen storage, whereas in skeletal muscle they play a permissive role for catecholamine-induced glycogenolysis and/or inhibit insulin-stimulated glycogen synthesis. Moreover, glucocorticoids modulate the function of pancreatic α and β cells to regulate the secretion of glucagon and insulin, two hormones that play a pivotal role in the regulation of blood glucose levels. Overall, the major glucocorticoid effect on glucose homeostasis is to preserve plasma glucose for brain during stress, as transiently raising blood glucose is important to promote maximal brain function. In this chapter we will discuss the current understanding of the mechanisms underlying different aspects of glucocorticoid-regulated mammalian glucose homeostasis.

Keywords Glucocorticoids • Glucocorticoid receptor • Gluconeogenesis • Insulin • Glucose utilization • Glycogen • Pancreas • Glucose metabolism

Introduction

Glucocorticoids (GC) are stress hormones that play a key role in the regulation of mammalian glucose homeostasis. The name "glucocorticoids" originates from their profound effects on plasma glucose levels. GC regulate multiple aspects of glucose

T. Kuo, Ph.D. • A. McQueen, B.S. • T.-C. Chen, M.S.
Department of Nutritional Sciences and Toxicology, University of California Berkeley,
212 Morgan Hall, Berkeley, CA 94720, USA
e-mail: tk2592@cumc.columbia.edu; amcqueen@berkeley.edu; tcchen@berkeley.edu

J.-C. Wang, Ph.D. (✉)
Department of Nutritional Sciences and Toxicology, University of California Berkeley,
119 Morgan Hall, Berkeley, CA 94720-3104, USA
e-mail: walwang@berkeley.edu

J.-C. Wang, C. Harris (eds.), *Glucocorticoid Signaling*, Advances in Experimental Medicine and Biology 872, DOI 10.1007/978-1-4939-2895-8_5

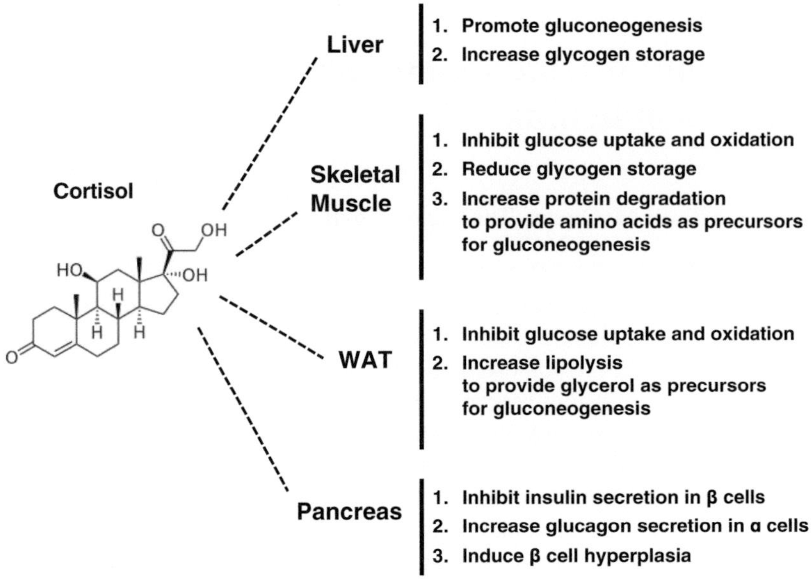

Fig. 5.1 Glucocorticoid effects on glucose homeostasis. The effects of cortisol on glucose homeostasis in peripheral tissues

homeostasis (Fig. 5.1). First, GC promote hepatic gluconeogenesis [1, 2] and reduce glucose uptake and utilization in skeletal muscle and white adipose tissue (WAT) [3, 4]. These effects are critical for metabolic adaptation during stress, such as fasting/starvation, when plasma glucose needs to be preserved because it is the brains' primary energy source, and transiently raising blood glucose is important to promoting maximal brain functions [5]. Insulin, a hormone secreted from pancreatic β cells, exerts opposite effects on these physiological processes by inhibiting hepatic gluconeogenesis and promoting glucose utilization in skeletal muscle and WAT. Thus, to exert their responses, GC need to antagonize insulin actions. These effects are critical during stress, which in short term does not affect or even enhances glucose tolerance. However, chronic GC exposure results in hyperglycemia and insulin resistance [3, 4, 6]. Second, GC exert tissue-specific effects on glycogen metabolism. In liver, GC increase glycogen storage, whereas in skeletal muscle GC play a permissive role for catecholamine-induced glycogenolysis or inhibit insulin-stimulated glycogen synthesis [7–9]. Third, GC modulate insulin and glucagon secretion from pancreas. GC treatment increases plasma glucagon levels [10, 11], whereas the effects of GC on insulin secretion are complex [12–16]. GC induce pancreas islet hyperplasia *in vivo* that leads to hyperinsulinemia [12, 17–19] and have been shown to exert cytotoxic effects on β cells [20, 21]. Overall, in this chapter we will discuss the current understanding of the mechanisms underlying these distinct aspects of GC-regulated glucose homeostasis.

Gluconeogenesis

The gluconeogenic pathway generates glucose from non-carbohydrate substrates (Fig. 5.2) [22, 23]. Gluconeogenesis mainly occurs in liver, though kidney and intestine are also contributors. The major gluconeogenic precursors are lactate, glycerol and gluconeogenic amino acids, such as alanine. Both lactate and alanine can be converted to pyruvate, which is then carboxylated to oxaloacetate (OAA) by pyruvate carboxylase (PC) [24, 25], a step that occurs in mitochondria. OAA needs to be converted to malate to be shuttled to the cytoplasm where it is converted back to OAA. Cytosolic phosphoenolpyruvate carboxykinase (PCK1) [26, 27] then catalyzes OAA to phosphoenolpyruvate (PEP), which then enters the gluconeogenic pathway. There is also a mitochondrial form of PCK1 (m-PCK1) that can directly convert OAA to PEP in mitochondria, which is then transported to cytoplasm and participates in gluconeogenesis. Recent studies suggest that m-PCK1 could also play a role in gluconeogenesis [28–31].

Conversion of PEP to fructose-1,6-phosphate (F1,6BP) requires five enzymatic steps that are essentially the reverse of glycolysis (Fig. 5.2). The "bifunctional" enzyme, phosphofructokinase 2 (PFK2)/fructose bisphosphatase 2 (FBPase2) (a.k.a. PFKFB1) [32, 33], plays a critical role in the switch between gluconeogenesis and glycolysis. PFKFB1 regulates the production of fructose 2,6 bisphosphate (F2,6BP), which is an allosteric activator of phosphofructokinase 1 (PFK1), an enzyme in the glycolytic pathway. When circulating glucose levels are low, such as during fasting and starvation, glucagon inactivates PFK2, which allows FBPase2 activity to be favored. This results in decreased production of F2,6BP, and reduced glycolysis and enhanced gluconeogenesis. Fructose 1,6-bisphosphatase (FBP1) converts F1,6BP to fructose-6-phosphate (F6P) [34], which is then converted to glucose-6-phosphate (G6P). G6P enters the endoplasmic reticulum (ER), where the enzyme glucose-6-phosphatase (G6PC) [35–37] converts G6P to glucose (Fig. 5.2). Notably, distinct gluconeogenic amino acids can be converted to specific intermediates in the tricarboxylic acid (TCA) cycle. These TCA cycle intermediates are converted to OAA to enter the gluconeogenic cycle. Glycerol enters gluconeogenesis through conversion to dihydroxyacetone phosphate (DHAP), which can then be metabolized to glycerol-3-phosphate (G3P) or directly to F1,6BP (Fig. 5.2) thus entering the gluconeogenic cycle above PCK1.

The positive effect of GC on hepatic gluconeogenesis is well established and has been under extensive study for several decades. Injecting GC into humans [38, 39] and rodents increases hepatic gluconeogenesis. In addition, GC are important for other hormones to activate gluconeogenesis. Although glucagon is regarded as a major hormone that activates gluconeogenesis during fasting, fasting-induced gluconeogenesis is reduced in adrenalectomized mice, an effect that is restored by treating mice with GC [40–42]. In fact, glucagon-, epinephrine-, or cyclic AMP (cAMP)-induced gluconeogenesis are all attenuated in adrenalectomized mice. Giving GC to adrenalectomized mice restores the ability of these hormones to induce gluconeogenesis. Thus, GC play a "permissive" role promoting the optimal ability of

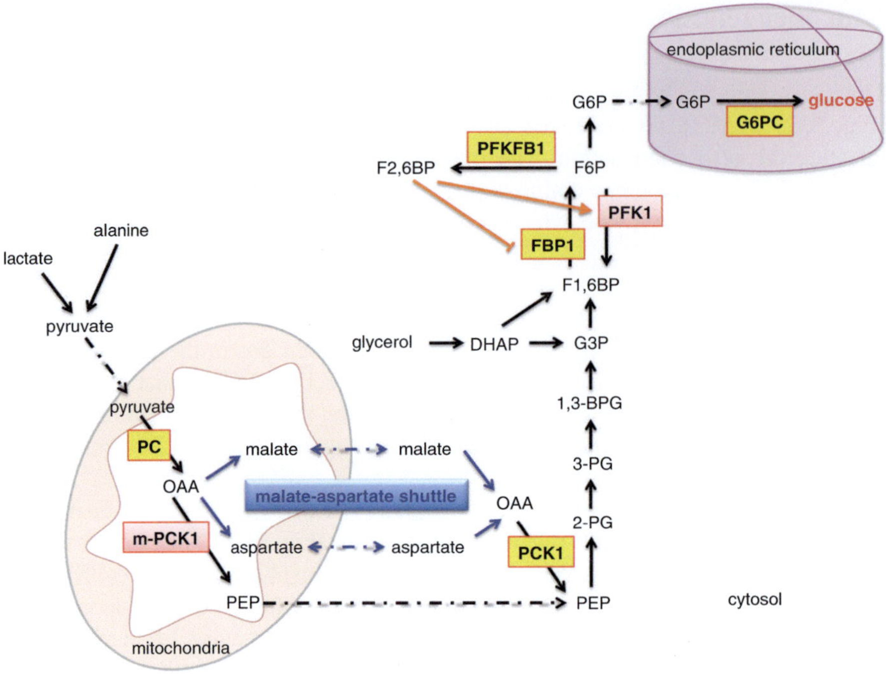

Fig. 5.2 Gluconeogenic pathway in hepatocytes. Lactate and alanine are converted to pyruvate, which enters the mitochondria and is then converted to OAA by enzyme PC. Through malate-aspartate shuttle, OAA exits the mitochondria to form PEP. OAA can also be converted to PEP directly within the mitochondria. PEP then feeds into the gluconeogenic pathway. In addition, glycerol is metabolized to DHAP, which is then converted directly or indirectly through G3P to F1,6BP. The final product, glucose, is produced in the ER by enzyme G6PC. The key enzymes are boxed, with GR primary targets shown in *yellow*. *Abbreviation*: *OAA* oxaloacetate, *PEP* phosphoenolpyruvate, *DHAP* dihydroxyacetone phosphate, *G3P* glyceraldehyde-3-phosphate, *F1,6BP* fructose-1,6-bisphosphate, *2-PG* 2-phosphoglycerate, *3-PG* 3-phosphoglycerate, *1,3-BPG* 1,3-bisphosphoglycerate, *G3P* glyceraldehyde-3-phosphate, *F1,6BP* fructose-1,6-bisphosphate, *F2,6BP* fructose-2,6-bisphosphate, *F6P* fructose-6-phosphate, and *G6P* glucose-6-phosphate. *Enzyme abbreviation*: *PC* pyruvate carboxylase, *m-PCK1* mitochondrial phosphoenolpyruvate carboxykinase, *PCK1* cytosolic phosphoenolpyruvate carboxykinase, *FBP1* fructose-1,6-bisphosphatase 1, *PFK1* phosphofructokinase 1, *PFKFB1* phosphofructokinase 2/fructose bisphosphatase 2, *G6PC* glucose-6-phosphatase catalytic subunit

these hormones in gluconeogenesis. GC also provide gluconeogenic precursors by promoting protein degradation in skeletal muscle to generate gluconeogenic amino acids [1, 4]. They also enhance lipolysis in WAT. This releases glycerol and fatty acids. Glycerol is used as a gluconeogenic precursor, whereas fatty acids provide energy to drive the gluconeogenic pathway [8, 43, 44].

GC promote gluconeogenesis mainly through activation of the transcription of genes encoding enzymes in the gluconeogenic pathway. The transcription of *PC, PCK1, FBP1, PFKFB1, G6PC* and *G6P transporter* (*SLC37A4*) are all stimulated by GC (Fig. 5.2). GC convey their signals mainly through an intracellular receptor, the glucocorticoid receptor (GR). Before binding to GC, GR resides in cytoplasm and is associated with Hsp90 chaperone complex. Upon binding to GC, GR dissociates from Hsp90 chaperone complex and enters the nucleus, where the GR is recruited to specific genomic sequences, called glucocorticoid response elements (GREs). The GR:GRE association could be a direct GR:DNA interaction or an indirect association through other DNA-binding transcription factors. In any case, once GR associates with a specific GRE, it recruits a number of transcriptional coregulators. While they do not bind to DNA directly, these coregulators are able to assist the GR in modulation of the transcriptional rate of nearby genes through distinct mechanisms: altering chromatin structure, inducing histone modifications; recruiting RNA polymerase II containing basal transcription machinery; and modulating transcriptional elongation. Notably, most genomic GREs are "composite" GREs (also called glucocorticoid response units, GRUs) that consist of multiple cis-acting elements, which include binding sites for GR and other transcription factors (called accessory elements and accessory factors, respectively), to mediate a complete GC response. Because different DNA binding transcription factors other than GR are involved in the regulation of distinct GR primary target genes, the multi-protein transcriptional regulatory complex assembled on each GRE is likely distinct. Composite GREs could allow GC to differentially regulate distinct target genes in different cell types depending on the presence of cell type specific accessory factors and transcriptional coregulators. Another advantage of employing composite GREs is to allow specific cross-talk between GC and other signaling pathways at the accessory elements and their respective accessory factors. Thus, cross-talk does not need to occur directly through GR.

The GREs of *Pck1, Pfkfb1*, and *G6pc* genes have been identified. In particular, the mechanism of GR-regulated *Pck1* gene transcription has been extensively studied. By contrast, the mechanisms governing GC-activated *Pc* and *Fbp1* gene are unclear. Below we will discuss the mechanisms of GR-stimulated *Pck1, Pfkfb1*, and *G6pc* gene transcription.

PCK1

Rat *Pck1* gene contains two GREs, GRE1 and GRE2 (Fig. 5.3) [45]. GRE1 is located between −388 and −374 (relative to transcription start site, TSS), whereas GRE2 is located between and −367 and −354 in the *Pck1* gene promoter. When GRE1 or GRE2 is placed in front of TATA box in a synthetic reporter gene, neither mediates a GC response [46]. The combination of GRE1 and GRE2 also fails to confer GC-induced transcription [46]. In fact, both GRE1 and GRE2 bind GR very weakly *in vitro* [46]. However, in cooperation with other accessory elements on the

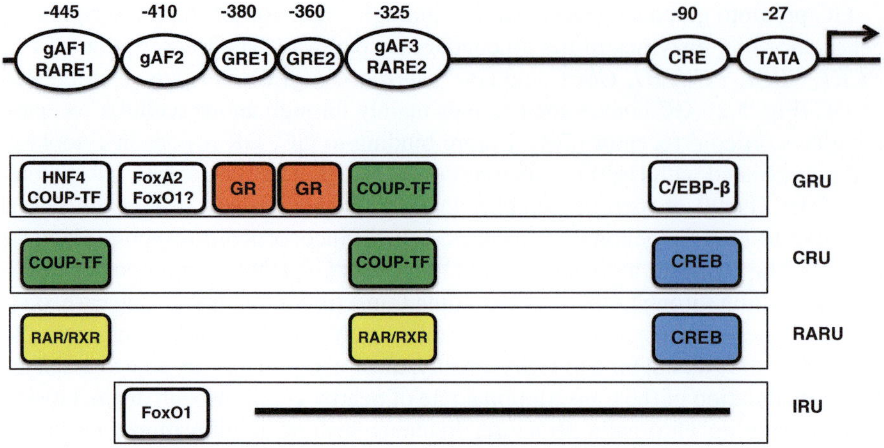

Fig. 5.3 Hormone response units in the PEPCK gene. Binding sites for various regulatory and transcription factors are shown in the *top row*, with the number indicating the center nucleotide position of each element with respect to the transcription start site. Four hormone-specific response units are drawn: proximal glucocorticoid response unit (GRU), cyclic AMP response unit (CRU), retinoic acid response unite (RARU), and insulin response unit (IRU). In the absence of the other hormones, the components of each response unit are depicted. These units interact functionally, cooperating or competing, to comprise the PEPCK promoter. Except for gAF2, DNA elements involved in IRU are not yet identified

Pck1 promoter, they confer a robust GC response. These accessory elements include gAF1 [45], gAF2 [45], gAF3 [47] and the cAMP response element (CRE) [48]. Both gAF1 (between −451 and −434 of rat *PCK1* promoter) and gAF2 (−416 and −407) are located 5′ from GRE1 and GRE2, whereas gAF3 and CRE are located in 3′ of these GREs (Fig. 5.3). Hepatic nuclear factor 4 (HNF4, NR2A1) and chicken ovalbumin upstream transcription factor (COUP-TF, NR2F2) bind to gAF1 and serve as accessory factors for a complete GC response [49]. The gAF1 element also serves as a retinoic acid response element (RARE). An all-trans retinoic acid receptor (RAR) and 9-cis retinoic acid receptor (RXR) heterodimer binds to gAF1 and confers retinoic acid (RA)-activated *Pck1* gene transcription [50]. RA has been shown to synergize with GC to stimulate *Pck1* gene transcription [51].

The gAF2 element binds to members of the forkhead box transcription factor family that include FoxA1 (also called hepatic nuclear factor 3 α, HNF3α), FoxA2 (hepatic nuclear factor 3 β, HNF3β), FoxO1 (FKHR) and FoxO3A (FKHRL1). FoxA2 have been shown to act as accessory factors for GR-regulated *Pck1* gene transcription *in vitro* [52]. Liver specific deletion of FoxO1 but not FoxO3A significantly reduces fasting-induced *Pck1* gene expression [53]. Because GC play an important role in fasting-induced *Pck1* gene transcription, these results suggest that FoxO1 may serve as an accessory factor for GR *in vivo*. The gAF2 element also serves as an insulin response sequence (IRS) that confers at least part of repressive effect of insulin on *Pck1* gene transcription [54, 55]. The ability of insulin to suppress *Pck1* gene expression is compromised in liver specific *FoxO1* knockout mice [53] or mice overexpressed dominant negative FoxO1 [56]. Notably, the ability of insulin to reduce *Pck1* gene expression is not affected in *FoxO3A* deletion mice [53]. These results support the key role of FoxO1 in the regulation of *Pck1* gene expression *in vivo*. The potential role of FoxA2 inhibiting insulin effects has been reported [57], although more studies are needed to confirm the importance of FoxA2 in insulin-suppressed *Pck1* gene expression.

Streptozotocin is a relatively specific pancreas β cell cytotoxin. Treatment of mice with streptozotocin induces a state that mimics type 1 diabetes. Circulating GC levels are increased in such animals and *Pck1* gene expression is augmented. In mice bearing a reporter gene containing a construct containing −2 kb rat *Pck1* gene promoter with a mutation at gAF2, streptozotocin-induced reporter gene expression is markedly reduced [58]. These results confirm the importance of the gAF2 element in GC-activated *Pck1* gene *in vivo*.

The gAF3 element (−337 and −321) binds to COUP-TF [47] and, like gAF1, also serves as a RARE that binds to RAR/RXR heterodimer [59]. The cAMP response element (CRE, between −90 and −82) is also required for a complete GC-stimulated *Pck1* gene transcription [48]. CCAAT enhancer binding protein β (C/EBPβ) binds to the CRE to mediate the accessory activity for the GC response [60].

Chromatin immunoprecipitation (ChIP) was used to monitor the recruitment of GR and various accessory factors to their respective binding sites in rat H4IIE hepatoma cells. GC treatment increases the recruitment of GR, FoxO1, FoxO3A and RNA polymerase II (Pol II) to the *Pck1* promoter [61]. FoxA2, C/EBPβ, HNF4 and COUP-TF occupy the *Pck1* promoter before GC treatment and their occupancy is not altered after GC treatment [61]. The recruitment of transcriptional coregulators, SRC-1, p300 and CREB binding protein (CBP), to the *Pck1* GRU is markedly increased by GC treatment [61]. This suggests that GC treatment initiates the assembly of multiprotein transcriptional regulatory complex on the *Pck1* promoter. Insulin treatment for just 3 min markedly decreases the GC-induced recruitment of GR, FoxO1, FoxO3A, FoxA2, SRC-1, p300, CBP, and Pol II, and only the occupancy of C/EBPβ, HNF-4 and COUP-TF remains unchanged [61]. Thus, insulin treatment rapidly disrupts the assembly of GC-induced transcriptional complex on the *Pck1* GRU. Analyzing epigenetic marks showed that the most significant change is the methylation at histone H3 arginine residue 17, which is significantly increased upon GC treatment and is abolished by insulin [61]. CARM1/PRMT4, is the histone methyltransferase that methylates histone H3 tail arginine 17 residue (H3R17) [62, 63]. CARM1 has been shown to serve as a transcriptional coactivator for GR. However, the occupancy of CARM1 on the *Pck1* promoter is not significantly changed upon GC or insulin treatment. One way to explain these results is that CARM1 is present on the *Pck1* GRU before GC treatment and the activity of CARM1 is modulated by insulin treatment. The activity of CARM1 is regulated by post-translational modifications [64]. Alternatively, it is possible that an unknown histone methyltransferase is involved in the elevation of H3R17 methylation on the *Pck1* GRU.

An analysis of the rat *Pck1* promoter in human hepatoma HepG2 cells identified two additional accessory elements (dAF1 and dAF2) that are involved in GC-stimulated *Pck1* gene transcription. The dAF1 element is located at −993 and has sequence similarity to the gAF1 element, whereas the dAF2 element, located at −1365, resembles more proximal gAF2 [65]. HNF4 and peroxisome proliferator activated receptor α (PPARα) and RXR heterodimer (PPARα/RXR) bind to dAF1, whereas FoxA1, FoxA2 and FoxO1 bind to dAF2 [65]. ChIP experiments showed that GC treatment increases FoxO1 and PPARα recruitment to the gAF2 and the dAF2, and HNF4 to the dAF1 in mouse liver [65]. Overexpressing PPARα or HNF4 synergizes with GR to activate a reporter gene harboring 2 kb of rat *Pck1* promoter [65]. A role for PPARα in GC-induced *Pck1* expression *in vivo* is consistent with the observation that GC-induced *Pck1* gene expression is markedly reduced in *Ppara* null mice [66]. Notably, streptozotocin-induced *Pck1* gene expression and blood glucose levels are reduced in transgenic mice that harbor a targeted ablation of the dAF1 in the *Pck1* gene promoter. These results support the importance of the dAF1 in GC-activated *Pck1* gene expression *in vivo* [65].

Overall, GC activate *Pck1* gene transcription through a complex GRU. Intriguingly, all accessory elements in the *Pck1* GRU are involved in the responses of other hormones that provide potential cross-talk between GC and other hormones, including glucagon, retinoic acid and insulin (Fig. 5.3). The complexity of GRU also allows various signaling pathways to fine tune the transcription levels of *Pck1* gene. Interestingly, GC repress the transcription of *Pck1* gene in adipocytes [67, 68] where the major metabolic role of *Pck1* is glyceroneogenesis [69, 70]. It is not entirely clear why *Pck1* GRU is not functional in adipocytes, although it is proposed that GR inhibits *Pck1* gene transcription through antagonizing C/EBP family of transcription factors in adipocytes [67, 68]. Nonetheless, the requirement of accessory factors to act with GR on "weak" GREs provides the flexibility for GC to regulate *Pck1* gene transcription in a tissue specific manner. While it is unknown how accessory factors participate in GR-regulated *Pck1* gene transcription, there are two potential mechanisms. First, accessory factors may aid in recruitment of transcriptional coregulators to the GRU to stimulate the transcription. Previous studies have shown that the transactivation domain of HNF4 and FoxA2 are required for their accessory activities [71]. When gAF1, gAF2 or gAF3 is replaced by the binding site of a yeast transcription factor Gal4, a fusion protein that consists of GAL4 DNA binding domain and a transcriptional coregulator, SRC1, is able to provide accessory activity [72]. Moreover, another transcriptional coregulator, peroxisome proliferator activated receptor γ coactivator-1α (PGC1α), has been shown to interact with HNF4 to participate in GC-activated *Pck1* gene transcription. PGC1α also interacts with and coactivates FoxO1 [73] that binds to the gAF2 element. But the role of PGC1α-FoxO1 interaction in GC-stimulated *Pck1* gene transcription has not been examined. Second, accessory factors can potentiate GR:GRE association. Using quantitative, real time equilibrium and stopped-flow fluorescence anisotropy measurements of nuclear protein-DNA interactions it was shown that GR binds to the *Pck1* GREs poorly. However, the presence of the gAF1 and the gAF2 elements markedly enhanced the association between GR and the *Pck1* GREs [74]. It is possible that the assembly of a multi-protein complex that includes GR, accessory factors and transcriptional coregulators enhances the association between GR and the two *Pck1* GREs.

G6PC

G6PC gene transcription is induced by GC, whereas insulin suppresses both basal and GC-activated *G6PC* gene transcription. Mouse *G6pc* also contains a complex GRU in the proximal promoter region. This GRU consists of three GREs, a CRE, a HNF4 binding site, a hepatic nuclear factor 1 (HNF1) binding site, and multiple FoxO/FoxA binding sites (FREs) [75]. GC significantly activate the expression of a reporter gene that contains the *G6pc* GRU. Mutation at any of these accessory elements in the *G6pc* GRU reporter gene reduces the GC response [75]. Mutations at FREs that bind FoxO1 and FoxO3A also reduce both the inhibitory effect of insulin and basal expression of *G6pc* gene [76]. In liver specific FoxO1 deletion mice the expression of *G6pc* in 18 h fasted mice is markedly lower than that of 18 h fasted wild type mice [53]. Moreover, the ability of insulin to suppress the expression of *G6pc* is abolished in liver specific FoxO1 deletion mice [53] or mice overexpressed dominant negative FoxO1 [56]. In contrast, liver specific FoxO3A deletion does not affect basal expression of *G6pc* and the ability of insulin to inhibit *G6pc* remains intact [53]. These results are reminiscent of the regulation of *Pck1* gene and suggests that FoxO1 plays a key role in the regulation of gluconeogenic genes *in vivo*. Also, similar to the regulation of *Pck1* gene, transcription coregulator PGC1α positively regulates basal *G6pc* gene transcription and enhances GC-stimulated *G6pc* gene transcription through interaction with HNF4 [77, 78]. FoxO1 and PGC1α appear to synergistically activate *G6pc* gene transcription [73, 79].

GC also activate the transcription of the G6P transporter (*SLC37A4*) gene, which encodes a protein that is responsible for shuttling G6P from the cytoplasm to the ER lumen. The mouse *Slc37a4* gene promoter contains a GRE [80, 81] and a FoxO1 binding site is identified nearby the GRE [81]. GC increase the activity of a luciferase reporter gene under the control of the *Slc37a4* gene promoter in 293 cells, whereas the mutation at the FoxO1 binding site reduces the ability of GC to potentiate this reporter gene activity [81]. Overexpression of FoxO1 in 293 cells potentiates the ability of GC to activate the reporter gene activity [81]. These results suggest that GC stimulate the *Slc37a4* gene through a GRU that contains at least a GRE and a FoxO1 binding site.

PFKFB1

Hepatic rat *Pfkfb1* gene transcription is stimulated by GC and a GRU has been identified in the intronic region of this gene. In addition to the GRE, this GRU consists of binding sites for FoxA2, hepatic nuclear factor 6 (HNF6, a.k.a. Onecut1), C/EBP and Nuclear factor 1 (NF1) [82]. Insulin antagonizes the stimulatory effect of GC on *Pfkfb1* gene [82, 83]. While insulin acts through PI3K and Akt to inhibit *Pck1* and *G6pc* gene expression [84, 85], this pathway is apparently not involved in the suppressive effect of insulin on GC-induced *Pfkfb1* gene. Instead, insulin activates the Jun N-terminal Kinase (JNK) pathway to inhibit GC-induced *Pfkfb1* gene expression [86].

Factors Regulating GC-Stimulated Hepatic Gluconeogenesis

Many other factors regulate gluconeogenesis by modulating GC signaling. Liver X receptor β (LXRβ) is involved in GC-activated *Pck1* gene expression. In *Lxrβ* null mice (*Lxr*β$^{-/-}$) GC-induced *Pck1* gene expression and glucose production are reduced, and GC-induced recruitment of GR to the *Pck1* GRE is impaired [87]. In contrast, in *Lxrα* null mice (*Lxra*$^{-/-}$) GC-regulated *Pck1* gene expression is not affected. Moreover, GC-stimulated expression of *tyrosine aminotransferase* (*Tat*) gene, which encodes the enzyme that converts tyrosine to 4-hydroxyphenolpyruvate, is not affected in *Lxrb*$^{-/-}$ mice. Thus, the role of LXRβ in GC action is relatively specific to the *Pck1* gene. The mechanisms governing LXRβ action in GC-activated *Pck1* gene transcription are unclear, especially in view of another study, which showed that treating hepatoma cells with LXR ligands suppresses GC-stimulated *Pck1* and *G6pc* gene expression [88]. Microarray analyses showed that LXR ligands affect only a subset of GC-regulated genes. Both gel shift and ChIP experiments suggest that an LXRα/RXRα heterodimer competes with GR for binding at rat *G6pc* GRE [88]. In agreement with its effects on GC-induced gluconeogenic gene expression is the observation that treating rats with an LXR ligand attenuates GC-augmented plasma glucose levels [88]. In summary, unliganded LXRβ is required for maximal GC-induced *Pck1* gene transcription and is necessary for GC-induced recruitment of GR to the *Pck1* GRE. By contrast, LXR ligands suppress GC-activated gluconeogenic gene transcription by inhibiting the recruitment of GR to the GREs of these genes.

The expression of transcription factor ying yang 1 (YY1) is increased upon fasting and in insulin resistant state [89]. Overexpression of YY1 in mouse liver increases gluconeogenesis [89]. In contrast, the deletion of YY1 in mouse liver results in hypoglycemia. YY1 potentiates gluconeogenesis through the increase of hepatic GR expression, which in turn augments the expression of gluconeogenic genes [89].

The expression of farnesoid x receptor (FXR, NR1H4), a bile acid receptor, is also increased during fasting [90]. *Fxr* null mice become hypoglycemic during fasting and have a reduced glucose production after a pyruvate challenge and a decreased expression of *Pck1* and *G6pc* [90]. The treatment of fasted mice with an FXR ligand, 6α-ethylchenodeoxycholic acid (6E-CDCA), increases hepatic glucose production and *Pck1* and *G6pc* gene expression. These effects are not observed in *Fxr* null mice, nor are they seen in fed mice [90]. 6E-CDCA elevates the expression of GR. By contrast, the expression of GR is decreased in *Fxr* null mice. Reducing GR expression in liver abolishes 6E-CDCA-induced glucose production and the expression of *Pck1* and *G6pc* [90]. Thus, FXR activation elevates GR expression, which in turn results in enhanced gluconeogenesis.

Ubiquitin-specific protease 2 (USP2) expression is induced by fasting by both GC and glucagon, and by PGC1α overexpression [91]. Overexpression of Usp2 in mouse liver increases glucose production and exacerbates high fat diet-induced glucose intolerance, whereas the reduction of Usp2 expression in mouse liver improves systemic glycemic control [91]. Usp2 induces the expression of 11β-hydroxysteroid dehydrogenase type 1 (11β-HSD1), the enzyme that converts the inactive GC,

11-DHC (rodents) or cortisone (humans), to active corticosterone (rodents) or cortisol (humans) in liver. Thus, Usp2 increases hepatic gluconeogenesis by increasing the active GC levels in hepatocytes. Usp2 is a ubiquitin specific protease. How Usp2 increases the expression of 11β-HSD1 is unclear.

Transforming growth factor β (TGFβ) decreases hepatic gluconeogenesis. TGFβ increases the expression of SMAD6, which directly associates with the N-terminus of GR [92]. When SMAD6 associates with GRE-bound GR, it recruits histone deacetylase 3 (HDAC3) which antagonizes the acetylation of histone H3 and H4 on genomic regions near the GREs [92]. This results in an inhibition of the transactivation activity of GR. Overexpression of SMAD6 in liver reduces GC-induced *Pck1* transcription and blood glucose levels, which mimics TGFβ effects [92].

GR associates with the Hsp90-containing chaperone complex in the cytoplasm, an interaction that induces a GR conformation favorable for binding GC. The Hsp90-containing chaperone complex also plays a role in the translocation of GR into the nucleus. Acetylation of the lysine 294 residue of Hsp90 reduces its interaction with GR [93]. Therefore, keeping lysine 294 in a deacetylated state is critical for the GR-Hsp90 interaction. Histone deacetylase 6 (HDAC6) deacetylates lysine 294 of Hsp90. Ablation of *Hdac6* results in a decreased GR response due to defective GR translocation into the nucleus [93]. In *Hdac6* knockout mice, GC-induced hepatic *Pck1* and *G6pc* gene expression and glucose production are reduced [93]. GC-induced glucose production in primary hepatocytes isolated from *Hdac6* knockout mice is markedly lower than in primary hepatocytes isolated from wild type mice [93]. Notably, *Hdac6* ablation generally reduces GR signaling. Not surprisingly, other GR-regulated processes, including GC-induced insulin resistance and lipolysis in adipocytes, are also affected when *Hdac6* is deleted.

Glucose Utilization

GC inhibit glucose utilization by reducing both glucose uptake and glucose oxidation in skeletal muscle and WAT, two major tissues involved in insulin-responsive glucose utilization [94, 95]. These GC effects counteract those of insulin, which promote glucose uptake, glycolysis and glucose oxidation. These result in a transient increase of circulating glucose, which is considered beneficial during stress [5]. In both mouse and human myotubes GC reduce insulin-stimulated glucose uptake [96–98] by attenuating insulin-induced GLUT4 translocation to the cell membrane. By contrast, reduced GC signaling improves insulin sensitivity and glucose utilization in mouse and human skeletal muscle. Circulating GC levels are increased in genetically obese *ob/ob*, *db/db* and lipotrophic *A-ZIP/F-1* [99] mice as compared to wild type. These mice are insulin resistant, but adrenalectomy improves insulin-stimulated muscle glucose disposal [100, 101]. In high fat diet-induced obese mice, adrenalectomy or treatment with the GR antagonist, RU-486, improves insulin sensitivity and increases glucose utilization in skeletal muscle. Moreover,

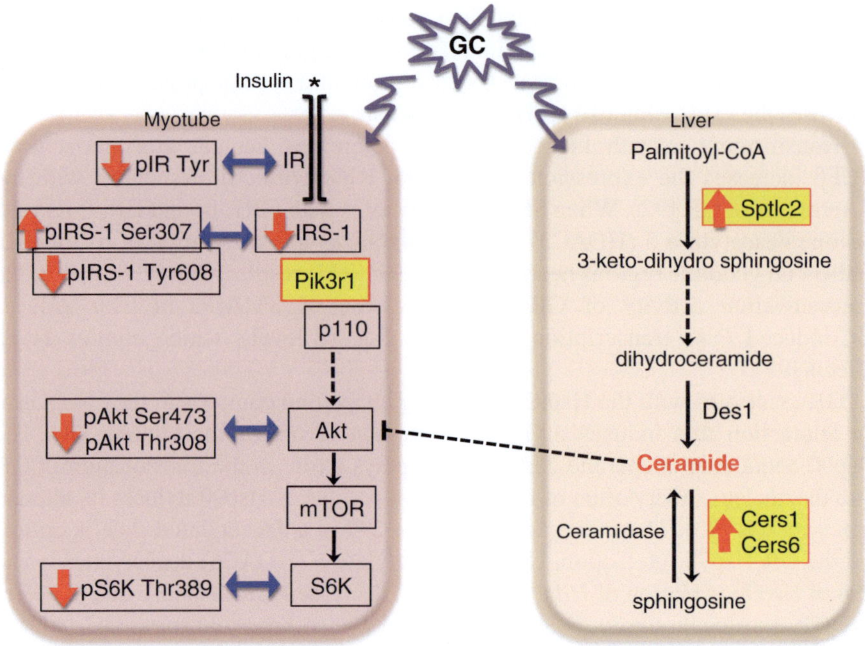

Fig. 5.4 Models of glucocorticoid-regulated insulin action. Mechanisms of glucocorticoid-induced insulin resistance are depicted. In myotubes, glucocorticoids (GC) decrease the tyrosine phosphorylation of insulin receptor (IR) and the expression of IRS1. They increase the serine 307 phosphorylation while decrease the tyrosine 608 phosphorylation of IRS1. GC also increase the expression of Pik3r1, which results in decreased activity of Akt and p70 S6 kinase (S6K). In the liver, GC increase the gene expression of enzymes involved in ceramide synthesis, including Sptlc2, Cers1 and Cers6, which results in increased levels of ceramides. These ceramides then interfere with insulin signaling. The GC-regulated genes are shown in *yellow*

11β-HSD1-specific inhibitors, which reduce corticosterone levels in various tissues, improve insulin sensitivity and skeletal muscle glucose utilization in animal models of diabetes or insulin resistance [96].

The ability of GC to inhibit glucose uptake and glucose oxidation in skeletal muscle is due to, at least in part, a direct effect of GCs on myotubes. Treating cultured myotubes with GC inhibits insulin-stimulated glucose utilization [96, 102]. One mechanism by which GC reduce glucose utilization is the inhibition of insulin signaling [96, 103, 104]. Insulin binds to and activates the cell-surface insulin receptor (IR) tyrosine kinase, which in turn phosphorylates the members of insulin receptor substrate (IRS) protein family [105] (Fig. 5.4). Tyrosine-phosphorylated IRS proteins associate with the IR and initiate downstream signaling events [105] (Fig. 5.4). Mice treated with GCs have reduced levels of tyrosine-phosphorylated IR and total IRS-1 proteins in skeletal muscle [96], and the activities of phosphoinositide-3-kinase (PI3K) and Akt, two signaling molecules downstream of IRS-1, are decreased [96, 106, 107] (Fig. 5.4).

The ability of GC to inhibit glucose utilization in myotubes requires protein synthesis. A list of potential GR primary target genes that can suppress insulin action has been identified in mouse C2C12 myotubes [108]. Among these genes, the role of *Pik3r1* (a.k.a. *p85α*) in the GC response was examined *in vitro. Pik3r1* encodes the regulatory subunit of PI3K, which binds to activate IRS1 (Fig. 5.4) through its SRC homology 2 (SH2) domain to bring the catalytic subunit of PI3K, Pik3ca (a.k.a. p110), to the plasma membrane [109, 110]. Pik3ca then catalyzes the conversion of phosphatidylinositol (4,5)-bisphosphate (PIP2) to phosphatidylinositol (3,4,5)-bisphosphate (PIP3) [109, 110]. PIP3 anchors the protein kinase Akt protein kinase family to the plasma membrane and thus initiates downstream signaling events [110]. Pik3r1 is a key component in the insulin signaling pathway. But monomeric Pik3r1 is thought to compete with the Pik3r1/Pik3ca heterodimer to interact with IRS-1 to suppress insulin action [111, 112]. In addition, Pik3r1 is required for the maximal activity of phosphatase and tensin homolog (PTEN) [113], which antagonizes PI3K activity. In C2C12 mouse myotubes, GC treatment reduces the activity of several components of the insulin signaling pathway. These GC effects are markedly decreased in C2C12 myotubes that have reduced Pik3r1 expression [108]. Thus, *Pik3r1* is a potential GR primary target gene in the mediation of the suppressive effect of GC on glucose utilization in skeletal muscle, though this notion needs to be confirmed *in vivo.* Interestingly, global heterozygous deletion of *Pik3r1* gene in mice has improved whole body insulin sensitivity [114, 115]. Elevated expression of *PIK3R1* is found in patients with insulin resistance [116]. Notably, *Pik3r1* is likely not the only GR target gene that mediating suppressive effects of GC in insulin response, and additional GR primary target genes could also participate in this process.

GC also modulate insulin sensitivity and glucose utilization through the generation of specific lipid mediators. GC treatment increases ceramide levels in mouse liver and portal circulation [117]. These ceramides cause hepatic insulin resistance [118]. They are also delivered to skeletal muscle so one might expect systemic effects. In fact, hyperinsulinemic-euglycemic clamp studies in mice show that GC decrease the glucose infusion rate required to maintain euglycemia, prevent insulin-inhibited hepatic glucose output, and inhibit 2-deoxyglucose uptake into skeletal muscle, all evidence of reduced insulin sensitivity, whereas mice pretreat with myriocin, an inhibitor of serine palmitoyaltransferase (Sptlc1 and Sptlc2), an enzyme in the ceramide synthetic pathway, have reduced GC-induced insulin resistance in skeletal muscle and hepatic glucose output (Fig. 5.4) [117]. Notably, GC augment the expression of enzymes in the ceramide synthetic and metabolic pathway in liver, such as Sptlc2, ceramide synthase 1 (Cers1, a.k.a. Lass1), and ceramide synthase 6 (Cers6, a.k.a. Lass6) (Fig. 5.4) [117]. However, it is not yet clear whether the genes encoding these enzymes are primary targets of GR signaling.

In addition to increased ceramide levels in liver, GC-promoted lipolysis in WAT could impair whole body glucose homeostasis. Acipimox, an inhibitor of lipolysis in adipocytes, improves whole body glucose homeostasis in human subjects treated with GC [119]. Although it is not clear how GC-induced lipolysis affects glucose homeostasis, it is likely that the fatty acids generated from lipolysis are mobilized to skeletal muscle and liver and converted to lipid mediators, such as diacylglycerol (DAG) and ceramides, that cause insulin resistance.

Notably, Brennan-Speranza et. al. also reported that GC inhibit the expression of osteocalcin, a secreted protein from bone that reduce adiposity and hepatic steatosis, to decrease insulin sensitivity.

GC also inhibit glucose oxidation by stimulating the expression of several members of the pyruvate dehydrogenase kinase family (PDK). PDK regulates glucose oxidation by inhibiting the pyruvate dehydrogenase complex that converts pyruvate to acetyl-CoA [120]. Among the PDK family, *PDK4* is a GR primary target gene; GREs have been identified in human *PDK4* and rat *Pdk4* genes [121, 122]. FoxO1 binding sites have been identified near the human *PDK4* gene GRE and they are required for the maximal induction of *PDK4* gene transcription by GC. These FREs also mediate the inhibitory response of insulin on *PDK4* gene transcription [122]. For the rat *Pdk4* gene, an FRE located approximately 6 kb away from the GREs is thought to participate in both the insulin and GC responses [121]. Thus, the mechanisms governing the transcriptional regulation of the *PDK4* gene by GC appear to be similar to GC-activated *Pck1* and *G6pc* gene transcription discussed above.

In addition to skeletal muscle, GC reduce glucose uptake and glucose oxidation in many other tissues. GC inhibit insulin-stimulated glucose uptake in both mouse 3T3-L1 and primary adipocytes. The mechanisms governing these GC effects are mostly unknown. Overexpression of *dual specificity protein phosphatase 1* (*Dusp1*, a.k.a. MAP kinase phosphatase 1, *Mkp1*), a primary GR target gene, inhibits insulin-stimulated glucose uptake in 3T3-L1 adipocytes [123]. However, the exact role of Dusp1 in GC-induced insulin resistance in adipocytes has not been established. Most reports indicate that GC inhibit glucose uptake by antagonizing the insulin response, though the direct inhibition of glucose transporter 4 (Glut4) trafficking process by GC in 3T3-L1 adipocytes has also been reported [124]. *Pik3r1* expression is also increased by GC in adipocytes [125]. Thus, Pik3r1 may also participate in the GC-inhibited insulin response in adipocytes. Studies of human adipocytes found that GC inhibit insulin-stimulated glucose uptake and signaling in omental but not subcutaneous adipocytes [126]. In fact, studies in human primary subcutaneous adipocytes show that GC pre-treatment potentiates insulin-stimulated glucose uptake [102, 127]. This suggests that GC affect insulin signaling in a depot-specific manner in humans.

The ability of GC to inhibit glucose oxidation has been linked to GC-induced apoptosis in leukemia cells. GC inhibit the expression of glucose transporter 1 (GLUT1) that results in a decreased glucose uptake into leukemia cells [128]. GC also suppress glucose uptake and oxidation in certain regions of brain, such as the hypothalamus and hippocampus [129–131]. The exact mechanisms of these GC effects on these cell types are mostly unclear.

Glycogen Metabolism

GC regulate glycogen metabolism in a tissue-specific manner. In liver the administration of GC to fasted mice increases liver glycogen content [40, 132]. GC induce the activity of glycogen synthase [7, 133, 134]. Glycogen synthase activity is regulated by post-translational modification. Protein kinase A (PKA) and glycogen synthase kinase 3 (GSK3) phosphorylates and inactivates glycogen synthase [135–137]. In contrast, protein phosphatase 1 (PP1) dephosphorylates glycogen synthase, which potentiates its activity [138, 139]. The active form of glycogen phosphorylase, glycogen phosphorylase *a*, breaks down glycogen to glucose units. Glycogen phosphorylase *a* also inhibits the dephosphorylation of glycogen synthase by PP1. PKA can also inhibit PP1 activity. Data suggest that the activation of glycogen synthase phosphatase, PP1, by GC [140, 141], though the exact the mechanism of this activation is unclear.

Epinephrine plays a key role in skeletal muscle glycogenolysis. This effect is blunted in adrenalectomized rats [142]. The stimulatory effects of epinephrine on muscle glycogen phosphorylase and phosphorylase kinase are all attenuated in the skeletal muscle of adrenalectomized rats. PP1 dephosphorylates glycogen phosphorylase and phosphorylase kinase and suppresses their activities. PP1 activity is increased in skeletal muscle of adrenalectomized rats and the ability of epinephrine to inhibit PP1 is reduced in the skeletal muscle of adrenalectomized rats. Cortisol treatment in adrenalectomized rats restores normal epinephrine effects through the activation of phosphorylase kinase and glycogen phosphorylase and the inhibition of PP1. The induction of *PDK4* gene transcription by GR may contribute to GC-regulated glycogen metabolism in myotubes. One report shows that a reduction of the expression of PDK4 in human primary myotubes diminishes GC-repressed glycogen synthesis [143]. Notably, GC also inhibits insulin-stimulated glycogen synthesis and the activity of glycogen synthase [9]. These GC effects are mainly due to their ability to reduce insulin signaling in skeletal muscle. Interestingly, one report shows that glycogen storage is actually increased in soleus muscle by dexamethasone treatment despite a decrease of glycogen synthesis [9]. The mechanisms underlying these phenotypes are not clear.

In cardiac muscle, adrenalectomy also blocks epinephrine-induced glycogenolysis [144]. GC treatment, however, has been shown to facilitate glycogen storage in cardiac muscle [145]. GC increase AMP-activated protein kinase (AMPK) activity that leads to the elevation of glucose uptake, the induction of glycogen synthase, and the reduction of glycogen phosphorylase. These contribute to an augmentation of glycogen content in cardiac muscle. Brain stores certain amounts of glycogen, though the levels are lower than in skeletal muscle or liver. Astrocytes store most of the brain glycogen, whose levels are increased by norepinephrine. In primary astrocyte culture norepinephrine induces glycogen synthesis, an effect which is suppressed by GC [146].

GC Effects on Pancreas

As an endocrine organ, the pancreas secretes several hormones that include insulin (from β cells in the islets of Langerhans), glucagon (from α cells), and somatostatin (from δ cells). Insulin secreted during the fed state promotes glucose uptake and utilization, and inhibits gluconeogenesis. In contrast, glucagon secreted during fasting stimulates gluconeogenesis and glycogenolysis. Somatostatin suppresses both insulin and glucagon secretion.

The effect of GC on insulin secretion is an area of vigorous research; however, the direct mechanistic relationship between GC and β cell function *in vivo* remains elusive. This is largely due to the difficulty of separating the indirect effects of GC-induced peripheral insulin resistance from the direct actions of GC on pancreatic β cells. Hypercortisolism induces insulin resistance in tissues such as liver, adipose and skeletal muscle, as discussed above. In murine models and human studies, this GC-induced insulin resistance can lead to compensatory β cell hyperplasia and hyperinsulinemia, and consequent normoglycemia. However, long-term GC treatment that exceeds the β cell compensatory capacity leads to insufficient insulin secretion, hyperglycemia, and eventually Type 2 diabetes [15, 17, 147–150]. Overall, the effects of GC on β cell function are dependent on the length and dosage of the treatment, the experimental animal model under investigation, as well as the susceptibility of the strain exposed [151]. It is important to note that short-term effects of GC on β cells are likely reversible [152]. To begin to dissect the direct *in vivo* effects of GC on β cells, a mouse model of insulin promoter-driven GR overexpression was generated [14]. These mice show glucose intolerance due to blunted insulin secretion.

Insulin secretion is mainly triggered by a postprandial increase in the concentration of plasma glucose, which enters pancreatic β cells through GLUT2 transporters. Glucokinase phosphorylates glucose to produce glucose-6-phosphate, which enters the glycolytic cycle. Through mitochondrial β-oxidation, the ratio of ATP to ADP increases, which leads to the closure of ATP-sensitive potassium channels and plasma membrane depolarization. The increase in cellular electrical conductivity drives the opening of voltage-sensitive calcium ion channels. The consequent influx of calcium ions eventually leads to exocytosis of insulin-containing granules.

Compared to islets isolated from GC-treated animals, isolated islets directly treated with GC present opposite results. First, synthetic GC dexamethasone promotes posttranslational degradation of GLUT2 and decreases glucose-stimulated insulin secretion (GSIS) in isolated rat pancreatic islets [153]. Second, dexamethasone inhibits glucokinase mRNA expression in RIN cells, a rat pancreatic islet tumor cell line [154]. GC have also been reported to suppress the gene expression of insulin [155] and pancreatic and duodenal homeobox 1 (Pdx1) in β cells [156]. Pdx1 is a transcription factor essential for pancreatic development and β cell maturation. Third, in *ob/ob* mice, GC enhance the activity of glucose-6-phosphatase, which dephosphorylates glucose-6-phosphate to produce glucose, resulting in a futile cycle [157]. Fourth, in the INS-1 rat pancreatic β cell line, the transcription of voltage-gated potassium channel Kv1.5 is upregulated by dexamethasone treatment [158]. This results in an increase in the repolarizing outward current and reduces the influx of calcium ions, which compromises GSIS. On the contrary, islets purified from GC-treated rats display enhanced insulin secretion, likely due to compensatory hyperinsulinemia [159–161], and have improved glucose sensitivity and oxidative metabolism [162]. Furthermore, transmission electron microscopy shows an increased amount of docked secretory granules in GC-treated β cells [148].

GC also regulates the expression of genes that modulate insulin secretion. The blunting of GSIS by dexamethasone is significantly restored in islets isolated from *serum and glucocorticoid-induced kinase (Sgk1)* knockout mice compared to wild type [158]. *Sgk1* is a well-established GR primary target gene. Moreover, GC treatment of MIN6 mouse insulinoma cells and mouse islets augments the expression of corticotropin-releasing factor receptor type 2α (CRFR2α) and inhibits the expression of CRFR type 1 (CRFR1) [163]. CRFR1 potentiates both glucose-induced insulin secretion *in vitro* and *in vivo* and the proliferation of neonatal rat β cells [164].

GC can stimulate β cell proliferation in vivo; however, GC exposure could also have cytotoxic effects on mouse isolated pancreatic islets of Langerhans and MIN6 cells [165]. Interestingly, increased intracellular cAMP levels have previously been shown to attenuate dexamethasone-induced β cell death using exendin-4 [20] and forskolin [166]. Co-treatment with RU486, an antagonist of GC, abolished these GC effects. Mitogen-activated protein kinases (MAPK) such as p38 MAPK and JNK regulate apoptosis. In MIN6 cells, inhibition of p38 MAPK reduces glucocorticoid-induced apoptosis [167]. This may be due to decreased phosphorylation of mouse GR at serine residue 220 by p38 MAPK. Phosphorylation of GR at serine 220 is positively associated with GR transcriptional activity [168, 169]. Moreover, in isolated islets, inhibition of p38 MAPK decreases glucocorticoid-induced formation of cleaved caspase 3, which plays a key role in the execution phase of apoptosis [167]. Thus, p38 MAPK is required for glucocorticoid-induced cytotoxicity. In contrast, JNK signaling dampens glucocorticoid-induced cytotoxicity, as inhibition of JNK potentiates the cytotoxic effect of GC in MIN6 cells [167]. A plausible explanation is that JNK is responsible for phosphorylating GR at serine 234, which is negatively associated with the transcriptional activity of GR [170].

Reactive oxygen species (ROS) may also mediate glucocorticoid-induced cytotoxicity [171]. Thioredoxin-interacting protein (Txnip) could exert its pro-apoptotic effect by blocking the activity of thioredoxin, which is involved in a major pro-survival thiol-reducing pathway in cells. In *db/db* mice, overexpression of thioredoxin suppresses the progression of hyperglycemia, likely through the prevention of the reduction of Pdx1 and V-maf musculoaponeurotic fibrosarcoma oncogene homologue A (MafA) transcription factors [172]. GC induces Txnip expression in human and mouse islets, and in MIN6 cells [165]. Overexpressing Txnip mimicked pro-apoptotic effect of GC, while knocking down Txnip partially rescued this phenotype in MIN6 cells. Furthermore, thioredoxin overexpression protected MIN6 cells from GC-induced cytotoxicity [165]. Interestingly, GC-induced Txnip is dependent on the presence of p38 MAPK. Studies in leukemia and other cell types have shown that Txnip is a primary GR target gene [173].

In addition to insulin secretion, GC has been shown to modulate glucagon secretion. GC treatment causes both hyperinsulinemia and hyperglucagonemia. Rodents treated with GC have unaltered insulin/glucagon ratio from the fasted state to the fed state. They tend to have increased α cell mass, and show impaired high glucose-suppressed glucagon secretion. This hyperglucagonemia leads to hyperglycemia through the activation of hepatic glycogenolysis and gluconeogenesis [174]. Notably, co-treatment of GC and a glucagon receptor antagonist on rats resulted in normoglycemia [11]. The mechanisms underlying the effects of GC on glucagon secretion, however, are unclear.

Conclusion and Future Directions

The physiological responses of GC that modulate glucose homeostasis have been well documented. Significant progress has been made to understand these GC actions in the last three decades. However, the precise manner by which GC regulates glucose homeostasis is still unclear. The key to understanding the mechanisms of GC-regulated glucose homeostasis is the identification of GR primary target genes that mediate GC actions, and understanding how these primary target genes are transcriptionally regulated by GR in various processes, including gluconeogenesis, glucose uptake and utilization, glycogen metabolism, and pancreatic endocrine secretions. For transcriptional regulation, GC-activated *Pck1* gene transcription has received the most attention. Nevertheless, further studies are needed to understand how different accessory factors coordinate with GR and transcriptional coregulators to activate *Pck1* gene transcription. Analyzing the transcriptional regulatory mechanisms of these GC-regulated gluconeogenic genes should allow the identification of common molecular features in GR-stimulated gluconeogenesis. *In vivo*, GC actions are affected by other hormones and environmental cues. Understanding the cross-talk and integration of these signals should be the focus of future work.

The emergence of advanced techniques and technologies during the last decade have facilitated the characterization of molecular features of GC-regulated glucose homeostasis. Using high throughput DNA sequencing technology, genomic GR binding regions, transcriptional coregulator occupancy sites, and global chromatin structure changes were identified. The discovery of GC-regulated non-coding RNA and enhancer RNA, using genomic run-on sequencing (GRO-seq) and RNA sequencing (RNA-seq), unveiled another layer of glucose homeostasis regulation [175–178]. Clustered Regularly Interspaced Short Palindromic Repeats (CRISPR) genome editing technology can now be used to efficiently ablate a GR primary target gene, and modify a particular residue of GRE to study its role in the natural chromosomal context [179–183]. These new approaches in combination with refined genetic, physiological and biochemical tools will play a critical role in improving our understanding of GC-regulated biology in metabolic tissues.

Ultimately, the goal of the studying the basic mechanisms of GC-regulated glucose homeostasis is to develop a pharmacotherapy targeting GC signaling for treating metabolic diseases. Reducing GC signaling *in vivo* improves insulin sensitivity and decreases plasma glucose levels, which would largely benefit type 2 diabetes patients. However, reducing GC actions globally is not ideal. For example, the body's inflammatory status would be elevated with reduced GC signaling. In fact, as inflammation promotes insulin resistance, the anti-inflammatory activity of GR could also improve insulin sensitivity. Overall, identifying approaches to dissociate metabolic and anti-inflammatory actions of GR could provide novel ways to treat metabolic disease.

Acknowledgment We thank Drs. Daryl Granner and Richard O'Brien for their insightful comments and suggestions for this chapter. The Wang laboratory is supported by NIH DK83591.

References

1. Exton JH. Regulation of gluconeogenesis by glucocorticoids. Monogr Endocrinol. 1979;12: 535–46.
2. Kraus-Friedmann N. Hormonal regulation of hepatic gluconeogenesis. Physiol Rev. 1984;64:170–259.
3. Di Dalmazi G, Pagotto U, Pasquali R, Vicennati V. Glucocorticoids and type 2 diabetes: from physiology to pathology. J Nutr Metab. 2012;2012:525093. doi:10.1155/2012/525093.
4. Kuo T, Harris CA, Wang JC. Metabolic functions of glucocorticoid receptor in skeletal muscle. Mol Cell Endocrinol. 2013;380:79–88. doi:10.1016/j.mce.2013.03.003.
5. Charmandari E, Tsigos C, Chrousos G. Endocrinology of the stress response. Annu Rev Physiol. 2005;67:259–84. doi:10.1146/annurev.physiol.67.040403.120816.
6. Andrews RC, Walker BR. Glucocorticoids and insulin resistance: old hormones, new targets. Clin Sci (Lond). 1999;96:513–23.
7. Stalmans W, Laloux M. Glucocorticoids and hepatic glycogen metabolism. Monogr Endocrinol. 1979;12:517–33.
8. Exton JH, Friedmann N, Wong EH, Brineaux JP, Corbin JD, Park CR. Interaction of glucocorticoids with glucagon and epinephrine in the control of gluconeogenesis and glycogenolysis in liver and of lipolysis in adipose tissue. J Biol Chem. 1972;247:3579–88.
9. Ruzzin J, Wagman AS, Jensen J. Glucocorticoid-induced insulin resistance in skeletal muscles: defects in insulin signalling and the effects of a selective glycogen synthase kinase-3 inhibitor. Diabetologia. 2005;48:2119–30. doi:10.1007/s00125-005-1886-0.
10. Wise JK, Hendler R, Felig P. Influence of glucocorticoids on glucagon secretion and plasma amino acid concentrations in man. J Clin Invest. 1973;52:2774–82. doi:10.1172/JCI107473.
11. Rafacho A, Goncalves-Neto LM, Santos-Silva JC, Alonso-Magdalena P, Merino B, Taboga SR, Carneiro EM, Boschero AC, Nadal A, Quesada I. Pancreatic alpha-cell dysfunction contributes to the disruption of glucose homeostasis and compensatory insulin hypersecretion in glucocorticoid-treated rats. PLoS One. 2014;9:e93531. doi:10.1371/journal.pone.0093531.
12. Beaudry JL, Riddell MC. Effects of glucocorticoids and exercise on pancreatic beta-cell function and diabetes development. Diabetes Metab Res Rev. 2012;28:560–73. doi:10.1002/dmrr.2310.
13. Longano CA, Fletcher HP. Insulin release after acute hydrocortisone treatment in mice. Metabolism. 1983;32:603–8.
14. Delaunay F, Khan A, Cintra A, Davani B, Ling ZC, Andersson A, Ostenson CG, Gustafsson J, Efendic S, Okret S. Pancreatic beta cells are important targets for the diabetogenic effects of glucocorticoids. J Clin Invest. 1997;100:2094–8. doi:10.1172/JCI119743.
15. Ogawa A, Johnson JH, Ohneda M, McAllister CT, Inman L, Alam T, Unger RH. Roles of insulin resistance and beta-cell dysfunction in dexamethasone-induced diabetes. J Clin Invest. 1992;90:497–504. doi:10.1172/JCI115886.
16. Dinneen S, Alzaid A, Miles J, Rizza R. Metabolic effects of the nocturnal rise in cortisol on carbohydrate metabolism in normal humans. J Clin Invest. 1993;92:2283–90.
17. Rafacho A, Cestari TM, Taboga SR, Boschero AC, Bosqueiro JR. High doses of dexamethasone induce increased beta-cell proliferation in pancreatic rat islets. Am J Physiol Endocrinol Metab. 2009;296:E681–9. doi:10.1152/ajpendo.90931.2008.
18. Morisset J, Jolicoeur L. Effect of hydrocortisone on pancreatic growth in rats. Am J Physiol. 1980;239:G95–8.
19. Like AA, Chick WL. Pancreatic beta cell replication induced by glucocorticoids in subhuman primates. Am J Pathol. 1974;75:329–48.

20. Ranta F, Avram D, Berchtold S, Dufer M, Drews G, Lang F, Ullrich S. Dexamethasone induces cell death in insulin-secreting cells, an effect reversed by exendin-4. Diabetes. 2006;55:1380–90.
21. Weinhaus AJ, Bhagroo NV, Brelje TC, Sorenson RL. Dexamethasone counteracts the effect of prolactin on islet function: implications for islet regulation in late pregnancy. Endocrinology. 2000;141:1384–93. doi:10.1210/endo.141.4.7409.
22. Pilkis SJ, el-Maghrabi MR, Claus TH. Hormonal regulation of hepatic gluconeogenesis and glycolysis. Annu Rev Biochem. 1988;57:755–83. doi:10.1146/annurev.bi.57.070188.003543.
23. Pilkis SJ, Granner DK. Molecular physiology of the regulation of hepatic gluconeogenesis and glycolysis. Annu Rev Physiol. 1992;54:885–909. doi:10.1146/annurev.ph.54.030192.004321.
24. Jitrapakdee S, Wallace JC. Structure, function and regulation of pyruvate carboxylase. Biochem J. 1999;340(Pt 1):1–16.
25. Menefee AL, Zeczycki TN. Nearly 50 years in the making: defining the catalytic mechanism of the multifunctional enzyme, pyruvate carboxylase. FEBS J. 2014;281:1333–54. doi:10.1111/febs.12713.
26. Hanson RW, Garber AJ. Phosphoenolpyruvate carboxykinase. I. Its role in gluconeogenesis. Am J Clin Nutr. 1972;25:1010–21.
27. Hanson RW, Patel YM. Phosphoenolpyruvate carboxykinase (GTP): the gene and the enzyme. Adv Enzymol Relat Areas Mol Biol. 1994;69:203–81.
28. Mendez-Lucas A, Duarte JA, Sunny NE, Satapati S, He T, Fu X, Bermudez J, Burgess SC, Perales JC. PEPCK-M expression in mouse liver potentiates, not replaces, PEPCK-C mediated gluconeogenesis. J Hepatol. 2013;59:105–13. doi:10.1016/j.jhep.2013.02.020.
29. Stark R, Guebre-Egziabher F, Zhao X, Feriod C, Dong J, Alves TC, Ioja S, Pongratz RL, Bhanot S, Roden M, Cline GW, Shulman GI, Kibbey RG. A role for mitochondrial phosphoenolpyruvate carboxykinase (PEPCK-M) in the regulation of hepatic gluconeogenesis. J Biol Chem. 2014;289:7257–63. doi:10.1074/jbc.C113.544759.
30. Arinze IJ, Garber AJ, Hanson RW. The regulation of gluconeogenesis in mammalian liver. The role of mitochondrial phosphoenolpyruvate carboxykinase. J Biol Chem. 1973;248:2266–74.
31. Cheng SC, Cheng RH. A mitochondrial phosphoenolpyruvate carboxykinase from rat brain. Arch Biochem Biophys. 1972;151:501–11.
32. Rider MH, Bertrand L, Vertommen D, Michels PA, Rousseau GG, Hue L. 6-phosphofructo-2-kinase/fructose-2,6-bisphosphatase: head-to-head with a bifunctional enzyme that controls glycolysis. Biochem J. 2004;381:561–79. doi:10.1042/BJ20040752.
33. Rousseau GG, Hue L. Mammalian 6-phosphofructo-2-kinase/fructose-2,6-bisphosphatase: a bifunctional enzyme that controls glycolysis. Prog Nucleic Acid Res Mol Biol. 1993;45: 99–127.
34. Marcus F, Rittenhouse J, Gontero B, Harrsch PB. Function, structure and evolution of fructose-1,6-bisphosphatase. Arch Biol Med Exp. 1987;20:371–8.
35. Burchell A, Waddell ID. The molecular basis of the hepatic microsomal glucose-6-phosphatase system. Biochim Biophys Acta. 1991;1092:129–37.
36. Hutton JC, O'Brien RM. Glucose-6-phosphatase catalytic subunit gene family. J Biol Chem. 2009;284:29241–5. doi:10.1074/jbc.R109.025544.
37. van Schaftingen E, Gerin I. The glucose-6-phosphatase system. Biochem J. 2002;362:513–32.
38. Schneiter P, Tappy L. Kinetics of dexamethasone-induced alterations of glucose metabolism in healthy humans. Am J Physiol. 1998;275:E806–13.
39. Tounian P, Schneiter P, Henry S, Delarue J, Tappy L. Effects of dexamethasone on hepatic glucose production and fructose metabolism in healthy humans. Am J Physiol. 1997;273:E315–20.
40. Exton JH, Miller TB, Harper SC, Park CR. Carbohydrate metabolism in perfused livers of adrenalectomized and steroid-replaced rats. Am J Physiol. 1976;230:163–70.
41. Exton JH, Park CR. Control of gluconeogenesis in the perfused liver of normal and adrenalectomized rats. J Biol Chem. 1965;240:955–7.
42. Sistare FD, Haynes Jr RC. Acute stimulation by glucocorticoids of gluconeogenesis from lactate/pyruvate in isolated hepatocytes from normal and adrenalectomized rats. J Biol Chem. 1985;260:12754–60.

43. Exton JH, Mallette LE, Jefferson LS, Wong EH, Friedmann N, Miller Jr TB, Park CR. The hormonal control of hepatic gluconeogenesis. Recent Prog Horm Res. 1970;26:411–61.
44. Menon RK, Sperling MA. Carbohydrate metabolism. Semin Perinatol. 1988;12:157–62.
45. Imai E, Stromstedt PE, Quinn PG, Carlstedt-Duke J, Gustafsson JA, Granner DK. Characterization of a complex glucocorticoid response unit in the phosphoenolpyruvate carboxykinase gene. Mol Cell Biol. 1990;10:4712–9.
46. Scott DK, Stromstedt PE, Wang JC, Granner DK. Further characterization of the glucocorticoid response unit in the phosphoenolpyruvate carboxykinase gene. The role of the glucocorticoid receptor-binding sites. Mol Endocrinol. 1998;12:482–91.
47. Scott DK, Mitchell JA, Granner DK. The orphan receptor COUP-TF binds to a third glucocorticoid accessory factor element within the phosphoenolpyruvate carboxykinase gene promoter. J Biol Chem. 1996;271:31909–14.
48. Imai E, Miner JN, Mitchell JA, Yamamoto KR, Granner DK. Glucocorticoid receptor-cAMP response element-binding protein interaction and the response of the phosphoenolpyruvate carboxykinase gene to glucocorticoids. J Biol Chem. 1993;268:5353–6.
49. Hall RK, Sladek FM, Granner DK. The orphan receptors COUP-TF and HNF-4 serve as accessory factors required for induction of phosphoenolpyruvate carboxykinase gene transcription by glucocorticoids. Proc Natl Acad Sci U S A. 1995;92:412–6.
50. Hall RK, Scott DK, Noisin EL, Lucas PC, Granner DK. Activation of the phosphoenolpyruvate carboxykinase gene retinoic acid response element is dependent on a retinoic acid receptor/coregulator complex. Mol Cell Biol. 1992;12:5527–35.
51. Wang XL, Herzog B, Waltner-Law M, Hall RK, Shiota M, Granner DK. The synergistic effect of dexamethasone and all-trans-retinoic acid on hepatic phosphoenolpyruvate carboxykinase gene expression involves the coactivator p300. J Biol Chem. 2004;279:34191–200. doi:10.1074/jbc.M403455200.
52. Wang JC, Stromstedt PE, O'Brien RM, Granner DK. Hepatic nuclear factor 3 is an accessory factor required for the stimulation of phosphoenolpyruvate carboxykinase gene transcription by glucocorticoids. Mol Endocrinol. 1996;10:794–800.
53. Zhang K, Li L, Qi Y, Zhu X, Gan B, DePinho RA, Averitt T, Guo S. Hepatic suppression of Foxo1 and Foxo3 causes hypoglycemia and hyperlipidemia in mice. Endocrinology. 2012;153:631–46.
54. O'Brien RM, Lucas PC, Forest CD, Magnuson MA, Granner DK. Identification of a sequence in the PEPCK gene that mediates a negative effect of insulin on transcription. Science. 1990;249:533–7.
55. Forest CD, O'Brien RM, Lucas PC, Magnuson MA, Granner DK. Regulation of phosphoenolpyruvate carboxykinase gene expression by insulin. Use of the stable transfection approach to locate an insulin responsive sequence. Mol Endocrinol. 1990;4:1302–10. doi:10.1210/mend-4-9-1302.
56. Nakae J, Kitamura T, Silver DL, Accili D. The forkhead transcription factor Foxo1 (Fkhr) confers insulin sensitivity onto glucose-6-phosphatase expression. J Clin Invest. 2001;108: 1359–67.
57. Wolfrum C, Asilmaz E, Luca E, Friedman JM, Stoffel M. Foxa2 regulates lipid metabolism and ketogenesis in the liver during fasting and in diabetes. Nature. 2004;432:1027–32. doi:10.1038/nature03047.
58. Lechner PS, Croniger CM, Hakimi P, Millward C, Fekter C, Yun JS, Hanson RW. The use of transgenic mice to analyze the role of accessory factor two in the regulation of phosphoenolpyruvate carboxykinase (GTP) gene transcription during diabetes. J Biol Chem. 2001;276:22675–9. doi:10.1074/jbc.M102422200.
59. Scott DK, Mitchell JA, Granner DK. Identification and characterization of the second retinoic acid response element in the phosphoenolpyruvate carboxykinase gene promoter. J Biol Chem. 1996;271:6260–4.
60. Yamada K, Duong DT, Scott DK, Wang JC, Granner DK. CCAAT/enhancer-binding protein beta is an accessory factor for the glucocorticoid response from the cAMP response element in the rat phosphoenolpyruvate carboxykinase gene promoter. J Biol Chem. 1999;274:5880–7.

61. Hall RK, Wang XL, George L, Koch SR, Granner DK. Insulin represses phosphoenolpyruvate carboxykinase gene transcription by causing the rapid disruption of an active transcription complex: a potential epigenetic effect. Mol Endocrinol. 2007;21:550–63. doi:10.1210/me.2006-0307.
62. Schurter BT, Koh SS, Chen D, Bunick GJ, Harp JM, Hanson BL, Henschen-Edman A, Mackay DR, Stallcup MR, Aswad DW. Methylation of histone H3 by coactivator-associated arginine methyltransferase 1. Biochemistry. 2001;40:5747–56.
63. Lee DY, Teyssier C, Strahl BD, Stallcup MR. Role of protein methylation in regulation of transcription. Endocr Rev. 2005;26:147–70.
64. Feng Q, He B, Jung SY, Song Y, Qin J, Tsai SY, Tsai MJ, O'Malley BW. Biochemical control of CARM1 enzymatic activity by phosphorylation. J Biol Chem. 2009;284:36167–74. doi:10.1074/jbc.M109.065524.
65. Cassuto H, Kochan K, Chakravarty K, Cohen H, Blum B, Olswang Y, Hakimi P, Xu C, Massillon D, Hanson RW, Reshef L. Glucocorticoids regulate transcription of the gene for phosphoenolpyruvate carboxykinase in the liver via an extended glucocorticoid regulatory unit. J Biol Chem. 2005;280:33873–84. doi:10.1074/jbc.M504119200.
66. Bernal-Mizrachi C, Weng S, Feng C, Finck BN, Knutsen RH, Leone TC, Coleman T, Mecham RP, Kelly DP, Semenkovich CF. Dexamethasone induction of hypertension and diabetes is PPAR-alpha dependent in LDL receptor-null mice. Nat Med. 2003;9:1069–75.
67. Beale EG, Forest C, Hammer RE. Regulation of cytosolic phosphoenolpyruvate carboxykinase gene expression in adipocytes. Biochimie. 2003;85:1207–11.
68. Olswang Y, Blum B, Cassuto H, Cohen H, Biberman Y, Hanson RW, Reshef L. Glucocorticoids repress transcription of phosphoenolpyruvate carboxykinase (GTP) gene in adipocytes by inhibiting its C/EBP-mediated activation. J Biol Chem. 2003;278:12929–36. doi:10.1074/jbc.M300263200.
69. Hanson RW, Reshef L. Glyceroneogenesis revisited. Biochimie. 2003;85:1199–205.
70. Cadoudal T, Leroyer S, Reis AF, Tordjman J, Durant S, Fouque F, Collinet M, Quette J, Chauvet G, Beale E, Velho G, Antoine B, Benelli C, Forest C. Proposed involvement of adipocyte glyceroneogenesis and phosphoenolpyruvate carboxykinase in the metabolic syndrome. Biochimie. 2005;87:27–32. doi:10.1016/j.biochi.2004.12.005.
71. Wang JC, Stromstedt PE, Sugiyama T, Granner DK. The phosphoenolpyruvate carboxykinase gene glucocorticoid response unit: identification of the functional domains of accessory factors HNF3 beta (hepatic nuclear factor-3 beta) and HNF4 and the necessity of proper alignment of their cognate binding sites. Mol Endocrinol. 1999;13:604–18.
72. Stafford JM, Waltner-Law M, Granner DK. Role of accessory factors and steroid receptor coactivator 1 in the regulation of phosphoenolpyruvate carboxykinase gene transcription by glucocorticoids. J Biol Chem. 2001;276:3811–9. doi:10.1074/jbc.M009389200.
73. Puigserver P, Rhee J, Donovan J, Walkey CJ, Yoon JC, Oriente F, Kitamura Y, Altomonte J, Dong H, Accili D, Spiegelman BM. Insulin-regulated hepatic gluconeogenesis through FOXO1-PGC-1alpha interaction. Nature. 2003;423:550–5. doi:10.1038/nature01667.
74. Stafford JM, Wilkinson JC, Beechem JM, Granner DK. Accessory factors facilitate the binding of glucocorticoid receptor to the phosphoenolpyruvate carboxykinase gene promoter. J Biol Chem. 2001;276:39885–91.
75. Vander Kooi BT, Onuma H, Oeser JK, Svitek CA, Allen SR, Vander Kooi CW, Chazin WJ, O'Brien RM. The glucose-6-phosphatase catalytic subunit gene promoter contains both positive and negative glucocorticoid response elements. Mol Endocrinol. 2005;19:3001–22.
76. Onuma H, Vander Kooi BT, Boustead JN, Oeser JK, O'Brien RM. Correlation between FOXO1a (FKHR) and FOXO3a (FKHRL1) binding and the inhibition of basal glucose-6-phosphatase catalytic subunit gene transcription by insulin. Mol Endocrinol. 2006;20:2831–47. doi:10.1210/me.2006-0085.
77. Yoon JC, Puigserver P, Chen G, Donovan J, Wu Z, Rhee J, Adelmant G, Stafford J, Kahn CR, Granner DK, Newgard CB, Spiegelman BM. Control of hepatic gluconeogenesis through the transcriptional coactivator PGC-1. Nature. 2001;413:131–8.

78. Boustead JN, Stadelmaier BT, Eeds AM, Wiebe PO, Svitek CA, Oeser JK, O'Brien RM. Hepatocyte nuclear factor-4 alpha mediates the stimulatory effect of peroxisome proliferator-activated receptor gamma co-activator-1 alpha (PGC-1 alpha) on glucose-6-phosphatase catalytic subunit gene transcription in H4IIE cells. Biochem J. 2003;369:17–22. doi:10.1042/BJ20021382.
79. Matsumoto M, Pocai A, Rossetti L, Depinho RA, Accili D. Impaired regulation of hepatic glucose production in mice lacking the forkhead transcription factor Foxo1 in liver. Cell Metab. 2007;6:208–16.
80. Hiraiwa H, Chou JY. Glucocorticoids activate transcription of the gene for the glucose-6-phosphate transporter, deficient in glycogen storage disease type 1b. DNA Cell Biol. 2001;20:447–53. doi:10.1089/104454901316976073.
81. Kallwellis-Opara A, Zaho X, Zimmermann U, Unterman TG, Walther R, Schmoll D. Characterization of cis-elements mediating the stimulation of glucose-6-phosphate transporter promoter activity by glucocorticoids. Gene. 2003;320:59–66.
82. Pierreux CE, Urso B, De Meyts P, Rousseau GG, Lemaigre FP. Inhibition by insulin of glucocorticoid-induced gene transcription: involvement of the ligand-binding domain of the glucocorticoid receptor and independence from the phosphatidylinositol 3-kinase and mitogen-activated protein kinase pathways. Mol Endocrinol. 1998;12:1343–54. doi:10.1210/mend.12.9.0172.
83. Lemaigre FP, Lause P, Rousseau GG. Insulin inhibits glucocorticoid-induced stimulation of liver 6-phosphofructo-2-kinase/fructose-2,6-bisphosphatase gene transcription. FEBS Lett. 1994;340:221–5.
84. Sutherland C, O'Brien RM, Granner DK. Phosphatidylinositol 3-kinase, but not p70/p85 ribosomal S6 protein kinase, is required for the regulation of phosphoenolpyruvate carboxykinase (PEPCK) gene expression by insulin. Dissociation of signaling pathways for insulin and phorbol ester regulation of PEPCK gene expression. J Biol Chem. 1995;270:15501–6.
85. Liao J, Barthel A, Nakatani K, Roth RA. Activation of protein kinase B/Akt is sufficient to repress the glucocorticoid and cAMP induction of phosphoenolpyruvate carboxykinase gene. J Biol Chem. 1998;273:27320–4.
86. De Los Pinos E, Fernandez De Mattos S, Joaquin M, Tauler A. Insulin inhibits glucocorticoid-stimulated L-type 6-phosphofructo-2-kinase/fructose-2,6-bisphosphatase gene expression by activation of the c-Jun N-terminal kinase pathway. Biochem J. 2001;353:267–73.
87. Patel R, Patel M, Tsai R, Lin V, Bookout AL, Zhang Y, Magomedova L, Li T, Chan JF, Budd C, Mangelsdorf DJ, Cummins CL. LXRbeta is required for glucocorticoid-induced hyperglycemia and hepatosteatosis in mice. J Clin Invest. 2010;121:431–41.
88. Nader N, Ng SS, Wang Y, Abel BS, Chrousos GP, Kino T. Liver X receptors regulate the transcriptional activity of the glucocorticoid receptor: implications for the carbohydrate metabolism. PLoS One. 2012;7:e26751.
89. Lu Y, Xiong X, Wang X, Zhang Z, Li J, Shi G, Yang J, Zhang H, Ning G, Li X. Yin Yang 1 promotes hepatic gluconeogenesis through upregulation of glucocorticoid receptor. Diabetes. 2013;62:1064–73. doi:10.2337/db12-0744.
90. Renga B, Mencarelli A, D'Amore C, Cipriani S, Baldelli F, Zampella A, Distrutti E, Fiorucci S. Glucocorticoid receptor mediates the gluconeogenic activity of the farnesoid X receptor in the fasting condition. FASEB J. 2012;26:3021–31. doi:10.1096/fj.11-195701.
91. Molusky MM, Li S, Ma D, Yu L, Lin JD. Ubiquitin-specific protease 2 regulates hepatic gluconeogenesis and diurnal glucose metabolism through 11beta-hydroxysteroid dehydrogenase 1. Diabetes. 2012;61:1025–35. doi:10.2337/db11-0970.
92. Ichijo T, Voutetakis A, Cotrim AP, Bhattachryya N, Fujii M, Chrousos GP, Kino T. The Smad6-histone deacetylase 3 complex silences the transcriptional activity of the glucocorticoid receptor: potential clinical implications. J Biol Chem. 2005;280:42067–77. doi:10.1074/jbc.M509338200.
93. Winkler R, Benz V, Clemenz M, Bloch M, Foryst-Ludwig A, Wardat S, Witte N, Trappiel M, Namsolleck P, Mai K, Spranger J, Matthias G, Roloff T, Truee O, Kappert K, Schupp M, Matthias P, Kintscher U. Histone deacetylase 6 (HDAC6) is an essential modifier of

glucocorticoid-induced hepatic gluconeogenesis. Diabetes. 2012;61:513–23. doi:10.2337/db11-0313.
94. Ferrannini E, Simonson DC, Katz LD, Reichard Jr G, Bevilacqua S, Barrett EJ, Olsson M, DeFronzo RA. The disposal of an oral glucose load in patients with non-insulin-dependent diabetes. Metabolism. 1988;37:79–85.
95. DeFronzo RA, Tripathy D. Skeletal muscle insulin resistance is the primary defect in type 2 diabetes. Diabetes Care. 2009;32 Suppl 2:S157–63.
96. Morgan SA, Sherlock M, Gathercole LL, Lavery GG, Lenaghan C, Bujalska IJ, Laber D, Yu A, Convey G, Mayers R, Hegyi K, Sethi JK, Stewart PM, Smith DM, Tomlinson JW. 11beta-hydroxysteroid dehydrogenase type 1 regulates glucocorticoid-induced insulin resistance in skeletal muscle. Diabetes. 2009;58:2506–15.
97. Dimitriadis G, Leighton B, Parry-Billings M, Sasson S, Young M, Krause U, Bevan S, Piva T, Wegener G, Newsholme EA. Effects of glucocorticoid excess on the sensitivity of glucose transport and metabolism to insulin in rat skeletal muscle. Biochem J. 1997;321(Pt 3):707–12.
98. Weinstein SP, Wilson CM, Pritsker A, Cushman SW. Dexamethasone inhibits insulin-stimulated recruitment of GLUT4 to the cell surface in rat skeletal muscle. Metabolism. 1998;47:3–6.
99. Moitra J, Mason MM, Olive M, Krylov D, Gavrilova O, Marcus-Samuels B, Feigenbaum L, Lee E, Aoyama T, Eckhaus M, Reitman ML, Vinson C. Life without white fat: a transgenic mouse. Genes Dev. 1998;12:3168–81.
100. Ohshima K, Shargill NS, Chan TM, Bray GA. Effects of dexamethasone on glucose transport by skeletal muscles of obese (ob/ob) mice. Int J Obes. 1989;13:155–63.
101. Haluzik M, Dietz KR, Kim JK, Marcus-Samuels B, Shulman GI, Gavrilova O, Reitman ML. Adrenalectomy improves diabetes in A-ZIP/F-1 lipoatrophic mice by increasing both liver and muscle insulin sensitivity. Diabetes. 2002;51:2113–8.
102. Gathercole LL, Bujalska IJ, Stewart PM, Tomlinson JW. Glucocorticoid modulation of insulin signaling in human subcutaneous adipose tissue. J Clin Endocrinol Metab. 2007;92:4332–9.
103. Pivonello R, De Leo M, Vitale P, Cozzolino A, Simeoli C, De Martino MC, Lombardi G, Colao A. Pathophysiology of diabetes mellitus in Cushing's syndrome. Neuroendocrinology. 2010;92 Suppl 1:77–81.
104. Schakman O, Gilson H, Thissen JP. Mechanisms of glucocorticoid-induced myopathy. J Endocrinol. 2008;197:1–10.
105. Lee YH, White MF. Insulin receptor substrate proteins and diabetes. Arch Pharm Res. 2004;27:361–70.
106. Saad MJ, Folli F, Kahn JA, Kahn CR. Modulation of insulin receptor, insulin receptor substrate-1, and phosphatidylinositol 3-kinase in liver and muscle of dexamethasone-treated rats. J Clin Invest. 1993;92:2065–72.
107. Giorgino F, Almahfouz A, Goodyear LJ, Smith RJ. Glucocorticoid regulation of insulin receptor and substrate IRS-1 tyrosine phosphorylation in rat skeletal muscle in vivo. J Clin Invest. 1993;91:2020–30.
108. Kuo T, Lew MJ, Mayba O, Harris CA, Speed TP, Wang JC. Genome-wide analysis of glucocorticoid receptor-binding sites in myotubes identifies gene networks modulating insulin signaling. Proc Natl Acad Sci U S A. 2012;109:11160–5. doi:10.1073/pnas.1111334109.
109. Backer JM. The regulation of class IA PI 3-kinases by inter-subunit interactions. Curr Top Microbiol Immunol. 2010;346:87–114. doi:10.1007/82_2010_52.
110. Cantley LC. The phosphoinositide 3-kinase pathway. Science. 2002;296:1655–7. doi:10.1126/science.296.5573.1655.
111. Barbour LA, Shao J, Qiao L, Leitner W, Anderson M, Friedman JE, Draznin B. Human placental growth hormone increases expression of the p85 regulatory unit of phosphatidylinositol 3-kinase and triggers severe insulin resistance in skeletal muscle. Endocrinology. 2004;145:1144–50.
112. Draznin B. Molecular mechanisms of insulin resistance: serine phosphorylation of insulin receptor substrate-1 and increased expression of p85alpha: the two sides of a coin. Diabetes. 2006;55:2392–7.

113. Chagpar RB, Links PH, Pastor MC, Furber LA, Hawrysh AD, Chamberlain MD, Anderson DH. Direct positive regulation of PTEN by the p85 subunit of phosphatidylinositol 3-kinase. Proc Natl Acad Sci U S A. 2010;107:5471–6. doi:10.1073/pnas.0908899107.
114. Terauchi Y, Tsuji Y, Satoh S, Minoura H, Murakami K, Okuno A, Inukai K, Asano T, Kaburagi Y, Ueki K, Nakajima H, Hanafusa T, Matsuzawa Y, Sekihara H, Yin Y, Barrett JC, Oda H, Ishikawa T, Akanuma Y, Komuro I, Suzuki M, Yamamura K, Kodama T, Suzuki H, Yamamura K, Kodama T, Suzuki H, Koyasu S, Aizawa S, Tobe K, Fukui Y, Yazaki Y, Kadowaki T. Increased insulin sensitivity and hypoglycaemia in mice lacking the p85 alpha subunit of phosphoinositide 3-kinase. Nat Genet. 1999;21:230–5. doi:10.1038/6023.
115. Mauvais-Jarvis F, Ueki K, Fruman DA, Hirshman MF, Sakamoto K, Goodyear LJ, Iannacone M, Accili D, Cantley LC, Kahn CR. Reduced expression of the murine p85alpha subunit of phosphoinositide 3-kinase improves insulin signaling and ameliorates diabetes. J Clin Invest. 2002;109:141–9.
116. Bandyopadhyay GK, Yu JG, Ofrecio J, Olefsky JM. Increased p85/55/50 expression and decreased phosphotidylinositol 3-kinase activity in insulin-resistant human skeletal muscle. Diabetes. 2005;54:2351–9.
117. Holland WL, Brozinick JT, Wang LP, Hawkins ED, Sargent KM, Liu Y, Narra K, Hoehn KL, Knotts TA, Siesky A, Nelson DH, Karathanasis SK, Fontenot GK, Birnbaum MJ, Summers SA. Inhibition of ceramide synthesis ameliorates glucocorticoid-, saturated-fat-, and obesity-induced insulin resistance. Cell Metab. 2007;5:167–79.
118. Chavez JA, Summers SA. A ceramide-centric view of insulin resistance. Cell Metab. 2012;15:585–94. doi:10.1016/j.cmet.2012.04.002.
119. Tappy L, Randin D, Vollenweider P, Vollenweider L, Paquot N, Scherrer U, Schneiter P, Nicod P, Jequier E. Mechanisms of dexamethasone-induced insulin resistance in healthy humans. J Clin Endocrinol Metab. 1994;79:1063–9.
120. Sugden MC, Holness MJ. Recent advances in mechanisms regulating glucose oxidation at the level of the pyruvate dehydrogenase complex by PDKs. Am J Physiol Endocrinol Metab. 2003;284:E855–62. doi:10.1152/ajpendo.00526.2002.
121. Connaughton S, Chowdhury F, Attia RR, Song S, Zhang Y, Elam MB, Cook GA, Park EA. Regulation of pyruvate dehydrogenase kinase isoform 4 (PDK4) gene expression by glucocorticoids and insulin. Mol Cell Endocrinol. 2010;315:159–67. doi:10.1016/j.mce.2009.08.011.
122. Kwon HS, Huang B, Unterman TG, Harris RA. Protein kinase B-alpha inhibits human pyruvate dehydrogenase kinase-4 gene induction by dexamethasone through inactivation of FOXO transcription factors. Diabetes. 2004;53:899–910.
123. Bazuine M, Carlotti F, Tafrechi RS, Hoeben RC, Maassen JA. Mitogen-activated protein kinase (MAPK) phosphatase-1 and -4 attenuate p38 MAPK during dexamethasone-induced insulin resistance in 3T3-L1 adipocytes. Mol Endocrinol. 2004;18:1697–707.
124. Sakoda H, Ogihara T, Anai M, Funaki M, Inukai K, Katagiri H, Fukushima Y, Onishi Y, Ono H, Fujishiro M, Kikuchi M, Oka Y, Asano T. Dexamethasone-induced insulin resistance in 3T3-L1 adipocytes is due to inhibition of glucose transport rather than insulin signal transduction. Diabetes. 2000;49:1700–8.
125. Yu CY, Mayba O, Lee JV, Tran J, Harris C, Speed TP, Wang JC. Genome-wide analysis of glucocorticoid receptor binding regions in adipocytes reveal gene network involved in triglyceride homeostasis. PLoS One. 2010;5:e15188.
126. Hazlehurst JM, Gathercole LL, Nasiri M, Armstrong MJ, Borrows S, Yu J, Wagenmakers AJ, Stewart PM, Tomlinson JW. Glucocorticoids fail to cause insulin resistance in human subcutaneous adipose tissue in vivo. J Clin Endocrinol Metab. 2013;98:1631–40. doi:10.1210/jc.2012-3523.
127. Gathercole LL, Morgan SA, Bujalska IJ, Stewart PM, Tomlinson JW. Short- and long-term glucocorticoid treatment enhances insulin signalling in human subcutaneous adipose tissue. Nutr Diabetes. 2011;1:e3. doi:10.1038/nutd.2010.3.
128. Buentke E, Nordstrom A, Lin H, Bjorklund AC, Laane E, Harada M, Lu L, Tegnebratt T, Stone-Elander S, Heyman M, Soderhall S, Porwit A, Ostenson CG, Shoshan M, Tamm KP, Grander D. Glucocorticoid-induced cell death is mediated through reduced glucose metabolism in lymphoid leukemia cells. Blood Cancer J. 2011;1:e31. doi:10.1038/bcj.2011.27.

129. de Leon MJ, McRae T, Rusinek H, Convit A, De Santi S, Tarshish C, Golomb J, Volkow N, Daisley K, Orentreich N, McEwen B. Cortisol reduces hippocampal glucose metabolism in normal elderly, but not in Alzheimer's disease. J Clin Endocrinol Metab. 1997;82:3251–9. doi:10.1210/jcem.82.10.4305.
130. Horner HC, Packan DR, Sapolsky RM. Glucocorticoids inhibit glucose transport in cultured hippocampal neurons and glia. Neuroendocrinology. 1990;52:57–64.
131. Cherian AK, Briski KP. Effects of adrenalectomy on neuronal substrate fuel transporter and energy transducer gene expression in hypothalamic and hindbrain metabolic monitoring sites. Neuroendocrinology. 2010;91:56–63. doi:10.1159/000264919.
132. Mersmann HJ, Segal HL. Glucocorticoid control of the liver glycogen synthetase-activating system. J Biol Chem. 1969;244:1701–4.
133. Ray PD, Foster DO, Lardy HA. Mode of action of glucocorticoids. I. Stimulation of gluconeogenesis independent of synthesis de novo of enzymes. J Biol Chem. 1964;239:3396–400.
134. De Wulf H, Hers HG. The stimulation of glycogen synthesis and of glycogen synthetase in the liver by glucocorticoids. Eur J Biochem. 1967;2:57–60.
135. Huang TS, Krebs EG. Amino acid sequence of a phosphorylation site in skeletal muscle glycogen synthetase. Biochem Biophys Res Commun. 1977;75:643–50.
136. Rylatt DB, Aitken A, Bilham T, Condon GD, Embi N, Cohen P. Glycogen synthase from rabbit skeletal muscle. Amino acid sequence at the sites phosphorylated by glycogen synthase kinase-3, and extension of the N-terminal sequence containing the site phosphorylated by phosphorylase kinase. Eur J Biochem. 1980;107:529–37.
137. Pugazhenthi S, Khandelwal RL. Regulation of glycogen synthase activation in isolated hepatocytes. Mol Cell Biochem. 1995;149–150:95–101.
138. Brady MJ, Nairn AC, Saltiel AR. The regulation of glycogen synthase by protein phosphatase 1 in 3T3-L1 adipocytes. Evidence for a potential role for DARPP-32 in insulin action. J Biol Chem. 1997;272:29698–703.
139. Printen JA, Brady MJ, Saltiel AR. PTG, a protein phosphatase 1-binding protein with a role in glycogen metabolism. Science. 1997;275:1475–8.
140. Laloux M, Stalmans W, Hers HG. On the mechanism by which glucocorticoids cause the activation of glycogen synthase in mouse and rat livers. Eur J Biochem. 1983;136:175–81.
141. Vanstapel F, Dopere F, Stalmans W. The role of glycogen synthase phosphatase in the glucocorticoid-induced deposition of glycogen in foetal rat liver. Biochem J. 1980;192: 607–12.
142. Green GA, Chenoweth M, Dunn A. Adrenal glucocorticoid permissive regulation of muscle glycogenolysis: action on protein phosphatase(s) and its inhibitor(s). Proc Natl Acad Sci U S A. 1980;77:5711–5.
143. Salehzadeh F, Al-Khalili L, Kulkarni SS, Wang M, Lonnqvist F, Krook A. Glucocorticoid-mediated effects on metabolism are reversed by targeting 11 beta hydroxysteroid dehydrogenase type 1 in human skeletal muscle. Diabetes Metab Res Rev. 2009;25:250–8. doi:10.1002/dmrr.944.
144. Miller TB, Exton JH, Park CR. A block in epinephrine-induced glycogenolysis in hearts from adrenalectomized rats. J Biol Chem. 1971;246:3672–8.
145. Puthanveetil P, Rodrigues B. Glucocorticoid excess induces accumulation of cardiac glycogen and triglyceride: suggested role for AMPK. Curr Pharm Des. 2013;19:4818–30.
146. Baltrons MA, Agullo L, Garcia A. Dexamethasone up-regulates a constitutive nitric oxide synthase in cerebellar astrocytes but not in granule cells in culture. J Neurochem. 1995;64: 447–50.
147. Binnert C, Ruchat S, Nicod N, Tappy L. Dexamethasone-induced insulin resistance shows no gender difference in healthy humans. Diabetes Metab. 2004;30:321–6.
148. Rafacho A, Marroqui L, Taboga SR, Abrantes JL, Silveira LR, Boschero AC, Carneiro EM, Bosqueiro JR, Nadal A, Quesada I. Glucocorticoids in vivo induce both insulin hypersecretion and enhanced glucose sensitivity of stimulus-secretion coupling in isolated rat islets. Endocrinology. 2010;151:85–95. doi:10.1210/en.2009-0704.

149. Besse C, Nicod N, Tappy L. Changes in insulin secretion and glucose metabolism induced by dexamethasone in lean and obese females. Obes Res. 2005;13:306–11. doi:10.1038/oby.2005.41.
150. Fransson L, Franzen S, Rosengren V, Wolbert P, Sjoholm A, Ortsater H. beta-Cell adaptation in a mouse model of glucocorticoid-induced metabolic syndrome. J Endocrinol. 2013;219:231–41. doi:10.1530/JOE-13-0189.
151. Jeong IK, Oh SH, Kim BJ, Chung JH, Min YK, Lee MS, Lee MK, Kim KW. The effects of dexamethasone on insulin release and biosynthesis are dependent on the dose and duration of treatment. Diabetes Res Clin Pract. 2001;51:163–71.
152. van Raalte DH, Nofrate V, Bunck MC, van Iersel T, Elassaiss Schaap J, Nassander UK, Heine RJ, Mari A, Dokter WH, Diamant M. Acute and 2-week exposure to prednisolone impair different aspects of beta-cell function in healthy men. Eur J Endocrinol. 2010;162:729–35. doi:10.1530/EJE-09-1034.
153. Gremlich S, Roduit R, Thorens B. Dexamethasone induces posttranslational degradation of GLUT2 and inhibition of insulin secretion in isolated pancreatic beta cells. Comparison with the effects of fatty acids. J Biol Chem. 1997;272:3216–22.
154. Borboni P, Porzio O, Magnaterra R, Fusco A, Sesti G, Lauro R, Marlier LN. Quantitative analysis of pancreatic glucokinase gene expression in cultured beta cells by competitive polymerase chain reaction. Mol Cell Endocrinol. 1996;117:175–81.
155. Goodman PA, Medina-Martinez O, Fernandez-Mejia C. Identification of the human insulin negative regulatory element as a negative glucocorticoid response element. Mol Cell Endocrinol. 1996;120:139–46.
156. Sharma S, Jhala US, Johnson T, Ferreri K, Leonard J, Montminy M. Hormonal regulation of an islet-specific enhancer in the pancreatic homeobox gene STF-1. Mol Cell Biol. 1997;17:2598–604.
157. Khan A, Ostenson CG, Berggren PO, Efendic S. Glucocorticoid increases glucose cycling and inhibits insulin release in pancreatic islets of ob/ob mice. Am J Physiol. 1992;263:E663–6.
158. Ullrich S, Berchtold S, Ranta F, Seebohm G, Henke G, Lupescu A, Mack AF, Chao CM, Su J, Nitschke R, Alexander D, Friedrich B, Wulff P, Kuhl D, Lang F. Serum- and glucocorticoid-inducible kinase 1 (SGK1) mediates glucocorticoid-induced inhibition of insulin secretion. Diabetes. 2005;54:1090–9.
159. Negrato CA, Jovanovic L, Rafacho A, Tambascia MA, Geloneze B, Dias A, Rudge MV. Association between different levels of dysglycemia and metabolic syndrome in pregnancy. Diabetol Metab Syndr. 2009;1:3. doi:10.1186/1758-5996-1-3.
160. Rafacho A, Quallio S, Ribeiro DL, Taboga SR, Paula FM, Boschero AC, Bosqueiro JR. The adaptive compensations in endocrine pancreas from glucocorticoid-treated rats are reversible after the interruption of treatment. Acta Physiol. 2010;200:223–35. doi:10.1111/j.1748-1716.2010.02146.x.
161. Rafacho A, Abrantes JL, Ribeiro DL, Paula FM, Pinto ME, Boschero AC, Bosqueiro JR. Morphofunctional alterations in endocrine pancreas of short- and long-term dexamethasone-treated rats. Horm Metab Res. 2011;43:275–81. doi:10.1055/s-0030-1269896.
162. Rafacho A, Giozzet VA, Boschero AC, Bosqueiro JR. Functional alterations in endocrine pancreas of rats with different degrees of dexamethasone-induced insulin resistance. Pancreas. 2008;36:284–93. doi:10.1097/MPA.0b013e31815ba826.
163. Huising MO, Pilbrow AP, Matsumoto M, van der Meulen T, Park H, Vaughan JM, Lee S, Vale WW. Glucocorticoids differentially regulate the expression of CRFR1 and CRFR2alpha in MIN6 insulinoma cells and rodent islets. Endocrinology. 2011;152:138–50. doi:10.1210/en.2010-0791.
164. Huising MO, van der Meulen T, Vaughan JM, Matsumoto M, Donaldson CJ, Park H, Billestrup N, Vale WW. CRFR1 is expressed on pancreatic beta cells, promotes beta cell proliferation, and potentiates insulin secretion in a glucose-dependent manner. Proc Natl Acad Sci U S A. 2010;107:912–7. doi:10.1073/pnas.0913610107.
165. Reich E, Tamary A, Sionov RV, Melloul D. Involvement of thioredoxin-interacting protein (TXNIP) in glucocorticoid-mediated beta cell death. Diabetologia. 2012;55:1048–57. doi:10.1007/s00125-011-2422-z.

166. Avram D, Ranta F, Hennige AM, Berchtold S, Hopp S, Haring HU, Lang F, Ullrich S. IGF-1 protects against dexamethasone-induced cell death in insulin secreting INS-1 cells independent of AKT/PKB phosphorylation. Cell Physiol Biochem. 2008;21:455–62. doi:10.1159/000129638.
167. Fransson L, Rosengren V, Saha TK, Grankvist N, Islam T, Honkanen RE, Sjoholm A, Ortsater H. Mitogen-activated protein kinases and protein phosphatase 5 mediate glucocorticoid-induced cytotoxicity in pancreatic islets and beta-cells. Mol Cell Endocrinol. 2014;383:126–36. doi:10.1016/j.mce.2013.12.010.
168. Bodwell JE, Webster JC, Jewell CM, Cidlowski JA, Hu JM, Munck A. Glucocorticoid receptor phosphorylation: overview, function and cell cycle-dependence. J Steroid Biochem Mol Biol. 1998;65:91–9.
169. Ismaili N, Garabedian MJ. Modulation of glucocorticoid receptor function via phosphorylation. Ann N Y Acad Sci. 2004;1024:86–101. doi:10.1196/annals.1321.007.
170. Rogatsky I, Logan SK, Garabedian MJ. Antagonism of glucocorticoid receptor transcriptional activation by the c-Jun N-terminal kinase. Proc Natl Acad Sci U S A. 1998;95: 2050–5.
171. Roma LP, Oliveira CA, Carneiro EM, Albuquerque GG, Boschero AC, Souza KL. N-acetylcysteine protects pancreatic islet against glucocorticoid toxicity. Redox Rep. 2011;16:173–80. doi:10.1179/1351000211Y.0000000006.
172. Yamamoto M, Yamato E, Toyoda S, Tashiro F, Ikegami H, Yodoi J, Miyazaki J. Transgenic expression of antioxidant protein thioredoxin in pancreatic beta cells prevents progression of type 2 diabetes mellitus. Antioxid Redox Signal. 2008;10:43–9. doi:10.1089/ars.2007.1586.
173. Wang Z, Rong YP, Malone MH, Davis MC, Zhong F, Distelhorst CW. Thioredoxin-interacting protein (txnip) is a glucocorticoid-regulated primary response gene involved in mediating glucocorticoid-induced apoptosis. Oncogene. 2006;25:1903–13. doi:10.1038/sj.onc.1209218.
174. Quesada I, Tuduri E, Ripoll C, Nadal A. Physiology of the pancreatic alpha-cell and glucagon secretion: role in glucose homeostasis and diabetes. J Endocrinol. 2008;199:5–19. doi:10.1677/JOE-08-0290.
175. Everett LJ, Lazar MA. Cell-specific integration of nuclear receptor function at the genome. Wiley Interdiscip Rev Syst Biol Med. 2013;5:615–29. doi:10.1002/wsbm.1231.
176. Hah N, Danko CG, Core L, Waterfall JJ, Siepel A, Lis JT, Kraus WL. A rapid, extensive, and transient transcriptional response to estrogen signaling in breast cancer cells. Cell. 2011;145:622–34. doi:10.1016/j.cell.2011.03.042.
177. Hah N, Kraus WL. Hormone-regulated transcriptomes: lessons learned from estrogen signaling pathways in breast cancer cells. Mol Cell Endocrinol. 2014;382:652–64. doi:10.1016/j.mce.2013.06.021.
178. Madsen JG, Schmidt SF, Larsen BD, Loft A, Nielsen R, Mandrup S. iRNA-seq: computational method for genome-wide assessment of acute transcriptional regulation from total RNA-seq data. Nucleic Acids Res. 2015;43(6):e40. doi:10.1093/nar/gku1365.
179. Jinek M, East A, Cheng A, Lin S, Ma E, Doudna J. RNA-programmed genome editing in human cells. eLife. 2013;2:e00471. doi:10.7554/eLife.00471.
180. Mali P, Yang L, Esvelt KM, Aach J, Guell M, DiCarlo JE, Norville JE, Church GM. RNA-guided human genome engineering via Cas9. Science. 2013;339:823–6. doi:10.1126/science.1232033.
181. Cong L, Ran FA, Cox D, Lin S, Barretto R, Habib N, Hsu PD, Wu X, Jiang W, Marraffini LA, Zhang F. Multiplex genome engineering using CRISPR/Cas systems. Science. 2013;339: 819–23. doi:10.1126/science.1231143.
182. Cho SW, Kim S, Kim JM, Kim JS. Targeted genome engineering in human cells with the Cas9 RNA-guided endonuclease. Nat Biotechnol. 2013;31:230–2. doi:10.1038/nbt.2507.
183. Yang H, Wang H, Jaenisch R. Generating genetically modified mice using CRISPR/Cas-mediated genome engineering. Nat Protoc. 2014;9:1956–68. doi:10.1038/nprot.2014.134.

Chapter 6
How Do Glucocorticoids Regulate Lipid Metabolism?

Roldan M. de Guia and Stephan Herzig

Abstract Glucocorticoids (GCs) and their cognate, intracellular receptor, the glucocorticoid receptor (GR) have been characterized as critical checkpoints in the hormonal control of energy homeostasis in mammals. Whereas physiological levels of GCs are required for proper metabolic control, aberrant GC action has been linked to a variety of severe metabolic diseases, including type 2 diabetes and obesity. As a member of the nuclear receptor superfamily of transcription factors, the GR translocates into the cell nucleus upon GC binding where it serves as a transcriptional regulator of distinct GC-responsive target genes that are in many cases associated with lipid regulatory pathways and thereby intricately control both physiological and pathophysiological systemic lipid homeostasis. Thus, this chapter focuses on the current knowledge of GC/GR function in lipid handling and its implications for systemic metabolic dysfunction.

Keywords Glucocorticoid receptor • Glucocorticoid hormones • Liver • Adipose tissue • Lipid metabolism • Lipoprotein metabolism • Cholesterol metabolism

Lipid Metabolism at a Glance

The maintenance of metabolic homeostasis depends on the highly regulated interaction of nutritional, neural, and endocrine stimuli. The continuous cycle of nutrient acquisition from the ingested food and the depletion of nutrients involves the coordination of several biomolecular processes at all levels of the animal's biology. Each of the major metabolic organs of the body—the liver, muscles, adipose tissues, intestines, pancreas, and the brain—depends on the major food groups of carbohydrates, proteins, and fats to support various physiologic processes of the organism [1, 2].

R.M. de Guia, Ph.D. • S. Herzig, Ph.D. (✉)
Department of Molecular Metabolic Control, German Cancer Research Center, Center for Molecular Biology and University Hospital Heidelberg, Im Neuenheimer Feld 280, Heidelberg 69120, Germany
e-mail: s.herzig@dkfz.de

J.-C. Wang, C. Harris (eds.), *Glucocorticoid Signaling*, Advances in Experimental Medicine and Biology 872, DOI 10.1007/978-1-4939-2895-8_6

Compared to carbohydrates and proteins, lipids, due to their amphipathic or non-polar nature, are digested, absorbed, and processed in the body differently. Lipids in food require emulsification by liver-derived bile salts in the digestive tract to facilitate pancreas-derived lipase action and intestinal absorption. Cholesterol esters and re-formed triglycerides in the intestinal enterocytes are packaged into lipoproteins, the chylomicrons (Fig. 6.1), which drain via the lymphatics before emptying into the blood vessels [1, 2]. Triglycerides in lipoproteins are acted upon by lipoprotein lipases (LPL) which are tethered to the walls of capillaries by proteoglycans and GPIHBP1 [3, 4]. Fatty acid then move into the cells with the help of transport proteins (fatty acid transport proteins, FATP; fatty acid translocase, FAT/CD36; caveolin-1 and fatty acid-binding protein, FABP) [5]. Cholesterol esters inside lipoproteins are transported into the cell via receptor-mediated processes involving the low-density lipoprotein receptor (LDLR), lipolysis-stimulated lipoprotein receptor (LSR), LDLR-related protein (LRP), and the scavenger receptor (SRB1). The majority of the fatty acid transporters are located on the surface of adipocytes, hepatocytes, and muscle cells, while most lipoprotein receptors are expressed by hepatocytes [6].

Apart from chylomicrons, four other types of lipoprotein particles are recognized to be involved in lipid transport (Fig. 6.1). They differ in size and density which

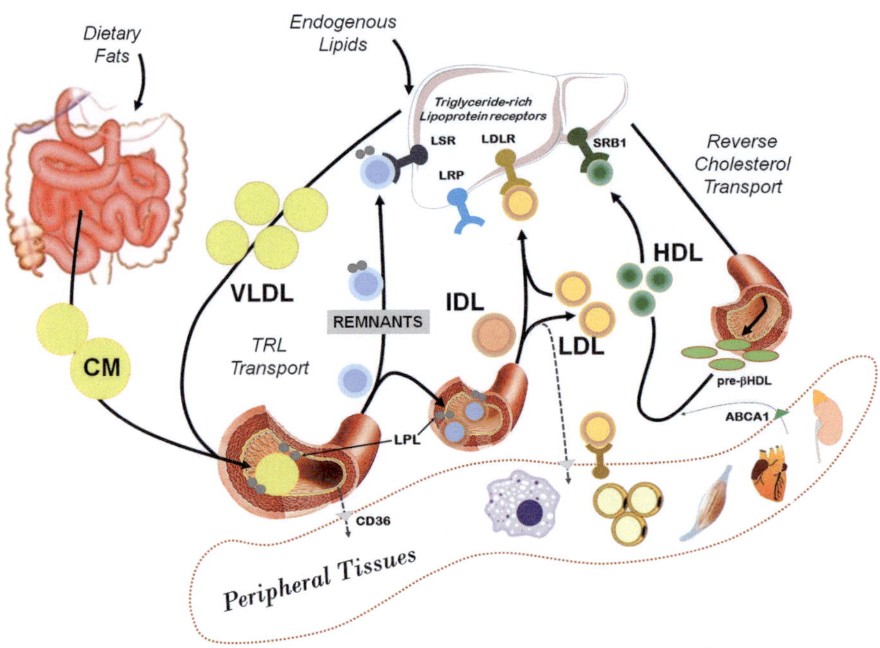

Fig. 6.1 The lipoprotein pathway. *ABCA1* ATP-binding cassette transporter A1, *CM* chylomicrons, *HDL* high-density lipoproteins, *IDL* intermediate density lipoproteins, *LDLR* LDL receptor, *LPL* lipoprotein lipase, *LDL* low-density lipoproteins, *LRP* LDL receptor-related protein, *LSR* lipolysis-stimulated lipoprotein receptor, *SRB1* Scavenger receptor class B member 1, *TRL* triglyceride-rich lipoproteins, *VLDL* very-low-density-lipoproteins

depend on the varying lipid and apolipoprotein contents of each particle [6]. Triglycerides and cholesterol esters comprise the core of the particle while amphipathic phospholipids, unesterified cholesterol, and proteins are found in the outer shell. Apolipoproteins function by activating or inhibiting enzymes important in the transport process and by serving as ligands for cell surface receptors [6]. ApoB and apoE, for example, can be recognized by the LDLR and LSR receptors while ApoCII activates lipoprotein lipase that hydrolyzes triglycerides [7]. Dietary (exogenous) lipids are transported through chylomicrons while endogenous lipids from the liver are transported by very-low density lipoproteins (VLDL) which can give rise to intermediate density (IDL), low-density (LDL), and remnant lipoprotein particles. High density lipoproteins (HDL) are involved primarily in the transport of cholesterol from the peripheral tissues to the liver [2, 6].

The triglycerides and fatty acids being transported from the liver come from different sources: (1) from the chylomicron remnants, (2) *de novo* lipogenic pathway (DNL), and (3) fatty acids from lipolysis in adipocytes [1, 8]. Cholesterol, on the other hand, is provided from (1) the diet via chylomicron remnants, (2) *de novo* synthesis of cholesterol, and (3) from other lipoproteins internalized by the liver [9].

In the cell, fatty acids can be converted into triglycerides or used as source of energy during starvation through mitochondrial β-oxidation. Triglycerides can be packaged into lipoprotein particles (liver) or stored in depots (adipose tissues). In β-oxidation, long-chain fatty acids (LCFAs) are channeled across the mitochondrial membrane, with the help of carnitine palmitoyltransferase-I (CPT-I), and acetyl-CoAs are generated and extracted for energy via the Kreb's cycle and electron transport chain. Fatty acids can also be used for the synthesis of eicosanoids (prostaglandins, thromboxanes, leukotrienes), glycerophospholipids (plasmalogens and phosphatidates), and sphingolipids (sphingomyelin, cerebrosides, gangliosides) [10]. Cholesterol is then used as component of biomembranes and for the synthesis of bile acids (cholic, glycocholic, taurocholic, lithocholic acids), steroid hormones (androgens, progestins, estrogens, glucocorticoids, and mineralocorticoids), and secosteroids (vitamin D) [11].

Insulin and counter-regulatory hormones, which include the glucocorticoids (GCs), play major roles in keeping the balance of metabolites in the cell by activating or deactivating key enzymes involved in different anabolic and catabolic pathways [12]. Furthermore, inter-organ nutrient cross-talk depends primarily on the hormonal signals perceived by the cells. These are true for glucose, amino acid, and lipid metabolism. Below, we present collective information on how GCs regulate lipid metabolism in adipocytes, the liver, and other tissues. Most of the enzymes that are regulated by GCs in lipid pathways are in fact the same ones that are dysregulated in chronic conditions of GC excess i.e. Metabolic and Cushing's syndromes.

GCs and Adipose Tissue

The adipose tissues are the major fat depot of the body. Excess triglycerides that cannot be used by the body as well as excess glucose can be stored as fats in white adipose tissues [2]. Here, triglycerides remain until needed in times of nutrient deprivation.

Abdominal and visceral adiposity are now regarded as an important criteria contributing to the collective dysfunctional phenotypes of the Metabolic Syndrome [13]. Interestingly, this abnormal deposition of fats in the body is likewise observed in Cushings syndrome where the individual suffers from hyperactive HPA axis and hypercortisolemia [14, 15]. However, one key distinction between metabolic syndrome and Cushing syndrome is that patients with Metabolic Syndrome have normal or increased amounts of peripheral adipose and patients with Cushings syndrome typically have decreased peripheral adipose. This partially overlapping phenotype between Metabolic and Cushings syndrome demonstrates potential important role of GCs in adipose tissue physiology.

Effects of GCs on Adipocyte Differentiation

GCs have been reported to possess pro-adipogenic function. This was realized in transgenic mice over-expressing 11beta (β)-HSD1 in adipocytes thus increasing active local GC levels. In these mice, a significant increase in abdominal fat was observed whereas peripheral fats were less affected [16]. Furthermore, the specific elevation of GCs in adipose tissues resulted in the manifestation of major phenotypes of the metabolic syndrome such as abdominal obesity, glucose intolerance, and hypertriglyceridemia. The adipocyte hypertrophic phenotype observed in these animals also results in decreased levels of adiponectin, an insulin-sensitizing adipokine, and increased local and systemic levels of the cytokine tumor necrosis factor-alpha (TNF-α), a marker of insulin resistance [16–19].

Apart from the adipocyte hypertrophic phenotype of mice over-expressing 11β (beta)-HSD1, GCs have been shown to promote differentiation *in vitro* [20]. Cortisol and dexamethasone, a synthetic glucocorticoid, have been widely used as a component of the adipogenic induction cocktail for adipose stromal cells, primary and 3T3-L1 pre-adipocytes [21, 22]. This function of GCs can be attributed to both transcriptional and non-transcriptional action of the GR. The GR has been reported to act on specific histone deacetylase 1 complex for degradation by the 26S proteasome, thus promoting the expression of transcription factors (i.e. C/EBP-alpha (α) and STAT5) necessary for the differentiation process [23, 24]. In primary adipocytes isolated from 11β (beta)-HSD1-over-expressing mice, peroxisome-proliferator activated receptor—gamma (PPARγ), an important regulator of lipogenesis and adipogenesis, has been reported to be up-regulated [25–27]. In other recent reports where the GR function was rendered deficient pharmacologically or genetically [28], adipogenesis was shown to be inhibited *in vitro* which may depend on GR-induced expression of genes such as Krüppel-like factor 15 (KLF15) [29]. Since elevated 11β (beta)-HSD1 activity has been reported in both human and mouse obesity, these are therefore, indicative of the possible contribution of GC/GR signaling in the development of the Metabolic Syndrome by regulating the expression of transcriptional regulators and complexes [30–32].

Effects of GCs on Adipose Tissue Lipolysis

During fasting and starvation, fatty acids and glycerol are release from their storage form as triglycerides in adipose tissue fat depots. Lipases are involved in catalyzing the hydrolysis of the ester bonds between fatty acids and glycerol. Adipose triglyceride lipase (ATGL/desnutrin) hydrolyzes triglyceride into diacylglyceride (DAG) and fatty acid. DAG is in turn hydrolyzed by hormone-sensitive lipase (HSL) into monoacylglyceride (MAG) and another molecule of fatty acid. The final hydrolysis step that completely liberates fatty acids and glycerol is catalyzed by monoglyceride lipase (MGL). Under normal, basal (fasted) physiologic state, GCs promote lipolysis in adipose tissues by inducing activity of all three lipases and reducing LPL activity [33–37]. This lipolytic function of GCs is more pronounce in peripheral adipose than in abdominal depots where adipogenic and lipogenic programs are favored [38, 39]. GCs are also known to induce expression of FoxO transcription factors which are known regulators of the lipases [40, 41].

In addition to the possible direct GR regulation [34] or indirect effects via FoxO, GCs can also affect lipolysis by acting on upstream regulators. Cyclic adenosine monophosphate (cAMP)–protein kinase A (PKA) pathway which is stimulated by catecholamines via G-protein coupled, β (beta)-adrenergic receptor activates HSL and the lipid droplet surface protein perilipin which facilitates fatty acid release by promoting ATGL and HSL function [42, 43]. This process is actually inhibited by insulin via phosphoinositide-3 kinase (PI3K)-protein kinase B (PKB/Akt) signaling which activates the cAMP-hydrolyzing enzyme, phosphodiesterase 3B (PDE3B) [42, 43]. The lipolytic action of catecholamines via beta (β)-adrenergic stimulation occurs rapidly compared to glucocorticoids which takes several hours [44]. It has been postulated that GCs increase intracellular cAMP levels by (1) increasing PKA activity, (2) inhibiting PDE3B expression and/or (3) by inducing expression of angiopoietin-like 4 (ANGPTL4) [44–46].

ANGPTL4 is a hypoxia-induced, secreted glycoprotein that can interact with proteoglycans of the extracellular matrix [47, 48]. It has been reported that GCs promote secretion of ANGPTL4 from the liver and white adipose tissues [49]. ANGPTL4 in turn activates adipose tissue lipolysis and inhibits LPL activity resulting in elevation of free fatty acids in the circulation [50]. Furthermore, ANGPTL4 expression is found to be highly induced during fasting, when the levels of GCs are high, and is inhibited by treatment with RU486, a GR antagonist. [51] Mice lacking ANGPTL4 have reduced fasting-induced WAT lipolysis further demonstrating the possible role of the GC/GR signaling in the process [51]. In line with this, a conserved GRE is present in the rat, mouse, and human ANGPTL4 gene locus [45]. The exact mechanism by which GC-dependent, ANGPTL4-induced lipolysis works remains to be elucidated, but the main hypothesis points to cAMP activation which was observed to be deficient in WAT of ANGPTL4 knockout mice [45]. The importance of this protein in adipose tissue biology, nevertheless, is justifiable as increased ANGPTL4 expression is demonstrated to be associated with elevated fatty acids in both Type 1 and Type 2 diabetes mouse models [52]. This process is also implicated in fatty acid redistribution from adipose tissue to the liver which in turn increases triglyceride synthesis in hepatocytes [33].

Studies on the lipolytic action of GCs in humans have been largely inconsistent across different steroidal preparations, length of GC administration, distinct fat depots, and characteristics of the study population itself. To mention a few, in Cushing's patients [53, 54] and individuals who received chronic GC treatments [55–57], the rate of non-esterified fatty acid (NEFA) turnover is unaffected. This potential anti-lipolytic activity of GCs is also observed *in vitro* in 3T3-L1 cells treated with high concentration of corticosterone (>1 μM). The effect though, is reversed upon removal of the GC from the medium favoring once more lipolysis [35]. In other studies in humans involving infusion of fatty acid stable-isotope and acute physiological hypercortisolemia, an increase in blood NEFA levels is observed [58, 59]. This is probably due to an increased lipolysis from subcutaneous fat depots when insulin levels were normal [58].

With regards to fatty acid oxidation, there has been no concrete report on how GCs could possibly affect this pathway. Furthermore, no study has examined the role of adipose tissue GR using genetic gain- or loss-of-function models which could further provide insight on how GCs regulate lipid metabolism.

Effects of GCs on *de novo* Lipogenesis and (β)-Oxidation

Fatty acids can be synthesized from non-lipid materials, like glucose, via the de novo lipogenic pathway (DNL). DNL takes place both in the liver and the adipose tissues with synthesized fatty acids stored as triglycerides as cytosolic lipid droplets [60]. In adipocytes, TG is made by successive acylation and dephosphorylation of a glycerol-3-phosphate backbone. Specifically, glycerol-3-phosphate is acylated by GPAT resulting in lysophosphatidic acid which is acylated by AGPAT to form phosphatidic acid. Phosphatidic acid is dephosphosphorylated by phosphatidic acid phosphatase, otherwise known as lipins to yield diacylglycerol. In the final step DAG is acylated by DGAT to form triglyceride. There are several isoforms of each enzyme class. Interestingly, some, but not all isoforms are induced by glucocorticoids in mice [34].

In contrast to lipolysis, lipogenesis is activated by insulin in times of nutrient excess. One might expect that in this case, GCs might be inhibitory but experimental evidence said otherwise. Despite the low levels in the fed state, GCs have been shown to potentiate the lipogenic function of insulin by regulating the expression of some genes in the pathway [61]. This synergistic response in lipogenesis is shown to be important also in adrenalectomised rats [62] and human adipocytes [63].

The synthesis of fatty acids from glucose starts with acetyl-CoA, the product of the action of pyruvate dehydrogenase on the final metabolite of the glycolytic pathway. Acetyl-CoA carboxylase (ACC) then converts acetyl-CoA to malonyl-CoA which is in turn the substrate for fatty acid synthase (FASN), a rate-determining enzyme of lipogenesis. GCs have been reported, both *in vivo* and *in vitro*, to regulate the expression of ACC and FASN [44, 61, 64, 65]. GR-binding regions in or nearby the genes coding for these enzymes have been identified in both mouse and human genome [34, 63, 66]. In one study, mRNA levels of ACC and FASN in a human adipocyte cell line are shown to be induced by GC treatment [64]. The lipogenic program, however, is decreased and could only be increased by addition of insulin which supports previous reports on the GC-insulin synergism [62, 63].

GCs are known to affect the oxidation of fatty acids in both adipose tissues and the liver. In the former, one proposed mechanism is via the induction of expression of Tribbles-homologue 3 (TRB3) by GCs. TRB3 expression is induced in adipose tissues during fasting and over-expression of the protein resulted in increased fatty acid oxidation and protection of mice against diet-induced obesity [67]. In contrast to pro-lipogenic effect of GCs, TRB3 has been shown to promote ubiquitin-mediated degradation of ACC explaining how the oxidative pathway could have been favored [67]. In a similar report but in chronic lymphocytic leukemia (CLL) cells where resistance has been observed with GC treatments [68], dexamethasone is found to induce the expression of PPAR-α (alpha) and δ (delta) [69]. PPARα and δ are known to regulate fatty acid oxidation [70]. Whether the same mechanism works in adipose tissues and the liver remains to be investigated. Nevertheless, all these reports prove that GCs play an important role in the regulation of lipid homeostasis in adipose tissues.

The Hepatic Functions of GCs

The liver is a central metabolic organ involved in the control of mammalian glucose and lipid homeostasis [71]. Genome-wide analysis of GC-regulated target gene networks has shown that the GR controls many aspects of hepatic energy metabolism. More than 50 genes seem to be direct, regulatory targets of GC action. In many cases, the GR functionally interacts with other transcription factors to control specific genetic networks in the liver [72], among which only few have been characterized.

Control of Hepatic Triglyceride Metabolism and VLDL by GCs

Besides ACC and FASN, other enzymes involved in lipogenesis and TG synthesis have been reported to be regulated by GCs in the liver. Stearoyl-CoA desaturase (SCD1/2), glycerol-3-phosphate acyltransferase (GPAT3/4), 1-acylglycerol-3-phosphate acyltransferase (AGPAT2), lipin 1 (LPIN1), and diacylglyceride acyltransferase (DGAT1/2) are all implicated to be regulated directly or indirectly by the GR [73–75]. As in adipose tissues, GC/GR pro-lipogenic activity in the liver seems to require insulin signaling [74, 76]. This process is possibly important in the development of hepatic steatosis prior to the onset of insulin resistance in Metabolic Syndrome [8]. Hepatic steatosis is, in fact, a prominent negative side-effect of long-term GCs treatment [39]. In accordance to this, liver specific GR loss-of-function reduces TG levels in normal [77] as well as diabetic [78] mice and reduces accumulation of TG in liver during liver regeneration [28].

Several mechanisms have been proposed on how GCs promote TG synthesis and hepatic steatosis one being the inhibition of the hairy and enhancer of split 1 (HES1) repressor by GCs [78]. GR loss-of-function manipulations in the liver and hepatocytes resulted in up-regulation of HES1 that increases expression of pancreatic lipase (PNL) and pancreatic lipase-related protein (PLRP2), proteins involved in the "lipase arm" of GR-dependent TG metabolism. Both PNL and PLRP2 are repressed by GC-mediated HES1 inhibition thereby explaining hepatic triglyceride accumulation. Furthermore, liver HES1 expression is repressed in mouse models of hepatic steatosis and overexpressing HES1 in these models partially reverses the steatotic phenotype [78]. Another possible mechanism by which GCs control triglyceride turnover in the liver is via carboxyesterase 1d (CES1D/triacylglycerol hydrolase, TGH). TGH is believed to be involved in triglyceride lipolysis of fats droplets in hepatocytes [75]. Just like PNL and PLRP2, TGH mRNA levels are repressed upon dexamethasone treatment which could be explained by GC-induced destabilization of the mRNA at the 3′-UTR [75].

The composite function of the GR in concert with other transcription factors which could be a repressor or activator adds additional layer to the role of GCs in controlling hepatic lipid metabolism. The liver X receptor (LXR), for example, is known to interact with the GR [79] and it has been reported that mice lacking LXR-β (beta) has reduced GC-induced triglyceride accumulation [80]. The same phenotype is observed in mice where MED1 is specifically inhibited in the liver ($Med1\Delta^{liver}$) [81]. MED1 is a co-activator subunit of the Mediator complex which in turn participates in RNA Pol II-dependent transcription [82]. In dexamethasone-treated $Med1\Delta^{liver}$ mice, levels of fatty acid oxidation genes—short-chain and medium-chain acyl-CoA dehydrogenase (SCAD & MCAD)—are normalized compared to wild type mice where the genes are repressed by GC treatment [81]. The mechanism could also involve PPARγ which is known to be induced by GCs [27]. In a follow-up study in the same $Med1\Delta^{liver}$ mice where PPARγ is inhibited with an adenovirus, mice are protected from high-fat diet induced hepatic steatosis with impaired induction of lipogenic genes, adiponectin, and lipid droplet-associated genes [83].

Apart from potential direct effects on the enzymes of DNL and TG synthesis, GCs can regulate expression of secreted factors that can likewise affect lipid load in hepatocytes. As mentioned above, GCs induce secretion of ANGPTL4 from the liver and adipose tissues. ANGPTL4, with its pro-lipolytic activity, increases NEFA being released in the circulation which can serve as substrate for TG synthesis in the liver. Indeed, ANGPTL4 null mice are protected from GC-induced hepatic steatosis and hyperlipidemia [45]. Another intriguing recent study has shown that osteoblast specific GR knockout reduces the negative metabolic side effects of chronic GC treatments probably involving the bone secretion of osteocalcin or the bone γ-carboxyglutamate protein (BGALP) which has been shown to inhibit adiposity and hepatic steatosis [84].

The liver, aside from being at the receiving end for NEFA, can transport lipids to peripheral tissues via VLDL. The assembly and secretion of VLDL is a complicated process that depends not only on substrate availability but also on co-translational regulation of ApoB and the microsomal triglyceride transfer protein (MTTP); re-uptake of under-lipidated, immature particles; and the status of the intracellular secretory pathway [85]. Both *in vitro* and *in vivo* studies done on primary hepatocytes and dexamethasone-treated mice, respectively, showed increased VLDL triglycerides and plasma concentration of HDL particles [86–88]. Promotion of VLDL secretion by GCs is possibly due to increased production and stabilization of ApoB together with increased triglyceride synthesis [89]. Similarly, increased VLDL-TG production resulting in hypertriglyceridemia is observed in rats treated with dexamethasone for 2 weeks [90]. The rats also had impaired adipose tissue LPL activity which contributes to hypertriglyceridemia. In human cases of Cushing's syndrome, the rate of VLDL production, and consequently VLDL and LDL levels, is found to be elevated and peripheral clearance is unaffected [91]. Besides VLDL secretion, the receptor-mediated clearance of the remnants of triglyceride-rich lipoproteins (TRLs: chylomicrons, VLDL, IDL, and LDL) is also possibly regulated by GCs. It has been reported that the levels of these remnant particles are elevated in conditions of metabolic syndrome where hypertriglyceridemia and HPA hyperactivity are likewise observed [7, 92]. The major remnant receptors include the lipolysis-stimulated lipoprotein receptor (LSR), LDL receptor (LDLR), LDLR-related protein (LRP), and heparan sulfate proteoglycans (HSPGs) [7, 93]. With the exception of LDLR which has been shown to be down-regulated by GC treatment [94], no other studies have been done on how GC or the GR can affect receptor-mediated clearance of TRL remnant particles that can then impact on systemic triglyceride levels.

Effects of GCs on Cholesterol and Bile Acid Homeostasis

Cholesterol is an important lipid metabolite serving as a component of biomembranes and as precursor for the synthesis of steroids. Cholesterol transport in the body is largely dependent on LDL and HDL particles. Both interact with its respective receptors enabling endocytosis of the entire particle. Besides VLDL and TRL remnants, GCs are also shown to elevate HDL in the plasma of rats which can be attributed to the inhibition of hepatic lipase (HL) and activation of lecithin-cholesterol acyltransferase (LCAT) activity [95]. Similar results are reported in dexamethasone-treated livers of adrenalectomized rats [88], in overweight and obese individuals [96], and persons who received prednisone for 1 month as part of rheumatic disease management [97]. Furthermore, GCs are shown to increase transcriptional activity of apolipoprotein A-I (ApoA1), the major protein component of HDL particles which is important for the transport of cholesterol from the peripheral tissues to the liver [98]. A dose-dependent effect of dexamethasone is also observed in the hepatocyte binding capacity of HDL in dexamethasone-treated adrenalectomized rats [99]. The HDL scavenger receptor (SCRAB1/SRB1) promotes glucocorticoid production by the adrenal gland [100] particularly in response to inflammation and bacterial endotoxins [101]. Furthermore, adrenal-specific SRB1 deficiency resulted into glucocorticoid insufficiency and lower VLDL and LDL levels suggesting potential cross-talk between the liver and adrenal glands [102]. In case of LDL, reports on the effects of GCs on its metabolism have been elusive. The negative effects of GCs on hepatic LDLR levels [94] enables one to speculate the pathophysiologic consequences of elevated GC levels but the limited, often contradicting, reports warrant further studies.

In the field of cholesterol handling by the liver, hepatocyte-specific GR loss-of-function reduced the serum hypercholesterolemia in obese mice [78]. This is also accompanied by higher hepatic expression of the sterol regulatory element binding protein 2 (SREBP2) and increased liver cholesterol levels suggesting functional role of the GR in hepatic cholesterol homeostasis. In another study, as well as Addison's patients, serum bile acids were elevated in the fasted state and systemic bile acid homeostasis was disrupted upon the fasted-fed transition [103]. This abnormality in bile acid homeostasis can be attributed to lower expression of the major basolateral hepatocyte bile acid transporter, Na^+-taurocholate transport protein (NTCP/SLC10A1) resulting in faulty trans-hepatic bile acid recycling [103]. The gene coding for NTCP is shown to be a direct GR target based on avidin-biotin DNA-binding (ABCD) and chromatin immunoprecipitation (ChIP) assays and studies in mice with mutant GR where DNA-binding and dimerization function was lost (GR^{dim}) [103, 104]. The impaired bile acid uptake likewise exacerbated development of gallstones in mice with hepatocyte-specific GR knockdown [103]. Furthermore, the mice did not gain as much weight as the control owing to reduced fat mass which could be due to lower feed efficiency brought by ineffective lipid digestion and absorption [103]. The findings though, are contradictory to what is observed in Cushing's patients who have elevated serum bile acid and in a separate study where liver-specific GR knockdown results in lower serum bile acid due to inhibition of FXR activity which then blunts bile acid synthesis [105]. Further studies are therefore required to clarify these discrepancies.

GC-Dependent Lipid Processes in Muscles and Macrophages

With the exception of macrophages and smooth muscles, little attention has been given to how glucocorticoids affect lipid metabolism in other tissues. In muscles, most studies have focused on the effects of GCs on protein metabolism. With regards to lipids, dexamethasone has been shown to promote intramuscular fat accumulation in chicken by AMPK inhibition and mTOR activation [106]. The process is aggravated by saturated fats and alleviated by unsaturated fatty acids. This is contradictory to what has been reported in C2C12 myotubes where dexamethasone treatment resulted into dose-dependent reduction of expression of lipogenic genes (FASN, ACC, and GPAT) which can be reversed by the GR antagonist RU486 [107]. It has also been reported that GCs can induce excessive NEFA oxidation that can lead into muscle insulin resistance [108] and perturbation of equilibrium between fats and glucose as energy sources [109]. In rats, GC-induced insulin resistance has been shown to increase serum insulin and NEFA levels and decrease oxidative phosphorylation by producing ATP from phosphocreatine [110]. How GC/GR signaling exactly orchestrates these entire events in skeletal muscles, with the possibility of inter-organ cross-talk, remains to be studied. In smooth muscles, dexamethasone has been demonstrated to increase the rate of cholesteryl ester formation in cultured human smooth muscle cells [111] and *in vivo* [112] which has implications on the atherogenic effect of GCs. But whether the mechanism of GC-induced lipid oxidation in muscle is similar as in fats and the liver and how GCs promote cholesteryl ester accumulation in smooth muscles remain to be investigated. In macrophages, increased GC-dependent, esterification of cholesterol is likewise observed [112, 113], which provide a unifying theme by which GCs can adversely affect atherogenesis. Furthermore, it has been demonstrated also in macrophages that dexamethasone can decrease both mRNA and protein levels of the cholesterol transporter ATP-binding cassette transporter-A1 (ABCA1) and that apoA1-mediated cholesterol efflux is impeded [113] which would further support the hypothesis of GC-associated cardiovascular risk.

Concluding Remarks and Outlook

Apart from the peripheral functions of the GC/GR axis in metabolic control (Fig. 6.2), recent studies have established further ties between GC/GR signaling and regulatory functions of the central nervous system, osteoblasts as well as intestinal cells in energy homeostasis and lipid handling.

Together, these studies have substantially broadened the spectrum of GC target organs in energy homeostasis and revealed unexpected communication routes between peripheral and/or central organ compartments. GC/GR-dependent endocrine control beyond the classical GC-mediated pathways therefore represents a largely unexplored area of research that holds the promise for exciting discoveries in endocrine circuitry in the future. It can be anticipated that recent advances in –OMICS technologies and systems biology will provide substantial support for our further understanding of endocrine GC/GR signaling and its impact on metabolic health and disease.

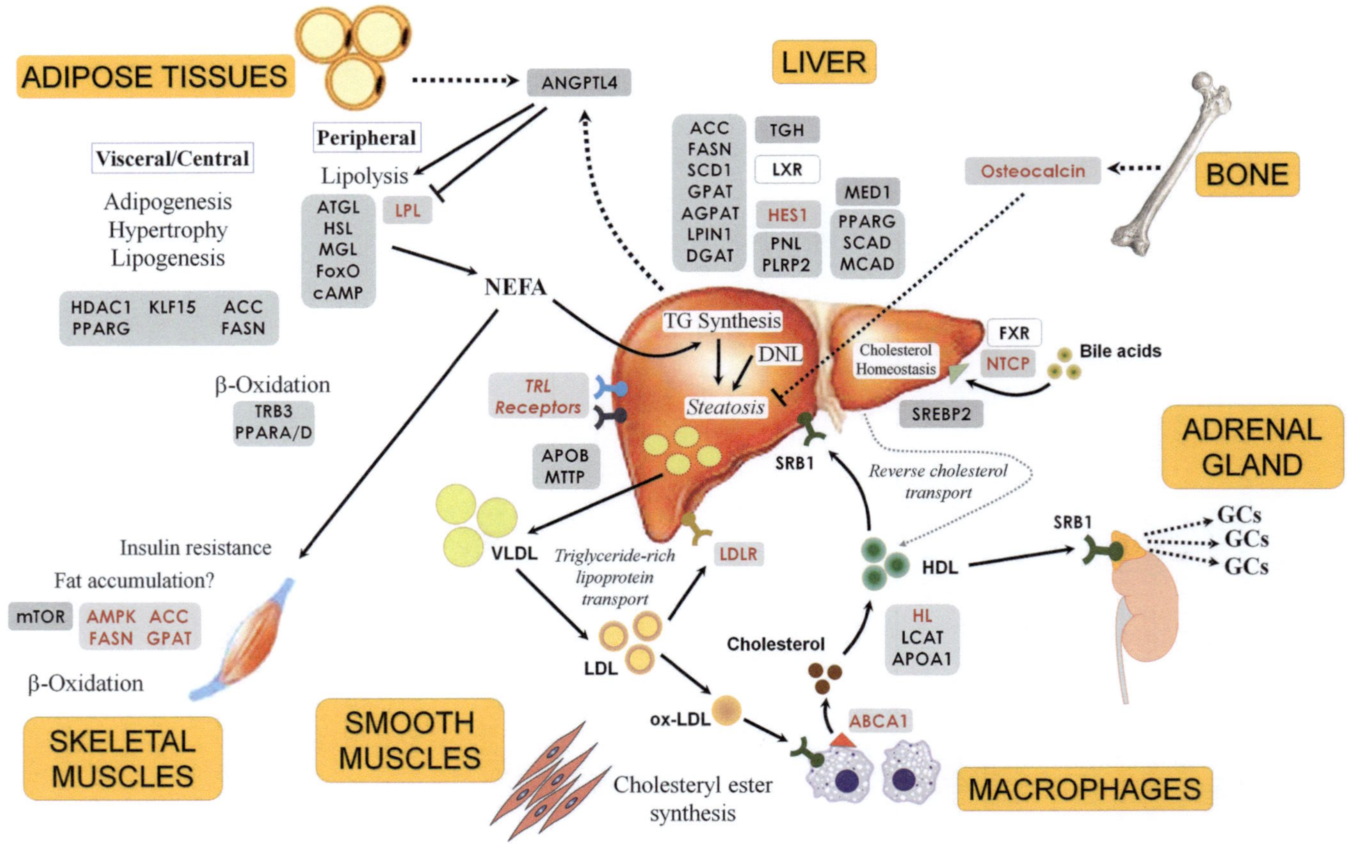

Fig. 6.2 GC/GR targets in lipid metabolism. Overview of lipid metabolic processes regulated by the GC/GR signaling in adipose tissues, the liver, bone, muscles, macrophages, and adrenal glands. Genes regulated by GCs in each metabolic process are listed in square-round boxes: genes in *black* texts are up-regulated; genes in *red* texts are inhibited by GCs

References

1. Nguyen P, Leray V, Diez M, et al. Liver lipid metabolism. J Anim Physiol Anim Nutr (Berl). 2008;92(3):272–83.
2. Werner A, Kuipers F, Verkade HJ. Fat absorption and lipid metabolism in cholestasis. Landes Biosci; 2000-2013.
3. Mead JR, Irvine SA, Ramji DP. Lipoprotein lipase: structure, function, regulation, and role in disease. J Mol Med (Berl). 2002;80(12):753–69.
4. Young SG, Davies BS, Voss CV, Gin P, Weinstein MM, Tontonoz P, Reue K, Bensadoun A, Fong LG, Beigneux AP. GPIHBP1, an endothelial cell transporter for lipoprotein lipase. J Lipid Res. 2011;52(11):1869–84.
5. Williams KJ, Fisher EA. Globular warming: how fat gets to the furnace. Nat Med. 2011;17(2):157–9.
6. Ginsberg HN. Lipoprotein physiology in nondiabetic and diabetic states. Relationship to atherogenesis. Diabetes Care. 1991;14(9):839–55.
7. Narvekar P, Berriel Diaz M, Krones-Herzig A, et al. Liver-specific loss of lipolysis-stimulated lipoprotein receptor triggers systemic hyperlipidemia in mice. Diabetes. 2009;58(5): 1040–9.
8. Choi SH, Ginsberg HN. Increased very low density lipoprotein (VLDL) secretion, hepatic steatosis, and insulin resistance. Trends Endocrinol Metab. 2011;22(9):353–63.
9. Huff MW. Dietary cholesterol, cholesterol absorption, postprandial lipemia and atherosclerosis. Can J Clin Pharmacol. 2003;10 Suppl A:26A–32A.
10. Mathews CK, van Holde KE, Ahern KG. Biochemistry. 3rd ed. Upper Saddle River, NJ: Prentice Hall; 1999.
11. Russell DW. Cholesterol biosynthesis and metabolism. Cardiovasc Drugs Ther. 1992;6(2): 103–10.
12. Desvergne B, Michalik L, Wahli W. Transcriptional regulation of metabolism. Physiol Rev. 2006;86(2):465–514.
13. Grundy SM, Cleeman JI, Daniels SR, et al. Diagnosis and management of the metabolic syndrome: an American Heart Association/National Heart, Lung, and Blood Institute scientific statement. Curr Opin Cardiol. 2006;21(1):1–6.
14. Arnaldi G, Mancini T, Tirabassi G, Trementino L, Boscaro M. Advances in the epidemiology, pathogenesis, and management of Cushing's syndrome complications. J Endocrinol Invest. 2012;35(4):434–48.
15. Anagnostis P, Athyros VG, Tziomalos K, Karagiannis A, Mikhailidis DP. Clinical review: the pathogenetic role of cortisol in the metabolic syndrome: a hypothesis. J Clin Endocrinol Metab. 2009;94(8):2692–701.
16. Masuzaki H, Paterson J, Shinyama H, et al. A transgenic model of visceral obesity and the metabolic syndrome. Science. 2001;294(5549):2166–70.
17. Viengchareun S, Zennaro MC, Pascual-Le Tallec L, Lombes M. Brown adipocytes are novel sites of expression and regulation of adiponectin and resistin. FEBS Lett. 2002;532(3): 345–50.
18. Fasshauer M, Klein J, Neumann S, Eszlinger M, Paschke R. Hormonal regulation of adiponectin gene expression in 3T3-L1 adipocytes. Biochem Biophys Res Commun. 2002;290(3):1084–9.
19. Masuzaki H, Yamamoto H, Kenyon CJ, et al. Transgenic amplification of glucocorticoid action in adipose tissue causes high blood pressure in mice. J Clin Invest. 2003;112(1): 83–90.
20. Gregoire FM, Smas CM, Sul HS. Understanding adipocyte differentiation. Physiol Rev. 1998;78(3):783–809.
21. Pantoja C, Huff JT, Yamamoto KR. Glucocorticoid signaling defines a novel commitment state during adipogenesis in vitro. Mol Biol Cell. 2008;19(10):4032–41.

22. Hauner H, Schmid P, Pfeiffer EF. Glucocorticoids and insulin promote the differentiation of human adipocyte precursor cells into fat cells. J Clin Endocrinol Metab. 1987;64(4):832–5.
23. Wiper-Bergeron NWD, Pope L, Schild-Poulter C, Hache RJ. Stimulation of preadipocyte differentiation by steroid through targeting of an HDAC1 complex. EMBO J. 2003;22(9):2135–45.
24. Floyd ZE, Stephens JM. STAT5A promotes adipogenesis in nonprecursor cells and associates with the glucocorticoid receptor during adipocyte differentiation. Diabetes. 2003;52(2):308–14.
25. Liu Y, Sun WL, Sun Y, Hu G, Ding GX. Role of 11-beta-hydroxysteroid dehydrogenase type 1 in differentiation of 3T3-L1 cells and in rats with diet-induced obesity. Acta Pharmacol Sin. 2006;27(5):588–96.
26. Liu Y, Yan C, Wang Y, et al. Liver X receptor agonist T0901317 inhibition of glucocorticoid receptor expression in hepatocytes may contribute to the amelioration of diabetic syndrome in db/db mice. Endocrinology. 2006;147(11):5061–8.
27. Wu Z, Bucher NL, Farmer SR. Induction of peroxisome proliferator-activated receptor gamma during the conversion of 3T3 fibroblasts into adipocytes is mediated by C/EBPbeta, C/EBPdelta, and glucocorticoids. Mol Cell Biol. 1996;16(8):4128–36.
28. Shteyer E, Liao Y, Muglia LJ, Hruz PW, Rudnick DA. Disruption of hepatic adipogenesis is associated with impaired liver regeneration in mice. Hepatology. 2004;40(6):1322–32.
29. Asada M, Rauch A, Shimizu H, et al. DNA binding-dependent glucocorticoid receptor activity promotes adipogenesis via Kruppel-like factor 15 gene expression. Lab Invest. 2011;91(2):203–15.
30. Rask E, Olsson T, Soderberg S, et al. Tissue-specific dysregulation of cortisol metabolism in human obesity. J Clin Endocrinol Metab. 2001;86(3):1418–21.
31. Livingstone DE, Jones GC, Smith K, et al. Understanding the role of glucocorticoids in obesity: tissue-specific alterations of corticosterone metabolism in obese Zucker rats. Endocrinology. 2000;141(2):560–3.
32. Baudrand R, Carvajal CA, Riquelme A, et al. Overexpression of 11beta-hydroxysteroid dehydrogenase type 1 in hepatic and visceral adipose tissue is associated with metabolic disorders in morbidly obese patients. Obes Surg. 2010;20(1):77–83.
33. Peckett AJ, Wright DC, Riddell MC. The effects of glucocorticoids on adipose tissue lipid metabolism. Metabolism. 2011;60(11):1500–10.
34. Yu CY, Mayba O, Lee JV, et al. Genome-wide analysis of glucocorticoid receptor binding regions in adipocytes reveal gene network involved in triglyceride homeostasis. PLoS One. 2010;5(12):e15188.
35. Campbell JE, Peckett AJ, D'Souza AM, Hawke TJ, Riddell MC. Adipogenic and lipolytic effects of chronic glucocorticoid exposure. Am J Physiol Cell Physiol. 2011;300(1):C198–209.
36. Slavin BG, Ong JM, Kern PA. Hormonal regulation of hormone-sensitive lipase activity and mRNA levels in isolated rat adipocytes. J Lipid Res. 1994;35(9):1535–41.
37. Villena JA, Roy S, Sarkadi-Nagy E, Kim KH, Sul HS. Desnutrin, an adipocyte gene encoding a novel patatin domain-containing protein, is induced by fasting and glucocorticoids: ectopic expression of desnutrin increases triglyceride hydrolysis. J Biol Chem. 2004;279(45):47066–75.
38. Ebbert JO, Jensen MD. Fat depots, free fatty acids, and dyslipidemia. Nutrients. 2013;5(2):498–508.
39. Vegiopoulos A, Herzig S. Glucocorticoids, metabolism and metabolic diseases. Mol Cell Endocrinol. 2007;275(1–2):43–61.
40. Wang JC, Gray NE, Kuo T, Harris CA. Regulation of triglyceride metabolism by glucocorticoid receptor. Cell Biosci. 2012;2(1):19.
41. Macfarlane DP, Forbes S, Walker BR. Glucocorticoids and fatty acid metabolism in humans: fuelling fat redistribution in the metabolic syndrome. J Endocrinol. 2008;197(2):189–204.
42. Carmen GY, Victor SM. Signalling mechanisms regulating lipolysis. Cell Signal. 2006;18(4):401–8.

43. Holm C. Molecular mechanisms regulating hormone-sensitive lipase and lipolysis. Biochem Soc Trans. 2003;31(Pt 6):1120–4.
44. Xu C, He J, Jiang H, et al. Direct effect of glucocorticoids on lipolysis in adipocytes. Mol Endocrinol. 2009;23(8):1161–70.
45. Koliwad SK, Kuo T, Shipp LE, et al. Angiopoietin-like 4 (ANGPTL4, fasting-induced adipose factor) is a direct glucocorticoid receptor target and participates in glucocorticoid-regulated triglyceride metabolism. J Biol Chem. 2009;284(38):25593–601.
46. van Raalte DH, Brands M, Serlie MJ, et al. Angiopoietin-like protein 4 is differentially regulated by glucocorticoids and insulin in vitro and in vivo in healthy humans. Exp Clin Endocrinol Diabetes. 2012;120(10):598–603.
47. Le Jan S, Amy C, Cazes A, et al. Angiopoietin-like 4 is a proangiogenic factor produced during ischemia and in conventional renal cell carcinoma. Am J Pathol. 2003;162(5):1521–8.
48. Cazes A, Galaup A, Chomel C, et al. Extracellular matrix-bound angiopoietin-like 4 inhibits endothelial cell adhesion, migration, and sprouting and alters actin cytoskeleton. Circ Res. 2006;99(11):1207–15.
49. Mandard S, Zandbergen F, van Straten E, et al. The fasting-induced adipose factor/angiopoietin-like protein 4 is physically associated with lipoproteins and governs plasma lipid levels and adiposity. J Biol Chem. 2006;281(2):934–44.
50. Lichtenstein L, Berbee JF, van Dijk SJ, et al. Angptl4 upregulates cholesterol synthesis in liver via inhibition of LPL- and HL-dependent hepatic cholesterol uptake. Arterioscler Thromb Vasc Biol. 2007;27(11):2420–7.
51. Gray NE, Lam LN, Yang K, Zhou AY, Koliwad S, Wang JC. Angiopoietin-like 4 (Angptl4) protein is a physiological mediator of intracellular lipolysis in murine adipocytes. J Biol Chem. 2012;287(11):8444–56.
52. Mizutani N, Ozaki N, Seino Y, et al. Reduction of insulin signaling upregulates angiopoietin-like protein 4 through elevated free fatty acids in diabetic mice. Exp Clin Endocrinol Diabetes. 2012;120(3):139–44.
53. Birkenhager JC, Timmermans HA, Lamberts SW. Depressed plasma FFA turnover rate in Cushing's syndrome. J Clin Endocrinol Metab. 1976;42(1):28–32.
54. Saunders J, Hall SE, Sonksen PH. Glucose and free fatty acid turnover in Cushing's syndrome. J Endocrinol Invest. 1980;3(3):309–11.
55. Miyoshi H, Shulman GI, Peters EJ, Wolfe MH, Elahi D, Wolfe RR. Hormonal control of substrate cycling in humans. J Clin Invest. 1988;81(5):1545–55.
56. Gravholt CH, Dall R, Christiansen JS, Moller N, Schmitz O. Preferential stimulation of abdominal subcutaneous lipolysis after prednisolone exposure in humans. Obes Res. 2002; 10(8):774–81.
57. Johnston DG, Gill A, Orskov H, Batstone GF, Alberti KG. Metabolic effects of cortisol in man—studies with somatostatin. Metabolism. 1982;31(4):312–7.
58. Djurhuus CB, Gravholt CH, Nielsen S, et al. Effects of cortisol on lipolysis and regional interstitial glycerol levels in humans. Am J Physiol Endocrinol Metab. 2002;283(1):E172–7.
59. Samra JS, Clark ML, Humphreys SM, MacDonald IA, Bannister PA, Frayn KN. Effects of physiological hypercortisolemia on the regulation of lipolysis in subcutaneous adipose tissue. J Clin Endocrinol Metab. 1998;83(2):626–31.
60. Ferramosca A, Zara V. Modulation of hepatic steatosis by dietary fatty acids. World J Gastroenterol. 2014;20(7):1746–55.
61. Hillgartner FB, Salati LM, Goodridge AG. Physiological and molecular mechanisms involved in nutritional regulation of fatty acid synthesis. Physiol Rev. 1995;75(1):47–76.
62. Williams BH, Berdanier CD. Effects of diet composition and adrenalectomy on the lipogenic responses of rats to starvation-refeeding. J Nutr. 1982;112(3):534–41.
63. Wang Y, Jones Voy B, Urs S, et al. The human fatty acid synthase gene and de novo lipogenesis are coordinately regulated in human adipose tissue. J Nutr. 2004;134(5):1032–8.
64. Gathercole LL, Morgan SA, Bujalska IJ, Hauton D, Stewart PM, Tomlinson JW. Regulation of lipogenesis by glucocorticoids and insulin in human adipose tissue. PLoS One. 2011; 6(10):e26223.

65. Sul HS, Wang D. Nutritional and hormonal regulation of enzymes in fat synthesis: studies of fatty acid synthase and mitochondrial glycerol-3-phosphate acyltransferase gene transcription. Annu Rev Nutr. 1998;18:331–51.
66. Lu Z, Gu Y, Rooney SA. Transcriptional regulation of the lung fatty acid synthase gene by glucocorticoid, thyroid hormone and transforming growth factor-beta 1. Biochim Biophys Acta. 2001;1532(3):213–22.
67. Qi L, Heredia JE, Altarejos JY, et al. TRB3 links the E3 ubiquitin ligase COP1 to lipid metabolism. Science. 2006;312(5781):1763–6.
68. Hala M, Hartmann BL, Bock G, Geley S, Kofler R. Glucocorticoid-receptor-gene defects and resistance to glucocorticoid-induced apoptosis in human leukemic cell lines. Int J Cancer. 1996;68(5):663–8.
69. Tung S, Shi Y, Wong K, et al. PPARalpha and fatty acid oxidation mediate glucocorticoid resistance in chronic lymphocytic leukemia. Blood. 2013;122(6):969–80.
70. Grygiel-Gorniak B. Peroxisome proliferator-activated receptors and their ligands: nutritional and clinical implications—a review. Nutr J. 2014;13:17.
71. van den Berghe G. The role of the liver in metabolic homeostasis: implications for inborn errors of metabolism. J Inherit Metab Dis. 1991;14(4):407–20.
72. Le Phuc P, Friedman JR, Schug J, et al. Glucocorticoid receptor-dependent gene regulatory networks. PLoS Genet. 2005;1(2):e16.
73. Legrand P, Catheline D, Hannetel JM, Lemarchal P. Stearoyl-CoA desaturase activity in primary culture of chicken hepatocytes. Influence of insulin, glucocorticoid, fatty acids and cordycepin. Int J Biochem. 1994;26(6):777–85.
74. Dich J, Bro B, Grunnet N, Jensen F, Kondrup J. Accumulation of triacylglycerol in cultured rat hepatocytes is increased by ethanol and by insulin and dexamethasone. Biochem J. 1983;212(3):617–23.
75. Dolinsky VW, Douglas DN, Lehner R, Vance DE. Regulation of the enzymes of hepatic microsomal triacylglycerol lipolysis and re-esterification by the glucocorticoid dexamethasone. Biochem J. 2004;378(Pt 3):967–74.
76. Mangiapane EH, Brindley DN. Effects of dexamethasone and insulin on the synthesis of triacylglycerols and phosphatidylcholine and the secretion of very-low-density lipoproteins and lysophosphatidylcholine by monolayer cultures of rat hepatocytes. Biochem J. 1986; 233(1):151–60.
77. Opherk C, Tronche F, Kellendonk C, et al. Inactivation of the glucocorticoid receptor in hepatocytes leads to fasting hypoglycemia and ameliorates hyperglycemia in streptozotocin-induced diabetes mellitus. Mol Endocrinol. 2004;18(6):1346–53.
78. Lemke U, Krones-Herzig A, Berriel Diaz M, et al. The glucocorticoid receptor controls hepatic dyslipidemia through Hes1. Cell Metab. 2008;8(3):212–23.
79. Nader N, Ng SS, Wang Y, Abel BS, Chrousos GP, Kino T. Liver x receptors regulate the transcriptional activity of the glucocorticoid receptor: implications for the carbohydrate metabolism. PLoS One. 2012;7(3):e26751.
80. Patel R, Patel M, Tsai R, et al. LXRbeta is required for glucocorticoid-induced hyperglycemia and hepatosteatosis in mice. J Clin Invest. 2011;121(1):431–41.
81. Jia Y, Viswakarma N, Fu T, et al. Conditional ablation of mediator subunit MED1 (MED1/PPARBP) gene in mouse liver attenuates glucocorticoid receptor agonist dexamethasone-induced hepatic steatosis. Gene Expr. 2009;14(5):291–306.
82. Zhang Y, Xiaoli, Zhao X, Yang Y. The mediator complex and lipid metabolism. J Biochem Pharmacol Res. 2013;1(1):51–5.
83. Bai L, Jia Y, Viswakarma N, et al. Transcription coactivator mediator subunit MED1 is required for the development of fatty liver in the mouse. Hepatology. 2011;53(4):1164–74.
84. Brennan-Speranza TC, Henneicke H, Gasparini SJ, et al. Osteoblasts mediate the adverse effects of glucocorticoids on fuel metabolism. J Clin Invest. 2012;122(11):4172–89.
85. Sundaram M, Yao Z. Recent progress in understanding protein and lipid factors affecting hepatic VLDL assembly and secretion. Nutr Metab (Lond). 2010;7:35.

86. Martin-Sanz P, Vance JE, Brindley DN. Stimulation of apolipoprotein secretion in very-low-density and high-density lipoproteins from cultured rat hepatocytes by dexamethasone. Biochem J. 1990;271(3):575–83.
87. Duerden JM, Bartlett SM, Gibbons GF. Long-term maintenance of high rates of very-low-density-lipoprotein secretion in hepatocyte cultures. A model for studying the direct effects of insulin and insulin deficiency in vitro. Biochem J. 1989;263(3):937–43.
88. Cole TG, Wilcox HG, Heimberg M. Effects of adrenalectomy and dexamethasone on hepatic lipid metabolism. J Lipid Res. 1982;23(1):81–91.
89. Wang CN, Hobman TC, Brindley DN. Degradation of apolipoprotein B in cultured rat hepatocytes occurs in a post-endoplasmic reticulum compartment. J Biol Chem. 1995;270(42): 24924–31.
90. Bagdade JD, Yee E, Albers J, Pykalisto OJ. Glucocorticoids and triglyceride transport: effects on triglyceride secretion rates, lipoprotein lipase, and plasma lipoproteins in the rat. Metabolism. 1976;25(5):533–42.
91. Taskinen MR, Nikkila EA, Pelkonen R, Sane T. Plasma lipoproteins, lipolytic enzymes, and very low density lipoprotein triglyceride turnover in Cushing's syndrome. J Clin Endocrinol Metab. 1983;57(3):619–26.
92. Chan DC, Watts GF, Barrett PH, Mamo JC, Redgrave TG. Markers of triglyceride-rich lipoprotein remnant metabolism in visceral obesity. Clin Chem. 2002;48(2):278–83.
93. Foley EM, Gordts PL, Stanford KI, et al. Hepatic remnant lipoprotein clearance by heparan sulfate proteoglycans and low-density lipoprotein receptors depend on dietary conditions in mice. Arterioscler Thromb Vasc Biol. 2013;33(9):2065–74.
94. Hazra A, Pyszczynski NA, DuBois DC, Almon RR, Jusko WJ. Modeling of corticosteroid effects on hepatic low-density lipoprotein receptors and plasma lipid dynamics in rats. Pharm Res. 2008;25(4):769–80.
95. Jansen H, van Tol A, Auwerx J, Skretting G, Staels B. Opposite regulation of hepatic lipase and lecithin: cholesterol acyltransferase by glucocorticoids in rats. Biochim Biophys Acta. 1992;1128(2–3):181–5.
96. Wang X, Magkos F, Patterson BW, Reeds DN, Kampelman J, Mittendorfer B. Low-dose dexamethasone administration for 3 weeks favorably affects plasma HDL concentration and composition but does not affect very low-density lipoprotein kinetics. Eur J Endocrinol. 2012;167(2):217–23.
97. Ettinger WH, Klinefelter HF, Kwiterovitch PO. Effect of short-term, low-dose corticosteroids on plasma lipoprotein lipids. Atherosclerosis. 1987;63(2–3):167–72.
98. Taylor AH, Raymond J, Dionne JM, et al. Glucocorticoid increases rat apolipoprotein A-I promoter activity. J Lipid Res. 1996;37(10):2232–43.
99. Bocharov AV, Huang W, Vishniakova TG, et al. Glucocorticoids upregulate high-affinity, high-density lipoprotein binding sites in rat hepatocytes. Metabolism. 1995;44(6):730–8.
100. Temel RE, Trigatti B, DeMattos RB, Azhar S, Krieger M, Williams DL. Scavenger receptor class B, type I (SR-BI) is the major route for the delivery of high density lipoprotein cholesterol to the steroidogenic pathway in cultured mouse adrenocortical cells. Proc Natl Acad Sci U S A. 1997;94(25):13600–5.
101. Cai L, Ji A, de Beer FC, Tannock LR, van der Westhuyzen DR. SR-BI protects against endotoxemia in mice through its roles in glucocorticoid production and hepatic clearance. J Clin Invest. 2008;118(1):364–75.
102. Hoekstra M, van der Sluis RJ, Van Eck M, Van Berkel TJ. Adrenal-specific scavenger receptor BI deficiency induces glucocorticoid insufficiency and lowers plasma very-low-density and low-density lipoprotein levels in mice. Arterioscler Thromb Vasc Biol. 2013;33(2): e39–46.
103. Rose AJ, Berriel Diaz M, Reimann A, et al. Molecular control of systemic bile acid homeostasis by the liver glucocorticoid receptor. Cell Metab. 2011;14(1):123–30.
104. Eloranta JJ, Jung D, Kullak-Ublick GA. The human Na+-taurocholate cotransporting polypeptide gene is activated by glucocorticoid receptor and peroxisome proliferator-activated

receptor-gamma coactivator-1alpha, and suppressed by bile acids via a small heterodimer partner-dependent mechanism. Mol Endocrinol. 2006;20(1):65–79.

105. Lu Y, Zhang Z, Xiong X, et al. Glucocorticoids promote hepatic cholestasis in mice by inhibiting the transcriptional activity of the farnesoid X receptor. Gastroenterology. 2012;143(6): 1630–40. e1638.
106. Wang X, Wei D, Song Z, Jiao H, Lin H. Effects of fatty acid treatments on the dexamethasone-induced intramuscular lipid accumulation in chickens. PLoS One. 2012;7(5):e36663.
107. Morgan S, Gathercole L, Stewart S, Smith D, Tomlinson J. Impact of glucocorticoids upon lipogenesis and β-oxidation in skeletal muscle. Paper presented at Society for Endocrinology BES 20102010; Manchester, UK.
108. Guillaume-Gentil C, Assimacopoulos-Jeannet F, Jeanrenaud B. Involvement of non-esterified fatty acid oxidation in glucocorticoid-induced peripheral insulin resistance in vivo in rats. Diabetologia. 1993;36(10):899–906.
109. Venkatesan N, Davidson MB, Hutchinson A. Possible role for the glucose-fatty acid cycle in dexamethasone-induced insulin antagonism in rats. Metabolism. 1987;36(9):883–91.
110. Dumas JF, Bielicki G, Renou JP, et al. Dexamethasone impairs muscle energetics, studied by (31)P NMR, in rats. Diabetologia. 2005;48(2):328–35.
111. Petrichenko IE, Daret D, Kolpakova GV, Shakhov YA, Larrue J. Glucocorticoids stimulate cholesteryl ester formation in human smooth muscle cells. Arterioscler Thromb Vasc Biol. 1997;17(6):1143–51.
112. Stein O, Dabach Y, Hollander G, Ben-Naim M, Halperin G, Stein Y. Dexamethasone impairs cholesterol egress from a localized lipoprotein depot in vivo. Atherosclerosis. 1998;137(2): 303–10.
113. Ayaori M, Sawada S, Yonemura A, et al. Glucocorticoid receptor regulates ATP-binding cassette transporter-A1 expression and apolipoprotein-mediated cholesterol efflux from macrophages. Arterioscler Thromb Vasc Biol. 2006;26(1):163–8.

Chapter 7
Glucocorticoids and Skeletal Muscle

Sue C. Bodine and J. David Furlow

Abstract Glucocorticoids are known to regulate protein metabolism in skeletal muscle, producing a catabolic effect that is opposite that of insulin. In many catabolic diseases, such as sepsis, starvation, and cancer cachexia, endogenous glucocorticoids are elevated contributing to the loss of muscle mass and function. Further, exogenous glucocorticoids are often given acutely and chronically to treat inflammatory conditions such as asthma, chronic obstructive pulmonary disease, and rheumatoid arthritis, resulting in muscle atrophy. This chapter will detail the nature of glucocorticoid-induced muscle atrophy and discuss the mechanisms thought to be responsible for the catabolic effects of glucocorticoids on muscle.

Keywords Muscle atrophy • Protein synthesis • Proteolysis • Ubiquitin proteasome pathway • Gene transcription

Introduction

Skeletal muscle comprises ~40 % of body mass and is a major target of glucocorticoids (GC), whose actions primarily alter protein and glucose metabolism. Glucocorticoids are released endogenously in response to a variety of physically and psychologically stressful conditions including fasting, illness and exercise. The effect of elevated levels of circulating GCs on skeletal muscle is a decrease in

S.C. Bodine, Ph.D. (✉)
Department of Neurobiology, Physiology and Behavior, University of California, Davis, One Shields Avenue, Davis, CA 95616, USA

Department of Physiology and Membrane Biology, University of California, Davis, One Shields Avenue, Davis, CA 95616, USA
e-mail: scbodine@ucdavis.edu

J.D. Furlow, Ph.D.
Department of Neurobiology, Physiology and Behavior, University of California, Davis, One Shields Avenue, Davis, CA 95616, USA
e-mail: jdfurlow@ucdavis.edu

J.-C. Wang, C. Harris (eds.), *Glucocorticoid Signaling*, Advances in Experimental Medicine and Biology 872, DOI 10.1007/978-1-4939-2895-8_7

protein synthesis and an increase in proteolysis in order to generate an increase in free amino acids that can serve as substrates for gluconeogenesis in the liver and help to maintain blood glucose levels. Thus, acute increases in glucocorticoids are beneficial in that they maintain circulating glucose levels to support brain function under stressful conditions. However, when circulating glucocorticoid levels are elevated chronically due to a pathological condition or as a course of medical treatment the effects on skeletal muscle are negative causing loss of mass and weakness, as well as insulin resistance. In this chapter, our discussion will concentrate on the effects of elevated glucocorticoids on skeletal muscle mass and function, as the effects on skeletal muscle metabolism are discussed in detail in other chapters of this book.

Glucocorticoid Induced Muscle Atrophy

Glucocorticoids are known to have catabolic effects on skeletal muscle, either as an endocrine hormone released in response to stress or as a drug given to treat inflammation. Elevated endogenous glucocorticoids are a major component of Cushing's syndrome, as well as other diseases such as sepsis and cancer cachexia [1]. In addition, millions of adults are treated with glucocorticoid therapy for a variety of acute and chronic inflammatory conditions such as asthma, chronic obstructive pulmonary disease, rheumatoid arthritis, and neurological conditions such as spinal cord injury, brain tumors, and myasthenia gravis [2], and in an apparent paradox, muscular dystrophy (see below). In many cases, treatment with glucocorticoids can exceed 5 years leading to a variety of side effects including muscle wasting and weakness.

To demonstrate the catabolic effects of glucocorticoids on muscle, many studies have utilized healthy juvenile and adult rodents, and delivered a variety of synthetic glucocorticoids (e.g. dexamethasone, methylprednisolone, prednisolone, and triamcinolone acetonide) at relatively high doses on a daily basis for days to weeks. The effect of excess glucocorticoids on muscle mass is well characterized, and in general, is dependent on the fiber type composition of the muscle. Muscle fibers are classified into types based on their myosin heavy chain composition (I, IIa, IIx, or IIb) and metabolic properties (SO, slow oxidative; FOG, fast oxidative glycolytic; and FG, fast glycolytic) [3] (Fig. 7.1). Chronic, excess levels of GCs cause a loss of muscle mass that is greatest in muscles composed predominantly of type II, fast twitch fibers [4, 5]. At the individual fiber level, the loss of fiber cross-sectional area is greatest in the fast-twitch glycolytic fibers (FG, type IIb or IIx) and least in the slow-twitch oxidative fibers (SO, type I) [6–8]. In rat muscles of mixed fiber type composition, such as the medial gastrocnemius, atrophy usually follows the order (from highest to lowest) of FG>FOG>SO [6, 9]. Moreover, there can be regional differences in the degree of atrophy of the fast glycolytic fibers, in that the FG fibers in the predominantly "white" region of the medial gastrocnemius atrophy more than the FG fibers in the predominantly "red" region [7].

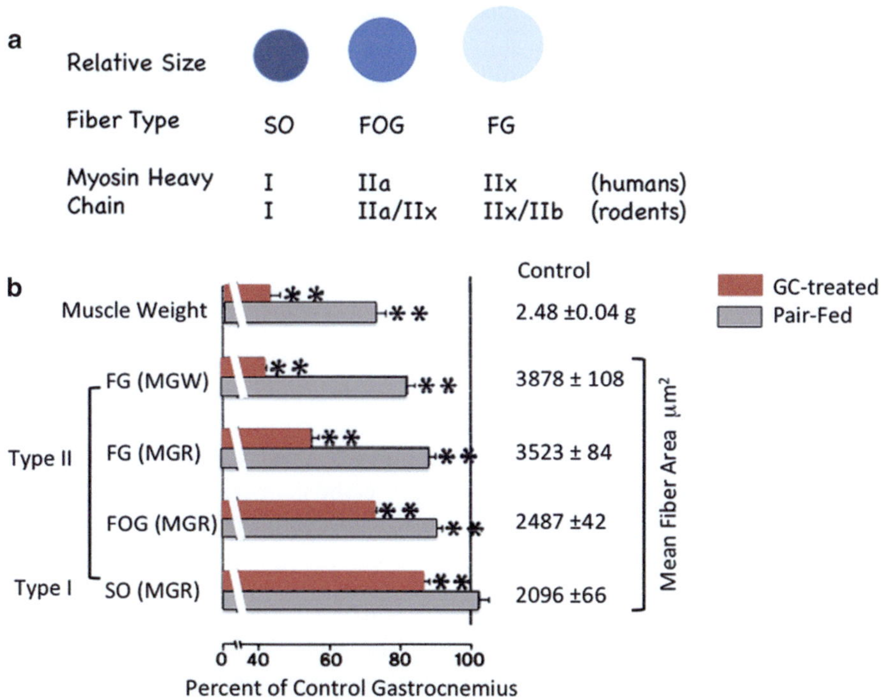

Fig. 7.1 *Fiber types and glucocorticoid-induced atrophy*. (**a**) Skeletal muscles are composed of fibers that express different myosin heavy chain isoforms (MHC) and have varying metabolic properties. There are four MHCs expressed in rodent skeletal muscles (I, IIa, IIx and IIb), while there are only three MHCs expressed in human skeletal muscle (I, IIa, and IIx). The metabolic profiles of fibers vary with respect to their capacity for oxidative metabolism and glycolysis. Fibers can be classified into three types based on their myosin ATPase activity, oxidative capacity and, glycolytic capacity: *SO* slow oxidative, *FOG* fast oxidative glycolytic, and *FG* fast glycolytic. The cross-sectional area of muscle fibers vary with the largest fibers being the FG fibers and the smallest being the SO fibers. (**b**) Data taken from Gardiner et al. [6] demonstrating fiber type specific atrophy following 6 weeks of glucocorticoid treatment (1 mg/kg triamcinoloneacetonide-21 phosphate) in male rats. Gastrocnemius muscle weights and mean areas of each fiber type in GC-treated and pair-fed rats are expressed as a percentage of control values. The control means ± SE are given to the right of the bars. MGR and MGW refer to regions of the medial gastrocnemius muscle containing high percentages of "*R*ed" (i.e. oxidative) and "*W*hite" (i.e. non-oxidative) fibers. Gardiner, PF Montanaro, G Simpson, DR Edgerton, VR Effects of glucocorticoid treatment and food restriction on rat hindlimb muscles. The American journal of physiology.1980; 238(2): E124–30

The extent to which oxidative fibers are affected by exogenous glucocorticoids depends in part on the dose and type of synthetic glucocorticoid used, and also on the specific muscle. For example, the fluorinated synthetic glucocorticoids (dexamethasone and triamcinolone acetonide) are more potent atrophy-inducing agents than the nonfluorinated preparations (prednisolone and predisone) [10, 11]. Additional factors to be taken into consideration when studying the effects of glucocorticoids on muscle mass are the sex of the animal and the animal species. In general, rats are more sensitive to the atrophy-inducing effects of glucocorticoids than mice. Whereas a dose of <0.5 mg/kg of dexamethasone will induce significant muscle atrophy in the rat over a 7-day period, doses in the range of 1–3 mg/kg are required to induce atrophy in the mouse. Furthermore, male and female rodents show different sensitivities to glucocorticoids, with male rodents requiring higher doses than female rodents to achieve similar amounts of atrophy [12]. The mechanisms responsible for the sex differences are unclear, but may relate to the ability of testosterone to protect against the catabolic effects of glucocorticoids [13, 14, 15].

Food restriction and starvation, which cause an increase in circulating glucocorticoids, also induce fiber-type specific atrophy [6, 8]. Supraphysiologic doses of synthetic glucocorticoids can cause a reduction in food intake, especially in rats, and thus the atrophy inducing effects of glucocorticoids are often compared to pair-fed animals (i.e. food intake matched to the glucocorticoid-treated animals). In general, the degree of body weight loss, or muscle atrophy, observed in response to pair-feeding is significantly less than that observed in response to glucocorticoid treatment [6, 7]. Thus, the mechanisms regulating muscle atrophy under glucocorticoid-treatment and food deprivation may overlap, but likely are not identical. Furthermore, glucocorticoids are thought to increase food intake in humans and still cause muscle atrophy. One indication that this may be the case is the observation that deletion of the E3 ubiquitin ligase, MuRF1, can spare muscle fibers from atrophy induced by glucocorticoid-treatment, but not nutritional deprivation [12].

Synthetic glucocorticoids are used to treat a variety of inflammatory diseases in humans and can induce muscle atrophy in both the acute and chronic phase of treatment. Glucocorticoid-induced myopathies are most often associated with fluorinated glucocorticoids, but can occur with all formulations depending on dose, length of treatment, and underlying diseases [11]. The risk of acquiring a glucocorticoid-induced myopathy is higher in the elderly and in patients with cancer, respiratory distress syndrome, and those who are physically inactive and in negative nitrogen balance at the onset of glucocorticoid treatment [1, 16]. Acute muscle atrophy associated with glucocorticoid treatment often occurs in the intensive care setting where high doses of glucocorticoids are given to patients who are immobilized, often are mechanically ventilated, given paralyzing agents, and have additional complications such as nutritional deficiencies and sepsis.

Acute muscle wasting associated with high doses of glucocorticoids can affect both respiratory and limb muscles [1, 17]. In humans, chronic glucocorticoid-induced weakness and wasting is generally associated with proximal muscles, especially the pelvic girdle muscles, but can also affect distal muscles [11]. Glucocorticoid-induced atrophy in humans is fiber type specific, as observed in animal models, being greatest in the type IIx, glycolytic, non-oxidative fibers [18, 19]. It should be noted that in human limb muscles the fast glycolytic fibers express the type IIx myosin heavy chain and not IIb, which is not found in human limb muscles (Fig. 7.1). Type II specific atrophy is also observed in diseases with elevated endogenous glucocorticoids, such as Cushing's syndrome and sepsis [20].

Glucocorticoids and Muscle Function

Excess glucocorticoids also cause a decrease in muscle strength that parallels the loss of fiber cross-sectional area. For example, patients with Cushing's syndrome and patients receiving glucocorticoid treatment have decreases in muscle strength [20–22]. On physical exam, a patient with chronic glucocorticoid excess (Cushing syndrome) will have difficulty rising from a chair without use of the hands, a test of proximal muscle function. The use of glucocorticoids as treatment of an inflammatory disease can often exacerbate the deleterious effects of the disease on muscle strength. For example, measurement of isometric and isokinetic muscle strength in rheumatoid arthritis patients found significantly greater decreases in strength in those patients treated with prednisone compared to those patients who received no treatment [23]. Furthermore, glucocorticoid-mediated muscle atrophy in patients with chronic obstructive pulmonary disease can negatively affect their pulmonary function due to decrease muscle (bellows) function.

Decreases in muscle force output are also found in animal disease models associated with increases in endogenous glucocorticoids, such as sepsis and critical care illness [24–26]. In some animals, exogenous glucocorticoid treatment produces comparable decreases in both muscle contractile force and fiber cross-sectional [9, 27]. Interestingly, while treatment of rodents with no underlying disease with a synthetic glucocorticoid, such as dexamethasone, causes a loss of muscle mass, especially in predominantly fast-twitch muscle, there is often no measurable decrease in maximum muscle force output leading to an increase in specific force output (i.e., force per cross-sectional area) [12, 28, 29]. This increase in specific force is apparent in limb muscles, especially fast-twitch muscles, but not in the diaphragm of juvenile rodents given synthetic glucocorticoids [8, 30]. In rodents, the mechanism responsible for the increase in specific force is unclear, but does not appear to be related to a change in sarcolemmal excitability or excitation-contraction coupling [29].

Glucocorticoids and Muscular Dystrophy

Duchenne muscular dystrophy (DMD) is the most prevalent muscular dystrophy occurring in 1 out of 3500 live male births, and caused by mutations in the dystrophin gene [31]. DMD is a severe progressive neuromuscular disease that presents with gait disturbances between the ages of 3 and 5 years and loss of ambulation by 12. There are no cures for DMD, but the use of glucocorticoid therapy (prednisone, prednisolone and deflazacort) has been shown to improve muscle strength during the first 6 months of usage followed by stabilization of the course of the disease for up to 2 years, after which there is a decline in function. There are many potential adverse effects of glucocorticoid therapy including osteoporosis, excessive weight gain, and behavioral abnormalities [32]. While there appear to be short-term benefits of glucocorticoid therapy in DMD patients, the long-term functional benefits are unclear [31]. A study performed on mdx mice showed that daily administration of prednisolone resulted in early (initial 50 days) improvements in muscle strength and motor coordination, but these benefits were lost after 100 days of continuous treatment [33]. Further, there was a deterioration of cardiac function and increased fibrosis of the heart with prolonged glucocorticoid treatment [34].

The mechanism by which glucocorticoids improve muscle strength in DMD patients is unknown, and thought to occur through a suppression of inflammation. DMD patients often show pseudohypertrophy of muscles, especially in the calf, which is thought to be due to inflammation. However, other immunosuppressive drugs that reduce inflammatory infiltrates in the muscles of DMD patients do not improve strength, as seen with prednisone. In mdx mice, treatment with prednisolone increases specific force in muscle while having no affect on muscle fiber size [35, 36]. The increase in specific force output is similar to what is seen in normal mice and rats given glucocorticoids. The mechanism by which glucocorticoids improve force output in DMD and mdx is still unclear and may be distinct from the anti-inflammatory actions of the glucocorticoids.

Glucocorticoids and Protein Turnover

It is well established that the overall effect of supraphysiologic levels of glucocorticoids on skeletal muscle is catabolic resulting in a loss of muscle mass, fiber cross-sectional, and force output. Loss of muscle protein occurs as the result of an inbalance between protein synthesis and breakdown, with net breakdown exceeding synthesis. The response of skeletal muscle to catabolic doses of exogenous glucocorticoids is both a decrease in protein synthesis and an increase in protein degradation. The extent to which protein synthesis and degradation are altered by exogenously delivered glucocorticoids is dependent on both the type of steroid used and the dose given.

Alteration in Protein Synthesis: The earliest changes in muscle protein metabolism following glucocorticoid administration appear to be related primarily to a decrease in the rate of protein synthesis, rather than an increase in protein breakdown. Protein synthesis is the process by which mRNA is translated on ribosomes into amino acids and can be divided into three distinct phases: initiation, elongation and termination [37]. Translation initiation is a key rate limiting process that is controlled at two key steps: (1) modulation of the binding of the initiator methionyl-tRNA (met-tRNAi) to the 40S ribosomal subunit to form the 43S preinitiation complex, and (2) the binding of the mRNA to the 43S preinitiation complex to form the 48S preinitiation complex. Regulation of these translation initiation steps is controlled through the phosphorylation of initiation factors and protein kinases that are controlled by signaling pathways downstream of phosphatidylinositol 3-kinase (PI3K) [38] (Fig. 7.2). In mammalian muscle, muscle mass is critically regulated by activation of the Ser/Thr protein kinase, the mammalian (or mechanical) target of rapamycin (mTOR),

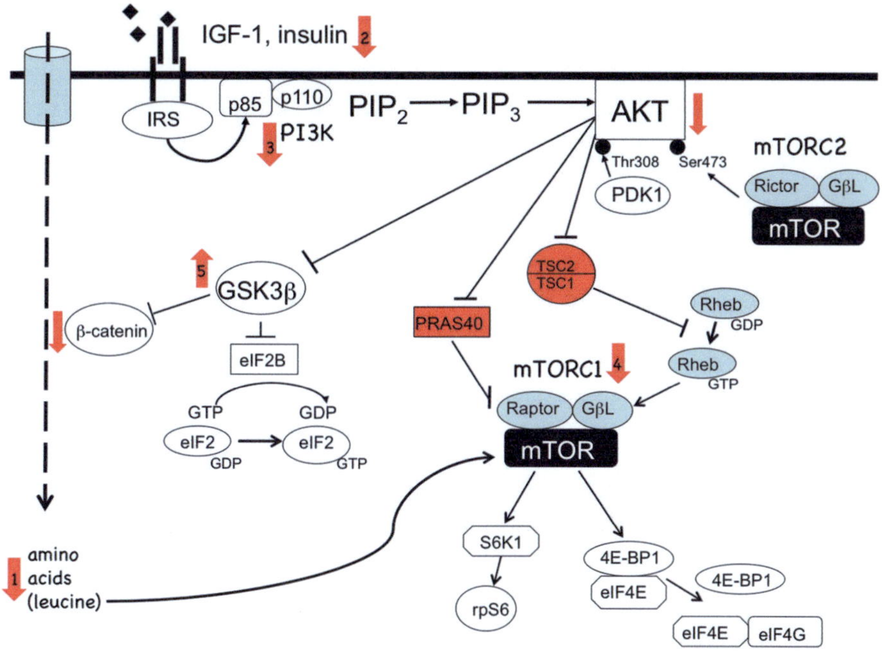

Fig. 7.2 Signaling pathways downstream of insulin and insulin-like growth factor 1 (IGF-1) that are affected by glucocorticoids. The mammalian target of rapamycin complex 1 (mTORC1) is activated by growth factors such as insulin and IGF-1 and branch chain amino acids such as leucine, leading to increases in mRNA translation and protein synthesis. Glucocorticoids potentially inhibit protein synthesis through multiple mechanisms including (1) a decrease in amino acid uptake, (2) a decrease in insulin and IGF1 levels (3) a decrease in the activity of phosphatidylinositol 3-kinase (PI3K) leading to a decrease in Protein Kinase B/Akt activity, (4) a decrease in mTORC1 activity leading to a decrease in the phosphorylation and activation of S6K1 and 4E-BP1, and (5) an increase in GSK3β activity leading to a decrease in β-catenin and possibly inhibition of eIF2B

and its downstream targets the ribosomal protein S6 kinase (S6K1/p70^{s6k}) and the translational repressor eukaryotic initiation factor 4E-binding protein 1 (4E-BP1). In muscle, mTOR can be found in at least two distinct complexes: the rapamycin-sensitive mTOR complex 1 (mTORC1) and the rapamycin-insensitive mTOR complex 2 (mTORC2) [39]. Activation of mTORC1 is critical for the regulation of muscle mass and is regulated by many pathways including growth factor signaling through PI3K/Protein Kinase B (PKB/Akt), amino acids, and mechanical loading [40, 41].

In response to moderate to high doses of glucocorticoids the synthesis of muscle proteins is significantly reduced in both humans and rodents [10, 42, 43]. Glucocorticoids delivered acutely at low doses do not produce significant decreases in basal protein synthesis, but do attenuate the ability of branched chain amino acids and insulin to activate protein translation [44, 45]. Chronic levels of low doses of glucocorticoids could induce muscle loss through anabolic resistance to elevated levels of insulin and amino acids as would occur following a meal.

The decrease in protein synthesis that occurs in response to glucocorticoids could be related to a number of mechanisms. One possible mechanism is through a decrease in the ability of muscle fibers to uptake amino acids from extracellular fluids [46, 47]. Further, decreases in ribosome content, as well as, decreases in the efficiency of protein translation have been measured following glucocorticoid treatment and reflect a decrease in the capacity for protein synthesis. Initial studies revealed both a decrease in the number of polyribosomes, and a reduced ability to incorporate amino acids into proteins in fast-twitch muscle following glucocorticoid-treatment [48, 49]. Later studies by Rannels et al. revealed that the 55 % decrease in protein synthesis of rat fast-twitch muscles measured over 5 days of glucocorticoid treatment was due primarily to disaggregation of actively engaged polysomes through inhibition of peptide-chain initiation and partly to a ~15 % decrease in RNA content [43, 50]. Subsequent studies demonstrated that acute delivery (4 h) of glucocorticoids in rats induced rapid inhibition of protein synthesis (59 % decrease) independent of changes in total RNA, but through a mechanism that involved dephosphorylation of 4E-BP1 and S6K1, and stabilization of the inhibitory complex 4E-BP · eIF4E, leading to inhibition of the assembly of the cap-recognition complex, eIF4F (which is composed of the scaffolding protein eIF4G, the mRNA cap-recognition protein eIF4E and the ATP-dependent RNA helicase eIF4A) [51, 52]. Acute glucocorticoid treatment *in vivo* was not associated with a decrease in the rate of eIF2B-catalyzed guanine nucleotide exchange, which is a rate-limiting step in formation of the ternary complex (eIF2 · GTP, Met-tRNA$_i$, 40S ribosomal subunit, eIF1A and eIF3). The reduced phosphorylation of both 4E-BP1 and S6K1 suggested that glucocorticoids inhibit protein synthesis through inhibition of mTORC1.

Alterations to Protein Degradation: The inhibitory effect of glucocorticoids on protein synthesis were the first to be established, while the ability of glucocorticoids to increase protein degradation was debated for some time due to conflicting findings. Many early studies were able to measure a decrease in protein synthesis following hours to days of glucocorticoid treatment, but failed to measure a significant increase in proteolysis. Some of the conflicting proteolysis data was contributed to differences in the type and dose of glucocorticoid used, the muscles studied, and the nutritional status of the animals.

Goldberg (1969) was one of the first to measure an increase in proteolysis, as well as a decrease in protein synthesis, in the rat plantaris muscle (but not the soleus) following a high dose of cortisone acetate (10 mg/100 g body weight) that induced a 30 % and 50 % decrease in muscle mass 7 and 12 days following the initiation of GC treatment, respectively [4]. In 1979, Tomas et al. examined the effect of a range of corticosterone doses on rat skeletal muscle proteolysis as measured by 3-methyhistidine (N-tau-methylhistidine, 3MeH) excretion [53]. The excretion of 3-methylhistidine is often used to measure the breakdown of muscle since this amino acid is formed by the methylation of certain histidine residues in the myofibrillar proteins (actin and myosin) during proteolysis, and since it unable to be reincorporated into other proteins it is excreted in the urine [54]. Skeletal muscle is the major source of excreted 3MeH, with skin and intestine contributing a small amount [55]. Tomas et al. found that skeletal muscle proteolysis increased only when plasma glucocorticoid levels were significantly greater than resting levels, which occurred only at the highest doses of corticosterone treatment (5 and 10 mg/100 g body weight/day) and not at lower doses (0.2, 0.5 or 1.0 mg/100 g BW/day) [53]. The peak plasma corticosterone concentrations (40–80 μg/100 mL of plasma) that caused muscle proteolysis were in the range of corticosterone levels measured following burn injury and prolonged psychological/physical stress (50–100 μg/100 mL plasma).

These data revealed that in a fed state, glucocorticoids can increase muscle proteolysis but only if given in exceptionally high doses. In a fed state, low levels of glucocorticoids do not induce muscle breakdown due to the presence of insulin. However, in a fasted state, elevated levels of glucocorticoids and low insulin produce increases in muscle proteolysis [56, 57]. Glucocorticoids given directly to isolated muscles in a perfused bath or to myotubes in culture increase proteolysis, however, the effect is suppressed in the presence of insulin [58].

The importance of glucocorticoids in promoting muscle proteolysis under conditions of fasting and acidosis was demonstrated using adrenalectomized rats. Wing and Goldberg showed that the rate of protein degradation increased ~50 % in the extensor digitorum longus muscle of rats fasted for 24 h. In contrast, fasting of adrenalectomized rats induced a significantly smaller increase in muscle degradation (2.5-fold difference between fasted and fasted adrenalectomized rats). Administration of dexamethasone to fasted, adrenalectomized rats elevated the rate of protein breakdown to that observed in normal fasted rats thus demonstrating the role of glucocorticoids in regulating muscle proteolysis under conditions of fasting. Elevated levels of plasma glucocorticoids have also been shown to be required for the increase in muscle protein breakdown observed in response to metabolic acidosis [59, 60]. Glucocorticoids together with elevated cytokines also contribute to an increase in muscle proteolysis in various catabolic diseases such as sepsis, cancer and burns [57].

The Ubiquitin Proteasome Pathway and Glucocorticoids

Accelerated protein breakdown in skeletal muscle in response to glucocorticoids has been shown to be largely due to activation of the Ubiquitin Proteasome Pathway (UPP) [61, 62]. The UPP is an ATP-dependent proteolytic pathway made up of 3 classes of proteins involved in tagging proteins with the 76 amino acid protein ubiquitin, and the 26S proteasome, which recognizes polyubiquitinated proteins and degrades them into small peptides [63] (Fig. 7.3). Proteins can be monoubiquitinated, diubiquitinated, or polyubiquitinated via the actions of an E1 ubiquitin-activating enzyme, an E2 ubiquitin-conjugating enzyme, and an E3 ubiquitin ligase enzyme. To form a polyubiquitin chain, ubiquitin proteins are linked at one of their seven lysine residues [64]. Proteins containing a chain of at least four lysine-48 linked ubiquitin molecules are recognized and degraded by the proteasome [65].

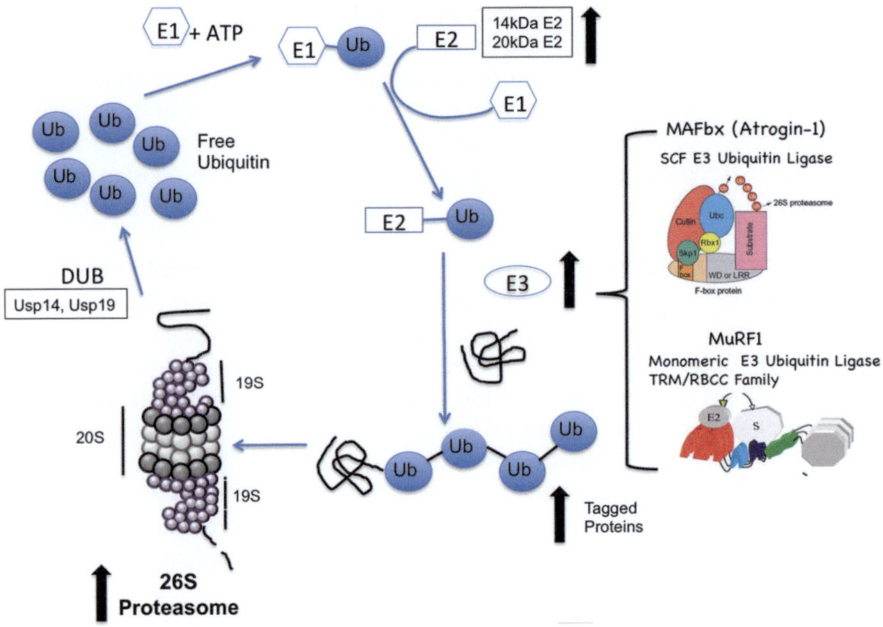

Fig. 7.3 The ubiquitin proteasome pathway (UPP). Excess glucocorticoids can lead to an increase in protein degradation through an increase in components of the UPP and an increase in the number of ubiquitin-tagged proteins. The process of ubiquitination is controlled by the ubiquitin-activating enzymes (E1), ubiquitin-conjugating enzymes (E2) and ubiquitin-protein ligases (E3). MuRF1 and MAFbx are both E3 ubiquitin ligases that control the ubiquitination of specific substrates. Ubiquitin can be added to a substrate as a chain of ubiquitins of variable length on a single lysine (polyubiquitination). Polyubiquitination of a substrate with an ubiquitin chain using K48 linkages generally results in proteasomal degradation. Deubiquitinating enzymes (DUB) associated with the proteasome or in the cytoplasm are responsible for disassembly of the polyubiquitin chains and maintenance of the pool of free ubiquitin

The 26S proteasome is a large (2-MDa) multi-subunit complex consisting of a 20S core particle and two 19S regulatory particles found at each end of the 20S core, and is responsible for the breakdown of the majority (60–70 %) of muscle proteins, especially under catabolic conditions. The 19S particle is involved in the recognition of polyubiquitinated proteins and functions to unfold these proteins and guide them into the 20S catalytic core [66]. The hydrolysis of ATP drives the unfolding of large globular proteins and translocation of the proteins into the 20S core. The 20S core proteasome is a cylindrical structure made up of two outer α rings and two inner β rings. Each alpha ring contains seven α subunits and each beta ring contain seven β subunits, with the proteolytic sites located on the inner surface of the β1, β2, and β5 subunits [67]. The β1 subunit possesses caspase-like proteolytic activity, the β2 subunit has trypsin-like proteolytic activity, and the β5 subunit has chymotrypsin-like proteolytic activity [67]. The substrate protein that enters the 20S core particle exits as small peptides ranging in size from 3 to 25 residues in length

[68]. These peptides are quickly digested into amino acids that can be reused to synthesize new proteins or metabolized to provide energy. Deubiquitinating enzymes (DUBs), associated with the 26S proteasome and also found in the cytoplasm, are responsible for disassembly of the polyubiquitin chain and release of free ubiquitin that can be reused to tag additional substrates.

The selection of proteins to be ubiquitinated and subsequently degraded by the proteasome is dependent on the specific E2 and E3 ligases present in the cell. Specificity of the process is determined largely by the E3 ligase, which binds the substrate and interacts with the E2 ligase to direct the transfer of ubiquitin to the substrates [69]. The specific E2-E3 pairings dictate the type of ubiquitin linkage and the length of the ubiquitin chains. In mammals, ~40 E2 ligases and over 600 E3 ligases have been identified. The E3 ligases can be classified into two major groups: the HECT (Homologous to the E6AP Carboxyl Terminus) E3 ligases and the RING (Really Interesting New Gene) E3 ubiquitin ligases [70]. The RING E3 ligases constitute the majority of E3s found in mammals and function as a scaffold that mediates the transfer of ubiquitin directly from the E2 ~ Ub to substrate.

Under catabolic conditions in which glucocorticoids are elevated, genes encoding components of the ubiquitin proteasome pathway and proteasome subunits have been shown to increase in skeletal muscle. An increase in ATP-dependent proteasome activity has been measured in both L6 and C2C12 myotubes exposed to dexamethasone *in vitro* and in hindlimb muscles of rats treated with dexamethasone [71, 72]. Interestingly, no increase in 20S or 26S proteasome activity was measured in hindlimb muscles of C57BL6 mice treated with dexamethasone for up to 14 days [12, 73]. In these mice, significant muscle atrophy occurred and was associated with a decrease in protein synthesis, as opposed to elevated proteolysis. The difference in the reported proteolytic response of rats versus mice to dexamethasone may be related to the nutritional status of the animals and the levels of plasma glucocorticoids achieved. In general, glucocorticoid-treatment in rats, is accompanied with a decrease in food consumption and rapid loss of muscle mass (especially when fluorinated glucocorticoids are used). In contrast, in mice no changes in food consumption are observed, at least at doses of 3–5 mg/kg of dexamethasone. Increases in proteasome-mediated degradation have been noted in catabolic disease states such as starvation, metabolic acidosis, sepsis, diabetes and cancer [74–80]. Associated with increases in proteasome activity are increases in the amount of ubiquitin-conjugates isolated from skeletal muscle. In addition, increases in the mRNA expression of ubiquitin, proteasome subunits and several E2 ligases ($E2_{14k}$, $E2_{17k}$) have been measured.

The E3 Ubiquitin Ligases: MuRF1 and MAFbx

A number of E3 ubiquitin ligases have been identified in skeletal muscle, but two muscle-specific E3s, Muscle RING Finger 1 (MuRF1) and Muscle Atrophy F-Box (MAFbx)/atrogin-1, have been shown to be transcriptionally increased under a variety of atrophy conditions including glucocorticoid treatment, as well as other catabolic conditions such as starvation, renal failure, cancer cachexia and diabetes [81]. Muscle RING Finger 1 (gene name:*Trim63*) is a simple RING E3 that belongs to a family of genes (MuRF1, 2, and 3) which represent a subgroup of the TRIM superfamily of multidomain ubiquitin E3 ligases characterized by the presence of the N-terminal tripartite motif: RING domain, zinc-finger B-box domain, and leucine-rich coiled-coil domain [82]. Muscle Atrophy F-Box (gene name:*FBX032*), also known as atrogin-1, is 1 of ~70F-box proteins and belongs to the cullin–RING E3 ligase (CRL) superfamily. MAFbx functions in a multi-subunit complex consisting of: SKP1 and a F-box protein which serve as the substrate adaptor module, a cullin protein (CUL1) that acts as a scaffold, and a RING finger protein (RBX1 or RBX2) that associates with CUL1 and recruits the ubiquitin charged E2s [83]. In resting skeletal muscle, MuRF1 and MAFbx are expressed at relatively low levels, but become transcriptionally upregulated in response to a variety of stressors, including glucocorticoids and cytokines [84].

The analysis of mice with null alleles in genes encoding either MuRF1 or MAFbx has provided evidence that both genes are key regulators of muscle atrophy. Baehr et al. examined the response of both male and female MuRF1 and MAFbx null mice to 14 days of dexamethasone-treatment and to 3 days of nutritional deprivation (80 % deficit) [12]. Glucocorticoid-induced muscle atrophy was significantly attenuated in both male and female MuRF1 KO mice, but not MAFbx KO mice. Muscle sparing was evident as measured by the maintenance of muscle mass, muscle fiber cross-sectional area, and muscle contractile force. In contrast, the deletion of MuRF1 or MAFbx did not result in muscle sparing in response to nutritional deprivation, even though glucocorticoids are elevated under conditions of starvation. These results and others [85] suggest that starvation activates additional glucocorticoid independent pathways that contribute to muscle atrophy.

Anti-anabolic Effects of Glucocorticoids

Suppression of protein synthesis in skeletal muscle by glucocorticoids can occur through a number of mechanisms. As mentioned previously, one mechanism by which glucocorticoids could decrease protein synthesis is related to its ability to inhibit the transport of amino acids into the muscle. Glucocorticoids can also inhibit protein synthesis by altering the activation level of specific signaling pathways (such as the PI3K/Akt/mTOR pathway), which mediate the anabolic actions of amino acids (primarily leucine), growth factors (such as insulin-like growth factor-1, IGF-1), and insulin.

As a key example, IGF-1 has been shown to have both an anabolic and an anti-catabolic role in skeletal muscle with the anabolic effects being mediated through its ability to increase protein synthesis through activation of the PI3K/Akt/mTOR pathways [86]. Under catabolic conditions and in response to excess glucocorticoids, circulating and muscle IGF-1 levels are often reduced [87, 88]. In young growing rats, a catabolic dose of dexamethasone has been shown to decrease muscle IGF-1 levels, as well as, decrease the phosphorylation levels (and thus activation) of both PKB/Akt and S6K1 [89, 90]. Evidence to support a role for decreased IGF-1 levels and reduced activation of PKB/Akt in glucocorticoid-induced muscle atrophy has come primarily from over-expression studies. Using *in vitro* myotube culture systems, IGF-1 has been shown to prevent glucocorticoid-induced loss of myotube diameter through the activation of PI3K/Akt and blockage of the nuclear translocation of the FOXO transcription factors and downregulation of the E3 ubiquitin ligases, MuRF1 and MAFbx/atrogin-1 [91, 92]. Moreover, *in vivo* systemic administration [93–95] and direct overexpression of IGF-1 into skeletal muscle [90] has been shown to attenuate glucocorticoid-induced muscle atrophy.

The attenuation of muscle atrophy in response to overexpression of either IGF-1 or constitutively active PKB/Akt is associated with an increase in mTOR signaling, a decrease in the nuclear translocation of FOXO1/3a, decreased expression of MuRF1 and MAFbx/atrogin-1 and an increase in β-catenin expression [96]. The responses to IGF1 and PKB/Akt-1 activation suggest that the attenuated atrophy is related to both an anabolic effect (maintenance of protein synthesis through activation of mTOR and its downstream targets) and anti-catabolic effect (reduced activation of the FOXO transcription factors and reduced expression of genes associated with elevated proteolysis, such as MuRF1, MAFbx and cathepsin-L). The increase in β-catenin expression in response to IGF-1 and PKB/Akt activation was interesting given that the transcription factor has been reported to play a role in skeletal muscle hypertrophy [97].

The amount of β-catenin in the cell is regulated by its phosphorylation and degradation by the proteasome. Phosphorylation of β-catenin is controlled by glycogen synthase kinase 3β (GSK-3β) whose activity in turn is inhibited by phosphorylation by PKB/Akt. In response to dexamethasone-treatment, a significant reduction in β-catenin protein levels were also reported, which is consistent with the hypothesis that GSK3β is activated in response to glucocorticoids leading to an increase in the phosphorylation and degradation of β-catenin [90]. An increase in β-catenin in response to IGF-1 or Akt activation is consistent with inactivation of GSK3β. Overexpression of a stable β-catenin protein, which is resistant to proteasomal degradation, in skeletal muscle is capable of blocking GC-induced muscle atrophy, thus providing some evidence that β-catenin may be an important regulator of the negative effects of glucocorticoids [90].

Independent of changes in growth factor availability, glucocorticoids can also decrease the activation of PI3K through mechanisms that modify the amount of insulin receptor substrate-1 (IRS-1) and the relative levels of the regulatory subunit of PI3K, i.e., p85α, and not the catalytic subunit, p110. Glucocorticoid-treatment consistently leads to a decrease in the amount of IRS-1 protein and inactivation of IRS-1 by serine-phosphorylation (ser307) *in vitro* in C_2C_{12} myotubes [98–100]. Acceleration of the degradation of IRS-1 could come about through multiple mechanisms involving either an increase in the amount of the tyrosine phosphatase, C1-Ten [101] or the E3 ubiquitin ligase, Cblb [98].

Another mechanism that can account for a decrease in PI3K activity following glucocorticoid treatment is the selective increase in the regulatory subunit of PI3K, p85α (a.k.a. Pik3r1). An increase in the regulatory subunit of PI3K (p85α), without a concomitant increase in the catalytic subunit (p110) occurs in response to glucocorticoids, and could contribute to the inhibition of PI3K activity [102, 103]. Overexpression of p85α *in vitro* results in a decrease in the diameter of C_2C_{12} myotubes [102]. In contrast, inhibition of p85α in C_2C_{12} myotubes, with small interfering RNA (siRNA), reduced the ability of glucocorticoids to suppress PI3K activity, inhibit protein synthesis and induce muscle atrophy. These data suggest that increased expression of p85α, the regulatory subunit of PI3K, is a major regulator of glucocorticoid-induced atrophy through the suppression of PI3K activity and protein synthesis.

Downstream of PI3K/Akt is the serine-threonine kinase, mTOR, which is a major regulator of skeletal muscle size [40]. As previously noted, mTORC1 activity, as measured by the phosphorylation of S6K1 and 4E-BP1, is suppressed in response to glucocorticoid treatment. Suppression of mTOR activity may be related to the inhibition of PI3K/Akt activation, but could also be a consequence of other mechanisms. Recently, it has been shown that suppression of mTORC1 in response to glucocorticoids is related to the up-regulation of genes such as REDD1 [104, 105]. The upregulation of REDD1 has been directly associated with a decrease in phosphorylation of the mTOR targets S6K1, 4E-BP1, and ULK1, as well as a decrease in protein synthesis. More recently it was demonstrated that mice with a deletion of REDD1 were resistant to the effects of dexamethasone [73]. The lack of REDD1 expression in skeletal muscle prevented dexamethasone-induced muscle atrophy and prevented the suppression of mTOR activity and protein synthesis. The mechanism underlying glucocorticoid-mediated mTOR inhibition has been unclear, however, recent studies suggest that inhibition is dependent on Akt/PRAS40 signaling [73]. It appears that overexpression of REDD1 leads to a decrease in Akt thr308 phosphorylation, which lessens the inhibition on PRAS40, allowing it to inhibit mTOR. An alternative hypothesis is that the REDD1 induced dephosphorylation of Akt on Thr308 leads to TSC1-TSC2 complex assembly and attenuation of Rheb-GTP and mTORC1 activation [106].

Molecular Mode of Action of Glucocorticoids in Skeletal Muscle

The Glucocorticoid Receptor and the "Classical" Genomic Actions of Glucocorticoids

As discussed in more detail in other chapters, the major means by which natural and synthetic GCs exert their actions in skeletal muscle as in other tissues is through the intracellular glucocorticoid receptor (GR), a member of the steroid receptor gene superfamily [107]. Although widely expressed in mammalian tissues, GR is relatively highly expressed in skeletal muscle and at higher levels than in more well-studied tissues such as the liver or adipose tissue. The GR is most closely related to the mineralocorticoid receptor (MR) that is best known for its role in sodium and potassium balance via its cognate hormone, aldosterone, in many vertebrates. However, the MR is expressed at much lower levels in skeletal muscle than other cell types. There is only one GR gene in mammals (GRα), although humans express a GRβ splicing isoform that appears to be expressed at low levels in skeletal muscle [108]. Internal ribosome entry may generate further diversity in the major GRα isoform (GR A-D), which alters target gene selectivity. Additional GR encoding genes and various alternately spliced isoforms have been reported in other vertebrates, most notably in fish [109]. In this chapter, for simplicity sake, we will refer to the skeletal muscle GRα generically as the GR.

The GR is expressed at much higher levels in fast-twitch glycolytic fibers versus slow-twitch oxidative fibers, which is consistent with the much greater sensitivity of fast twitch glycolytic fibers to the atrophy-inducing effects of glucocorticoids [110]. GR expression in muscle is not static, but is commonly down-regulated by its own cognate ligand [111] and also varies in a circadian manner with peak levels in the morning in rodents when endogenous glucocorticoids are low [112]. GR expression levels have also been reported to vary as a consequence of aging and obesity, although the significance of these expression changes has not been clearly established [113]. Skeletal muscle GR expression is also regulated by other nuclear receptors such as retinoic acid (RARs) [114], and estrogen receptor-related (ERR) receptors [115]; in the latter case, this has been linked to metabolic reprogramming by this orphan member of the gene superfamily.

The GR is a modular protein with interacting but separable domains that allow the versatile receptor to bind ligand, as well as, interact with specific DNA sequences, additional DNA-bound transcription factors, and transcriptional coregulatory proteins. As a result, activation of the GR by synthetic or natural ligands leads to a change in the expression of dozens to hundreds of genes in skeletal muscle, with the number and magnitude reported dependent on expression platform, dose and duration of exposure, and whether cultured myotubes or intact animals were used. The identified regulated genes in skeletal muscle include: (1) known, classical GR target genes that have been identified in many other tissues and (2) genes involved in substrate metabolism, insulin signaling, inflammation, the extracellular matrix, and angiogenesis as the most common functional clusters [102, 103, 116–118]. A subset of these genes, including their presumed mode of regulation by glucocorticoids, are highlighted in Table 7.1. In addition to protein encoding genes, potential roles for GR mediated transcriptional upregulation of microRNAs such as miR1 have been described [119].

In the unliganded state, the GR is retained in the cytoplasm as a complex with Hsp90 and other chaperones, and upon binding ligand releases these chaperones and exhibits nuclear translocation [107]. Recent chromatin immunoprecipitation experiments in C_2C_{12} myotubes identified multiple direct genomic targets of the GR in muscle [102]. A DNA sequence was commonly observed in the vicinity of both up and down-regulated primary response genes that was similar to a consensus

Table 7.1 Selected glucocorticoid regulated genes in skeletal muscle related to atrophy

Gene	Mode of regulation	Roles	References
MuRF1/TRIM63/RNF	GRE/FOXO1/KLF15 (direct[a])	E3 ubiquitin ligase	[84, 110, 134]
MAFbx/Atrogin-1/Fbxo32	FOXOs (indirect)	E3 ubiquitin ligase	[83, 84, 167]
PI3kr1	GRE (direct)	PI3 kinase subunit	[102]
REDD1/Ddit4	GRE (direct)	mTOR inhibitor	[73, 102, 104]
KLF15	GRE (direct)	Transcription factor	[110]
FOXOs	GRE (direct) (FOXO1[b])	Transcription factors	[128, 168]
Myostatin/GDF8	GRE (direct)	Muscle growth inhibitor	[169, 170]

[a]Both direct (via GRE) and indirect (via induced FOXO, KLF15 and other transcription factors) possible
[b]Only FOXO1 regulation by glucocorticoids has been investigated to date

glucocorticoid response element (GRE) recognized by ligand bound GR homodimers (5′ AGAACAnnnTGTTCT-3′, where n is any nucleotide in the three base pair spacer between the hexameric inverted repeat half sites), However, only a minority of these binding sites are found in the proximal promoter region of target genes (i.e. <5 kb upstream of the promoter); most GR binding sites are found more than 25 kb up or downstream of target gene transcription units or within introns. Glucocorticoids are also capable of direct transcriptional suppression of genes in many cell types, including skeletal muscle. A motif similar (but not identical) to the GRE described in up-regulated genes is enriched in skeletal muscle down-regulated genes [102], although protein–protein interactions with other transcription factors has traditionally been ascribed as the major mode of gene expression inhibition by glucocorticoids. For example, the GR is able to bind to NF-κB and AP-1 subunits as a major means of suppressing gene expression in fibroblasts (i.e. structural genes like collagen) and immune cells (i.e. pro-inflammatory cytokine genes) [120], but most of the work in skeletal muscle to date has focused on up-regulated genes.

To accomplish transcriptional regulation of its target genes, the GR must recruit additional transcriptional cofactors that serve to post-translationally modify chromatin or other transcription factors (primarily via acetylation and methylation), or otherwise collaborate to influence RNA polymerase II activity. For example, the GR both transcriptionally induces and physically associates with the histone acetyltransferase p300 and its paralog CBP [121]as well as the p300/CBP interacting protein Cited2 in skeletal muscle cells [122], in both sepsis and glucocorticoid excess models. Inhibition of global acetylation reduces muscle atrophy in cultured myotubes [123, 124], although the relative contribution of p300/CBP versus other GR interacting and non-interacting acetyltransferases/deacetylases is unknown at present. In addition to acetylation, the methyltransferase SMYD3 is also involved in glucocorticoid-induced atrophy *in vitro* [125], and interestingly is specifically induced by glucocorticoids in C_2C_{12} cells, whereas other SMYD family members are not. SMYD3 recruits the bromodomain containing, chromatin remodeling protein, BRD4, to specific muscle promoters activated during atrophy. A specific role in skeletal muscle for these transcriptional cofactors is proposed, however, they may simply be generally supporting GR transcriptional activity in myotubes given that HATs like p300/CBP are recruited to GR regulated promoters in many other cell types and target gene promoters. Indeed, SMYD3 also interacts with other steroid receptors as a mediator of steroid hormone action in several contexts [126].

While glucocorticoids directly up-regulate a number of genes implicated in skeletal muscle protein turnover and lipid/carbohydrate metabolism, multiple transcription factors are also induced by glucocorticoids in this tissue, serving to expand and integrate the hormone regulated transcriptional network in these cells. One important class of transcription factors discussed in previous sections is the forkhead domain containing class O transcription factors (FOXOs). FOXO1, 3 and 4 are all induced by exogenous glucocorticoids in skeletal muscle and are also consistently induced under a variety of conditions that elevate endogenous glucocorticoids [117, 127, 128]. Functional GREs may be present in at least one FOXO gene promoter [129]. FOXO transcription factors are important mediators of insulin action in multiple tissues, and are also involved in the transcriptional regulation of metabolic enzymes, response to oxidative stress, and autophagy [130, 131]. FOXO activity is inhibited via activation of Akt and phosphorylation at specific serine residues that inhibit FOXO nuclear localization. FOXO binding sites have been reported in multiple atrophy-associated and metabolic enzyme genes in skeletal muscle and inhibition of FOXO activity by growth factor addition or genetic means strongly suppresses muscle atrophy and associated gene expression [92, 132–135], suggesting that FOXOs may serve as key mediators of glucocorticoid-induced atrophy. FOXOs are also subject to acetylation by p300 in addition to phosphorylation, but in an isoform specific manner [136]. For example, nuclear translocation of FOXO1, the family member particularly linked to GR actions, is activated by p300-mediated acetylation, whereas FOXO3 transcriptional activity and localization are inhibited. Class I histone deacetylase activity supports muscle atrophy in disuse models presumably by targeting FOXO3a (at least in part) [137]; however, isoform specific FOXO activity in the context of glucocorticoid treatment models, where histone acetyltransferases are actively recruited by the GR to target gene promoters will need to be clarified.

In addition to FOXOs, several other glucocorticoid regulated transcription factors have been identified that may also play a key role in mediating the steroid's effects on muscle mass and function. For example, members of the C/EBP family, in particular C/EBPβ and δ, are up-regulated in cultured myotubes and *in vivo* [138] in response to glucocorticoids, and C/EBPβ may be necessary but not sufficient to support atrophy and associated gene expression in cultured myotubes [139]. Another transcription factor of note, Kruppel-like factor 15 (KLF15), is a strongly induced, direct GR target in multiple cell types, including skeletal muscle. This factor participates in regulation of previously described atrophy associated genes such as MuRF1 and MAFbx, on one hand, as well as a key enzyme involved in branch chained amino acid metabolism, BCAT2, on the other, resulting in compromised mTOR function and suppressed protein synthesis [110]. Lastly, overexpression of GILZ, a widely induced GR target gene that is strongly up-regulated in primary and C_2C_{12} myoblasts, inhibits myotube differentiation, thus providing a distinct pathway by which glucocorticoids can affect skeletal muscle mass beyond its effects in mature muscle fibers [140]. The function of GILZ is not fully understood since it may not be a transcription factor as originally believed based on its leucine zipper motif, but rather may act via interactions with several cytoplasmic signal transduction proteins [141].

Several other transcription factor genes have been identified as being glucocorticoid up- and down-regulated by microarray and candidate gene analysis. While these genes are less well characterized to date in terms of regulating muscle mass and function, they may be important in executing or fine-tuning the integrative response to glucocorticoid exposure in these cells in concert with the factors we have focused on so far. Despite significant progress in identifying key downstream transcription factors, the particular code that produces a skeletal muscle specific array of glucocorticoid regulated, functionally relevant genes has not been established. Presumably, this specificity depends on developmental programming of access to specific GREs that allows expression of the muscle specific glucocorticoid response, as has been demonstrated with specific "pioneering" factors such as FoxA isoforms in liver development [142]. While glucocorticoids regulate FOXO, C/EBP, and KLF15 transcription factors genes in many cell types, for example, the expression of a key muscle atrophy gene, MuRF1, is exclusive to heart and skeletal muscle. Even in this context, MuRF1 is induced to a greater degree and from a much lower baseline in skeletal muscle versus the heart, suggesting differential set points in MuRF1 gene activity in its two major sites of expression and action that are not well understood.

GR Interactions with Cytoplasmic Signal Transduction Proteins

While the previous section focused on the "classical" actions of the GR in skeletal muscle, there is growing evidence that the GR can also mediate the effects of glucocorticoids via specific protein-protein interactions with signal transduction machinery in the cytoplasm. For example, the GR can directly interact with and inhibit the activity of mitogen activated kinases (MAPK), IkB kinase, and PI3 kinase [143]. While most of these interactions have been studied in the context of immune system modulation, the best studied with respect to skeletal muscle mass is the direct physical interaction between GR and the p85 regulatory subunit of PI3 kinase. GR/p85 binding inhibits the p110 catalytic subunit interaction with IRS-1, thus contributing to subsequent downstream Akt inhibition, FOXO activation, and transcriptional regulation of atrophy-associated genes, among other consequences [144]. Elevated growth factor exposure, including insulin and IGF-1, can override the inhibitory PI3K/GR interaction; thus, atrophy stimulated by physiological glucocorticoid levels must be coupled with abrogated insulin signaling or decreased insulin levels as is observed in diabetes mellitus or starvation [144]. The importance of this aspect of glucocorticoid signaling in muscle will be discussed in more detail below.

In turn, the GR itself is phosphorylated by a variety of kinases, some of which have been implicated in the regulation of muscle mass and metabolism. Interestingly, while on one hand activated GSK3β has been implicated as an important mediator of glucocorticoid induced muscle atrophy [145], this kinase also directly phosphorylates the GR [146]. This results in increased activation of downstream targets such as GILZ; however, the effect of GSK3β phosphorylation of the GR appears to be highly target gene specific. Since these studies were conducted in nonmuscle cell lines, it will be interesting to examine GSK3β/GR interactions in the context of skeletal muscle atrophy in that it could serve as a selective feed forward mechanism to amply GR actions in the face of decreased insulin or other growth factor stimulation.

Potential Non-GR Mediated Effects of Glucocorticoids in Skeletal Muscle?

Despite considerable evidence for the bona fide GR in control of skeletal muscle mass and metabolism, nongenomic, non-GR mediated effects of glucocorticoids have been reported, particularly in the brain [147]. Rapid effects of the steroid appear in these cases to be mediated by a membrane bound receptor distinct from the GR itself [148]. While this response is still understudied relative to the classical, nuclear receptor pathway, a few reports of very rapid effects of glucocorticoids have appeared that support an alternative mode of glucocorticoid action in both cardiac and skeletal muscle [149–151]. Since the pharmacological profile of putative membrane receptors maybe distinct from the GR, it is noteworthy that particularly in cultured myotubes, very high doses of synthetic glucocorticoids are often used in order to observe the most robust effects on atrophy and associated gene expression [152, 153]. The GR equilibrium dissociation constant (K_d) for dexamethasone is very low (low nanomolar to subnanomolar range) yet concentrations of 100 μM are often reported as being required to observe maximal responses in these models. The necessity for very high doses of synthetic glucocorticoids in cultured myotubes has not adequately been explained to date.

Prereceptor Control of Glucocorticoid Actions

While the GR is critical for the actions of glucocorticoids in skeletal muscle, prereceptor control of access to high affinity ligands is another important regulatory point. Cells that express 11β hydroxysteroid dehydrogenase type I (11β HSD I) are able to convert the inactive cortisone and deoxycortisone to respectively to the active forms cortisol and corticosterone (the major endogenous glucocorticoid in man and rodent respectively), thereby increasing local glucocorticoid activity [154]. Interestingly, increased 11β-HSD type I activity in skeletal muscle has been linked to development of insulin resistance in skeletal muscle and as well as decreases in muscle mass and increases in atrophy associated gene expression [99]. Interestingly, 11β-HSD type I activity increases with age and obesity in multiple cell types, including skeletal muscle.

Non Cell Autonomous Actions of GCs on Skeletal Muscle Structure and Function

Most of the discussion thus far has focused on the direct actions of glucocorticoids in myofibers and myoblasts or satellite cells. However, systemic glucocorticoid exposure has important effects in several other cell types within skeletal muscle, in particular connective tissue and blood vessels. Glucocorticoids such as methylprednisolone acutely decrease blood flow to skeletal muscle, which correlates with decreased glucose up-take and systemic hypertension observed with prolonged glucocorticoid treatment [44, 155]. Transient up-regulation of transcripts related to vasoconstriction have been observed (e.g. endothelin-1) [156], presumably originating in associated blood vessel endothelial cells rather than in myofibers themselves. Likewise, in several gene expression profiling experiments, collagen gene expression is suppressed in intact muscle consistent with prior biochemical studies, presumably in fibroblasts supporting the muscle extracellular matrix.

Lastly, elevation of systemic glucocorticoids alters the levels of other circulating hormones and various metabolites that may indirectly impact muscle mass and function, in addition to metabolic adaptation and development of insulin resistance. In addition to effects on IGF1 described above, elevated glucocorticoids affect the levels of growth hormone, catecholamines, inflammatory cytokines, and other circulating factors which all may affect skeletal muscle structure and function indirectly [157–159]. In addition, central effects of glucocorticoids on food intake are well known [160], but this response varies among species. Furthermore, glucocorticoid excess causes hypogonadism, which could also indirectly affect muscle. Therefore, to examine the direct actions of glucocorticoids in skeletal muscle fibers themselves, it was important to develop experimental models with skeletal muscle specific reduction of or absent glucocorticoid signaling.

Genetic Approaches for GR Functional Analysis in Skeletal Muscle

As discussed in previous sections, the classical approach to establishing a role for glucocorticoids has been to challenge adrenalectomized animals or animals treated with the GR (and progesterone receptor) antagonist RU486. Since (a) the adrenal ablation or systemic pharmacological approaches are not specific enough to assess a direct role of the GR in muscle itself, and (b) germline knockout of the GR in mice is lethal around birth, a series of tissue specific knockout models have been developed in the intact organism (Table 7.2). These models allow for an analysis of cell autonomy in glucocorticoid induced atrophy as well as unequivocal identification of the GR as the bona fide target of excess glucocorticoids or elevated endogenous glucocorticoids in a range of atrophy models.

Three groups separately developed and published findings on muscle specific GR knockout mice (MGRKO), and challenged these mice with mostly different experimental models of atrophy. The first publication describing a MGRKO was used to investigate muscle atrophy caused by streptozotocin induced diabetes mellitus [144]. Mice with loxp sites flanking GR exon 2 were crossed with mice bearing a transgene with the muscle specific creatine kinase promoter driving cre recombinase expression in skeletal muscle (MCK-cre), although the promoter is also active at some level in the heart as well. Diabetes induced muscle atrophy and associated decreases in IRS-1 association with PI3 kinase was inhibited in the MGRKO mice. Interestingly, expression of a mutant GR that lacks the ability to translocate to the nucleus upon ligand binding in the MGRKO background at least partially rescues the atrophy and MAFbx/atrogin-1 induction in the diabetes model. Their data support a two step model for glucocorticoid induced atrophy in response to induced diabetes and starvation [144]: namely, insulin signaling must decrease or be otherwise impaired in order for the ligand activated GR to effectively further reduce PI3 kinase activity via a cytoplasmic interaction and activate

Table 7.2 Genetic approaches to studying glucocorticoid receptor activity in skeletal muscle atrophy

GR mutant	Atrophy model	References
GR dimerization mutant mice	Dexamethasone	[134]
GR floxed exon 2×MCK-cre mice	Diabetes/starvation	[144]
GR floxed exon 3×MCK-cre mice	Dexamethasone/starvation/denervation	[161]
GR floxed exon 2×MCK-cre mice	Cachexia/lipopolysaccharide/chemotherapy	[162, 163]

atrophy associated gene expression via reduced Akt activity and increased FOXO activity in the nucleus.

In a separate set of studies, Watson et al. [161] created a MGRKO mouse model (MGR^{e3}KO) by targeting a floxed exon 3 allele of the GR, again by crossing those mice with the MCK-cre transgene. These authors demonstrated that lack of the GR in skeletal muscle essentially prevents excess synthetic glucocorticoid induced atrophy and up-regulation of several atrophy associated genes including MuRF1 and MAFbx/atrogin-1. Up-regulation of atrophy-associated genes was blunted, but not completely prevented, in a nutritional deprivation model in the MGR^{e3}KO mice. Atrophy induced in a very different model, denervation of the sciatic nerve, was not different between wild type or MGR^{e3}KO mice nor was atrophy induced gene expression altered, demonstrating that the GR is not uniformly required for all conditions resulting in atrophy and it may be restricted to catabolic states [161]. These studies followed from prior work with a dimerization mutant GR expressing mouse line that surprisingly did not prevent excess glucocorticoid induced muscle atrophy, despite the presence of a near perfect palindromic GRE in the MuRF1 promoter [134]. MuRF1 induction was inhibited to a degree in the treated mice, but not enough to inhibit atrophy. Interestingly, dimerization mutant GRs alone fail to activate the palindromic GRE in the proximal MuRF1 promoter in transfection assays, but co-expression of FOXO1 re-establishes the strong synergistic activation of the promoter that is greater than observed upon expression of either the GR or FOXO1 on their own [134].

In a third mouse model, muscle specific disruption of the GR gene with the same strategy as in Hu et al. [144], and muscle atrophy was induced in a model of inflammation using systemic lipopolysaccharide (LPS) injection as well as a cancer cachexia model via implantation of Lewis lung carcinoma cells [162]. Muscle atrophy and induction of several associated muscle atrophy genes including MuRF1 and MAFbx/atrogin-1 were strongly abrogated in both challenges in exon 2 deleted MGRKO mice. The study also clarified the role of cytokines in inflammation and cancer induced atrophy with strong evidence that rather than acting directly on muscle itself, instead, increase circulating glucocorticoids that then in turn induce gene expression and subsequent atrophy via activation of the skeletal muscle GR [162]. Recently, this group also used the exon 2 deleted MGRKO model to demonstrate the requirement for skeletal muscle GR in atrophy induced by cytotoxic chemotherapeutic agents, again likely via increasing endogenous glucocorticoids that in turn induce muscle atrophy via the GR [163]. Taken together, the above studies clearly establish the critical central role of the skeletal muscle GR in regulating atrophy in rodent models of diabetes, excess glucocorticoid treatment, starvation, inflammation (as modeled by LPS exposure), cancer cachexia, and chemotherapy.

In addition to genetically modified mouse models, other approaches to investigate the role of the GR in skeletal muscle are emerging. Small interfering RNA (siRNA) directed against the GR inhibited glucocorticoid regulated atrophy and gene expression in cultured myotubes [164]. In an alternative intact vertebrate model, morpholinos directed against the zebrafish GR significantly affected proper muscle development [165], but atrophy has not been investigated in this system to date. The recent emergence of rapid genome editing technologies using TAL directed endonucleases (TALENs) or CRISPR/Cas9 [166], provides an even greater opportunity to specifically alter the GR coding sequence and/or its downstream effectors in a variety of cell based or in vivo models, including human stem cells or rats for example.

Summary of Glucocorticoid Regulated Processes in Skeletal Muscle Atrophy

The preponderance of evidence published to date is consistent with a central role for the GR, when bound by natural or synthetic glucocorticoids, in (a) interaction with cytoplasmic growth factor signal transduction machinery and (b) direct activation of genes that result in an interacting gene expression network leading to imbalances in protein synthesis and degradation pathways and ultimately muscle atrophy. A simplified model for the mechanisms underlying acute and chronic effects of glucocorticoids in initiating muscle atrophy is shown in Fig. 7.4. Under normal circumstances,

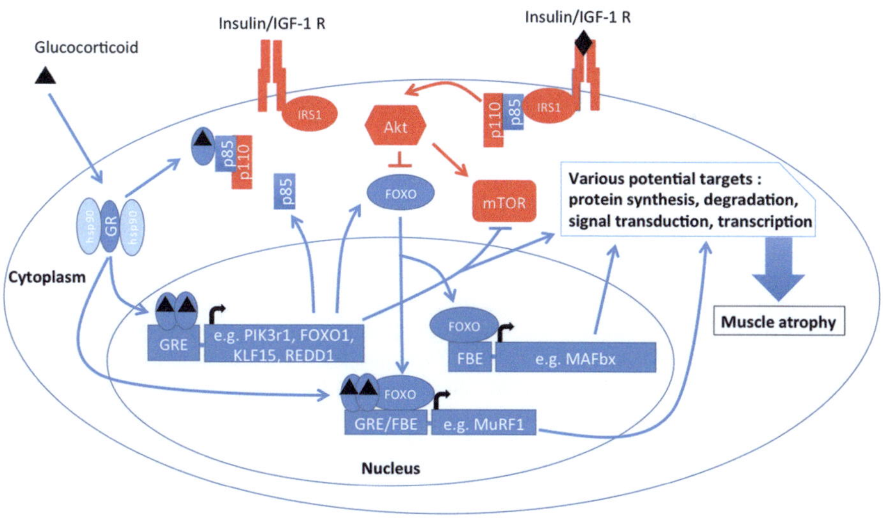

Fig. 7.4 Schematic for the mechanism underlying glucocorticoid induced skeletal muscle atrophy. More details are found in the text. *Red*, steps or proteins whose expression and/or activity are inhibited by glucocorticoids; *Blue*, steps or proteins whose expression or activity are induced by glucocorticoids. *GR* glucocorticoid receptor, *GRE* glucocorticoid response element, *FOXO* forkhead class O transcription factor, *FBE* forkhead transcription factors, *PIK3r1* phosphatidylinositol 3-phosphate kinase regulatory subunit 1 (PI3K p85), *p110* PI3K catalytic subunit

adequate insulin and growth factor activity keeps the pro-catabolic actions of normally fluctuating endogenous glucocorticoids in check via PI3 kinase, Akt, and mTOR activities. However, in cases of compromised insulin/IGF-1 levels or signal transduction (diabetes, starvation, inflammation), elevated endogenous glucocorticoids can acutely and directly interact with and inhibit PI3 kinase activity, decreasing downstream Akt activity and activating FOXO transcription factors that enter the nucleus and activate their pro-atrophy associated target genes. Excess synthetic glucocorticoid exposure, as the result of anti-inflammatory steroid use, may also overcome the brakes normally in place via insulin or IGF-1. The endogenous or synthetic ligand activated GR also translocates to the nucleus and directly activates atrophy associated genes via GREs; some of these targets are regulated in close collaboration with FOXO1 like MuRF1, and others (perhaps) more independently such as the regulatory subunit of PI3 kinase (p85), FOXOs themselves, and transcription factors or other effector genes like KLF15 and REDD1. Increased PI3 kinase p85 and FOXO expression with continued ligand occupation of the GR may then "feed-forward" to further inhibit Akt and downstream targets (including preventing inhibition of accumulating FOXOs), leading to an expanding network of downstream events and accelerated muscle loss seen in chronic glucocorticoid exposure.

References

1. Pereira RM, Freire de Carvalho J. Glucocorticoid-induced myopathy. Joint Bone Spine. 2011;78:41–4.
2. Overman RA, Yeh JY, Deal CL. Prevalence of oral glucocorticoid usage in the United States: a general population perspective. Arthritis Care Res (Hoboken). 2013;65:294–8.
3. Schiaffino S, Reggiani C. Fiber types in mammalian skeletal muscles. Physiol Rev. 2011;91:1447–531.
4. Goldberg AL. Protein turnover in skeletal muscle. II. Effects of denervation and cortisone on protein catabolism in skeletal muscle. J Biol Chem. 1969;244:3223–9.
5. Kelly FJ, Goldspink DF. The differing responses of four muscle types to dexamethasone treatment in the rat. Biochem J. 1982;208:147–51.
6. Gardiner PF, Montanaro G, Simpson DR, Edgerton VR. Effects of glucocorticoid treatment and food restriction on rat hindlimb muscles. Am J Physiol. 1980;238:E124–30.
7. Roy RR, Gardiner PF, Simpson DR, Edgerton VR. Glucocorticoid-induced atrophy in different fibre types of selected rat jaw and hind-limb muscles. Arch Oral Biol. 1983;28:639–43.
8. Dekhuijzen PN, Gayan-Ramirez G, Bisschop A, De Bock V, Dom R, Decramer M. Corticosteroid treatment and nutritional deprivation cause a different pattern of atrophy in rat diaphragm. J Appl Physiol. 1995;78:629–37.
9. Gardiner PF, Botterman BR, Eldred E, Simpson DR, Edgerton VR. Metabolic and contractile changes in fast and slow muscles of the cat after glucocorticoid-induced atrophy. Exp Neurol. 1978;62:241–55.
10. Bullock GR, Carter EE, Elliott P, Peters RF, Simpson P, White AM. Relative changes in the function of muscle ribosomes and mitochondria during the early phase of steroid-induced catabolism. Biochem J. 1972;127:881–92.
11. Anagnos A, Ruff RL, Kaminski HJ. Endocrine neuromyopathies. Neurol Clin. 1997;15: 673–96.

12. Baehr LM, Furlow JD, Bodine SC. Muscle sparing in muscle RING finger 1 null mice: response to synthetic glucocorticoids. J Physiol. 2011;589:4759–76.
13. Zhao W, Pan J, Zhao Z, Wu Y, Bauman WA, Cardozo CP. Testosterone protects against dexamethasone-induced muscle atrophy, protein degradation and MAFbx upregulation. J Steroid Biochem Mol Biol. 2008;110:125–9.
14. Ragnarsson O, Burt MG, Ho KK, Johannsson G. Effect of short-term GH and testosterone administration on body composition and glucose homoeostasis in men receiving chronic glucocorticoid therapy. Eur J Endocrinol. 2013;168:243–51.
15. Capaccio JA, Kurowski TT, Czerwinski SM, Chatterton Jr RT, Hickson RC. Testosterone fails to prevent skeletal muscle atrophy from glucocorticoids. J Appl Physiol. 1987;63: 328–34.
16. Batchelor TT, Taylor LP, Thaler HT, Posner JB, DeAngelis LM. Steroid myopathy in cancer patients. Neurology. 1997;48:1234–8.
17. Levin OS, Polunina AG, Demyanova MA, Isaev FV. Steroid myopathy in patients with chronic respiratory diseases. J Neurol Sci. 2014;338:96–101.
18. Decramer M, de Bock V, Dom R. Functional and histologic picture of steroid-induced myopathy in chronic obstructive pulmonary disease. Am J Respir Crit Care Med. 1996;153: 1958–64.
19. Hatakenaka M, Soeda H, Okafuji T, et al. Steroid myopathy: evaluation of fiber atrophy with T2 relaxation time—rabbit and human study. Radiology. 2006;238:650–7.
20. Khaleeli AA, Edwards RH, Gohil K, et al. Corticosteroid myopathy: a clinical and pathological study. Clin Endocrinol (Oxf). 1983;18:155–66.
21. Horber FF, Scheidegger JR, Grunig BE, Frey FJ. Thigh muscle mass and function in patients treated with glucocorticoids. Eur J Clin Invest. 1985;15:302–7.
22. Decramer M, Stas KJ. Corticosteroid-induced myopathy involving respiratory muscles in patients with chronic obstructive pulmonary disease or asthma. Am Rev Respir Dis. 1992;146:800–2.
23. Danneskiold-Samsoe B, Grimby G. Isokinetic and isometric muscle strength in patients with rheumatoid arthritis. The relationship to clinical parameters and the influence of corticosteroid. Clin Rheumatol. 1986;5:459–67.
24. Rossignol B, Gueret G, Pennec JP, et al. Effects of chronic sepsis on contractile properties of fast twitch muscle in an experimental model of critical illness neuromyopathy in the rat. Crit Care Med. 2008;36:1855–63.
25. Alamdari N, Toraldo G, Aversa Z, et al. Loss of muscle strength during sepsis is in part regulated by glucocorticoids and is associated with reduced muscle fiber stiffness. Am J Physiol Regul Integr Comp Physiol. 2012;303:R1090–9.
26. Supinski GS, Wang L, Song XH, Moylan JS, Callahan LA. Muscle-specific calpastatin overexpression prevents diaphragm weakness in cecal ligation puncture-induced sepsis. J Appl Physiol. 2014;117:921–9.
27. Robinson AJ, Clamann HP. Effects of glucocorticoids on motor units in cat hindlimb muscles. Muscle Nerve. 1988;11:703–13.
28. Gardiner PF, Edgerton VR. Contractile responses of rat fast-twitch and slow-twitch muscles to glucocorticoid treatment. Muscle Nerve. 1979;2:274–81.
29. Ruff RL, Martyn D, Gordon AM. Glucocorticoid-induced atrophy is not due to impaired excitability of rat muscle. Am J Physiol. 1982;243:E512–21.
30. Van Balkom RH, Zhan WZ, Prakash YS, Dekhuijzen PN, Sieck GC. Corticosteroid effects on isotonic contractile properties of rat diaphragm muscle. J Appl Physiol. 1997;83:1062–7.
31. Manzur AY, Kuntzer T, Pike M, Swan A. Glucocorticoid corticosteroids for Duchenne muscular dystrophy. Cochrane Database Syst Rev 2008;CD003725.
32. Hanaoka BY, Peterson CA, Horbinski C, Crofford LJ. Implications of glucocorticoid therapy in idiopathic inflammatory myopathies. Nat Rev Rheumatol. 2012;8:448–57.
33. Sali A, Guerron AD, Gordish-Dressman H, et al. Glucocorticoid-treated mice are an inappropriate positive control for long-term preclinical studies in the mdx mouse. PLoS One. 2012;7:e34204.

34. Janssen PM, Murray JD, Schill KE, et al. Prednisolone attenuates improvement of cardiac and skeletal contractile function and histopathology by lisinopril and spironolactone in the mdx mouse model of Duchenne muscular dystrophy. PLoS One. 2014;9:e88360.
35. Baltgalvis KA, Call JA, Nikas JB, Lowe DA. Effects of prednisolone on skeletal muscle contractility in mdx mice. Muscle Nerve. 2009;40:443–54.
36. Fisher I, Abraham D, Bouri K, Hoffman EP, Muntoni F, Morgan J. Prednisolone-induced changes in dystrophic skeletal muscle. FASEB J. 2005;19:834–6.
37. Hershey JW, Sonenberg N, Mathews MB. Principles of translational control: an overview. Cold Spring Harb Perspect Biol. 2012;4. pii: a011528.
38. Gordon BS, Kelleher AR, Kimball SR. Regulation of muscle protein synthesis and the effects of catabolic states. Int J Biochem Cell Biol. 2013;45:2147–57.
39. Ma XM, Blenis J. Molecular mechanisms of mTOR-mediated translational control. Nat Rev Mol Cell Biol. 2009;10:307–18.
40. Marcotte GR, West DW, Baar K. The molecular basis for load-induced skeletal muscle hypertrophy. Calcif Tissue Int. 2015;96(3):196–210.
41. Zoncu R, Efeyan A, Sabatini DM. mTOR: from growth signal integration to cancer, diabetes and ageing. Nat Rev Mol Cell Biol. 2011;12:21–35.
42. Millward DJ, Garlick PJ, Nnanyelugo DO, Waterlow JC. The relative importance of muscle protein synthesis and breakdown in the regulation of muscle mass. Biochem J. 1976;156: 185–8.
43. Rannels SR, Rannels DE, Pegg AE, Jefferson LS. Glucocorticoid effects on peptide-chain initiation in skeletal muscle and heart. Am J Physiol. 1978;235:E134–9.
44. Short KR, Nygren J, Bigelow ML, Nair KS. Effect of short-term prednisone use on blood flow, muscle protein metabolism, and function. J Clin Endocrinol Metab. 2004;89: 6198–207.
45. Liu Z, Li G, Kimball SR, Jahn LA, Barrett EJ. Glucocorticoids modulate amino acid-induced translation initiation in human skeletal muscle. Am J Physiol Endocrinol Metab. 2004;287: E275–81.
46. Mayer M, Shafrir E, Kaiser N, Milholland RJ, Rosen F. Interaction of glucocorticoid hormones with rat skeletal muscle: catabolic effects and hormone binding. Metabolism. 1976; 25:157–67.
47. Kostyo JL, Redmond AF. Role of protein synthesis in the inhibitory action of adrenal steroid hormones on amino acid transport by muscle. Endocrinology. 1966;79:531–40.
48. Young VR, Chen SC, Macdonald J. The sedimentation of rat skeletal-muscle ribosomes. Effect of hydrocortisone, insulin and diet. Biochem J. 1968;106:913–9.
49. Peters RF, Richardson MC, Small M, White AM. Some biochemical effects of triamcinolone acetonide on rat liver and muscle. Biochem J. 1970;116:349–55.
50. Rannels DE, Rannels SR, Li JB, Pegg AE, Morgan HE, Jefferson LS. Effects of glucocorticoids on peptide chain initiation in heart and skeletal muscle. Adv Myocardiol. 1980;1: 493–501.
51. Shah OJ, Kimball SR, Jefferson LS. Acute attenuation of translation initiation and protein synthesis by glucocorticoids in skeletal muscle. Am J Physiol Endocrinol Metab. 2000;278: E76–82.
52. Shah OJ, Kimball SR, Jefferson LS. Glucocorticoids abate p70(S6k) and eIF4E function in L6 skeletal myoblasts. Am J Physiol Endocrinol Metab. 2000;279:E74–82.
53. Tomas FM, Munro HN, Young VR. Effect of glucocorticoid administration on the rate of muscle protein breakdown in vivo in rats, as measured by urinary excretion of N taumethylhistidine. Biochem J. 1979;178:139–46.
54. Nagasawa T, Funabiki R. Quantitative determination of urinary N-tau-methylhistidine output as an index of myofibrillar protein degradation. J Biochem. 1981;89:1155–61.
55. Millward DJ, Bates PC. 3-Methylhistidine turnover in the whole body, and the contribution of skeletal muscle and intestine to urinary 3-methylhistidine excretion in the adult rat. Biochem J. 1983;214:607–15.

56. Mitch WE, Clark AS, May RC. Relationships between protein degradation and glucose metabolism in skeletal muscle. Prog Clin Biol Res. 1985;180:623–5.
57. Mitch WE, Goldberg AL. Mechanisms of muscle wasting. The role of the ubiquitin-proteasome pathway. N Engl J Med. 1996;335:1897–905.
58. Wing SS, Goldberg AL. Glucocorticoids activate the ATP-ubiquitin-dependent proteolytic system in skeletal muscle during fasting. Am J Physiol. 1993;264:E668–76.
59. Mitch WE, Medina R, Grieber S, et al. Metabolic acidosis stimulates muscle protein degradation by activating the adenosine triphosphate-dependent pathway involving ubiquitin and proteasomes. J Clin Invest. 1994;93:2127–33.
60. Price SR, Bailey JL, England BK. Necessary but not sufficient: the role of glucocorticoids in the acidosis-induced increase in levels of mRNAs encoding proteins of the ATP-dependent proteolytic pathway in rat muscle. Miner Electrolyte Metab. 1996;22:72–5.
61. Price SR, Mitch WE. Mechanisms stimulating protein degradation to cause muscle atrophy. Curr Opin Clin Nutr Metab Care. 1998;1:79–83.
62. Medina R, Wing SS, Haas A, Goldberg AL. Activation of the ubiquitin-ATP-dependent proteolytic system in skeletal muscle during fasting and denervation atrophy. Biomed Biochim Acta. 1991;50:347–56.
63. Lecker SH, Solomon V, Mitch WE, Goldberg AL. Muscle protein breakdown and the critical role of the ubiquitin-proteasome pathway in normal and disease states. J Nutr. 1999;129: 227S–37S.
64. Chau V, Tobias JW, Bachmair A, et al. A multiubiquitin chain is confined to specific lysine in a targeted short-lived protein. Science. 1989;243:1576–83.
65. D'Azzo A, Bongiovanni A, Nastasi T. E3 ubiquitin ligases as regulators of membrane protein trafficking and degradation. Traffic. 2005;6:429–41.
66. Tisdale MJ. The ubiquitin-proteasome pathway as a therapeutic target for muscle wasting. J Support Oncol. 2005;3:209–17.
67. Xie Y. Structure, assembly and homeostatic regulation of the 26S proteasome. J Mol Cell Biol. 2010;2:308–17.
68. Lecker SH, Goldberg AL, Mitch WE. Protein degradation by the ubiquitin-proteasome pathway in normal and disease states. J Am Soc Nephrol. 2006;17:1807–19.
69. Deshaies RJ, Joazeiro CA. RING domain E3 ubiquitin ligases. Annu Rev Biochem. 2009;78: 399–434.
70. Metzger MB, Hristova VA, Weissman AM. HECT and RING finger families of E3 ubiquitin ligases at a glance. J Cell Sci. 2012;125:531–7.
71. Auclair D, Garrel DR, Chaouki Zerouala A, Ferland LH. Activation of the ubiquitin pathway in rat skeletal muscle by catabolic doses of glucocorticoids. Am J Physiol. 1997;272: C1007–16.
72. Combaret L, Taillandier D, Dardevet D, et al. Glucocorticoids regulate mRNA levels for subunits of the 19S regulatory complex of the 26S proteasome in fast-twitch skeletal muscles. Biochem J. 2004;378:239–46.
73. Britto FA, Begue G, Rossano B, et al. REDD1 deletion prevents dexamethasone-induced skeletal muscle atrophy. Am J Physiol Endocrinol Metab. 2014;307:E983–93.
74. Price SR, Bailey JL, Wang X, et al. Muscle wasting in insulinopenic rats results from activation of the ATP-dependent, ubiquitin-proteasome proteolytic pathway by a mechanism including gene transcription. J Clin Invest. 1996;98:1703–8.
75. Tawa Jr NE, Odessey R, Goldberg AL. Inhibitors of the proteasome reduce the accelerated proteolysis in atrophying rat skeletal muscles. J Clin Invest. 1997;100:197–203.
76. Wing SS, Haas AL, Goldberg AL. Increase in ubiquitin-protein conjugates concomitant with the increase in proteolysis in rat skeletal muscle during starvation and atrophy denervation. Biochem J. 1995;307(Pt 3):639–45.
77. Tiao G, Hobler S, Wang JJ, et al. Sepsis is associated with increased mRNAs of the ubiquitin-proteasome proteolytic pathway in human skeletal muscle. J Clin Invest. 1997;99:163–8.

78. Voisin L, Breuille D, Combaret L, et al. Muscle wasting in a rat model of long-lasting sepsis results from the activation of lysosomal, Ca2+-activated, and ubiquitin-proteasome proteolytic pathways. J Clin Invest. 1996;97:1610–7.
79. Tiao G, Fagan JM, Samuels N, et al. Sepsis stimulates nonlysosomal, energy-dependent proteolysis and increases ubiquitin mRNA levels in rat skeletal muscle. J Clin Invest. 1994;94: 2255–64.
80. Baracos VE, DeVivo C, Hoyle DH, Goldberg AL. Activation of the ATP-ubiquitin-proteasome pathway in skeletal muscle of cachectic rats bearing a hepatoma. Am J Physiol. 1995;268: E996–1006.
81. Bodine SC, Baehr LM. Skeletal muscle atrophy and the E3 ubiquitin ligases MuRF1 and MAFbx/atrogin-1. Am J Physiol Endocrinol Metab. 2014;307:E469–84.
82. Centner T, Yano J, Kimura E, et al. Identification of muscle specific ring finger proteins as potential regulators of the titin kinase domain. J Mol Biol. 2001;306:717–26.
83. Gomes MD, Lecker SH, Jagoe RT, Navon A, Goldberg AL. Atrogin-1, a muscle-specific F-box protein highly expressed during muscle atrophy. Proc Natl Acad Sci U S A. 2001; 98:14440–5.
84. Bodine SC, Latres E, Baumhueter S, et al. Identification of ubiquitin ligases required for skeletal muscle atrophy. Science. 2001;294:1704–8.
85. Files DC, D'Alessio FR, Johnston LF, et al. A critical role for muscle ring finger-1 in acute lung injury-associated skeletal muscle wasting. Am J Respir Crit Care Med. 2012;185: 825–34.
86. Frost RA, Lang CH. Multifaceted role of insulin-like growth factors and mammalian target of rapamycin in skeletal muscle. Endocrinol Metab Clin North Am. 2012;41:297–322.
87. Gayan-Ramirez G, Vanderhoydonc F, Verhoeven G, Decramer M. Acute treatment with corticosteroids decreases IGF-1 and IGF-2 expression in the rat diaphragm and gastrocnemius. Am J Respir Crit Care Med. 1999;159:283–9.
88. Inder WJ, Jang C, Obeyesekere VR, Alford FP. Dexamethasone administration inhibits skeletal muscle expression of the androgen receptor and IGF-1–implications for steroid-induced myopathy. Clin Endocrinol (Oxf). 2010;73:126–32.
89. Schakman O, Gilson H, de Coninck V, et al. Insulin-like growth factor-I gene transfer by electroporation prevents skeletal muscle atrophy in glucocorticoid-treated rats. Endocrinology. 2005;146:1789–97.
90. Schakman O, Kalista S, Bertrand L, et al. Role of Akt/GSK-3beta/beta-catenin transduction pathway in the muscle anti-atrophy action of insulin-like growth factor-I in glucocorticoid-treated rats. Endocrinology. 2008;149:3900–8.
91. Sacheck JM, Ohtsuka A, McLary SC, Goldberg AL. IGF-1 stimulates muscle growth by suppressing protein breakdown and expression of atrophy-related ubiquitin-ligases, atrogin-1 and MuRF1. Am J Physiol Endocrinol Metab. 2004;287(4):E591–601.
92. Stitt TN, Drujan D, Clarke BA, et al. The IGF-1/PI3K/Akt pathway prevents expression of muscle atrophy-induced ubiquitin ligases by inhibiting FOXO transcription factors. Mol Cell. 2004;14:395–403.
93. Fournier M, Huang ZS, Li H, Da X, Cercek B, Lewis MI. Insulin-like growth factor I prevents corticosteroid-induced diaphragm muscle atrophy in emphysematous hamsters. Am J Physiol Regul Integr Comp Physiol. 2003;285:R34–43.
94. Chrysis D, Underwood LE. Regulation of components of the ubiquitin system by insulin-like growth factor I and growth hormone in skeletal muscle of rats made catabolic with dexamethasone. Endocrinology. 1999;140:5635–41.
95. Kanda F, Takatani K, Okuda S, Matsushita T, Chihara K. Preventive effects of insulin like growth factor-I on steroid-induced muscle atrophy. Muscle Nerve. 1999;22:213–7.
96. Schakman O, Kalista S, Barbe C, Loumaye A, Thissen JP. Glucocorticoid-induced skeletal muscle atrophy. Int J Biochem Cell Biol. 2013;45:2163–72.

97. Armstrong DD, Esser KA. Wnt/beta-catenin signaling activates growth-control genes during overload-induced skeletal muscle hypertrophy. Am J Physiol Cell Physiol. 2005;289: C853–9.
98. Nakao R, Hirasaka K, Goto J, et al. Ubiquitin ligase Cbl-b is a negative regulator for insulin-like growth factor 1 signaling during muscle atrophy caused by unloading. Mol Cell Biol. 2009;29:4798–811.
99. Morgan SA, Sherlock M, Gathercole LL, et al. 11beta-hydroxysteroid dehydrogenase type 1 regulates glucocorticoid-induced insulin resistance in skeletal muscle. Diabetes. 2009;58: 2506–15.
100. Zheng B, Ohkawa S, Li H, Roberts-Wilson TK, Price SR. FOXO3a mediates signaling cross-talk that coordinates ubiquitin and atrogin-1/MAFbx expression during glucocorticoid-induced skeletal muscle atrophy. FASEB J. 2010;24:2660–9.
101. Koh A, Lee MN, Yang YR, et al. C1-Ten is a protein tyrosine phosphatase of insulin receptor substrate 1 (IRS-1), regulating IRS-1 stability and muscle atrophy. Mol Cell Biol. 2013;33: 1608–20.
102. Kuo T, Lew MJ, Mayba O, Harris CA, Speed TP, Wang JC. Genome-wide analysis of glucocorticoid receptor-binding sites in myotubes identifies gene networks modulating insulin signaling. Proc Natl Acad Sci U S A. 2012;109:11160–5.
103. Furlow JD, Watson ML, Waddell DS, et al. Altered gene expression patterns in muscle ring finger 1 null mice during denervation- and dexamethasone-induced muscle atrophy. Physiol Genomics. 2013;45:1168–85.
104. Wang H, Kubica N, Ellisen LW, Jefferson LS, Kimball SR. Dexamethasone represses signaling through the mammalian target of rapamycin in muscle cells by enhancing expression of REDD1. J Biol Chem. 2006;281:39128–34.
105. Kumari R, Willing LB, Jefferson LS, Simpson IA, Kimball SR. REDD1 (regulated in development and DNA damage response 1) expression in skeletal muscle as a surrogate biomarker of the efficiency of glucocorticoid receptor blockade. Biochem Biophys Res Commun. 2011;412:644–7.
106. Dennis MD, Coleman CS, Berg A, Jefferson LS, Kimball SR. REDD1 enhances protein phosphatase 2A-mediated dephosphorylation of Akt to repress mTORC1 signaling. Sci Signal. 2014;7:ra68.
107. Kadmiel M, Cidlowski JA. Glucocorticoid receptor signaling in health and disease. Trends Pharmacol Sci. 2013;34:518–30.
108. Duma D, Jewell CM, Cidlowski JA. Multiple glucocorticoid receptor isoforms and mechanisms of post-translational modification. J Steroid Biochem Mol Biol. 2006;102:11–21.
109. Schaaf MJ, Champagne D, van Laanen IH, et al. Discovery of a functional glucocorticoid receptor beta-isoform in zebrafish. Endocrinology. 2008;149:1591–9.
110. Shimizu N, Yoshikawa N, Ito N, et al. Crosstalk between glucocorticoid receptor and nutritional sensor mTOR in skeletal muscle. Cell Metab. 2011;13:170–82.
111. Oakley RH, Cidlowski JA. Homologous down regulation of the glucocorticoid receptor: the molecular machinery. Crit Rev Eukaryot Gene Expr. 1993;3:63–88.
112. Yao Z, DuBois DC, Almon RR, Jusko WJ. Modeling circadian rhythms of glucocorticoid receptor and glutamine synthetase expression in rat skeletal muscle. Pharm Res. 2006;23: 670–9.
113. Witchel SF, DeFranco DB. Mechanisms of disease: regulation of glucocorticoid and receptor levels–impact on the metabolic syndrome. Nat Clin Pract Endocrinol Metab. 2006;2: 621–31.
114. Aubry EM, Odermatt A. Retinoic acid reduces glucocorticoid sensitivity in C2C12 myotubes by decreasing 11beta-hydroxysteroid dehydrogenase type 1 and glucocorticoid receptor activities. Endocrinology. 2009;150:2700–8.
115. Wang SC, Myers S, Dooms C, Capon R, Muscat GE. An ERRbeta/gamma agonist modulates GRalpha expression, and glucocorticoid responsive gene expression in skeletal muscle cells. Mol Cell Endocrinol. 2010;315:146–52.

116. Almon RR, DuBois DC, Yao Z, Hoffman EP, Ghimbovschi S, Jusko WJ. Microarray analysis of the temporal response of skeletal muscle to methylprednisolone: comparative analysis of two dosing regimens. Physiol Genomics. 2007;30:282–99.
117. Lecker SH, Jagoe RT, Gilbert A, et al. Multiple types of skeletal muscle atrophy involve a common program of changes in gene expression. FASEB J. 2004;18:39–51.
118. Carraro L, Ferraresso S, Cardazzo B, et al. Expression profiling of skeletal muscle in young bulls treated with steroidal growth promoters. Physiol Genomics. 2009;38:138–48.
119. Kukreti H, Amuthavalli K, Harikumar A, et al. Muscle-specific microRNA1 (miR1) targets heat shock protein 70 (HSP70) during dexamethasone-mediated atrophy. J Biol Chem. 2013;288:6663–78.
120. Busillo JM, Cidlowski JA. The five Rs of glucocorticoid action during inflammation: ready, reinforce, repress, resolve, and restore. Trends Endocrinol Metab. 2013;24:109–19.
121. Yang H, Menconi MJ, Wei W, Petkova V, Hasselgren PO. Dexamethasone upregulates the expression of the nuclear cofactor p300 and its interaction with C/EBPbeta in cultured myotubes. J Cell Biochem. 2005;94:1058–67.
122. Tobimatsu K, Noguchi T, Hosooka T, et al. Overexpression of the transcriptional coregulator Cited2 protects against glucocorticoid-induced atrophy of C2C12 myotubes. Biochem Biophys Res Commun. 2009;378:399–403.
123. Yang H, Wei W, Menconi M, Hasselgren PO. Dexamethasone-induced protein degradation in cultured myotubes is p300/HAT dependent. Am J Physiol Regul Integr Comp Physiol. 2007;292:R337-4.
124. Alamdari N, Aversa Z, Castillero E, Hasselgren PO. Acetylation and deacetylation—novel factors in muscle wasting. Metabolism. 2013;62:1–11.
125. Proserpio V, Fittipaldi R, Ryall JG, Sartorelli V, Caretti G. The methyltransferase SMYD3 mediates the recruitment of transcriptional cofactors at the myostatin and c-Met genes and regulates skeletal muscle atrophy. Genes Dev. 2013;27:1299–312.
126. Kim H, Heo K, Kim JH, Kim K, Choi J, An W. Requirement of histone methyltransferase SMYD3 for estrogen receptor-mediated transcription. J Biol Chem. 2009;284:19867–77.
127. Furuyama T, Kitayama K, Yamashita H, Mori N. Forkhead transcription factor FOXO1 (FKHR)-dependent induction of PDK4 gene expression in skeletal muscle during energy deprivation. Biochem J. 2003;375:365–71.
128. Cho JE, Fournier M, Da X, Lewis MI. Time course expression of Foxo transcription factors in skeletal muscle following corticosteroid administration. J Appl Physiol. 2010;108: 137–45.
129. Sun H, Gong Y, Qiu J, Chen Y, Ding F, Zhao Q. TRAF6 inhibition rescues dexamethasone-induced muscle atrophy. Int J Mol Sci. 2014;15:11126–41.
130. Webb AE, Brunet A. FOXO transcription factors: key regulators of cellular quality control. Trends Biochem Sci. 2014;39:159–69.
131. Gross DN, Wan M, Birnbaum MJ. The role of FOXO in the regulation of metabolism. Curr Diab Rep. 2009;9:208–14.
132. Zhao J, Brault JJ, Schild A, et al. FoxO3 coordinately activates protein degradation by the autophagic/lysosomal and proteasomal pathways in atrophying muscle cells. Cell Metab. 2007;6:472–83.
133. Mammucari C, Milan G, Romanello V, et al. FoxO3 controls autophagy in skeletal muscle in vivo. Cell Metab. 2007;6:458–71.
134. Waddell DS, Baehr LM, van den Brandt J, et al. The glucocorticoid receptor and FOXO1 synergistically activate the skeletal muscle atrophy-associated MuRF1 gene. Am J Physiol Endocrinol Metab. 2008;295:E785–97.
135. Reed SA, Sandesara PB, Senf SM, Judge AR. Inhibition of FoxO transcriptional activity prevents muscle fiber atrophy during cachexia and induces hypertrophy. FASEB J. 2012;26: 987–1000.

136. Senf SM, Sandesara PB, Reed SA, Judge AR. p300 Acetyltransferase activity differentially regulates the localization and activity of the FOXO homologues in skeletal muscle. Am J Physiol Cell Physiol. 2011;300:C1490–501.
137. Beharry AW, Sandesara PB, Roberts BM, Ferreira LF, Senf SM, Judge AR. HDAC1 activates FoxO and is both sufficient and required for skeletal muscle atrophy. J Cell Sci. 2014;127:1441–53.
138. Yang H, Mammen J, Wei W, et al. Expression and activity of C/EBPbeta and delta are upregulated by dexamethasone in skeletal muscle. J Cell Physiol. 2005;204:219–26.
139. Gonnella P, Alamdari N, Tizio S, Aversa Z, Petkova V, Hasselgren PO. C/EBPbeta regulates dexamethasone-induced muscle cell atrophy and expression of atrogin-1 and MuRF1. J Cell Biochem. 2011;112:1737–48.
140. Bruscoli S, Donato V, Velardi E, et al. Glucocorticoid-induced leucine zipper (GILZ) and long GILZ inhibit myogenic differentiation and mediate anti-myogenic effects of glucocorticoids. J Biol Chem. 2010;285:10385–96.
141. Ayroldi E, Riccardi C. Glucocorticoid-induced leucine zipper (GILZ): a new important mediator of glucocorticoid action. FASEB J. 2009;23:3649–58.
142. Kaestner KH. The FoxA factors in organogenesis and differentiation. Curr Opin Genet Dev. 2010;20:527–32.
143. Revollo JR, Cidlowski JA. Mechanisms generating diversity in glucocorticoid receptor signaling. Ann N Y Acad Sci. 2009;1179:167–78.
144. Hu Z, Wang H, Lee IH, Du J, Mitch WE. Endogenous glucocorticoids and impaired insulin signaling are both required to stimulate muscle wasting under pathophysiological conditions in mice. J Clin Invest. 2009;119:3059–69.
145. Fang CH, Li BG, James JH, et al. Protein breakdown in muscle from burned rats is blocked by insulin-like growth factor i and glycogen synthase kinase-3beta inhibitors. Endocrinology. 2005;146:3141–9.
146. Rubio-Patino C, Palmeri CM, Perez-Perarnau A, et al. Glycogen synthase kinase-3beta is involved in ligand-dependent activation of transcription and cellular localization of the glucocorticoid receptor. Mol Endocrinol. 2012;26:1508–20.
147. Sarabdjitsingh RA, Joels M, de Kloet ER. Glucocorticoid pulsatility and rapid corticosteroid actions in the central stress response. Physiol Behav. 2012;106:73–80.
148. Tasker JG, Di S, Malcher-Lopes R. Minireview: rapid glucocorticoid signaling via membrane-associated receptors. Endocrinology. 2006;147:5549–56.
149. Perez MH, Cormack J, Mallinson D, Mutungi G. A membrane glucocorticoid receptor mediates the rapid/non-genomic actions of glucocorticoids in mammalian skeletal muscle fibres. J Physiol. 2013;591:5171–85.
150. Kewalramani G, Puthanveetil P, Kim MS, et al. Acute dexamethasone-induced increase in cardiac lipoprotein lipase requires activation of both Akt and stress kinases. Am J Physiol Endocrinol Metab. 2008;295:E137–47.
151. Lee SR, Kim HK, Youm JB, et al. Non-genomic effect of glucocorticoids on cardiovascular system. Pflugers Arch. 2012;464:549–59.
152. Stitt TN, Drujan D, Clarke BA, et al. The IGF-1/PI3K/Akt pathway prevents expression of muscle atrophy-induced ubiquitin liganses by inhibiting FOXO transcription factors. Mol Cell. 2004;14:1–14.
153. Latres E, Amini AR, Amini AA, et al. Insulin-like growth factor-1 (IGF-1) inversely regulates atrophy-induced genes via the phosphatidylinositol 3-kinase/Akt/mammalian target of rapamycin (PI3K/Akt/mTOR) pathway. J Biol Chem. 2005;280:2737–44.
154. Chapman K, Holmes M, Seckl J. 11beta-hydroxysteroid dehydrogenases: intracellular gate-keepers of tissue glucocorticoid action. Physiol Rev. 2013;93:1139–206.
155. Short KR, Bigelow ML, Nair KS. Short-term prednisone use antagonizes insulin's anabolic effect on muscle protein and glucose metabolism in young healthy people. Am J Physiol Endocrinol Metab. 2009;297:E1260–8.

156. Almon RR, Dubois DC, Jin JY, Jusko WJ. Temporal profiling of the transcriptional basis for the development of corticosteroid-induced insulin resistance in rat muscle. J Endocrinol. 2005;184:219–32.
157. Patel R, Bookout AL, Magomedova L, et al. Glucocorticoids regulate the metabolic hormone FGF21 in a feed-forward loop. Mol Endocrinol. 2015;29(2):213–23.
158. Mazziotti G, Giustina A. Glucocorticoids and the regulation of growth hormone secretion. Nat Rev Endocrinol. 2013;9:265–76.
159. Wurtman RJ. Stress and the adrenocortical control of epinephrine synthesis. Metabolism. 2002;51:11–4.
160. la Fleur SE. The effects of glucocorticoids on feeding behavior in rats. Physiol Behav. 2006;89:110–4.
161. Watson ML, Baehr LM, Reichardt HM, Tuckermann JP, Bodine SC, Furlow JD. A cell-autonomous role for the glucocorticoid receptor in skeletal muscle atrophy induced by systemic glucocorticoid exposure. Am J Physiol Endocrinol Metab. 2012;302:E1210–20.
162. Braun TP, Grossberg AJ, Krasnow SM, et al. Cancer- and endotoxin-induced cachexia require intact glucocorticoid signaling in skeletal muscle. FASEB J. 2013;27:3572–82.
163. Braun TP, Szumowski M, Levasseur PR, et al. Muscle atrophy in response to cytotoxic chemotherapy is dependent on intact glucocorticoid signaling in skeletal muscle. PLoS One. 2014;9:e106489.
164. Zhao W, Qin W, Pan J, Wu Y, Bauman WA, Cardozo C. Dependence of dexamethasone-induced Akt/FOXO1 signaling, upregulation of MAFbx, and protein catabolism upon the glucocorticoid receptor. Biochem Biophys Res Commun. 2009;378:668–72.
165. Nesan D, Kamkar M, Burrows J, Scott IC, Marsden M, Vijayan MM. Glucocorticoid receptor signaling is essential for mesoderm formation and muscle development in zebrafish. Endocrinology. 2012;153:1288–300.
166. Segal DJ, Meckler JF. Genome engineering at the dawn of the golden age. Annu Rev Genomics Hum Genet. 2013;14:135–58.
167. Sandri M, Sandri C, Gilbert A, et al. Foxo transcription factors induce the atrophy-related ubiquitin ligase atrogin-1 and cause skeletal muscle atrophy. Cell. 2004;117:399–412.
168. Qin W, Pan J, Qin Y, Lee DN, Bauman WA, Cardozo C. Identification of functional glucocorticoid response elements in the mouse FoxO1 promoter. Biochem Biophys Res Commun. 2014;450:979–83.
169. Ma K, Mallidis C, Bhasin S, et al. Glucocorticoid-induced skeletal muscle atrophy is associated with upregulation of myostatin gene expression. Am J Physiol Endocrinol Metab. 2003;285:E363–71.
170. Qin J, Du R, Yang YQ, et al. Dexamethasone-induced skeletal muscle atrophy was associated with upregulation of myostatin promoter activity. Res Vet Sci. 2013;94:84–9.

Part III
Specific Effects of Glucocorticoids on Tissues

Chapter 8
Glucocorticoid-Induced Osteoporosis

Baruch Frenkel, Wendy White, and Jan Tuckermann

Abstract Osteoporosis is among the most devastating side effects of glucocorticoid (GC) therapy for the management of inflammatory and auto-immune diseases. Evidence from both humans and mice indicate deleterious skeletal effects within weeks of pharmacological GC administration, both related and unrelated to a decrease in bone mineral density (BMD). Osteoclast numbers and bone resorption are also rapidly increased, and together with osteoblast inactivation and decreased bone formation, these changes lead the fastest loss in BMD during the initial disease phase. Bone resorption then decreases to sub-physiological levels, but persistent and severe inhibition of bone formation leads to further bone loss and progressively increased fracture risk, up to an order of magnitude higher than that observed in untreated individuals. Bone forming osteoblasts are thus considered the main culprits in GC-induced osteoporosis (GIO). Accordingly, we focus this review primarily on deleterious effects on osteoblasts: inhibition of cell replication and function and acceleration of apoptosis. Mediating these adverse effects, GCs target pivotal regulatory mechanisms that govern osteoblast growth, differentiation and survival. Specifically, GCs inhibit growth factor pathways, including Insulin Growth Factors, Growth Hormone, Hepatocyte Growth/Scatter Factor and IL6-type cytokines. They also inhibit downstream kinases, including PI3-kinase and the MAP kinase ERK, the latter attributable in part to direct transcriptional stimulation of MAP kinase phosphatase 1. Most importantly, however, GCs inhibit the Wnt signaling pathway,

B. Frenkel, D.M.D., Ph.D. (✉)
Department of Orthopaedic Surgery, Keck School of Medicine,
Institute for Genetic Medicine, University of Southern California,
2250 Alcazar Street, CSC-240, Los Angeles, CA 90033, USA

Department of Biochemistry and Molecular Biology, Keck School of Medicine,
Institute for Genetic Medicine, University of Southern California,
2250 Alcazar Street, CSC-240, Los Angeles, CA 90033, USA
e-mail: frenkel@usc.edu

W. White, M.D.
Department of Diabetes and Endocrinology, Eisenhower Medical Center,
39000 Bob Hope Drive, Rancho Mirage, CA 92270, USA

J. Tuckermann
Institute for Comparative Molecular Endocrinology, University of Ulm, Helmholtzstrasse 8/1,
Ulm, D-89081, Germany

J.-C. Wang, C. Harris (eds.), *Glucocorticoid Signaling*, Advances in Experimental
Medicine and Biology 872, DOI 10.1007/978-1-4939-2895-8_8

which plays a pivotal role in osteoblast replication, function and survival. They transcriptionally stimulate expression of Wnt inhibitors of both the Dkk and Sfrp families, and they induce reactive oxygen species (ROS), which result in loss of ß-catenin to ROS-activated FoxO transcription factors. Identification of dissociated GCs, which would suppress the immune system without causing osteoporosis, is proving more challenging than initially thought, and GIO is currently managed by co-treatment with bisphosphonates or PTH. These drugs, however, are not ideally suited for GIO. Future therapeutic approaches may aim at GC targets such as those mentioned above, or newly identified targets including the Notch pathway, the AP-1/Il11 axis and the osteoblast master regulator RUNX2.

Keywords GIO • Proliferation • Osteoblast • Wnt • Notch • ERK • Akt • MKP-1 • FoxO • RUNX2

Glucocorticoid-Induced Osteoporosis (GIO)

The Clinical Problem

Since 1950's Nobel laureates P. Hench, E. Kendall and T. Reichstein introduced glucocorticoids (GCs) for the treatment of rheumatoid arthritis, these drugs have been widely used in the management of a myriad autoimmune and inflammatory diseases. GCs mainly act through the *NR3C1*-encoded glucocorticoid receptor (GR). Once bound, the activated GR translocates to the nucleus and regulates transcription of target genes either as a homodimer or a monomer that can interact with other transcription factors. Physicians frequently prescribe synthetic GCs to treat patients with advanced rheumatoid arthritis, asthma, multiple sclerosis, systemic lupus erythematosus, inflammatory bowel disease, and, in combination with other drugs, patients with hematologic and other malignancies and after organ transplantation. The high efficacy of synthetic GCs as immune suppressants is blemished, however, by deleterious side effects, in particular rapid bone loss leading to GC-induced osteoporosis (GIO). The GR is ubiquitously expressed, including in the various cell types present in bone, such as bone forming osteoblasts, matrix embedded osteocytes, and bone resorbing osteoclasts. GIO results from effects of GCs on all of these, as well as additional cell types, of which the effects on osteoblasts are generally considered the most important in mediating the progressive bone loss in patients chronically treated with GCs.

Following bone loss in post-menopausal women and in aging individuals of both sexes, GIO is the third most-common etiology of pathological bone loss, approaching by some estimates 20 % of all patients with osteoporosis [1]. Several large population-based and cohort studies have evaluated the relationship between oral GC and bone mineral density (BMD) as well as fracture risk, and the largest of these, the General Practice Research Database (GPRD) in the UK, has shown that even daily doses as low as 2.5 mg of prednisone are associated with increased risk for vertebral and hip fracture [2, 3]. BMD decreases as a function of cumulative exposure [4] and fracture risk depends additionally on the maximum daily dose [4]. The highest risk is observed in the spine, where the incidence of fracture compared to controls increases by more than fivefold with daily doses equivalent to or greater than 7.5 mg prednisolone [3]. Within the first year after treatment initiation, patients on oral GCs show an average of 54 % increased fracture risk, which far outpaces the decrease seen in BMD [3]. The apparent BMD-independent fracture risk may reflect effects of GCs on neurons and muscle cells, as well as microanatomical changes to the bone microarchitecture and/or material properties not captured by conventional imaging techniques [5]. Nonetheless, a decrease in BMD is readily detectable by the 12th week of GC administration and rapid bone loss continues through the 24th week of treatment. Thereafter, the BMD loss rate appears to slow, although fracture risk continues to accumulate [4, 6, 7]. Bone is lost from both the cortical and the trabecular compartments, and even though the latter is affected more severely, the trabecular thinning is usually not associated with perforations [4, 8, 9].

In addition to oral GC administration, many patients with severe obstructive respiratory diseases such as asthma and COPD are treated with high-dose inhaled corticosteroids. While there has been some controversy over whether or not inhaled GCs lead to an increased risk of fracture, or whether increased fracture rates were secondary to the underlying disease, the EOLO (Evaluation of Obstructive Lung Disease and Osteoporosis) Study group has recently shown that inhaled GCs at the highest doses, >1500 μg/day, have a 1.4-fold increased risk of fracture compared with controls and patients on lower doses of inhaled steroids. Additionally, univariate and multivariate analyses did not show any independent association between obstructive lung disease and fracture risk [10, 11].

Several recent reviews provide excellent coverage of clinical manifestations and management of GIO [12, 13]. When applicable, local GC administration should be preferred over systemic treatment, and chronic treatment with oral GCs should be considered a last resort. Once prescribed, oral GCs should be accompanied with prophylactic therapy to minimize deleterious side effects. This includes cases where high dose GCs are prescribed for short periods of time to control "flares" (exacerbations) of common inflammatory diseases because, as mentioned above, GC increase fracture risk within a few weeks of administration. Specialists treating autoimmune and inflammatory diseases often fail to take the necessary measures to prevent osteoporosis, and GIO-related fractures are frequently the basis for successful litigation [12]. The American College of Rheumatology advocates management of patients initiating ≥3-month treatment with a daily dose of ≥7.5 mg of prednisone equivalents with anti-resorptive therapy (e.g., bisphosphonates) to ameliorate GIO [14]. Recently, however, the bone anabolic peptide PTH(1-34) has been shown to counteract deleterious effects of GCs on osteoblasts *in vitro* and in mice [15], and a clinical trial suggested that it was in fact superior to anti-resorptive therapy for GIO [16]. The results of this clinical trial are consistent with the central role of osteoblasts in GIO (section "Cellular Mechanisms of GIO: Osteoblasts at the Center Stage"), and provide the prospect that further improvement of patient care may be achieved through better understanding of the underlying cellular and molecular disease mechanisms. Most of the published work on cellular and molecular mechanisms underlying GIO has focused on osteoblasts and their precursors, which is reviewed in detail in this chapter. We open, however, with a brief review of mouse models, which also highlight the less well-investigated roles for GCs in osteocytes and osteoclasts.

The Mouse as a Model

The adult mammalian skeleton undergoes continuous remodeling throughout life. Bone resorbing osteoclasts, bone forming osteoblasts and matrix-embedded osteocytes that derive from osteoblasts are the major cell types responsible for this process. Early studies with several animal models resulted in paradoxical observations, which impeded progress with in vivo investigation of GIO, but recent work shows that principal GIO mechanisms can be usefully modeled in mice of certain strains, including Swiss-Webster [17], Balb/c [18] and FVB/N [19]. Similar to the human disease, the chronic phase of decreased osteoblastogenesis and bone formation in these mouse models is preceded by an early phase dominated by exaggerated osteoclast-mediated bone resorption, resulting in the highest rates of bone loss early on after commencement of GC administration [17, 20].

Mechanistic investigation of GIO in vivo, including in mice, is limited because results of molecular analyses, for example gene expression data, are typically obtained at the tissue level and not from individual cell types at specific differentiation stages. Mouse genetics, however, has proven invaluable for functional assessment, in vivo, of the significance of various molecular aspects of GR signaling in particular cell types. For example, a direct role in GIO has been unequivocally assigned to osteoblasts using two mouse models where GC signaling was abrogated specifically in this cell type. In one model, GC signaling was ablated by knocking out the *GR* gene in cells that express Cre recombinase under the control of *Runx2* regulatory sequences [19]. In an earlier study, GC signaling in osteoblasts was abrogated by over-expression of the GC inactivating enzyme 11ß-HSD2 under the control of the Osteocalcin Gene 2 (*OG2*) promoter [21]. In both mouse models, administration of prednisolone resulted in less severe GIO compared to that observed in control mice, indicating that GC signaling in osteoblasts plays a critical role in GIO [19, 21] (section "Cellular Mechanisms of GIO: Osteoblasts at the Center Stage" below). Interestingly, the *OG2-HSD2* transgenic mice had no bone phenotype at the basal state [21]. In contrast, some skeletal deficiencies at the basal state were observed in transgenic mice expressing 11ß-HSD2 in osteoblasts under the control of the Collagen α1(I) promoter [22, 23] and in mice lacking the GR in the osteoblast lineage [19], indicating that endogenous GC signaling plays a bone anabolic role during early stages of osteoblast differentiation.

Another mouse model was used to assess the contribution of GR homodimerization to GIO. Classically, transcriptional stimulation in response to GCs occurs through binding of GR homodimers to palindromic GC response elements (GREs) at target gene enhancers, and for many years this was considered the predominant mechanism underlying GIO. A paradigm shift, however, was offered by administration of prednisolone to so-called GR^{dim} mice, harboring a GR mutant with an impaired dimerization interface [24, 25]. After 2 weeks of GC treatment, these mice had reduced osteoblast colony forming units (CFU-OBs) in the bone marrow, reduced osteoblast numbers on the bone surface, lower bone formation rates, and decreased bone mass, all similar to GC-treated wild type mice [19]. Thus, dimerization-independent mechanisms, such as binding of GR monomers to non-palindromic DNA response elements [26, 27], appear to be critical for the development of GIO. The significance of this paradigm shift to the future of GIO research is discussed in section "Glucocorticoids Without Osteoporosis?".

Finally, although this chapter addresses the contribution of osteoclastogenesis to GIO only briefly, such contribution appears quite sizable, especially at the early phase of GIO. Indeed, GCs promote osteoclast survival and function in vivo [20, 28] and the GC-induced bone loss (albeit without loss of bone strength) in the aforementioned OG2-11ß-HSD2 mice that lack GC signaling in osteoblasts [21] could result from persistent activation of osteoclasts in the presence of GCs. GC-stimulated bone resorption likely occurs through their receptors in cells of the osteoblast lineage (see section "Involvement of Cells Other than Osteoblasts in GIO"), although involvement of osteoclast GR in increased resorption has been suggested based on evidence from mice with conditional GR inactivation in the monocytic lineage [29, 30].

Cellular Mechanisms of GIO: Osteoblasts at the Center Stage

The multifaceted and complex mechanisms underlying GIO have been extensively reviewed [12, 13, 31–33]. Early anecdotal evidence suggested indirect effects of GCs on bone through their actions in the gonads and in calcium-regulating organs (kidney, intestine). However, more recent clinical observations and in vivo investigation of mouse models argue against such indirect effects as primary pathogenic mechanisms in GIO [12, 13, 44]. Instead, it is now widely accepted that GIO is caused primarily through direct effects of GCs in bone cells.

Bone loss in the chronic state of GIO is mostly attributable to decreased bone formation by osteoblasts [13], secondary to impaired osteoblast cell replication (section "Glucocorticoids Inhibit Osteoblast Cell Cycle" below), diminished osteoblast differentiation and function (section "Glucocorticoids Inhibit Osteoblast Differentiation and Function" below), and accelerated osteoblast and osteocyte apoptosis (section "Glucocorticoids Promote Osteoblast Apoptosis" below). Additional considerations will be briefly reviewed in the section "Involvement of Cells Other than Osteoblasts in GIO".

Glucocorticoids Inhibit Osteoblast Cell Cycle

Reports on GC-mediated inhibition of osteoblast proliferation *in vitro* date back to the 1970s [35]. Definitive in vivo evidence for inhibition of osteoblastic cell proliferation was demonstrated in GC-treated mice, where a dramatic decrease was observed in the number of bone marrow-derived CFU-Ob representing mesenchymal progenitors capable of bone formation [17, 19].

While acting as anti-mitogens in a variety of cell types, including fibroblasts, lymphocytes, hepatocytes, and lung alveolar cells, GCs engage different cell cycle regulatory mechanisms in a context-dependent manner. Even among osteoblast models, effects of GCs on cell cycle progression and the underlying molecular mechanisms vary as a function of the particular culture system and the differentiation stage. Treatment of mouse calvaria-derived osteoblasts with dexamethasone (dex) resulted in up to ~50 % reduction in the proportion of cells traversing through the active cell cycle phases (S/G2/M), but this inhibition occurred only at and after, not before, a well-defined developmental stage marked by a commitment to terminal differentiation [36, 37]. This differentiation stage-related anti-mitogenic effect of GCs was demonstrable in both the MC3T3-E1 immortalized cell line [36] and primary osteoblast cultures derived from newborn mouse calvariae [37], and in both cases inhibition of cell cycle progression was most strongly associated with suppression of cyclin A expression [36, 37]. In MC3T3-E1 cells, inhibition of cell cycle progression (as well as promotion of apoptosis) was also associated with activation of p53 [38]. In primary human osteoblast culture models, dex decreased thymidine incorporation into newly synthesized DNA and the proportion of cells traversing through the S/G2/M phases of the cell cycle [39]. In this system, as well as in human osteosarcoma cell lines, the anti-mitogenic effect of GCs is associated with either down-regulation of cell cycle stimulators such as CDK2, 4 and 6, cyclin D, c-Myc, and E2F-1, or upregulation of the cyclin-dependent inhibitors p21 and p27 [39, 40].

GC-mediated inhibition of thymidine incorporation was also demonstrated in the murine MBA-15.4 and the human MG-63 cell lines [41]. Co-treatment with sodium orthovanadate, a protein tyrosine phosphatase inhibitor, reversed the anti-mitogenic effect of dex, implicating tyrosine phosphatase(s) in the anti-mitogenic activity of GCs in these cells (see section "Additional Signaling Pathways"). Additional GC-regulated mechanisms involved in their anti-mitogenic effect in osteoblasts include the Wnt signaling pathway, Akt and MAPK signaling, as well as transcription factors such as FoxO3 (see section "Molecular Targets of Glucocorticoids in Osteoblasts"). Although the relative contribution of the aforementioned mechanisms to the inhibition of osteoblast replication remains to be determined, they all likely require the GC receptor (GR), as indicated by their sensitivity to GR antagonism by RU486 [36, 41].

Glucocorticoids Inhibit Osteoblast Differentiation and Function

Mesenchymal progenitors that reside in the bone marrow give rise to osteoblasts for the life-long process of filling eroded surfaces with new bone matrix. The early progenitors, commonly referred to as bone marrow stromal pluripotent cells, or mesenchymal stem cells, can select one of several developmental fates with appreciable plasticity, including a well documented reciprocal relationship between osteoblastogenesis and adipogenesis [42–47]. Many *in vitro* studies suggest that GCs influence this competitive cell fate choice, promoting the differentiation of bone marrow stromal cells into adipocytes at the expense of osteoblasts. Treatment of ST2 bone marrow-derived pluripotent cells with cortisol strongly stimulated the adipocytic genes PPARγ, C/EBPα, C/EBPδ and Adipsin [43] and inhibited RUNX2 (See also section "RUNX2"). Stimulation of C/EBPα and C/EBPδ was also observed in a microarray-based global gene expression study of dex-treated MC3T3-E1 pre-osteoblasts [48]. Similar to the commitment stage described in primary osteoblast cultures [37] and in MC3T3-E1 cells [49], GCs drive ST2 cells towards the adipocyte lineage and away from the osteoblastic cell fate only when administered early; once cells undergo commitment to the osteoblast lineage, the bone phenotype continues to develop normally even in the presence of GCs [50]. The idea that GIO is mediated in part by favoring differentiation of bone marrow pluripotent mesenchymal cells into adipocytes at the expense of osteoblasts is circumstantially supported by the increased marrow adiposity observed in GC-treated mice and humans [44], although direct in vivo evidence is lacking. Recent advances in the identification of skeletal stem cells may facilitate lineage tracing experiments to overcome this shortcoming [51, 52].

GCs at pharmacological concentrations suppress fundamental osteoblast functions, which can be partially traced to their influence, as described above, on cell fate at the osteoblast/adipocyte decision fork. Most significantly, GCs inhibit the biosynthesis of type I collagen, the predominant organic component of the bone matrix, in vitro [53] and in vivo [54]. Contributing to this inhibition, GCs first inhibit Procollagen α1(I) transcription; this was demonstrated in primary rat calvarial osteoblast cultures using nuclear run-off assays, where strong inhibition was evident within as little as two hours of treatment with 1 μM cortisol [55]. Secondly, GCs destabilize the Procollagen α1(I) transcript [55]. Thirdly, GCs inhibit collagen accumulation in a manner independent on Procollagen α1 mRNA; this was demonstrated, for example, by Sirius red-based assay of collagen accumulation in BMP2-treated MC3T3-E1 cultures. Co-treatment of these cultures with 0.1 or 1 μM dex resulted in a ~4-fold decline in the collagen accumulation rate even though Procollagen α1(I) mRNA levels did not decrease, but in fact increased [56]. The mRNA-independent decrease in collagen accumulation is attributable to inhibition of collagen translation, secretion, assembly, and/or accelerated collagen breakdown by GC-inducted collagenases [57].

Beside collagen accumulation, investigators in the GIO field often rely on assays of alkaline phosphatase (ALP) activity and mineral deposition to assess GC-mediated inhibition of osteoblast differentiation and function in culture models. Indeed, GCs at pharmacological concentrations inhibit both of these outcome parameters [19, 36, 53, 58–61]. The clinical relevance of ALP and mineralization assays to GIO, however, is not trivial because unlike inhibition of osteoblast proliferation and collagen synthesis, the results of ALP and mineralization assays can be species-dependent. In rodent osteoblast cultures, GCs at pharmacological levels usually inhibit ALP activity and mineralization, whereas physiological GC concentrations usually have stimulatory effects in these models [53, 62]. In human osteoblast cultures, on the other hand, GCs, often at low concentration, but sometimes even at pharmacological concentrations have been shown to stimulate ALP and mineral deposition [63, 64], and such stimulatory effects were also observed in murine cultures under specific experimental conditions [48, 65]. These observations initially led to a concept that GC might drive precocious osteoblast differentiation and thus exhaustion of a mesenchymal stem cell pool that otherwise continues to supply osteoblasts for bone formation throughout life. However, the paradoxical stimulation of osteoblast differentiation *in vitro* is typically observed at physiological rather than pharmacological GC concentrations [65], and most investigators no longer consider this a significant in vivo mechanism contributing to GIO. Among considerations arguing against this paradigm, regain of bone mass is observed shortly following GC withdrawal [6].

It is generally assumed that GC-mediated inhibition of ALP activity and mineralization, which are most reproducible in murine osteoblast cultures, capture a GC-sensitive phase of osteoblast differentiation. An alternative interpretation, however, is that some of the inhibitory effects of GCs in murine osteoblast cultures (e.g., on mineralization) are not as relevant to the human disease, and that other inhibitory effects (e.g., on cell replication and collagen synthesis) better model GIO in humans. Another cautionary note for interpreting results of ALP and mineralization assays in the context of GIO is that the two are not always coupled, and therefore inhibition of any one or even both of them by GCs may ultimately prove irrelevant to GIO. This is suggested, for example, by the good correlation between mineralization and the activity SMAD-BMPs, but not between mineralization and ALP activity in MC3T3-E1 cultures treated with dex along with recombinant BMP-2 or BMP-4 [67]. In addition ALP and mineralization assays can provide data that does not parallel collagen accumulation [68]. That cell proliferation and collagen accumulation assays are likely more relevant to GIO than mineralization assays is suggested by the lack of evidence for poorly mineralized osteoid in GIO in vivo. In section "Molecular Targets of Glucocorticoids in Osteoblasts", we review several signaling pathways adversely affected by exposure of osteoblast cultures to GCs; some of these pathways may ultimately provide the basis for assays most relevant to molecular mechanisms underlying GIO and to developing bone-sparing GC-based anti-inflammatory therapies. Some of these assays may capture mechanisms that reflect control of both osteoblast replication and maturation, two processes that classically represent opposite aspects of cell growth and differentiation, but may in fact operate hand-in-hand during development of the osteoblast phenotype [36, 69–71].

Glucocorticoids Promote Osteoblast Apoptosis

Many studies demonstrated GC-driven apoptosis of osteoblasts and osteocytes, both in vivo and *in vitro* [13, 33, 41, 72]. A milestone study invigorated this research avenue in 1998, showing by TUNEL staining increased apoptosis of osteoblasts and osteocytes in bone specimens from GC-treated mice and humans [17]. GC-mediated promotion of osteoblast apoptosis is a cell autonomous effect because apoptosis is not observed in transgenic mice with osteoblast-specific overexpression of the GC inactivating enzyme 11β-HSD2 [21].

Modeling osteoblast apoptosis in GIO, dex induced apoptosis in both the MC3T3-E1 and UMR-106 osteoblastic cell lines, and in both culture systems this was associated with activation of Caspase 3, a common downstream effector of multiple apoptotic signaling pathways [38, 73]. Apoptosis in the MC3T3-E1 culture model was also linked to activation of p53 [38]. GCs dose-dependently induced apoptosis in cultures of primary human osteoblasts derived from surgical bone chips, and this was attributable to increased mRNA and protein expression of Bak, as well as decreased mRNA and protein expression of Bcl-XL [39]. Annotations of genes differentially expressed in GC-treated versus control primary osteoblast cultures were highly enriched for apoptosis-related functions [74]. The same gene-set was also enriched for oxidative stress-related genes, hinting to one mechanism underlying GC-induced osteoblast apoptosis [74]. Related mechanisms of GC-induced apoptosis include inhibition of major survival pathways, such as Wnt, PI3K and ERK (section "Molecular Targets of Glucocorticoids in Osteoblasts"). Inhibition of these pathways by GCs leaves unopposed pro-apoptotic pathways that are activated by Fas, TNF, TRAIL [75] and Reactive Oxygen Species [ROS; sections "Pyk2, JNK and p66^{shc}" and "FoxO Proteins"].

Several open questions related to the role of osteoblast and osteocyte apoptosis in GIO should be noted. In the aforementioned milestone study [17], less than 1 % of murine osteoblasts in vivo were TUNEL-positive, and GCs increased this value threefold. The significance of these observations to the debilitation of bone formation, by up to ~80 %, in GIO remains a matter of debate. Under-estimation of the magnitude of basal and/or GC-induced apoptosis could result from technical difficulties with demonstration and quantitation of osteoblast apoptosis in vivo. Alternative techniques and/or approaches may shed light on this controversy. Furthermore, GC did not appear to induce apoptosis in osteoblasts during certain developmental stages [75] and in mice of certain strains [19]. Additionally, prevention of GC-induced osteoblast apoptosis did not rescue the low bone formation rates or the loss of spinal BMD in the aforementioned OG2-11ß-HSD2 mice (although the decrease in vertebral compression strength was prevented; see section "Involvement of Cells Other than Osteoblasts in GIO" below) [21]. Despite the uncertainty as to the relative contributions of effects of GCs on pre-osteoblast proliferation, impaired osteoblast function and increased apoptosis, the combined effect on all of these aspects together appears to create a "perfect storm" that leaves osteoblasts incapable of balancing bone resorption in GIO.

Involvement of Cells Other than Osteoblasts in GIO

Osteocytes. Not all osteoblasts undergo apoptosis after depositing new bone material at sites that have just been resorbed. Some give rise to flat lining cells that remain on the bone surface and many others incorporate into the newly formed matrix, where they continue to live for lengthy periods of time, contributing to the largest subpopulation of bone cells—the osteocytes. Through neuronal-like processes embedded in a highly interconnected canalicular system, osteocytes serve as biological relays, which stimulate osteoclasts, osteoblasts and their precursors in response to microdamage and mechanical loading. In fact, osteocytes are a major source of RANKL, a quintessential osteoclastogenic factor [77, 78], and their role in bone homeostasis is increasingly appreciated [79]. Thus, skeletal effects of GCs through osteocytes are both direct (as discussed immediately below) and indirect via osteoclastogenesis (see 'osteoclasts' thereafter).

There is significant evidence that high-dose GCs increase fracture risk not only by decreasing bone mass, but also by compromising bone material quality [13]. One of several explanations for this phenomenon entails GC-induced osteocyte apoptosis [17]. Contrasting osteocyte autophagy induced by GCs at physiological concentrations, which may protect these cells against stress [80], osteocyte apoptosis in response to high-dose GCs may deprive osteoclasts and osteoblasts the input, based on which they would otherwise respond to biomechanical needs. GC-induced osteocyte apoptosis is attributable to inhibition of survival mechanisms including Wnt signaling, Akt, and Pyk2 (section "Molecular Targets of Glucocorticoids in Osteoblasts" below). Additionally, recent evidence suggests that osteocytes directly modify the bone matrix in which they are embedded and that GCs interfere with a post-osteoblast mineralization process, whereby osteocytes regulate their immediate microenvironment [5]. This novel effect of GCs, hypomineralization of peri-osteocytic bone material, was demonstrated using a nanoindentation technique assisted by atomic force microscopy, and cannot be detected by conventional imaging or histmorphometric methods [5].

Osteoclasts. The early and most destructive phase of GIO is driven not only by the inhibition of osteoblastic bone formation as described above, but also by simultaneous stimulation of osteoclastic bone resorption. After prolonged treatment, however, bone resorption is suppressed to sub-physiological levels [28], contributing to the overall low bone turnover rates typical of GIO.

Stimulation of osteoclastogenesis by GCs at the early disease phase is attributable to both cell autonomous and paracrine mechanisms. Because osteoclastogenesis is a process that usually requires a few days for completion, the fast increase in osteoclast number observed immediately after GC administration has been attributable to extended life span of pre-existing osteoclasts. GC-mediated extension of the osteoclast life-span appears to be cell-autonomous because it occurred in isolated osteoclasts *in vitro* [28], and because osteoclast number was lower in wild type GC-treated mice compared to GC-treated transgenic mice over-expressing the GC-inactivating enzyme 11β-HSD2 is osteoclasts [30]. Additionally, *in vitro* data suggest enhancement of osteoblast-driven osteoclastogenesis by GCs, potentially contributing to the increased bone resorption observed in the early phase of GIO [29]. In particular, GCs dramatically suppress OPG mRNA and protein expression in human osteoblast cultures [18]. It remains to be clarified whether GC administration also stimulates bone resorption through osteocyte-borne RANKL [77, 78] and/or other osteoclastogenenic factors, and whether GC-induced osteocyte apoptosis facilitates the release of such factors to promote osteoclastogenesis [81].

The decreased osteoclastic bone resorption observed after prolonged GC treatment periods is generally considered secondary to attenuation of cell number and function in the osteoblast lineage as described above. There is intriguing evidence in vivo, however, suggesting that the GR in myeloid progenitor cells decreases the resorptive activity of osteoclasts, and that this is mediated by their failure to form cytoskeletal structures necessary for attachment to and resorption of the bone matrix [29].

Non-bone cells. In addition to mineralized tissue, the bone as an organ consists of soft tissues, such as cartilage, bone marrow, vessels and nerves. Potential contributions of these tissues to GIO are largely unknown. Furthermore, GCs may have indirect negative effects on bone through their actions in the neuroendocrine system, the gonads, the intestine, and the kidney (reviewed in [13]). Finally, adverse effects of GCs on motor function and muscle strength may contribute to fracture risk by increasing fall rates.

Molecular Targets of Glucocorticoids in Osteoblasts

The Wnt Signaling Pathway

Glucocorticoids Inhibit Wnt Signaling in Osteoblasts

As with several other cell types, Wnt signaling in osteoblasts increases cell proliferation by promoting cell cycle progression and by inhibiting apoptosis [82]. Because Wnt ligands are known to promote asymmetric stem cell division [83], they can be expected to allow early pre-osteoblast pool expansion to supply cells for the formative arm of bone remodeling, while preserving a mesenchymal stem cell pool in the bone marrow microenvironment. The central role of Wnt signaling in bone biology was initially indicated by the inactivating mutations in LRP5, coding a Wnt co-receptor, in patients with the familial bone disease osteoporosis-pseudoglioma [84]. The opposite, high bone mass (HBM), is observed in patients carrying LRP5 activating mutations [85, 86]. Alterations to many additional Wnt-related genes have been linked to bone mass control, and experiments manipulating such genes in mice solidified the notion that Wnt signaling plays a pivotal role in bone metabolism and bone mass control (reviewed in [82]). Accordingly, experimental stimulation of Wnt signaling in osteoblasts results in increased proliferation, increased expression of differentiation markers, decreased apoptosis, as well as attenuated osteoblast-driven osteoclastogenesis [82].

Bioinformatic analysis of global gene expression in bone tissue of GC-treated versus control mice linked the differentially expresed genes to the Wnt pathway, with indications for reduced signaling due, in part, to low expression of Wnt ligands [5]. GCs also decrease Wnt signaling in osteoblast cultures as indicated by decreased expression of Wnt targets, both endogenous genes and reporter constructs [37, 60, 87–90]. Mechanisms underlying inhibition of Wnt signaling in GIO are reviewed below and schematically summarized in Fig. 8.1.

Role of Dkks in GIO

DKK1, encoding dickkopf-1 (DKK1), a secreted antagonist of canonical WNT signaling, is one of few examples available to date for classical direct transcriptional stimulation by GCs in osteoblasts. Treatment of human osteoblasts with ≥10 nM dex resulted in robust stimulation of *DKK1* mRNA expression, attributable to a glucocorticoid response element (GRE) located 0.8 kb upstream of the *DKK1* transcription start site [91]. In a microarray study of primary human osteoblast cultures, *DKK1* was among the genes most strongly upregulated by GCs [74]. GCs also stimulated *Dkk1* expression in primary rat calvarial osteoblast cultures [92]. A role for GC-induced Dkk1 in inhibiting Wnt signaling in osteoblasts is indicated by reversal of the suppressive effect of dex on Wnt signaling upon addition of anti-Dkk1 antibody to the osteoblast culture medium [88].

Members of the Dkk family inhibit Wnt signaling through interaction with LRP5/LRP6 and with Kremen1/Kremen2 in the Wnt receptor complex [93, 94]. The critical role of DKKs in regulating Wnt signaling in osteoblasts and consequently controlling bone mass is demonstrated by the fact that the G171V mutation in LRP5 that leads to a HBM phenotype [85] is associated with abrogation of its binding to Dkk-1 [95, 96]. The HBM phenotype observed in patients with the G171V LRP5 mutation is therefore indicative of high sensitivity of the Wnt pathway and bone mass to reduced DKK1 activity upon LRP5 in osteoblasts. This is also demonstrated by the strong anabolic response and increased bone mass in DKK1 hypomorphic or heterozygous mice [97, 98]. The opposite effect, decreased bone mass, can therefore be expected to follow increased activity of DKK1 on osteoblasts' LRP5 in GIO. Indeed, GC-mediated suppression of Wnt/β-catenin signaling and osteoblast differentiation [86] was attenuated when stimulation of *Dkk1* was prevented by siRNA-mediated *Dkk1* knockdown [60]. In addition to the stimulation of Dkk1 in human and rat cells, GCs have also been shown to stimulate expression of the closely related *Dkk2* in primary mouse calvarial osteoblast cultures [37], potentially resulting in the observed inhibition of Wnt signaling in osteoblasts

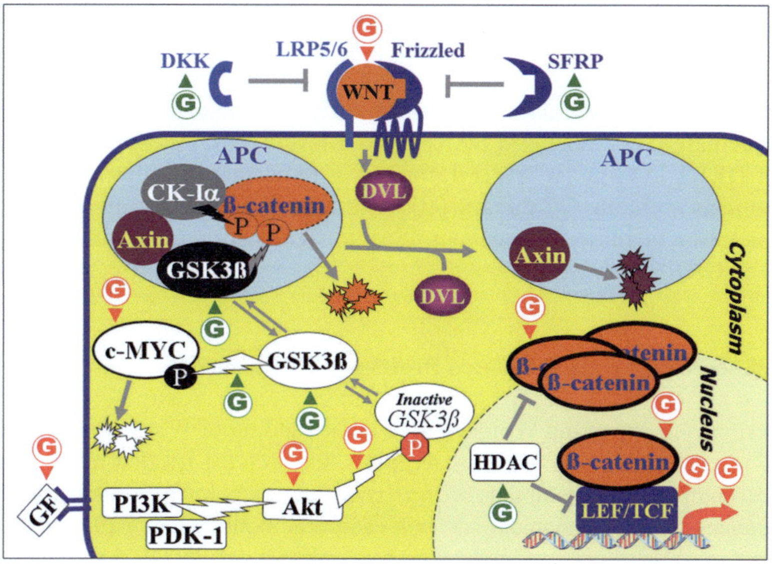

Fig. 8.1 GCs stimulate GSK3ß and inhibit Wnt signaling. Filled (*colored*) shapes represent components of the canonical Wnt pathway and empty (*white*) shapes represent signaling molecules that intersect with the Wnt pathway. GSK3ß is positioned at an interesection between the Wnt pathway and protein tyrosine kinase signaling. Lightning bolts indicate phosphorylation. Inhibitory and stimulatory effects of GCs are depicted by *red* and *green circled* G's, respectively. GCs inhibit growth factors (*GF*) and the downstream PI3K/Akt pathway. Consequently, the inhibitory phosphoryltion of GSK3ß on Ser^9 (octagon) is attenuated and GSK3ß is thus activated. Outside the canonical Wnt pathway, active GSK3ß phophorylates c-Myc on Thr^{58}, resulting in c-Myc degradation. Within the canonical Wnt pathway, GSK3ß phosphorylates ß-catenin, resulting in its degradation. This adds to increased rates of ß-catenin degradation due to stimulation of Wnt inhibitors of the SFRP and DKK families, as well as inhibition of Wnt ligands. Inactivation of the Wnt receptor complex stabilizes the ß-catenin destruction complex, where ß-catenin is phosphorylated, tagging it for degradation. Accumulation of ß-catenin is thus inhibited. Because ß-catenin is a critical coactivator for LEF/TCF transcription factors, their target genes (i.e., Wnt target genes) are suppressed. Additionally, GCs attenuate expression of some LEF/TCF transcription factors and stimulate that of HDAC proteins, which inhibit both ß-catenin and LEF/TCF. *Abbreviations*: *APC* adenomatous polyposis coli, CK-*Iα* casein kinase-Iα, *DVL* disheveled, *GSK* glycogen synthase kinase, *HDAC* histone deacetylase, *LRP5* low density lipoprotein-related protein 5, *SFRP* secreted freezled-related protein

through a similar mechanism, again leading to decreased osteoblast function and loss of bone mass.

Consistent with a potential role for DKK1 in GIO, its levels were elevated in the sera of children with 21-hydroxylase deficiency and chronically treated with high-dose glucocorticoids (10–25 mg/m^2 of hydrocortisone). The patients' sera inhibited osteoblast differentiation *in vitro* (although the inhibition—of ALP activity—was small), and the effect was ameliorated after administration of anti-DKK1 antibodies [99]. Additional data obtained from these patients suggested that the excess DKK1 could be partly biosynthesized by leukocytes, and that it played a role in regulating RANKL levels in these patients [99]. *Dkk1* mRNA was also elevated in mouse bones in vivo after 56 days of prednisolone treatment [20]; this study, however, raises the question of whether GC-mediated stimulation of *Dkk1* is a primary event in the mouse because the early time point in vivo (7-days) indicated decreased, not increased *DKK1* mRNA levels [20]. Further questioning the role of DKK1 in GIO, its levels decreased, not increased in a prospective study with patients initiating GC therapy [100]. Additional work is therefore needed, for example using conditional knockout mice, to rigorously test the potential role of DKK1 in GIO.

Role of Secreted Frizzled-Related Proteins (SFRPs) in GIO

The first step in activating Wnt signaling is the binding of Wnt ligands to frizzled family receptors. A group of decoy receptors, the secreted frizzled-related proteins (SFRPs) compete with membrane bound frizzled receptors for Wnt binding (Fig. 8.1), thus attenuating both canonical and non-canonical Wnt signaling [93, 101]. *Sfrp1* knockout mice have increased trabecular bone mineral density [102] and injection of rats with recombinant SFRP1 decreased bone mineral density [62]. Both canonical and non-canonical Wnt signaling in osteoblasts have been implicated in the regulation of osteoblast proliferation, differentiation and apoptosis by SFRP1 [102].

Dex at concentrations ≥ 0.1 μM dramatically stimulated *Sfrp1* mRNA expression in primary rat bone marrow stromal cell cultures, and this was independent of new protein synthesis [62]. Dex also stimulated *Sfrp1* expression in mouse primary calvarial osteoblast cultures [92]. Furthermore, siRNA knock down of SFRP1 led to increased β-catenin accumulation, enhanced Runx2 activity and high levels of ALP and osteocalcin expression, culminating in robust nodule formation even at high dex concentrations [62]. Hence, SFRP could serve as a therapeutic target, inhibition of which may ameliorate GIO.

Additional Wnt-Related GC Targets

Wnt ligands. GCs regulate the expression of some Wnt ligands, potentially contributing to inhibition Wnt signaling in osteoblasts. For example, corticosterone at 100 nM inhibited by ~50 % expression of Wnt 7b and Wnt 10b in mature green fluorescent protein (GFP)-expressing osteoblasts of Col2.3-GFP mice. Interestingly, 10 nM corticosterone had the opposite effect, potentially accounting for paradoxical anabolic effects often observed with low GC doses [92]. Loss of the autocrine/paracrine activity of Wnt ligands at high GC concentrations may amplify the aforementioned anti-Wnt effects of Dkk1, which were confirmed in the Col2.3-GFP-expressing cells [92].

GSK3β. GC-treated osteoblasts from both human and mouse origin display a decrease in the inhibitory phosphorylation of GSK3β on its Ser^9 residue, resulting in increased enzyme activity [46, 49]. The role of GSK3β in the anti-mitogenic effect of GCs was demonstrated by the rescue of cell cycle progression in GC-arrested MC3T3-E1 osteoblasts co-treated with lithium chloride, a GSK3β inhibitor [49]. GC-stimulated GSK3β attenuates cell cycle progression both by inhibiting ß-catenin/LEF-mediated transcription [87] and by phosphorylation of c-MYC on Thr^{58}, which marks the protein for proteasomal degradation [49]. The phosphorylation of GSK3β represents an important point of intersection between growth factor signaling and the canonical Wnt pathway (Fig. 8.1). Specifically, following the activation of PI3K by receptor tyrosine kinases [section "Akt"], Akt phosphorylates GSK3β's $serine^9$ residue, which results in loss of GSK3β activity upon its targets, such as β-catenin and c-Myc. Accordingly, pharmacological and molecular inhibition of PI3K/Akt in GC-treated MC3T3-E1 osteoblasts is associated with decreased phosphorylation of GSK3β's Ser^9 as well as c-MYC's Thr^{58} [49]. Thus, GC-mediated stimulation of the inhibitory kinase GSK3β results in (i) attenuation of β-catenin/LEF-driven transcription, adding to other inhibitory effects of GCs within the canonical Wnt pathway; and (ii) abrogation of GSK3β functions outside the canonical Wnt pathway (Fig. 8.1).

β-catenin. Ligand-bound GR has been shown to physically interact with β-catenin itself in U2OS/GR cells [103]. This could contribute to inhibition of LEF/TCF-mediated cyclin D transcription and to GIO in vivo, even though GCs did not inhibit cell cycle progression in the U2OS/GR cell culture model [103]. Additionally, GCs may inhibit Wnt signaling by translocating β-catenin from the cell nucleus to the cytoplasmic membrane, which is mediated though interactions of GR with calreticulin. Indeed, silencing of calreticulin abolished dex-mediated inhibition of cyclin D1 expression [104]. Finally, as will be described in section "FoxO Proteins", GCs interfere with canonical Wnt signaling at the level of β-catenin by generating reactive oxygen species, resulting in activation of FoxO transcription factors, which interact with β-catenin at the expense of LEF/TCF transcription factors.

Recent work suggests that GC-mediated suppression of Wnt/β-catenin signaling is mediated in part through inhibition of mir-29a [105]. In murine calvarial osteoblasts, both primary and MC3T3-E1 cells, mir-29a promotes bone phenotypic properties by suppressing expression of HDAC4, a β-catenin deacetylase [105]. GC-mediated downregulation of mir-29a, and the subsequent deacetylation and inactivation of ß-catenin by HDAC4 appear critical for suppression of the bone phenotype because anti-sense-mediated silencing of HDAC4 rendered the cultures resistant to GCs. Consistent with these findings, GC-mediated inhibition of cell cycle progression in MC3T3-E1 cultures was partially negated in the presence of the HDAC inhibitor trichostatin A [87].

LEF/TCF. Signals elicited by binding of Wnt ligands to their cell surface receptors ultimately lead to changes in gene expression by the binding of activated β-catenin to transcription factors of the LEF/TCF family (Fig. 8.1). In newborn mouse calvarial osteoblast cultures, 1 μM dex decreased the expression of *Lef1*, *Tcf1* and *Tcf4* (but not *Tcf3*) mRNA [37]. Interestingly, the effect of dex on *Lef1* and *Tcf1* expression depended on the developmental stage with respect to a commitment stage defined based on resistance that these cultures develop on day 6–7 to GC-mediated attenuation of mineral deposition. Specifically, dex inhibited *Lef1* only before the commitment stage, whereas the inhibition of Tcf1 was most robust after that stage [37].

Axin2. As discussed in section "Glucocorticoids Inhibit Osteoblast Differentiation and Function", GCs drive osteoblast precursors towards adipogenesis at the expense of osteogenesis [46, 90, 106]. In murine MC3T3-E1 pre-osteoblasts and ROB-C26 rat mesenchymal progenitor cells, this was attributable in part to a dex-mediated 3-fold increase in Axin2 mRNA expression [90, 107]. Indeed, dex also abrogated ß-catenin activation and this was no longer obvious after depletion of Axin2 in ROB-C26 cells [90]. Consistently, knockdown of Axin2 antagonized dex-mediated adipogenesis, although inhibition of ALP by dex persisted in Axin2-depleted ROB-C26 cultures [90].

Additional Signaling Pathways

In addition to the well documented role of the Wnt signaling pathway in bone pathophysiology in general, and GIO in particular, GCs affect several other pathways in osteoblasts, any of which may ultimately prove an effective target for therapeutic intervention. We briefly review here evidence for the involvement of Notch and BMP signaling, as well as several growth factor pathways, in GIO.

Notch Signaling

Glucocorticoids strongly stimulate transcription of *Notch1* and *Notch*2 in osteoblasts, resulting in several-fold increased mRNA expression within hours of treatment [108]. The activated Notch Intracellular Domain (NICD) is known to inhibit osteoblast differentiation by targeting RUNX2 both directly and indirectly [109, 110]. Although manipulation of Notch signaling in vivo results in a complex skeletal phenotype that depends on age, sex and bone tissue type [110–111], GC-mediated stimulation of Notch signaling likely plays an important role in GIO, which may be mediated in part by inhibition of RUNX2 [section "RUNX2"].

BMP Signaling

Comprehensive gene expression analysis in GC-arrested MC3T3-E1 osteoblast cultures indicated a threefold increase in the expression of *Follistatin* and *Dan* mRNAs, encoding inhibitors of BMP signaling [49]. In the same culture model, GCs also strongly inhibited *Bmp2* gene expression, and recombinant BMP2 reversed the inhibitory effects of GCs on mineral deposition, ALP activity, *osteocalcin* expression, as well as (transiently) cell cycle progression [56, 68]. These, however, remain indirect lines of evidence for a role that BMP signaling may play in GIO. In fact, dex did not inhibit the activity of a SMAD-BMP reporter in cultures of MC3T3-E1 cells [67], and some investigators even demonstrated stimulation of BMP signaling by GCs in osteoblasts [32]. Paradoxically, stimulation of BMP signaling by GCs may contribute to GIO through inhibition of Wnt signaling [112], although this conjuncture remains to be tested. Another interesting speculation is that GCs concomitantly stimulate and inhibit BMP signaling in a target gene-dependent manner. Be that as it may, global inhibition of BMP-SMAD signaling does not appear to occur in GIO, or at least not in the MC3T3-E1 culture model; high expression levels of other BMPs, in particular BMP-4, could have sustained activity of the BMP-SMAD reporter in the presence of GCs [67]. Still, decreased BMP2 levels may contribute to GIO when alternative osteogenic BMP genes are not expressed. Under such circumstances (and not in MC3T3-E1 cells), inhibition of *Bmp2* transcription, through regulatory sequences located >50-kb downstream of the transcription start site [67], might reduce BMP-SMAD signaling below a threshold necessary for development of the osteoblast phenotype.

Growth Factors

GCs inhibit the biosynthesis of hepatocyte growth factor (HGF, a.k.a. Scatter Factor) in human osteoblasts [113, 114], which could interrupt an autocrine mechanism whereby HGF stimulates osteoblast proliferation by binding to its c-Met receptor on the osteoblast membrane [115]. Indeed, inhibition of HGF signaling mimicked the anti-mitogenic effect of GCs in human osteoblast-like cultures, and GCs no longer inhibited cell proliferation in the presence of added recombinant HGF [114]. These results suggest that HGF may play an important role in GIO, although direct evidence to this effect is lacking.

Growth hormone (GH) and insulin-like growth factors (IGFs) influence bone metabolism through both endocrine and paracrine/autocrine mechanisms, of which the latter are more likely to play a role in GIO [116]. Indeed, osteoblast proliferation and collagen synthesis, probably the two most important functions inhibited in GIO, are stimulated by IGF-I and IGF-II, and GCs have been shown to inhibit IGF signaling at several levels. First, GCs inhibit IGF-I transcription and secretion in primary rat calvarial osteoblast cultures [117, 118], which again may be related to the promotion of the alternative, adipocyte cell fate decision. Indeed, the GC-induced adipogenic factors C/EBPα and C/EBPδ appear to bind *Igf-1* regulatory sequences and block transcriptional initiation or elongation [119]. Second, GCs inhibit expression of IGF-binding protein-3 (IGFBP-3) and IGFBP-5 in human osteoblast cultures [120]. The strong inhibition of *Igfbp5,* which occurs at the transcriptional level in rat calvarial osteoblast cultures [121], is of particular interest because IGFBP5 has an additional, autonomous effect on human osteoblast proliferation [122] and a selective bone anabolic effect in mice in vivo [123–125].

Downstream Kinases and Phosphatases

ERK

In addition to kinase activity directly associated with the receptors reviewed above, downstream kinases with roles in osteoblast growth and differentiation have been implicated in GIO. Chief among them, ERK is a central hub downstream of a variety of osteogenic stimuli elicited by interaction of growth factors and extracellular matrix proteins with their respective receptor tyrosine kinases and integrin receptors [126–129]. Activated ERK executes much of its function in the osteoblast nucleus, where it associates with specific DNA elements to activate key regulators of cell growth and differentiation, including the osteoblast master regulator RUNX2 [130, 131]. Accordingly, stimulation and inhibition of ERK results in enhancement and impediment, respectively, of osteoblast differentiation and bone formation in vivo and *in vitro* [41, 132, 133].

High dose GCs inhibit ERK signaling and its downstream effectors, and these inhibitory activities are similar to those observed after treatment of osteoblasts with the MEK/ERK inhibitor U0126 [41]. Both dex and U0126 decreased thymidine incorporation into newly synthesized DNA in serum- and TPA-stimulated MBA-15.4 and MG-63 osteoblastic cells [41]. Attenuated ERK activity in GC-treated osteoblasts is likely related to many aspects of GIO, including inhibition of osteoblast proliferation, differentiation and survival [75, 129]. Protection of ERK from GCs has the potential of partially reversing GIO [134].

Akt

Activation of the PI3K/Akt pathway by hormones and growth factors, including *PTH* and IGFs, is required for the differentiation of mesenchymal pluripotent cells into osteoblasts, as well as survival of the committed cells [135–137]. Mechanisms of action of Akt in regulating canonical signaling pathways such as Wnt (through GSK3β) and mTOR are well established [138, 139]. Additionally in osteoblasts, Akt promotes RUNX2 activity [140], in part through post-translational regulation of Smurf2 [141], and may also activate transcription factors directly within the cell nucleus [139]. GCs attenuate the activity of Akt in osteoblasts by decreasing growth factor availability [see section "Growth Factors"], and by compromising the cellular response to such growth factors [87], which is partially attributable to oxidative stress [89, 143].

Pyk2, JNK and p66^{shc}

In MLO-Y4 osteocyte-like cells, the two highly homologous kinases, FAK and Pyk2, play opposing roles with regard to interaction with the extracellular matrix (ECM). Whereas FAK promotes ECM attachment and cell survival, Pyk2 activation results in loss of cellular processes and anoikis, i.e., detachment-mediated programmed cell death [144]. GCs activates Pyk2 in a manner independent of either RNA or protein synthesis; the underlying mechanism of action, likely employing membrane-associated GR, involves stimulation of JNK [144]. Indeed, GCs no longer induce anoikis in Pyk2-depleted cells or in cells that express either an inactive Pyk2 mutant or a dominant negative JNK [143, 144]. GC-mediated detachment of cells from the ECM through the Pyk2/JNK axis may play a role in the apoptosis of both osteocytes and osteoblasts in vivo [17]. JNK also plays a role in GC-mediated ROS-driven apoptosis through activation of FoxO transcription factors [see section "FoxO Proteins"]. Indeed, GC-mediated FoxO-driven transcription and apoptosis was associated with activation of JNK, and both were severely compromised in fibroblasts derived from JNK1/2 double-knockout mouse embryos [89].

GCs also activated the $p66^{shc}$ kinase in bone in vivo and in cultured osteoblasts *in vitro*, leading to accumulation of reactive oxygen species (ROS) [89]. Indeed, GCs no longer stimulate ROS accumulation in osteoblast cultures in which $p66^{shc}$ is absent, or in which PKCß, the kinase that phosphorylates $p66^{shc}$, is pharmacologically inhibited [89]. ROS have many deleterious effects in osteoblasts. As described above they activate JNK, which results in apoptosis through FoxO-dependent and independent mechanisms, and they inhibit Wnt signaling through activation of FoxO transcription factors [section "FoxO Proteins" below] and inhibition of Akt [section "Akt" above].

MKP-1/DUSP

Protein tyrosine phosphatases (PTPs) have been strongly implicated in GC-mediated inhibition of ERK signaling and osteoblast function, as they are the dominant active phosphatase class in the lineage [41]. Inhibition of PTP with sodium orthovanadate restored ERK activity and osteoblast proliferation in dex-inhibited osteoblast cultures [134] and in methylprednisolone-treated Sprague–Dawley rats [145].

The main phosphatase implicated in GC-mediated inhibition of ERK is MAPK phosphatase-1 (MKP-1), also named dual specificity phosphatase-1 (DUSP1), which colocalizes with ERK and limits its effects on target gene transcription [146–149]. Although JNK and p38 can also be inactivated by MKP-1 [150] and even though MKP-3 is most efficient in inactivating ERK1/2 [148], studies in osteoblasts clearly indicate that GCs stimulate MKP-1 [41, 151, 152] while inhibiting MKP-3 [41], and that MKP-1 inactivates ERK [152]. In addition, knockdown of MKP-1 with siRNA in human MG-63 osteosarcoma cells prevented dex-mediated ERK dephosphorylation [152].

Stimulation of MKP-1 expression by GCs is rapid and robust. MKP-1 mRNA levels increased by >10-fold within 30 min of treatment of mouse MBA-15.4 and human MG-63 pre-osteoblasts with dex, an effect that lasted >24 h [41, 152]. This was associated with a similar >10-fold induction of MKP-1 protein, which precisely correlated with inhibition of ERK phosphorylation [41, 152]. GC-induced expression of MKP-1 was confirmed in a global microarray analysis of dex-treated

MC3T3-E1 osteoblast-like cells [48], as well as in fibroblast-like synoviocytes [147]. Whereas GR occupancy at the MKP-1 locus in osteoblasts has not been mapped systematically, it has been shown to associate, either directly or through a tethering mechanism, with a GRE-C/EBP composite element located ~1.3-kb upstream of the DUSP transcription start site in A549 human lung adenocarcinoma cells [153, 154]. GCs have also been shown to increase MKP-1 stability in mast cells and fibroblasts [151], although this mechanism is less likely relevant to bone cells because GCs no longer increase MKP-1 levels in cyclohexamide-treated MG-63 cells [152].

Does inhibition of MKP-1 offer a realistic approach for the management of GIO? Despite the pivotal roles of ERK in osteoblast growth and differentiation, side effects of global MKP-1 inhibition are more than likely to occur, which would necessitate specific targeting to bone. Additionally, the efficacy of prospective anti-MKP-1 approaches for GIO has been questioned by observations in MKP-1 knock-out mice treated for 28 days with methylprednisolone [155]. The absence of MKP-1 did not negate the GC-mediated decrease in bone formation, suggesting that inhibition of MKP-1 alone is insufficient for protection against GIO [155].

Transcription Factors

FoxO Proteins

The FoxO (forkhead box O) family, consisting of FoxO1, FoxO3a, FoxO4, and FoxO6 [156], play important roles in cellular responses to ROS as well as regulation of cell cycle progression and apoptosis [157]. As in other cells, FoxO family members defend osteoblasts against ROS [89, 158]. The balance between protective and deleterious effects of ROS and FoxO transcription factors is therefore key to for bone health [143]. GCs severely impair this balance by super-activating FoxO transcription factors. This results in the concomitant inhibition of Wnt signaling [143, 159, 160], thus compromising osteoblast proliferation and differentiation (see sections "Glucocorticoids Inhibit Osteoblast Cell Cycle"–"Glucocorticoids Promote Osteoblast Apoptosis").

Similar to the transcriptional and post-transcriptional regulation of FoxO3 in non-bone cells [161], GCs stimulate FoxO transcription factors in osteoblasts through several independent mechanisms. First, FoxO mRNA levels are upregulated by GCs. Indeed, FoxO3a and FoxO1a were two of the mRNAs most strongly upregulated in a microarray study of GC-treated primary human osteoblasts [74]. Second, GC-induced ROS stimulate the PKCβ/p66^{shc} axis, resulting in activation of JNK and the subsequent phosphorylation of FoxO [143] [see section "Pyk2, JNK and p66^{shc}"]. Third, GCs inhibit Akt [see section "Akt"], which results in the activation of FoxO proteins at the expense of LEF/TCF transcription factors [89].

When treated with pharmacologic doses of GCs, activated FoxO proteins bind and compete for a limited supply of ß-catenin [143, 162, 163]. Direct interaction between ß-catenin and FoxO3a was demonstrated by co-immunoprecipitation of assays in C2C12 cells [143]. Serving as a co-activator, ß-catenin stimulates expression of FoxO target genes at the expense of Wnt/TCF target genes [163]. Indeed, mimicry of GC treatment by over-expression of FoxO3a in Wnt3a-treated C2C12 cultures abrogated development of the osteoblast phenotype, illustrated by a decrease in ALP activity [143]. Furthermore, overexpression of ß-catenin partially overcame FoxO-mediated suppression of TCF/LEF-driven transcription, again suggesting that a limited pool of ß-catenin is shared for the activation of LEF/TCF and FoxO target genes [143]. Furthermore, unlike osteoblasts isolated from WT mice, Wnt3a-driven LEF/TCF activity was not inhibited by GCs in osteoblasts isolated from mice lacking FoxO1, FoxO2 and FoxO3, illustrating the critical role of FoxO proteins in GC-mediated inhibition of Wnt signaling [89].

AP-1

Much of the anti-inflammatory activity of GCs is attributable to both direct and indirect interactions between the GR and other transcription factors. Direct interactions occur both at cis-acting regulatory DNA elements and away from DNA. Indirect interactions involve, for example, regulation of phosphorylation and competition for common co-activators [164]. Perhaps the most important GR-interacting proteins in the context of immune suppression are AP1 (FOS/JUN) and NF-κB transcription factors. Of these, interaction with NF-kB does not appear to play a role in GIO because GCs suppress ALP activity in primary osteoblast cultures even when the cells are impaired for NF-κB activation [19]. In contrast, interactions of the GR with AP-1 family members, which are well documented in contexts other than osteoblasts [164–169], appear to play a role in GIO. This notion is consistent with the pivotal roles that FOS and JUN family members play in osteoblast growth and differentiation [170–172]. Indeed, GR^{dim} mice, in which classical transcriptional stimulation by GR dimers is impaired but inhibition of AP-1 is preserved, developed GIO [19]. Furthermore, abrogation of AP-1-depedent expression of IL-11 [174], a cytokine critical for bone formation in vivo [175], has been implicated in GIO, and administration of IL-11 restored ALP activity, mineralization, and RUNX2 expression in GC-inhibited murine primary calvarial osteoblast cultures [19]. Intriguingly, PTH administration overcame GC-mediated suppression of IL11 expression by stimulating Smad1 and the FOS family member delta FosB [176], a mechanism that could be exploited to combat GIO.

RUNX2

RUNX2 is the most powerful osteoblast lineage specifying factor known to date. Runx2 ablation in the mouse resulted in absence of osteoblasts and a general failure to form any mineralized tissue [177, 178]. Accordingly, manipulation of RUNX2 in cell culture models led to corresponding gain or loss of osteoblast phenotypic properties [179–181]. Because functional osteoblasts are needed not only for embryonic bone development, but also to balance the resorptive activity of osteoclasts at bone multicellular units (BMUs) throughout life, suppression of RUNX2 in adults can be expected to result in bone loss [182] [183]. Therefore, the inhibition of *Runx2* expression by GCs, observed in several osteoblast culture systems and in vivo, may play an important role in GIO [19, 43, 184, 185].

Because RUNX2 activity is strongly regulated post-translationally by covalent modifications and protein-protein interactions [186, 187], data on its mRNA and even protein expression levels can be misleading. It is therefore important to note that, in addition to inhibition RUNX2 expression, GCs have been reported to suppress the expression of RUNX2 targets, both endogenous genes such as osteocalcin and artificial constructs designed to specifically report on RUNX activity [61, 184]. Furthermore, a recent study demonstrated that GCs inhibit the activity of RUNX2 even when constitutively expressed from an exogenous lentiviral vector [61]. The implied post-translational inhibition of RUNX2, without a decrease in its mRNA or protein levels [61], is likely the primary effect leading secondarily to changes in the expression levels of endogenous *RUNX2* because endogenous *Runx2* is subjected to auto-regulation [188]. Indeed, exceptional cases where endogenous *Runx2* expression is not inhibited by GCs [185] may represent culture models, in which RUNX2 autoregulatory loops are not operative. Alternatively, inhibition of RUNX2 activity by GCs may be specific for differentiation stages represented by some and not other tissue culture models [56, 185].

Like several other steroid hormone receptors [189–191], the GR physically interacts with RUNX2 [61, 192], possibly inhibiting its DNA-binding and/or transcriptional activation activity. Such inhibition may vary depending on the target gene, including the topographic relationships between local sites occupied by RUNX2, GR and possibly other transcription factors [61, 193]. Additionally, GCs may inhibit RUNX2 indirectly, by suppressing Wnt [section "The Wnt Signaling Pathway"], ERK [section "ERK"] and Akt signaling [section "Akt"], all of which have been implicated in stimulating RUNX2 [130, 131, 140, 141, 194], or by stimulating Notch signaling, which inhibits RUNX2 [109, 110] [section "Notch Signaling"]. A comprehensive understanding of GC-mediated regulation of RUNX2 activity may ultimately lead to the development of novel therapeutic approaches for the management of GIO.

Additional Transcription Factors

The suppression of RUNX2 discussed in the previous section may constitute a critical mechanism underlying GC-mediated inhibition of osteocalcin, both a clinical marker of bone formation and a classical model for osteoblast-specific gene expression. The inhibition of osteocalcin expression by GCs, reproducibly observed both *in vitro* and in vivo, both in humans and mice, has been investigated for decades, with initial reports focusing on GR binding to osteocalcin proximal promoter elements [20, 195–201]. The inhibition of RUNX2 itself, however, is likely much more relevant to GIO than the inhibition of Osteocalcin, because Osteocalcin does not play any critical role in bone formation [202]. Still, an additional mechanism of osteocalcin transcriptional repression has been discovered using the MC3T3-E1 cell line, in which GCs do not inhibit *Runx2* [56, 185]. In these cells, GCs inhibit osteocalcin transcription by strongly repressing expression of *Krox20* [48, 203], which has been implicated in embryonal bone development in vivo [204]. Recent studies, however, have raised a doubt regarding the role of Krox20 in osteoblast suppression in GIO because its main function in the adult mouse skeleton in vivo appears to be inhibition of osteoclastogenesis and bone resorption [205, 206].

Microarray-assisted profiling of gene expression in GC-arrested MC3T3-E1 osteoblast cultures [48] confirmed the GC-mediated stimulation of the adipogenic regulators *C/EBPβ* and *C/EBPδ* and the inhibition *of Krox20* (see section "Glucocorticoids Inhibit Osteoblast Differentiation and Function" and previous paragraph, respectively). Together with *Krox20*, another zinc finger transcription factor gene, the *Kruppel-like factor 10* gene (*Klf10*; a.k.a *TGFβ-inducible growth response, or Tieg*), displayed the strongest suppression (6-fold) in the GC-arrested as compared to control cultures [48]. The relevance of these repressed transcription factor genes to GIO, as well as that of GC-stimulated transcription factors including *Klf 13, Period circadian clock 1 (Per1)* [48] and *Glucocorticoid-Inducible Leucine Zipper* (*Gilz*) [207], is less certain. Unexpectedly, some of the GC-upregulated genes play positive roles in osteoblast differentiation [207] and may explain paradoxical anabolic effects of GCs. Alternatively, these genes may play a role in GIO by abrogating a finely tuned circadian rhythm of gene expression [208, 209], and thus mediate the impact of GCs on proliferation and differentiation of osteoblasts.

Glucocorticoids Without Osteoporosis?

The current standard of care for GIO management is administration of bisphosphonates, which suppresses osteoclast activity. In contrast to high turnover osteoporosis (e.g., after estrogen loss), the use of bisphosphonates for patients undergoing long-term GC therapy is questionable, because it does not address osteoblast suppression and abrogation of bone formation, the hallmark of GIO. In fact, the outcome of bisphosphonate therapy for GIO is a further decrease of the bone turnover rate that is already low due to GC administration [15]. In this sense, intermittent treatment with recombinant PTH appears better suited for the management of GIO because it increases bone mass through stimulation of osteoblast function, directly counteracting adverse effects of GCs in osteoblasts [15, 210]. However, PTH therapy is costly, limited to 18–24 months, and requires daily subcutaneous injections, which together warrant efforts for the development of solutions more suitable for the prevention and treatment of GIO.

At least two further strategies can be envisioned for the development of bone-sparing GC-based therapy. One option is to develop GR ligands, which promote its anti-inflammatory properties without eliciting adverse effects in osteoblasts. This could be accomplished as a selective GR modulator monotherapy, or by taking advantage of bone targeting strategies [211]. Alternatively, rather than targeting GR itself, one can envision the development of combined therapies, whereby conventional GCs are administered along with an anabolic compound, preferably one that restores mechanisms that are compromised in osteoblasts and osteocytes exposed to GCs.

Dissociated Glucocorticoids

Transactivation versus Transrepression: A Model Too Simplistic for the Development of Osteoblast-Forgiving, Bone-Sparing GCs

A strategy used by pharmaceutical companies to reduce side effects of steroid therapy such as osteoporosis has been the design of ligands that avoid GR dimerization-dependent transactivation, but still transrepress inflammatory genes [212]. Limitations of this strategy, however, have quickly come to attention as compounds that specifically promote transrepression were not necessarily successful in avoiding bone loss [213]. Furthermore, such compounds demonstrated limited efficacy, and only a few have been tested in clinical trials [214]. Part of the problem with this approach was that first generation steroidal SGRMs were identified in *in vitro* high throughput screens, but appeared to be metabolized in vivo into non-dissociated compounds. It remains to be seen if second generation non-steroidal SGRMs will avoid such shortcomings.

The limited clinical success in identifying bone-sparing GR ligands also reflects under-appreciation of the complexity of GR-mediated transcriptional control. Basically, it was assumed that GR employs fundamentally different mechanisms in mediating immune suppression versus side effects. Specifically, an over-simplistic paradigm predominated, which associated side effects with GR homodimers at palindromic GREs, while ascribing the repression of pro-inflammatory genes to a tethering mechanism whereby the GR is recruited to pro-inflammatory genes by transcription factors such as NF-κB, AP-1 and IRF3 [215]. Recent progress, however, revealed a more complex picture. Specifically, chromatin immunoprecipitation (ChIP)-based techniques are currently allowing investigators to map transcription factor locations, as well as capture large-scale chromatin landscapes genome wide. The results highlight cell type-dependent GR-mediated mechanisms that are specific for activation states of cells and individual genes. Based on the emerging paradigm, the limited success of prior efforts is not surprising because both the immune suppression property and the side effects of GCs employ similar fundamental mechanisms of action.

According to the new paradigm, cell type-specific effects of GCs are based on the principle that GR binding is dependent to a large extent on chromatin accessibility, which itself is tissue-specific [216]. Furthermore, direct binding of GR monomers to DNA seems to be more common than initially anticipated and treatment with pharmacological GCs favors binding of GR dimers to classical palindromic DNA elements [26, 27]. Tissue-specific accessibility, in turn, depends on histone modifications and nucleosome positioning within gene regulatory regions [217]. Transcription factors such as C/EBP in liver and fat cells [218, 219], Stat3 in pituitary cells, and AP-1 in mammary epithelial cells [220] often determine accessibility. They are often lineage-specific master regulators, and serve as pioneering transcription factors. Thus, the view of interactions between GR and other transcription factors is changing: early studies of individual genes were mostly interpreted as tethering of the GR by other transcription factors via direct protein-protein interaction. The more recent, genome wide analyses, however, suggest a more complex picture. For example the combinatorial activation of GR and NF-κB leads to the creation of novel binding sites in addition to those occupied after activation of either GR alone or NF-κB alone [221]. Moreover NF-κB target genes sometimes bear GR DNA binding sites close to the NF-κB sites, leading to gene activation or repression dependent on the inflammatory state of innate immune cells [222]. These complex regulatory mechanisms explain, in retrospect, why early attempts to develop bone-sparing dissociating GR ligands were unsuccessful; they relied on transactivation/transrepression of a small number of genes and reporter constructs, which miss the big picture, where GR-mediated transcriptional control is heavily context-dependent. It greatly varies as a function of the individual target gene and it strongly depends on the cellular milieu, which itself depends on the cell type, on whether it is cycling or quiescent, and on the stage of differentiation.

New Requirements for Selective GR Agonists

New principles are sought after towards the identification of bone-sparing glucocorticoids, which do not depend on dissociating dimerization-dependent transactivation from dimerization-independent transrepression. Candidate dissociating ligands will have to be assessed for their bone-sparing property in primary cell systems. That such efforts may be fruitful is demonstrated by the discovery of the plant-derived GR ligand compound A (CpdA). CpdA does not compromise osteoblast differentiation [223]. In contrast to the classical ligand dex, CpdA does not antagonize AP-1-dependent IL-11 expression [223], a pivotal mechanism leading to inhibition of osteoblast differentiation [19]. Consequently, the anti-inflammatory activity of CpdA in arthritis [224, 225], experimental autoimmune encephalomyelitis (EAE) [226, 227] and asthma [225], which is attributable to inhibition of NF-κB [225] and thus decreased levels of cytokines such as IL-6 [223], is not associated with a decrease in bone mass [229]. Future cell-based screens may result in the discovery of additional GR ligands that spare osteoblasts, and some of these ligands may serve as lead compounds for the development of novel bone-sparing anti-inflammatory GC drugs.

Aiming at GR Targets

In addition to employing novel strategies for identification and optimization of novel GR ligands, parallel efforts are warranted towards the development of improved combination therapies for GIO. An optimal combination therapy would replace bisphosphonates or PTH with a treatment modality best suited to counteract pathogenic mechanisms of GIO. Several examples are provided below.

IL11 (interleukin-11), LIF1 (leukemia inhibitory factor 1). As described in section "AP-1", GCs inhibit AP-1-mediated stimulation of *Il11* in osteoblasts *in vitro* and in vivo. The GC-mediated decrease in IL11, as well as LIF1 expression, appears to significantly contribute to the inhibition of osteoblast differentiation [19, 176]. Supplementation of LIF and IL11 in animal models counteracts GIO at least partially (unpublished observations), consistent with data from transgenic mouse models indicating an important role for IL-11 in bone formation and bone mass control in vivo [230]. However, the safety profile of IL-11, as well as LIF, can be problematic; due to their pleiotropic effects, they might induce hematologic and other complications [231].

DKK1, sclerostin (*SOST*). We reviewed in section "Glucocorticoids Inhibit Wnt Signaling in Osteoblasts" evidence for the critical role of the Wnt pathway in osteoblast replication, differentiation and survival, and cited many lines of evidence implicating inhibition of Wnt signaling in GIO. One of the most appealing lines of evidence is the direct stimulation of Dkk1, a Wnt inhibitor, by GCs (section "Role of Dkks in GIO"). Regardless of whether or not GCs stimulate Dkk-1 expression in vivo as they do in vitro (see section "Role of Dkks in GIO"), restoration of Wnt signaling is an attractive option for the management of GIO. This can be achieved, for example, by neutralizing DKK1 using anti-DKK1 antibodies. Another potential approach to restore Wnt signaling in GC-treated patients is co-administration of antibodies against sclerostin, another Wnt antagonist. An advantage of the latter is that the sclerostin-coding gene *Sost* is primarily expressed in osteocytes [232], and its neutralization is therefore less likely to have undesirable extraskeletal effects. Encouraging results from a recent preclinical study, in which mice were co-treated with GCs and anti-sclerostin antibodies, warrant further efforts towards the development of Wnt-targeting strategies for the management of GIO [233].

microRNAs. Restoration of Wnt signaling in the presence of GCs may also be achieved by manipulation of microRNAs, an emerging therapeutic modality that remains to be exploited in the context of GIO. We reviewed in section "Additional Wnt-Related GC Targets" the suppression of HDAC4 by mir29-a and the consequential stimulation of β-catenin and the osteoblast phenotype. Because GCs inhibit mir29-a expression, its pharmacological stimulation appears an attractive avenue towards shielding osteoblasts from adverse effects of GCs.

Unbiased screens. With the advent of high throughput screening technologies, identification of lead compounds, steroidal or otherwise, for the management of GIO may be achieved in an agnostic fashion. Synthetic steroids or large chemical libraries can be screened in the presence of dex, for example, using cell-based assays that report on cellular features such as proliferation, differentiation and/or apoptosis [234]. Alternatively, the assay, typically fluorescent, may report on a distinct aspect of osteoblast differentiation. For example, the readout may reflect the activity of ALP, RUNX2, or a molecular pathway such as Notch or Wnt signaling [235].

Establishing such assays in primary mesenchymal stem cells, pre-osteoblastic cells or osteocytes may be challenging, but worth the effort. With recent advancements in the understanding of cellular and molecular mechanisms underlying GIO, the identification of target genes and compounds for combating this devastating iatrogenic disease is increasingly within reach.

Acknowledgements We thank Gillian H. Little, Michael R. Stallcup (University of Southern California) and Tamas Röszer (University of Ulm) for insightful comments. BF, who holds the J. Harold and Edna L. LaBriola Chair in Genetic Orthopaedic Research, acknowledges support from the National Institutes of Health (RO1 DK071122). JT acknowledges support from Deutsche Forschungsgemeinschaft Immunbone (SPP 1468 Tu220/6-2 INST 40/492 SFB 1149, Collaborative Research Centre 1149) and from KaroBioScience Foundation.

References

1. Soen S, Tanaka Y. Glucocorticoid-induced osteoporosis: skeletal manifestations of glucocorticoid use and 2004 Japanese Society for Bone and Mineral Research-proposed guidelines for its management. Mod Rheumatol. 2005;15:163–8.
2. van Staa TP, et al. A simple score for estimating the long-term risk of fracture in patients using oral glucocorticoids. QJM. 2005;98:191–8.
3. Van Staa TP, Leufkens HG, Abenhaim L, Zhang B, Cooper C. Use of oral corticosteroids and risk of fractures. J Bone Miner Res. 2000;15:993–1000.
4. van Staa TP, Leufkens HG, Cooper C. The epidemiology of corticosteroid-induced osteoporosis: a meta-analysis. Osteoporos Int. 2002;13:777–87.
5. Lane NE, et al. Glucocorticoid-treated mice have localized changes in trabecular bone material properties and osteocyte lacunar size that are not observed in placebo-treated or estrogen-deficient mice. J Bone Miner Res. 2006;21:466–76.
6. Laan RF, et al. Low-dose prednisone induces rapid reversible axial bone loss in patients with rheumatoid arthritis. A randomized, controlled study. Ann Intern Med. 1993;119:963–8.
7. McKenzie R, et al. Decreased bone mineral density during low dose glucocorticoid administration in a randomized, placebo controlled trial. J Rheumatol. 2000;27:2222–6.
8. Aaron JE, Francis RM, Peacock M, Makins NB. Contrasting microanatomy of idiopathic and corticosteroid-induced osteoporosis. Clin Orthop Relat Res. 1989;243:294–305.
9. Laan RF, et al. Differential effects of glucocorticoids on cortical appendicular and cortical vertebral bone mineral content. Calcif Tissue Int. 1993;52:5–9.
10. Maggi S, et al. Osteoporosis risk in patients with chronic obstructive pulmonary disease: the EOLO study. J Clin Densitom. 2009;12:345–52.
11. Gonnelli S, et al. Effect of inhaled glucocorticoids and beta(2) agonists on vertebral fracture risk in COPD patients: the EOLO study. Calcif Tissue Int. 2010;87:137–43.
12. Weinstein RS. Glucocorticoid-induced osteoporosis and osteonecrosis. Endocrinol Metab Clin North Am. 2012;41:595–611.
13. Canalis E, Mazziotti G, Giustina A, Bilezikian JP. Glucocorticoid-induced osteoporosis: pathophysiology and therapy. Osteoporos Int. 2007;18:1319–28.
14. Grossman JM, et al. American College of Rheumatology 2010 recommendations for the prevention and treatment of glucocorticoid-induced osteoporosis. Arthritis Care Res. 2010;62:1515–26.
15. Weinstein RS, Jilka RL, Almeida M, Roberson PK, Manolagas SC. Intermittent parathyroid hormone administration counteracts the adverse effects of glucocorticoids on osteoblast and osteocyte viability, bone formation, and strength in mice. Endocrinology. 2010;151:2641–9.

16. Gluer CC, et al. Comparative effects of teriparatide and risedronate in glucocorticoid-induced osteoporosis in men: 18-month results of the EuroGIOPs trial. J Bone Miner Res. 2013;28: 1355–68.
17. Weinstein RS, Jilka RL, Parfitt AM, Manolagas SC. Inhibition of osteoblastogenesis and promotion of apoptosis of osteoblasts and osteocytes by glucocorticoids. Potential mechanisms of their deleterious effects on bone. J Clin Invest. 1998;102:274–82.
18. Hofbauer LC, et al. Stimulation of osteoprotegerin ligand and inhibition of osteoprotegerin production by glucocorticoids in human osteoblastic lineage cells: potential paracrine mechanisms of glucocorticoid-induced osteoporosis. Endocrinology. 1999;140:4382–9.
19. Rauch A, et al. Glucocorticoids suppress bone formation by attenuating osteoblast differentiation via the monomeric glucocorticoid receptor. Cell Metab. 2010;11:517–31.
20. Yao W, et al. Glucocorticoid excess in mice results in early activation of osteoclastogenesis and adipogenesis and prolonged suppression of osteogenesis: a longitudinal study of gene expression in bone tissue from glucocorticoid-treated mice. Arthritis Rheum. 2008;58: 1674–86.
21. O'Brien CA, et al. Glucocorticoids act directly on osteoblasts and osteocytes to induce their apoptosis and reduce bone formation and strength. Endocrinology. 2004;145:1835–41.
22. Sher LB, et al. Transgenic expression of 11beta-hydroxysteroid dehydrogenase type 2 in osteoblasts reveals an anabolic role for endogenous glucocorticoids in bone. Endocrinology. 2004;145:922–9.
23. Yang M, et al. Col3.6-HSD2 transgenic mice: a glucocorticoid loss-of-function model spanning early and late osteoblast differentiation. Bone. 2010;47:573–82.
24. Reichardt HM, et al. Repression of inflammatory responses in the absence of DNA binding by the glucocorticoid receptor. EMBO J. 2001;20:7168–73.
25. Reichardt HM, et al. DNA binding of the glucocorticoid receptor is not essential for survival. Cell. 1998;93:531–41.
26. Lim HW et al. Genomic redistribution of GR monomers and dimers mediates transcriptional response to exogenous glucocorticoids in vivo. Genome Research 2015 May 8 [Epub ahead of print].
27. Schiller BJ, Chodankar R, Watson LC, Stallcup MR, Yamamoto KR. Glucocorticoid receptor binds half sites as a monomer and regulates specific target genes. Genome Biol. 2014;15:3181.
28. Weinstein RS, et al. Promotion of osteoclast survival and antagonism of bisphosphonate-induced osteoclast apoptosis by glucocorticoids. J Clin Invest. 2002;109:1041–8.
29. Kim HJ, et al. Glucocorticoids suppress bone formation via the osteoclast. J Clin Invest. 2006;116:2152–60.
30. Jia D, O'Brien CA, Stewart SA, Manolagas SC, Weinstein RS. Glucocorticoids act directly on osteoclasts to increase their life span and reduce bone density. Endocrinology. 2006; 147:5592–9.
31. Alesci S, De Martino MU, Ilias I, Gold PW, Chrousos GP. Glucocorticoid-induced osteoporosis: from basic mechanisms to clinical aspects. Neuroimmunomodulation. 2005;12:1–19.
32. Moutsatsou P, Kassi E, Papavassiliou AG. Glucocorticoid receptor signaling in bone cells. Trends Mol Med. 2012;18:348–59.
33. Baschant U, Lane NE, Tuckermann J. The multiple facets of glucocorticoid action in rheumatoid arthritis. Nat Rev Rheumatol. 2012;8:645–55.
34. Weinstein RS, et al. The skeletal effects of glucocorticoid excess override those of orchidectomy in mice. Endocrinology. 2004;145:1980–7.
35. Chen TL, Aronow L, Feldman D. Glucocorticoid receptors and inhibition of bone cell growth in primary culture. Endocrinology. 1977;100:619–28.
36. Smith E, et al. Glucocorticoids inhibit developmental stage-specific osteoblast cell cycle. Dissociation of cyclin A-cyclin-dependent kinase 2 from E2F4-p130 complexes. J Biol Chem. 2000;275:19992–20001.
37. Gabet Y, Noh T, Lee C, Frenkel B. Developmentally regulated inhibition of cell cycle progression by glucocorticoids through repression of cyclin A transcription in primary osteoblast cultures. J Cell Physiol. 2011;226:991–8.

38. Li H, et al. Glucocorticoid receptor and sequential P53 activation by dexamethasone mediates apoptosis and cell cycle arrest of osteoblastic MC3T3-E1 cells. PLoS One. 2012;7:e37030.
39. Chang JK, et al. Anti-inflammatory drugs suppress proliferation and induce apoptosis through altering expressions of cell cycle regulators and pro-apoptotic factors in cultured human osteoblasts. Toxicology. 2009;258:148–56.
40. Rogatsky I, Trowbridge JM, Garabedian MJ. Glucocorticoid receptor-mediated cell cycle arrest is achieved through distinct cell-specific transcriptional regulatory mechanisms. Mol Cell Biol. 1997;17:3181–93.
41. Engelbrecht Y, et al. Glucocorticoids induce rapid up-regulation of mitogen-activated protein kinase phosphatase-1 and dephosphorylation of extracellular signal-regulated kinase and impair proliferation in human and mouse osteoblast cell lines. Endocrinology. 2003;144: 412–22.
42. Lecka-Czernik B, et al. Inhibition of Osf2/Cbfa1 expression and terminal osteoblast differentiation by PPARgamma2. J Cell Biochem. 1999;74:357–71.
43. Pereira RC, Delany AM, Canalis E. Effects of cortisol and bone morphogenetic protein-2 on stromal cell differentiation: correlation with CCAAT-enhancer binding protein expression. Bone. 2002;30:685–91.
44. Rosen CJ, Bouxsein ML. Mechanisms of disease: is osteoporosis the obesity of bone? Nat Clin Pract Rheumatol. 2006;2:35–43.
45. Muruganandan S, Roman AA, Sinal CJ. Adipocyte differentiation of bone marrow-derived mesenchymal stem cells: cross talk with the osteoblastogenic program. Cell Mol Life Sci. 2009;66:236–53.
46. Naito M, Omoteyama K, Mikami Y, Takahashi T, Takagi M. Inhibition of Wnt/beta-catenin signaling by dexamethasone promotes adipocyte differentiation in mesenchymal progenitor cells, ROB-C26. Histochem Cell Biol. 2012;138:833–45.
47. Berendsen AD, Olsen BR. Osteoblast-adipocyte lineage plasticity in tissue development, maintenance and pathology. Cell Mol Life Sci. 2014;71:493–7.
48. Leclerc N, et al. Gene expression profiling of glucocorticoid-inhibited osteoblasts. J Mol Endocrinol. 2004;33:175–93.
49. Smith E, Coetzee GA, Frenkel B. Glucocorticoids inhibit cell cycle progression in differentiating osteoblasts via glycogen synthase kinase-3beta. J Biol Chem. 2002;277:18191–7.
50. Mikami Y, Lee M, Irie S, Honda MJ. Dexamethasone modulates osteogenesis and adipogenesis with regulation of osterix expression in rat calvaria-derived cells. J Cell Physiol. 2011;226:739–48.
51. Worthley et al. Gremlin 1 Identifies a Skeletal Stem Cell with Bone, Cartilage, and Reticular Stromal Potential Cell. 2015;160:269–284.
52. Chan et al. Identification and Specification of the Mouse Skeletal Stem Cell. Cell. 2015; 160:285–298.
53. Canalis E. Effect of glucocorticoids on type I collagen synthesis, alkaline phosphatase activity, and deoxyribonucleic acid content in cultured rat calvariae. Endocrinology. 1983;112: 931–9.
54. Harris C, et al. Large increases in adipose triacylglycerol flux in Cushingoid CRH-Tg mice are explained by futile cycling. Am J Physiol Endocrinol Metab. 2013;304:E282–93.
55. Delany AM, Gabbitas BY, Canalis E. Cortisol downregulates osteoblast alpha 1 (I) procollagen mRNA by transcriptional and posttranscriptional mechanisms. J Cell Biochem. 1995;57:488–94.
56. Luppen CA, et al. Brief bone morphogenetic protein 2 treatment of glucocorticoid-inhibited MC3T3-E1 osteoblasts rescues commitment-associated cell cycle and mineralization without alteration of Runx2. J Biol Chem. 2003;278:44995–5003.
57. Delany AM, Jeffrey JJ, Rydziel S, Canalis E. Cortisol increases interstitial collagenase expression in osteoblasts by post-transcriptional mechanisms. J Biol Chem. 1995;270: 26607–12.

58. Lian JB, et al. Species-specific glucocorticoid and 1,25-dihydroxyvitamin D responsiveness in mouse MC3T3-E1 osteoblasts: dexamethasone inhibits osteoblast differentiation and vitamin D down-regulates osteocalcin gene expression. Endocrinology. 1997;138:2117–27.
59. Chen TL, Fry D. Hormonal regulation of the osteoblastic phenotype expression in neonatal murine calvarial cells. Calcif Tissue Int. 1999;64:304–9.
60. Butler JS, et al. Silencing Dkk1 expression rescues dexamethasone-induced suppression of primary human osteoblast differentiation. BMC Musculoskelet Disord. 2010;11:210.
61. Koromila T, et al. Glucocorticoids antagonize RUNX2 during osteoblast differentiation in cultures of ST2 pluripotent mesenchymal cells. J Cell Biochem. 2014;115:27–33.
62. Wang FS, et al. Secreted frizzled-related protein 1 modulates glucocorticoid attenuation of osteogenic activities and bone mass. Endocrinology. 2005;146:2415–23.
63. Subramaniam M, et al. Glucocorticoid regulation of alkaline phosphatase, osteocalcin, and proto-oncogenes in normal human osteoblast-like cells. J Cell Biochem. 1992;50:411–24.
64. Cheng SL, Yang JW, Rifas L, Zhang SF, Avioli LV. Differentiation of human bone marrow osteogenic stromal cells in vitro: induction of the osteoblast phenotype by dexamethasone. Endocrinology. 1994;134:277–86.
65. Shalhoub V, et al. Glucocorticoids promote development of the osteoblast phenotype by selectively modulating expression of cell growth and differentiation associated genes. J Cell Biochem. 1992;50:425–40.
66. Ishida Y, Heersche JN. Glucocorticoid-induced osteoporosis: both in vivo and in vitro concentrations of glucocorticoids higher than physiological levels attenuate osteoblast differentiation. J Bone Miner Res. 1998;13:1822–6.
67. Luppen CA, Chandler RL, Noh T, Mortlock DP, Frenkel B. BMP-2 vs. BMP-4 expression and activity in glucocorticoid-arrested MC3T3-E1 osteoblasts: Smad signaling, not alkaline phosphatase activity, predicts rescue of mineralization. Growth Factors. 2008;26:226–37.
68. Luppen CA, Smith E, Spevak L, Boskey AL, Frenkel B. Bone morphogenetic protein-2 restores mineralization in glucocorticoid-inhibited MC3T3-E1 osteoblast cultures. J Bone Miner Res. 2003;18:1186–97.
69. Smith E, et al. Expression of cell cycle regulatory factors in differentiating osteoblasts: post-proliferative up-regulation of cyclins B and E. Cancer Res. 1995;55:5019–24.
70. Smith E, et al. Post-proliferative cyclin E-associated kinase activity in differentiated osteoblasts: inhibition by proliferating osteoblasts and osteosarcoma cells. J Cell Biochem. 1997;66:141–52.
71. Carcamo-Orive I, et al. Regulation of human bone marrow stromal cell proliferation and differentiation capacity by glucocorticoid receptor and AP-1 crosstalk. J Bone Miner Res. 2010; 25:2115–25.
72. Gu G, Hentunen TA, Nars M, Harkonen PL, Vaananen HK. Estrogen protects primary osteocytes against glucocorticoid-induced apoptosis. Apoptosis. 2005;10:583–95.
73. Liu Y, et al. Prevention of glucocorticoid-induced apoptosis in osteocytes and osteoblasts by calbindin-D28k. J Bone Miner Res. 2004;19:479–90.
74. Hurson CJ, et al. Gene expression analysis in human osteoblasts exposed to dexamethasone identifies altered developmental pathways as putative drivers of osteoporosis. BMC Musculoskelet Disord. 2007;8:12.
75. Conradie MM, et al. Vanadate prevents glucocorticoid-induced apoptosis of osteoblasts in vitro and osteocytes in vivo. J Endocrinol. 2007;195:229–40.
76. Zalavras C, Shah S, Birnbaum MJ, Frenkel B. Role of apoptosis in glucocorticoid-induced osteoporosis and osteonecrosis. Crit Rev Eukaryot Gene Expr. 2003;13:221–35.
77. Nakashima T, et al. Evidence for osteocyte regulation of bone homeostasis through RANKL expression. Nat Med. 2011;17:1231–4.
78. Xiong J, et al. Matrix-embedded cells control osteoclast formation. Nat Med. 2011;17: 1235–41.
79. Dallas SL, Prideaux M, Bonewald LF. The osteocyte: an endocrine cell … and more. Endocr Rev. 2013;34:658–90.
80. Jia J, et al. Glucocorticoid dose determines osteocyte cell fate. FASEB J. 2011;25:3366–76.

81. Gu G, Mulari M, Peng Z, Hentunen TA, Vaananen HK. Death of osteocytes turns off the inhibition of osteoclasts and triggers local bone resorption. Biochem Biophys Res Commun. 2005;335:1095–101.
82. Monroe DG, McGee-Lawrence ME, Oursler MJ, Westendorf JJ. Update on Wnt signaling in bone cell biology and bone disease. Gene. 2012;492:1–18.
83. Habib SJ, et al. A localized Wnt signal orients asymmetric stem cell division in vitro. Science. 2013;339:1445–8.
84. Gong Y, et al. LDL receptor-related protein 5 (LRP5) affects bone accrual and eye development. Cell. 2001;107:513–23.
85. Little RD, et al. A mutation in the LDL receptor-related protein 5 gene results in the autosomal dominant high-bone-mass trait. Am J Hum Genet. 2002;70:11–9.
86. Boyden LM, et al. High bone density due to a mutation in LDL-receptor-related protein 5. N Engl J Med. 2002;346:1513–21.
87. Smith E, Frenkel B. Glucocorticoids inhibit the transcriptional activity of LEF/TCF in differentiating osteoblasts in a glycogen synthase kinase-3beta-dependent and -independent manner. J Biol Chem. 2005;280:2388–94.
88. Ohnaka K, Tanabe M, Kawate H, Nawata H, Takayanagi R. Glucocorticoid suppresses the canonical Wnt signal in cultured human osteoblasts. Biochem Biophys Res Commun. 2005;329:177–81.
89. Almeida M, Han L, Ambrogini E, Weinstein RS, Manolagas SC. Glucocorticoids and tumor necrosis factor alpha increase oxidative stress and suppress Wnt protein signaling in osteoblasts. J Biol Chem. 2011;286:44326–35.
90. Naito M, Mikami Y, Takagi M, Takahashi T. Up-regulation of Axin2 by dexamethasone promotes adipocyte differentiation in ROB-C26 mesenchymal progenitor cells. Cell Tissue Res. 2013;354:761–70.
91. Ohnaka K, Taniguchi H, Kawate H, Nawata H, Takayanagi R. Glucocorticoid enhances the expression of dickkopf-1 in human osteoblasts: novel mechanism of glucocorticoid-induced osteoporosis. Biochem Biophys Res Commun. 2004;318:259–64.
92. Mak W, Shao X, Dunstan CR, Seibel MJ, Zhou H. Biphasic glucocorticoid-dependent regulation of Wnt expression and its inhibitors in mature osteoblastic cells. Calcif Tissue Int. 2009;85:538–45.
93. Cruciat CM, Niehrs C. Secreted and transmembrane wnt inhibitors and activators. Cold Spring Harb Perspect Biol. 2013;5:a015081.
94. Bafico A, Liu G, Yaniv A, Gazit A, Aaronson SA. Novel mechanism of Wnt signalling inhibition mediated by Dickkopf-1 interaction with LRP6/Arrow. Nat Cell Biol. 2001;3:683–6.
95. Patel MS, Karsenty G. Regulation of bone formation and vision by LRP5. N Engl J Med. 2002;346:1572–4.
96. Harada S, Rodan GA. Control of osteoblast function and regulation of bone mass. Nature. 2003;423:349–55.
97. Morvan F, et al. Deletion of a single allele of the Dkk1 gene leads to an increase in bone formation and bone mass. J Bone Miner Res. 2006;21:934–45.
98. MacDonald BT, et al. Bone mass is inversely proportional to Dkk1 levels in mice. Bone. 2007;41:331–9.
99. Brunetti G, et al. High dickkopf-1 levels in sera and leukocytes from children with 21-hydroxylase deficiency on chronic glucocorticoid treatment. Am J Physiol Endocrinol Metab. 2013;304:E546–54.
100. Gifre L, et al. Effect of glucocorticoid treatment on Wnt signalling antagonists (sclerostin and Dkk-1) and their relationship with bone turnover. Bone. 2013;57:272–6.
101. Jones SE, Jomary C. Secreted frizzled-related proteins: searching for relationships and patterns. Bioessays. 2002;24:811–20.
102. Bodine PV, et al. The Wnt antagonist secreted frizzled-related protein-1 is a negative regulator of trabecular bone formation in adult mice. Mol Endocrinol. 2004;18:1222–37.
103. Takayama S, Rogatsky I, Schwarcz LE, Darimont BD. The glucocorticoid receptor represses cyclin D1 by targeting the Tcf-beta-catenin complex. J Biol Chem. 2006;281:17856–63.

104. Olkku A, Mahonen A. Calreticulin mediated glucocorticoid receptor export is involved in beta-catenin translocation and Wnt signalling inhibition in human osteoblastic cells. Bone. 2009;44:555–65.
105. Ko J-Y, et al. MicroRNA-29a ameliorates glucocorticoid-induced suppression of osteoblast differentiation by regulating β-catenin acetylation. Bone. 2013;57:468–75.
106. Nuttall ME, Patton AJ, Olivera DL, Nadeau DP, Gowen M. Human trabecular bone cells are able to express both osteoblastic and adipocytic phenotype: implications for osteopenic disorders. J Bone Miner Res. 1998;13:371–82.
107. Hayashi K, et al. BMP/Wnt antagonists are upregulated by dexamethasone in osteoblasts and reversed by alendronate and PTH: potential therapeutic targets for glucocorticoid-induced osteoporosis. Biochem Biophys Res Commun. 2009;379:261–6.
108. Pereira RM, Delany AM, Durant D, Canalis E. Cortisol regulates the expression of Notch in osteoblasts. J Cell Biochem. 2002;85:252–8.
109. Engin F, et al. Dimorphic effects of Notch signaling in bone homeostasis. Nat Med. 2008;14:299–305.
110. Hilton MJ, et al. Notch signaling maintains bone marrow mesenchymal progenitors by suppressing osteoblast differentiation. Nat Med. 2008;14:306–14.
111. Zanotti S, Canalis E. Notch1 and Notch2 expression in osteoblast precursors regulates femoral microarchitecture. Bone. 2014;62:22–8.
112. Kamiya N, et al. BMP signaling negatively regulates bone mass through sclerostin by inhibiting the canonical Wnt pathway. Development. 2008;135:3801–11.
113. Skrtic S, Ohlsson C. Cortisol decreases hepatocyte growth factor levels in human osteoblast-like cells. Calcif Tissue Int. 2000;66:108–12.
114. Tsunashima Y, Kondo A, Matsuda T, Togari A. Hydrocortisone inhibits cellular proliferation by downregulating hepatocyte growth factor synthesis in human osteoblasts. Biol Pharm Bull. 2011;34:700–3.
115. Grano M, et al. Hepatocyte growth factor is a coupling factor for osteoclasts and osteoblasts in vitro. Proc Natl Acad Sci U S A. 1996;93:7644–8.
116. Giustina A, Mazziotti G, Canalis E. Growth hormone, insulin-like growth factors, and the skeleton. Endocr Rev. 2008;29:535–59.
117. McCarthy TL, Centrella M, Canalis E. Cortisol inhibits the synthesis of insulin-like growth factor-I in skeletal cells. Endocrinology. 1990;126:1569–75.
118. Delany AM, Canalis E. Transcriptional repression of insulin-like growth factor I by glucocorticoids in rat bone cells. Endocrinology. 1995;136:4776–81.
119. Delany AM, Durant D, Canalis E. Glucocorticoid suppression of IGF I transcription in osteoblasts. Mol Endocrinol. 2001;15:1781–9.
120. Chevalley T, Strong DD, Mohan S, Baylink D, Linkhart TA. Evidence for a role for insulin-like growth factor binding proteins in glucocorticoid inhibition of normal human osteoblast-like cell proliferation. Eur J Endocrinol. 1996;134:591–601.
121. Gabbitas B, Pash JM, Delany AM, Canalis E. Cortisol inhibits the synthesis of insulin-like growth factor-binding protein-5 in bone cell cultures by transcriptional mechanisms. J Biol Chem. 1996;271:9033–8.
122. Andress DL, Birnbaum RS. Human osteoblast-derived insulin-like growth factor (IGF) binding protein-5 stimulates osteoblast mitogenesis and potentiates IGF action. J Biol Chem. 1992;267:22467–72.
123. Richman C, Baylink DJ, Lang K, Dony C, Mohan S. Recombinant human insulin-like growth factor-binding protein-5 stimulates bone formation parameters in vitro and in vivo. Endocrinology. 1999;140:4699–705.
124. Miyakoshi N, et al. Evidence that IGF-binding protein-5 functions as a growth factor. J Clin Invest. 2001;107:73–81.
125. Salih DA, et al. Insulin-like growth factor-binding protein-5 induces a gender-related decrease in bone mineral density in transgenic mice. Endocrinology. 2005;146:931–40.
126. Chaudhary LR, Avioli LV. Activation of extracellular signal-regulated kinases 1 and 2 (ERK1 and ERK2) by FGF-2 and PDGF-BB in normal human osteoblastic and bone marrow stromal

cells: differences in mobility and in-gel renaturation of ERK1 in human, rat, and mouse osteoblastic cells. Biochem Biophys Res Commun. 1997;238:134–9.
127. Xiao G, et al. Bone morphogenetic proteins, extracellular matrix, and mitogen-activated protein kinase signaling pathways are required for osteoblast-specific gene expression and differentiation in MC3T3-E1 cells. J Bone Miner Res. 2002;17:101–10.
128. Xiao G, Jiang D, Gopalakrishnan R, Franceschi RT. Fibroblast growth factor 2 induction of the osteocalcin gene requires MAPK activity and phosphorylation of the osteoblast transcription factor, Cbfa1/Runx2. J Biol Chem. 2002;277:36181–7.
129. Greenblatt MB, Shim JH, Glimcher LH. Mitogen-activated protein kinase pathways in osteoblasts. Annu Rev Cell Dev Biol. 2013;29:63–79.
130. Xiao G, et al. MAPK pathways activate and phosphorylate the osteoblast-specific transcription factor, Cbfa1. J Biol Chem. 2000;275:4453–9.
131. Li Y, Ge C, Franceschi RT. Differentiation-dependent association of phosphorylated extracellular signal-regulated kinase with the chromatin of osteoblast-related genes. J Bone Miner Res. 2010;25:154–63.
132. Lai CF, et al. Erk is essential for growth, differentiation, integrin expression, and cell function in human osteoblastic cells. J Biol Chem. 2001;276:14443–50.
133. Ge C, Xiao G, Jiang D, Franceschi RT. Critical role of the extracellular signal-regulated kinase-MAPK pathway in osteoblast differentiation and skeletal development. J Cell Biol. 2007;176:709–18.
134. Hulley PA, Gordon F, Hough FS. Inhibition of mitogen-activated protein kinase activity and proliferation of an early osteoblast cell line (MBA 15.4) by dexamethasone: role of protein phosphatases. Endocrinology. 1998;139:2423–31.
135. Xian L, et al. Matrix IGF-1 maintains bone mass by activation of mTOR in mesenchymal stem cells. Nat Med. 2012;18:1095–101.
136. Yamamoto T, et al. Parathyroid hormone activates phosphoinositide 3-kinase-Akt-Bad cascade in osteoblast-like cells. Bone. 2007;40:354–9.
137. Mukherjee A, Rotwein P. Akt promotes BMP2-mediated osteoblast differentiation and bone development. J Cell Sci. 2009;122:716–26.
138. Liang J, Slingerland JM. Multiple roles of the PI3K/PKB (Akt) pathway in cell cycle progression. Cell Cycle. 2003;2:339–45.
139. Hennessy BT, Smith DL, Ram PT, Lu Y, Mills GB. Exploiting the PI3K/AKT pathway for cancer drug discovery. Nat Rev Drug Discov. 2005;4:988–1004.
140. Fujita T, et al. Runx2 induces osteoblast and chondrocyte differentiation and enhances their migration by coupling with PI3K-Akt signaling. J Cell Biol. 2004;166:85–95.
141. Choi YH, et al. Akt enhances Runx2 protein stability by regulating Smurf2 function during osteoblast differentiation. FEBS J. 2014;281(16):3656–66.
142. Borgatti P, et al. Translocation of Akt/PKB to the nucleus of osteoblast-like MC3T3-E1 cells exposed to proliferative growth factors. FEBS Lett. 2000;477:27–32.
143. Almeida M, Han L, Martin-Millan M, O'Brien CA, Manolagas SC. Oxidative stress antagonizes Wnt signaling in osteoblast precursors by diverting beta-catenin from T cell factor- to forkhead box O-mediated transcription. J Biol Chem. 2007;282:27298–305.
144. Plotkin LI, Manolagas SC, Bellido T. Glucocorticoids induce osteocyte apoptosis by blocking focal adhesion kinase-mediated survival. Evidence for inside-out signaling leading to anoikis. J Biol Chem. 2007;282:24120–30.
145. Hulley PA, Conradie MM, Langeveldt CR, Hough FS. Glucocorticoid-induced osteoporosis in the rat is prevented by the tyrosine phosphatase inhibitor, sodium orthovanadate. Bone. 2002;31:220–9.
146. Clark AR, Lasa M. Crosstalk between glucocorticoids and mitogen-activated protein kinase signalling pathways. Curr Opin Pharmacol. 2003;3:404–11.
147. Toh ML, Yang Y, Leech M, Santos L, Morand EF. Expression of mitogen-activated protein kinase phosphatase 1, a negative regulator of the mitogen-activated protein kinases, in rheumatoid arthritis: up-regulation by interleukin-1beta and glucocorticoids. Arthritis Rheum. 2004;50:3118–28.

148. Camps M, Nichols A, Arkinstall S. Dual specificity phosphatases: a gene family for control of MAP kinase function. FASEB J. 2000;14:6–16.
149. Slack DN, Seternes OM, Gabrielsen M, Keyse SM. Distinct binding determinants for ERK2/p38alpha and JNK map kinases mediate catalytic activation and substrate selectivity of map kinase phosphatase-1. J Biol Chem. 2001;276:16491–500.
150. Franklin CC, Kraft AS. Conditional expression of the mitogen-activated protein kinase (MAPK) phosphatase MKP-1 preferentially inhibits p38 MAPK and stress-activated protein kinase in U937 cells. J Biol Chem. 1997;272:16917–23.
151. Kassel O, et al. Glucocorticoids inhibit MAP kinase via increased expression and decreased degradation of MKP-1. EMBO J. 2001;20:7108–16.
152. Horsch K, et al. Mitogen-activated protein kinase phosphatase 1/dual specificity phosphatase 1 mediates glucocorticoid inhibition of osteoblast proliferation. Mol Endocrinol. 2007; 21:2929–40.
153. Johansson-Haque K, Palanichamy E, Okret S. Stimulation of MAPK-phosphatase 1 gene expression by glucocorticoids occurs through a tethering mechanism involving C/EBP. J Mol Endocrinol. 2008;41:239–49.
154. Shipp LE, et al. Transcriptional regulation of human dual specificity protein phosphatase 1 (DUSP1) gene by glucocorticoids. PLoS One. 2010;5:e13754.
155. Conradie MM, et al. MKP-1 knockout does not prevent glucocorticoid-induced bone disease in mice. Calcif Tissue Int. 2011;89:221–7.
156. Katoh M, Katoh M. Human FOX gene family (Review). Int J Oncol. 2004;25:1495–500.
157. Kenyon C. The plasticity of aging: insights from long-lived mutants. Cell. 2005;120: 449–60.
158. Houstis N, Rosen ED, Lander ES. Reactive oxygen species have a causal role in multiple forms of insulin resistance. Nature. 2006;440:944–8.
159. Essers MA, et al. FOXO transcription factor activation by oxidative stress mediated by the small GTPase Ral and JNK. EMBO J. 2004;23:4802–12.
160. Iyer S, et al. FOXOs attenuate bone formation by suppressing Wnt signaling. J Clin Invest. 2013;123:3409–19.
161. Lutzner N, Kalbacher H, Krones-Herzig A, Rosl F. FOXO3 is a glucocorticoid receptor target and regulates LKB1 and its own expression based on cellular AMP levels via a positive autoregulatory loop. PLoS One. 2012;7:e42166.
162. Essers MA, et al. Functional interaction between beta-catenin and FOXO in oxidative stress signaling. Science. 2005;308:1181–4.
163. Hoogeboom D, et al. Interaction of FOXO with beta-catenin inhibits beta-catenin/T cell factor activity. J Biol Chem. 2008;283:9224–30.
164. De Bosscher K, Vanden Berghe W, Haegeman G. The interplay between the glucocorticoid receptor and nuclear factor-kappaB or activator protein-1: molecular mechanisms for gene repression. Endocr Rev. 2003;24:488–522.
165. Jonat C, et al. Antitumor promotion and antiinflammation: down-modulation of AP-1 (Fos/Jun) activity by glucocorticoid hormone. Cell. 1990;62:1189–204.
166. Yang-Yen HF, et al. Transcriptional interference between c-Jun and the glucocorticoid receptor: mutual inhibition of DNA binding due to direct protein-protein interaction. Cell. 1990;62:1205–15.
167. Subramaniam N, Cairns W, Okret S. Studies on the mechanism of glucocorticoid-mediated repression from a negative glucocorticoid response element from the bovine prolactin gene. DNA Cell Biol. 1997;16:153–63.
168. Miner JN, Yamamoto KR. The basic region of AP-1 specifies glucocorticoid receptor activity at a composite response element. Genes Dev. 1992;6:2491–501.
169. Tuckermann JP, et al. The DNA binding-independent function of the glucocorticoid receptor mediates repression of AP-1-dependent genes in skin. J Cell Biol. 1999;147:1365–70.
170. McCabe LR, et al. Developmental expression and activities of specific fos and jun proteins are functionally related to osteoblast maturation: role of Fra-2 and Jun D during differentiation. Endocrinology. 1996;137:4398–408.

171. Sabatakos G, et al. Overexpression of DeltaFosB transcription factor(s) increases bone formation and inhibits adipogenesis. Nat Med. 2000;6:985–90.
172. Wagner EF. Functions of AP1 (Fos/Jun) in bone development. Ann Rheum Dis. 2002;61 Suppl 2:ii40–2.
173. Roohk DJ, et al. Dexamethasone-mediated changes in adipose triacylglycerol metabolism are exaggerated, not diminished, in the absence of a functional GR dimerization domain. Endocrinology. 2013;154:1528–39.
174. Matsumoto T, Kuriwaka-Kido R, Kondo T, Endo I, Kido S. Regulation of osteoblast differentiation by interleukin-11 via AP-1 and Smad signaling. Endocr J. 2012;59:91–101.
175. Sims NA, et al. Interleukin-11 receptor signaling is required for normal bone remodeling. J Bone Miner Res. 2005;20:1093–102.
176. Kuriwaka-Kido R, et al. Parathyroid hormone (1-34) counteracts the suppression of interleukin-11 expression by glucocorticoid in murine osteoblasts: a possible mechanism for stimulating osteoblast differentiation against glucocorticoid excess. Endocrinology. 2013;154: 1156–67.
177. Komori T, et al. Targeted disruption of Cbfa1 results in a complete lack of bone formation owing to maturational arrest of osteoblasts [see comments]. Cell. 1997;89:755–64.
178. Otto F, et al. Cbfa1, a candidate gene for cleidocranial dysplasia syndrome, is essential for osteoblast differentiation and bone development [see comments]. Cell. 1997;89:765–71.
179. Banerjee C, et al. Runt homology domain proteins in osteoblast differentiation: AML3/CBFA1 is a major component of a bone-specific complex. J Cell Biochem. 1997;66:1–8.
180. Ducy P, Zhang R, Geoffroy V, Ridall AL, Karsenty G. Osf2/Cbfa1: a transcriptional activator of osteoblast differentiation [see comments]. Cell. 1997;89:747–54.
181. Komori T. Signaling networks in RUNX2-dependent bone development. J Cell Biochem. 2011;112:750–5.
182. Ducy P, et al. A Cbfa1-dependent genetic pathway controls bone formation beyond embryonic development. Genes Dev. 1999;13:1025–36.
183. Estrada K, et al. Genome-wide meta-analysis identifies 56 bone mineral density loci and reveals 14 loci associated with risk of fracture. Nat Genet. 2012;44:491–501.
184. Chang DJ, et al. Reduction in transforming growth factor beta receptor I expression and transcription factor CBFa1 on bone cells by glucocorticoid. J Biol Chem. 1998;273:4892–6.
185. Prince M, et al. Expression and regulation of Runx2/Cbfa1 and osteoblast phenotypic markers during the growth and differentiation of human osteoblasts. J Cell Biochem. 2001;80:424–40.
186. Bae SC, Lee YH. Phosphorylation, acetylation and ubiquitination: the molecular basis of RUNX regulation. Gene. 2006;366:58–66.
187. Li X, Decker M, Westendorf JJ. TEThered to Runx: novel binding partners for runx factors. Blood Cells Mol Dis. 2010;45:82–5.
188. Drissi H, et al. Transcriptional autoregulation of the bone related CBFA1/RUNX2 gene. J Cell Physiol. 2000;184:341–50.
189. Paredes R, et al. Bone-specific transcription factor Runx2 interacts with the 1alpha,25-dihydroxyvitamin D3 receptor to up-regulate rat osteocalcin gene expression in osteoblastic cells. Mol Cell Biol. 2004;24:8847–61.
190. Khalid O, et al. Modulation of Runx2 activity by estrogen receptor-alpha: implications for osteoporosis and breast cancer. Endocrinology. 2008;149:5984–95.
191. Baniwal SK, et al. Repression of Runx2 by androgen receptor (AR) in osteoblasts and prostate cancer cells: AR binds Runx2 and abrogates its recruitment to DNA. Mol Endocrinol. 2009;23:1203–14.
192. Ning YM, Robins DM. AML3/CBFalpha1 is required for androgen-specific activation of the enhancer of the mouse sex-limited protein (Slp) gene. J Biol Chem. 1999;274:30624–30.
193. Little GH, et al. Differential effects of RUNX2 on the androgen receptor in prostate cancer: synergistic stimulation of a gene set exemplified by SNAI2 and subsequent invasiveness. Cancer Res. 2014;74:2857–68.

194. Gaur T, et al. Canonical WNT signaling promotes osteogenesis by directly stimulating Runx2 gene expression. J Biol Chem. 2005;280:33132–40.
195. Ekenstam E, Stalenheim G, Hallgren R. The acute effect of high dose corticosteroid treatment on serum osteocalcin. Metabolism. 1988;37:141–4.
196. Heinrichs AA, et al. Identification of multiple glucocorticoid receptor binding sites in the rat osteocalcin gene promoter. Biochemistry. 1993;32:11436–44.
197. Cosman F, Nieves J, Herbert J, Shen V, Lindsay R. High-dose glucocorticoids in multiple sclerosis patients exert direct effects on the kidney and skeleton. J Bone Miner Res. 1994;9:1097–105.
198. Morrison N, Eisman J. Role of the negative glucocorticoid regulatory element in glucocorticoid repression of the human osteocalcin promoter. J Bone Miner Res. 1993;8:969–75.
199. Morrison NA, et al. 1,25-dihydroxyvitamin D-responsive element and glucocorticoid repression in the osteocalcin gene. Science. 1989;246:1158–61.
200. Stromstedt PE, Poellinger L, Gustafsson JA, Carlstedt-Duke J. The glucocorticoid receptor binds to a sequence overlapping the TATA box of the human osteocalcin promoter: a potential mechanism for negative regulation. Mol Cell Biol. 1991;11:3379–83.
201. Meyer T, Gustafsson JA, Carlstedt-Duke J. Glucocorticoid-dependent transcriptional repression of the osteocalcin gene by competitive binding at the TATA box. DNA Cell Biol. 1997;16:919–27.
202. Ducy P, et al. Increased bone formation in osteocalcin-deficient mice. Nature. 1996;382:448–52.
203. Leclerc N, Noh T, Khokhar A, Smith E, Frenkel B. Glucocorticoids inhibit osteocalcin transcription in osteoblasts by suppressing Egr2/Krox20-binding enhancer. Arthritis Rheum. 2005;52:929–39.
204. Levi G, et al. Defective bone formation in Krox-20 mutant mice. Development. 1996;122:113–20.
205. Gabet Y, et al. Gender-specific control of peak bone mass by the Wnt pathway: androgen signaling protects against lef1 haploinsufficiency-induced bone loss. Bone. 2008;42:S50 (meeting abstract).
206. Kim HJ, et al. Early growth response 2 negatively modulates osteoclast differentiation through upregulation of Id helix-loop-helix proteins. Bone. 2012;51:643–50.
207. Shi X, et al. A glucocorticoid-induced leucine-zipper protein, GILZ, inhibits adipogenesis of mesenchymal cells. EMBO Rep. 2003;4:374–80.
208. Fu L, Patel MS, Karsenty G. The circadian modulation of leptin-controlled bone formation. Prog Brain Res. 2006;153:177–88.
209. Wu X, et al. Circadian mechanisms in murine and human bone marrow mesenchymal stem cells following dexamethasone exposure. Bone. 2008;42:861–70.
210. Jilka RL, et al. Increased bone formation by prevention of osteoblast apoptosis with parathyroid hormone. J Clin Invest. 1999;104:439–46.
211. Whyte MP, et al. Enzyme-replacement therapy in life-threatening hypophosphatasia. N Engl J Med. 2012;366:904–13.
212. Schäcke H, Berger M, Rehwinkel H, Asadullah K. Selective glucocorticoid receptor agonists (SEGRAs): novel ligands with an improved therapeutic index. Mol Cell Endocrinol. 2007;275:109–17.
213. Belvisi MG, et al. Therapeutic benefit of a dissociated glucocorticoid and the relevance of in vitro separation of transrepression from transactivation activity. J Immunol. 2001;166:1975–82.
214. Schäcke H, et al. Characterization of ZK 245186, a novel, selective glucocorticoid receptor agonist for the topical treatment of inflammatory skin diseases. Br J Pharmacol. 2009;158:1088–103.
215. Kassel O, Herrlich P. Crosstalk between the glucocorticoid receptor and other transcription factors: molecular aspects. Mol Cell Endocrinol. 2007;275:13–29.

216. John S, et al. Chromatin accessibility pre-determines glucocorticoid receptor binding patterns. Nat Genet. 2011;43:264–8.
217. Miranda TB, Morris SA, Hager GL. Complex genomic interactions in the dynamic regulation of transcription by the glucocorticoid receptor. Mol Cell Endocrinol. 2013;380:16–24.
218. Grøntved L, et al. C/EBP maintains chromatin accessibility in liver and facilitates glucocorticoid receptor recruitment to steroid response elements. EMBO J. 2013;32:1568–83.
219. Siersbæk R, et al. Extensive chromatin remodelling and establishment of transcription factor 'hotspots' during early adipogenesis. EMBO J. 2011;30:1459–72.
220. Biddie SC, et al. Transcription factor AP1 potentiates chromatin accessibility and glucocorticoid receptor binding. Mol Cell. 2011;43:145–55.
221. Rao NAS, et al. Coactivation of GR and NFKB alters the repertoire of their binding sites and target genes. Genome Res. 2011;21:1404–16.
222. Uhlenhaut NH, et al. Insights into negative regulation by the glucocorticoid receptor from genome-wide profiling of inflammatory cistromes. Mol Cell. 2013;49(1):158–71.
223. Rauch A, et al. An anti-inflammatory selective glucocorticoid receptor modulator preserves osteoblast differentiation. FASEB J. 2011;25:1323–32.
224. Rauner M, et al. Effects of the selective glucocorticoid receptor modulator compound A on bone metabolism and inflammation in male mice with collagen-induced arthritis. Endocrinology. 2013;154(10):3719–28.
225. De Bosscher K, et al. Selective modulation of the glucocorticoid receptor can distinguish between transrepression of NF-κB and AP-1. Cell Mol Life Sci. 2013. doi:10.1007/s00018-013-1367-4.
226. Wüst S, et al. Peripheral T cells are the therapeutic targets of glucocorticoids in experimental autoimmune encephalomyelitis. J Immunol. 2008;180:8434–43.
227. Van Loo G, et al. Antiinflammatory properties of a plant-derived nonsteroidal, dissociated glucocorticoid receptor modulator in experimental autoimmune encephalomyelitis. Mol Endocrinol. 2010;24(2):310–22.
228. Reber LL, et al. A dissociated glucocorticoid receptor modulator reduces airway hyperresponsiveness and inflammation in a mouse model of asthma. J Immunol. 2012;188(7):3478–87.
229. Thiele S, et al. Selective glucocorticoid receptor modulation maintains bone mineral density in mice. J Bone Miner Res. 2012;27:2242–50.
230. Sims NA, Walsh NC. GP130 cytokines and bone remodelling in health and disease. BMB Rep. 2010;43:513–23.
231. Metcalf D, Nicola NA, Gearing DP. Effects of injected leukemia inhibitory factor on hematopoietic and other tissues in mice. Blood. 1990;76:50–6.
232. van Bezooijen RL, ten Dijke P, Papapoulos SE, Lowik CW. SOST/sclerostin, an osteocyte-derived negative regulator of bone formation. Cytokine Growth Factor Rev. 2005;16:319–27.
233. Marenzana M, et al. Sclerostin antibody treatment enhances bone strength but does not prevent growth retardation in young mice treated with dexamethasone. Arthritis Rheum. 2011;63:2385–95.
234. Bickle M. The beautiful cell: high-content screening in drug discovery. Anal Bioanal Chem. 2010;398:219–26.
235. Borchert KM, et al. High-content screening assay for activators of the Wnt/Fzd pathway in primary human cells. Assay Drug Develop Technol. 2005;3:133–41.

Chapter 9
Effects of Glucocorticoids in the Immune System

Emmanuel Oppong and Andrew C.B. Cato

Abstract Glucocorticoids (GCs) are steroid hormones with widespread effects. They control intermediate metabolism by stimulating gluconeogenesis in the liver, mobilize amino acids from extra hepatic tissues, inhibit glucose uptake in muscle and adipose tissue, and stimulate fat breakdown in adipose tissue. They also mediate stress response. They exert potent immune-suppressive and anti-inflammatory effects particularly when administered pharmacologically. Understanding these diverse effects of glucocorticoids requires a detailed knowledge of their mode of action. Research over the years has uncovered several details on the molecular action of this hormone, especially in immune cells. In this chapter, we have summarized the latest findings on the action of glucocorticoids in immune cells with a view of identifying important control points that may be relevant in glucocorticoid therapy.

Keywords Glucocorticoid receptor • Inflammation • Signaling pathways • Immune cells

Introduction

The human body is constantly exposed to a host of pathogens that disturbs its proper function and survival. These pathogens, which are mainly microorganisms, gain access to the body through infections or through tissue injuries. However, the body is able to defend itself by making use of cells, tissues and organs which function to protect it from harm. Collectively, the network of these interacting cells, tissues and organs and the mechanisms for protection, form the immune system.

Without a proper functioning immune system, exposure of the human body to simple infections can be detrimental. The first line of defense by the immune system

---

E. Oppong, Ph.D. (✉) • A.C.B. Cato
Institute of Toxicology and Genetics, Karlsruhe Institute of Technology,
Hermann-von-Helmholtz-Platz 1, Eggenstein-Leopoldshafen, Karlsruhe,
Baden Württemberg 76344, Germany
e-mail: Emmanuel.oppong@kit.edu

J.-C. Wang, C. Harris (eds.), *Glucocorticoid Signaling*, Advances in Experimental Medicine and Biology 872, DOI 10.1007/978-1-4939-2895-8_9

is the innate immune system [1]. It is a nonspecific response and it acts rapidly. Components of this system are the physical, chemical and cellular barriers that offer protection against an invading pathogen [2]. These include the skin, the mucous lining of the digestive tract, the respiratory and the urogenital systems. Certain fluids found in the body such as tears, sweat, saliva, urine and gastric juice contain components that are not conducive for microbial growth and these form chemical barriers to infection [2]. The cellular aspect of the innate immune system is made up of cells that are programmed to recognize and destroy microorganisms. They act by engulfing or releasing toxic contents to destroy the invading organisms. In addition to the rapid innate immune system, there is a second line of defense, termed adaptive immune system, which has a slow onset of action and is very specific in the type of immune response it initiates [2, 3]. With regards to this type of immune protection, the primary response to an exposure of antigen elicits the production of immunoglobulins. These immunoglobulins subsequently initiate a robust response when the same antigen is later encountered.

In its quest to protect the body from injurious agents, the immune system triggers reactions that may damage tissues and organs in the body. This response, which alters the normal physiological function of various tissues and organs, occurs through the release of mediators such as cytokines and chemokines as well as prostaglandins and is termed inflammation [4, 5]. Inflammation seeks to establish tissue homeostasis, but it is also associated with deleterious effects such as pain, swelling, itching, redness and heat that do more harm than good. The inflammatory response involves a complex interplay of cells of the immune system. While some cells function to detect and alert the immune system of an impending danger, others are involved in terminating the response [5, 6].

As inflammation is a “harmful” process, there are efforts to attenuate or control it by steroids. Glucocorticoids are steroidal compounds that influence both arms of the immune system at multiple levels [7, 8]. Their effective action in terminating inflammatory responses has made them a gold standard in anti-inflammatory therapy. Dexamethasone, prednisolone methylprednisolone, triamcinolone and betamethasone are some of the synthetic glucocorticoid analogs that have been successfully used to manage various inflammatory diseases, ever since Hench and his colleagues first reported the anti-inflammatory effect of the natural hormone, cortisol, in patients with rheumatoid arthritis in the early 1950s [9–11].

The actions of glucocorticoids are mediated by the glucocorticoid receptor (GR), a ligand inducible transcription factor (TF). In target cells, the hormone activated receptor diminishes inflammation by either inhibiting the activity of pro-inflammatory TFs, terminating signaling pathways or upregulating the expression of anti-inflammatory proteins. Over the years, studies carried out in different immune cell types have revealed novel insights into the mechanism of action of glucocorticoids in the immune system. Thus, the objective of this chapter is to give an overview of various mechanisms used by glucocorticoids to modulate the behavior and function of the different immune cells.

Cells of the Immune System

The white blood cells, also referred to as leucocytes, form the class of cells involved in the defense against pathogens. They are derived from specialized progenitor cells in the bone marrow and they differentiate and mature into other cell types under the influence of different cytokines and transcription factors [12]. Based on the presence or absence of granules in their cytoplasm, leucocytes can be classified as either granulocytes or agranulocytes respectively [13]. Granulocytes have multi-lobed nuclei and are made up of neutrophils, basophils, eosinophils and mast cells. They can be distinguished from each other microscopically by the appearance of characteristic granules upon staining with hematoxylin and eosin as well as by the expression of cell type specific genes. Each cell type plays a specific role during an injury, infection or inflammation. For example, at inflammatory sites, local signals generated in response to chemokines and cytokines act as chemoattractants for neutrophils. These cells are recruited to the inflammatory sites where they exhibit very potent phagocytic and microbicidal activity which is characterized by the activation of myeloperoxidase that contributes to clearance and resolution of the inflammatory response [14–16]. Eosinophils are involved in protection against parasites and helminth infections [17, 18] and mast cells mediate allergic responses. These latter cells are characterized by the presence of cytoplasmic granules which consist of histamine, heparin and the proteases tryptase and chymase. When activated by allergens, they degranulate and release the contents of their cytoplasmic granules and they also increase the expression of cytokine genes [19, 20]. These events contribute to the symptoms of allergic and inflammatory reactions that are characteristic of mast cells. Basophils form a class of granulocytes with a rather unknown physiological function although they are thought to have a role in host defense just like other immune cells [21].

The second major class of immune cells is made up of agranulocytes. They are single nucleated cells that consist of monocytes and lymphocytes. Monocytes migrate from the blood and differentiate into either macrophages or dendritic cells. Macrophages are large, irregularly shaped cells which contain numerous vacuoles that are used for phagocytosis of cellular debris from tissue necrosis and invading microorganisms [22]. In contrast, dendritic cells are star shaped cells whose main functions are to capture, process and present antigens to B and T lymphocytes to prime and boost immunity [23, 24]. B lymphocytes are involved in antigen-specific antibody production and T lymphocytes are involved in cell-mediated immunity and cytokine production. Nearly all these immune cells are sensitive to the immune suppressive and anti-inflammatory actions of glucocorticoids and a summary of their response to the hormone is presented in Table 9.1.

Table 9.1 Summary of the effects of glucocorticoids on immune cells

Immune cell	Glucocorticoid action	References
Neutrophil	Increased survival and accumulation of cells at inflammatory sites	[92–94]
Basophil	Induction of apoptosis, rapid non-genomic inhibition of cell activation	[110, 111]
Eosinophil	No reported action	
Dendritic cells	Inhibition of cell differentiation, reduced T cell stimulatory capacity, increased endocytic activity, induction of apoptosis, inhibition of pro-inflammatory cytokines, inhibition of cell migration	[63, 64, 112–115]
Macrophages	Activation of JNK and p38 MAPK are inhibited, expression of pro inflammatory genes are attenuated, contact hypersensitivity is suppressed	[57, 61, 62]
Mast cells	Activation of FcεRI dependent signaling is attenuated, histamine and calcium release are downregulated, Phosphatases such as DUSP1 are upregulated	(see Oppong et al., 2013 for review) [37]
T lymphocytes	Apoptosis is induced, release of pro-inflammatory cytokines are inhibited, Non genomic inhibition of LCK and Fyn, rapid phosphorylation of ZAP70	[71, 72, 78–80, 83–86]
B lymphocytes	Induction of apoptosis, production of immunoglobulins is induced, suppression of cell activation, proliferation and differentiation	[116–118]

Mechanism of Glucocorticoid Action

The actions of glucocorticoids are mediated by the glucocorticoid receptor (GR), a cytoplasmic modular protein that consist of a central DNA binding domain, a N-terminal domain that contains transactivation function and a C-terminal ligand binding domain that allows for recruitment of coregulators [25]. Upon hormone binding, a conformational change is induced in the receptor that allows it to translocate to the nucleus where it binds chromatin and induces or represses the expression of target genes [26]. To facilitate binding of the GR to chromatin, an "open" chromatin conformation is established by the transcription factor, activation protein-1 (AP-1) that allows the receptor to recognize its target sequences [27]. The hormone activated receptor then homodimerizes and binds in a head-to-head manner to glucocorticoid response elements (GREs), that are made up of two hexameric half motifs separated by three base pairs [28, 29]. Subsequently, the GR recruits coactivator proteins (e.g., steroid receptor) coactivator 1, 2, 3 (SRC1, SRC2 and SRC3) with histone acetyltransferase activity or chromatin remodeling complexes to the promoters of the target genes to enhance gene activation [30, 31]. The GR also downregulates gene expression. In this case, the liganded receptor is tethered as a monomer to DNA bound TFs such as AP-1 or nuclear factor kappa-light-chain-enhancer of activated B cells (NF-κB) to inhibit their action [32]. The GR uses another mode of action to bind as monomers to negative glucocorticoid response elements (nGRE) in a head to tail conformation to inhibit gene expression [33, 34]. In this case, the consensus sequence for the nGREs consists of two inverted repeats separated by one base pair. Herein, the receptor assembles a repression complex via its association with corepressor proteins such as nuclear receptor co-repressor (NCoR) and histone deacetylases (HDACs) to inhibit gene expression [33, 34]. At certain DNA binding sites, termed "composite GRE", the binding motifs for the GR and other TFs are side by side such that the two proteins influence each other's activity to synergistically activate or repress gene expression [35]. All these mechanisms are classical actions of the GR, also referred to as its genomic mechanisms that are used in regulating immune cell function.

In addition, the hormone also elicits rapid non-genomic actions which are thought to arise from the plasma cell membrane. These actions of glucocorticoids involve modulation of signaling pathways, and they occur within seconds to minutes and are insensitive to transcriptional and translation inhibitors [36]. Mechanistically, the rapid actions of glucocorticoids are not clearly understood and are therefore described by three theories. First, they are thought to emanate from a membrane localized GR (mGR) that is distinct from the classical receptor. Second they are described as merely coming from receptor independent effects originating from hormone-mediated changes in membrane fluidity and third they are thought to arise from the classical intracellular GR that is recruited to the plasma membrane (for reviews see Stahl and Buttegreit 2010 and Oppong et al., 2013) [37, 38]. Over the years, different methods have been used to demonstrate the localization and action of the mGR. One of the early approaches was the use of a monoclonal antibody conjugated to fluorescein isothiocyanate to show the presence of mGR on S-49 lymphoma cells [39]. From these and later studies, it was postulated that they may be different from the classical receptor [40–44]. However, the identity of this receptor still remains obscure, as it has not been cloned. The second theory on glucocorticoid-mediated changes in membrane fluidity could only be achieved with high concentrations of the hormone which are different from the reported rapid effects with physiological concentrations of the hormone [45]. Thus, it is more likely that the classical GR is recruited to the plasma membrane to mediate the rapid action of the hormone. However, the membrane-localized classical GR represents a very small population of the intracellular receptor and this has made its detection challenging. Recently, antibody-conjugated magnetofluorescent liposomes were used to demonstrate membrane bound GR in peripheral monocytes [46]. This technique has the advantage of amplifying fluorescence signal up to 1000-fold to detect membrane bound GR [47–49]. Another technique that has been used to demonstrate membrane-bound GR in both monocytes and T cells is termed Fluorescence Amplification by Sequential Employment of Reagents (FASER) [50]. With this method, the fluorescence signal from cells labeled with antibody conjugated to a fluorochrome is amplified by the sequential addition of two reagents: a fluorochrome specific activator and an enhancer. In yet another technique using mast cells, allergens were patterned on a glass surface, to immobilize and activate the mast cells [51]. This approach was used to visualize in real time the recruitment of a GR tagged with GFP to the site of interaction of the allergen with the cell [52]. Further analyses using photobleaching techniques revealed slower dynamics of the receptor at the membrane of the activated mast cells compared to non-activated cells, indicating possible interaction of the GR with membrane bound components in the activated cells [52]. With these brief descriptions of the pathways and methods used for studying glucocorticoid action in immune cells, we will now take a look at the effect of the hormone on individual immune cells.

Glucocorticoid Modulation of Immune Cell Action

Macrophages

Macrophages act as scavenger cells of the immune system since they clear cellular debris and invading microorganisms by phagocytosis [22]. They are activated by bacteria components such as lipopolysaccharide (LPS) or endotoxin that bind to toll like receptors (TLRs) located on their surfaces. This triggers intracellular signaling cascades such as the MAP kinases (p38 MAPK, ERK and JNK) resulting in enhanced transcription of pro-inflammatory cytokine genes. Glucocorticoids interfere with these signaling pathways to dampen the inflammatory responses. For example, pretreatment of bone marrow derived macrophages with the synthetic glucocorticoid, dexamethasone, following TLR activation by LPS, resulted in the suppression of phosphorylation of JNK and p38 MAP kinase. However, phosphorylation of ERK was unaffected [53], indicating a differential action of glucocorticoids on the MAPK pathway in these cells. Additionally, pro-inflammatory cytokines and chemokines such as interferon gamma (IFNγ), interleukin 1α and interleukin 1β (IL-1α and IL-1β) were inhibited at both the mRNA and protein levels [53]. The anti-inflammatory action of glucocorticoids in macrophages was also reported to occur through an upregulation of the expression of dual specificity phosphatase one, DUSP1, a negative regulator of the MAPKs. Ablation of DUSP1 impaired the anti-inflammatory action of glucocorticoids in these cells [53, 54].

The ligand activated GR also negatively regulates pro-inflammatory gene expression through crosstalk with TFs such as AP-1 and NF-κB at the promoter of target genes [35]. Downregulation of the activities of these TFs by the GR was reported to involve the coregulator protein GRIP1, that was previously identified as a coactivator of the GR [55, 56]. GRIP1, possesses a unique domain that confers repression to target genes when it is bound by the GR and it is through this region that the negative regulation by the GR is achieved [57]. Conventional GRIP1 knock-out mice show defects in normal growth of the adrenal glands and glucocorticoid secretion as well as impaired metabolic and reproductive functions [58–60]. However to further clarify the contribution of GRIP1 to the anti-inflammatory action of glucocorticoids, macrophage-specific GRIP1 knockout mice were generated. Genome wide transcriptome analyses of bone marrow derived macrophages isolated from these mice, showed that the expression of approximately forty percent of the LPS-induced inflammatory genes repressed by glucocorticoids in the wide-type were not repressed by glucocorticoids in the macrophage-specific knockout mice [61]. Some of the pro-inflammatory cytokine genes that showed impaired inhibition by glucocorticoids in the knockout cells are IL-1α, IL-1β, TNFα and CCL4 [61]. Together these results confirm that GRIP1 is indeed an essential mediator of the negative action of GR in macrophages.

The role of macrophages as targets for the anti-inflammatory action of glucocorticoids has also been investigated in animal models of immune diseases. For example, contact hypersensitivity (CHS) induced by the allergens, 2, 4-dinitrofluorobenezene or oxazolone, which are characterized by swollen ears, was not suppressed by glucocorticoids in mice with a macrophage specific GR knockout although control mice showed glucocorticoid inhibition [62]. Furthermore, in the CHS model, mice expressing a dimerization defective GR were also unresponsive to glucocorticoids [62]. These mice exhibited increased accumulation of macrophages in the swollen ear [62], indicating that macrophages are important mediators of the anti-inflammatory action of glucocorticoids and that dimerization of the GR is necessary for the anti-inflammatory effect of the hormone.

Dendritic Cells

Dendritic cells (DCs) are antigen presenting cells that differentiate from precursor monocytes [23]. They exist either as freshly differentiated immature cells or terminally differentiated mature cells in resident tissues such as the spleen and lymph node. In response to tissue damage or injury, immature DCs capture invading pathogen, process them and present them to T cells for initiation of an immune response [23]. In general, immature DCs have increased antigen capturing and processing capability but low T cell stimulatory potential [23]. On the contrary, mature DCs are characterized by low antigen uptake capacity, increased ability of antigen presentation and are effective T-cell stimulators [24]. Dexamethasone interferes with the lifecycle of DCs. *In vitro* maturation of DCs by TNFα or CD40 ligand to express costimulatory molecules such as MHC I, MHC II and CD80 is impaired by dexamethasone and as such glucocorticoid treated DCs exhibit reduced T-cell stimulatory capacity and production of cytokines [63, 64]. Thus, it is likely that by keeping DCs in an immature state, glucocorticoids may enhance the clearance of invading pathogens and microorganisms. Other effects of glucocorticoids on DCs are presented in Table 9.1.

T Lymphocytes

T Lymphocytes, also referred to as T cells are immune cells that arise from lymphoid progenitors in the bone marrow and mature in the thymus [65]. They are classified as naïve, effector or memory cells based on their function during the immune response. Naïve T cells are nonactivated, short-lived cells that die when no antigen is encountered. However, upon activation, naïve T cells differentiate into distinct effector T cell types (Th1, Th2, Th17 and Tregs cells) based on their pattern of cytokine production [66, 67]. For example, IL-12, IFN-α and IFN-γ induce differentiation of Th1 phenotype to secrete IL-2, TNF-α and IFN-γ whereas IL-4 drives the Th2 to produce IL-4, IL-10 and IL-13 [68, 69]. Generally T effector cells disappear after removal of the antigenic agent. However, a subpopulation remains to form a third class of cells, termed memory T cells [70]. Upon reexposure to the same antigen, memory T lymphocytes are activated to mount a rapid and potent immune response. Nearly all the subtypes of T lymphocytes are sensitive to the actions of glucocorticoids, which result in changes in their survival, differentiation or function. Glucocorticoids also influence the pattern of cytokines that drive the differentiation of T cells into the distinct subtypes. These effects of the hormone have been USE extensively instead of intensively reviewed by Elenkov 2004 and Flammer et al. 2011 [71, 72] and readers are referred to these articles for further information on this topic.

One of the mechanisms by which the immune system elicits its protective role is by inducing programmed cell death, also termed apoptosis, of the different immune cells [73]. The Bcl-2 family of proteins which contain up to four Bcl-2 homology domains (BH1-BH4) mediate programmed cell death [74]. While some members of this class of proteins protect against cell death, other members initiate cellular apoptosis. In T lymphocytes, apoptosis is induced either by the intrinsic pathway which is usually activated by cellular stress or by the extrinsic pathway induced by aggregation of surface receptors such as Fas [73, 75]. The intrinsic pathway is initiated by a subset of proapoptotic proteins that contains only the BH3 domain such as Bid, Bad and Blk. These proteins trigger cell death by either activating other proapoptotic bcl-2 family members (Bax and Bak) or by inhibiting the antiapoptotic proteins Mcl, Bcl-2 and Bcl-x [73]. Consequently, cytochrome c is released from the mitochondria and an apoptosome consisting of cytochrome c, protease activating factor 1 (Apaf-1) and caspase 9 activate effector caspases that lead to cell death [76, 77]. In the extrinsic pathway, activation of cell surface "death" receptors recruits caspase 8 which in turn activates downstream effectors such as caspase 3 to trigger apoptosis of the cells [73]. Alternatively, caspase 8 can also cleave the BH3-only propaptotic member, Bid, resulting in the release of cytochrome c and initiation of apoptosis as described above [73, 74]. Glucocorticoids potently induce T cell death via both mechanisms. Studies documenting this action of the hormone have been reviewed by Distelhorst 2002, Ashwell et al. 2000 and Herold et al. 2006 [78–80].

The effects of glucocorticoids in T cells described above are not the only action of the hormone in these cells. There are reports that glucocorticoids also exert non-genomic actions in these cells by modulating the activities of kinases. Several kinases are activated when the T cell receptor (TCR) expressed on

T lymphocytes are cross-linked. Some of these kinases include the Src-like kinases, LCK and FYN [81, 82]. In an analysis of the kinome of human CD4+ T lymphocytes using a peptide array that consists of about 1176 kinase substrates, Lowenberg et al. 2005 showed that glucocorticoids differentially altered the phosphorylation of these substrates with about 116 kinases showing a rapid increase or decrease in phosphorylation following glucocorticoid treatment [83]. Importantly, within minutes of hormone administration, the activities of LCK and FYN were downregulated causing them to dissociate from the TCR [83, 84]. In another study, tyrosine phosphorylation of ZAP-70, an adapter molecule in the TCR pathway was rapidly enhanced by glucocorticoid treatment [85, 86]. These rapid actions of the hormone in T cells were all shown to require membrane localization of the GR.

Neutrophils

Neutrophils are immune cells that are rapidly recruited to inflammatory sites for phagocytosis and destruction of invading organisms [87]. After their phagocytic action, they undergo apoptosis and are subsequently cleared from the inflamed tissue by macrophages. The removal of neutrophils is valuable to the resolution of inflammation and return to tissue homeostasis [88]. However, when neutrophils persist for long periods, they accumulate and release cytotoxic serine proteases such as neutrophil elastase and metalloproteinases which further aggravate the inflammatory process [89–91]. Long-lived neutrophils at inflammatory sites are therefore undesirable. Intriguingly, glucocorticoids have been shown to increase the accumulation and survival of neutrophils [92, 93] with the half-lives of glucocorticoid treated neutrophils doubling compared to non-treated controls [92]. This glucocorticoid-mediated prolongation of neutrophil survival involves induced expression of anti-apoptotic proteins such as members of the Bcl-2 family of proteins and/or suppression of apoptotic signaling pathways [94]. Thus the action of glucocorticoids on neutrophils maybe detrimental since the hormone enhances the ability of neutrophils to survive and to secrete pro-inflammatory mediators [95]. This notion is supported by the finding that glucocorticoids are ineffective in neutrophilic inflammation [94]. This pro-inflammatory effect of neutrophils is therefore thought to contribute to the clinical disorder of glucocorticoid resistance [94, 96]. Besides, glucocorticoids also increase the concentration of neutrophils in the blood by causing demargination of these cells from the walls of the blood vessels [97].

Mast Cells

Mast cells are effector cells of allergic and inflammatory reactions. They express on their surface IgE receptor (FcεRI), TLRs, mast/stem cell growth factor receptor (SCFR), (also referred to as c-kit receptor) and IgG receptor that are involved in the activation of the cells. However, most studies exploring the behavior and function of these cells have concentrated on the FcεRI pathway since this is the main signaling pathway of mast cells. The FcεRI consists of α, β and γ subunits that play distinct roles during activation of mast cells (Fig. 9.1). The α subunit contains an extracellular domain that is important for IgE binding while the β and γ domains contain immunoreceptor tyrosine-based activation motifs (ITAMs) for signal amplification and propagation respectively [98, 99]. Allergen-mediated aggregation of FcεRI triggers a chain of signaling events where Lyn, a proximal Src-like kinase, is recruited to phosphorylate the ITAMs of the β and γ subunits. This leads to the recruitment of spleen tyrosine kinase (Syk) which further transduces the signal to downstream targets [100, 101] (Fig. 9.1). Signal transduction in mast cells can also occur through an alternative pathway involving Fyn kinase, Grb2- associated binder 2 (Gab2) and phosphatidylinositol 3-kinase (PI3K) [102] (Fig. 9.1). All these signaling events result in three main downstream effects: degranulation, cytokine/chemokine gene expression and arachidonic acid and eicosanoid production (Fig. 9.1). Glucocorticoids inhibit mast cell activation at different levels in this signaling cascade as recently described (Oppong et al [37] and Fig. 9.1). For example, following long-term treatment (16–24 h), glucocorticoids downregulate the activity of Erk1/2 [103], which could have an impact on their effect on arachidonic release and cytokine gene expression. Activation of PI3K and the release of calcium from intracellular store as well as degranulation are other examples of markers of mast cell activation that are inhibited by long-term glucocorticoid treatment of mast cells [37, 104].

Glucocorticoids also use rapid non-genomic actions to modulate mast cell action [105, 106]. Here, it has been shown that following allergen-mediated crosslinking of FcεRI of mast cells, the GR is rapidly recruited to the plasma membrane (Fig. 9.1) and this process is further transiently enhanced by glucocorticoid treatment [52]. It therefore appears that the membrane-localized GR may serve as an early sensor for the hormone to amplify its main genomic action [52]. In kinetic studies of hormone-dependent nuclear translocation of the GR, a transient delay in the mobility of the GR was observed in allergen-activated compared to non-activated mast cells [52]. This lag phase in nuclear translocation of the GR was suggested to have arisen from the time the receptor spent at the plasma membrane. The function of the membrane localized-GR was therefore postulated to modulate cell signaling pathways that could eventually be relevant to the genomic action of the receptor [52], a definitive proof for this mechanism is however lacking.

One of the downstream targets for this rapid action of glucocorticoids is the rapid- and transient- increase in FcεRI-dependent phosphorylation of Erk1 and 2 [52]. Along with the long-term action on phosphorylation of Erk-1/2, glucocorticoids therefore exert a dual effect on Erk phosphorylation in mast cells. The significance of these two regulatory pathways is not fully understood. However it is likely that the rapid upregulation of phosphorylation of the Erk-1/2 may prepare the cell for the later events that lead to the downregulation of the kinase activity. The latter negative action of the glucocorticoid is however thought to be the basis for the anti-allergic and anti-inflammatory actions of glucocorticoids since Erk1/2 regulates several of the pro-inflammatory pathways in mast cells that are downregulated by glucocorticoids.

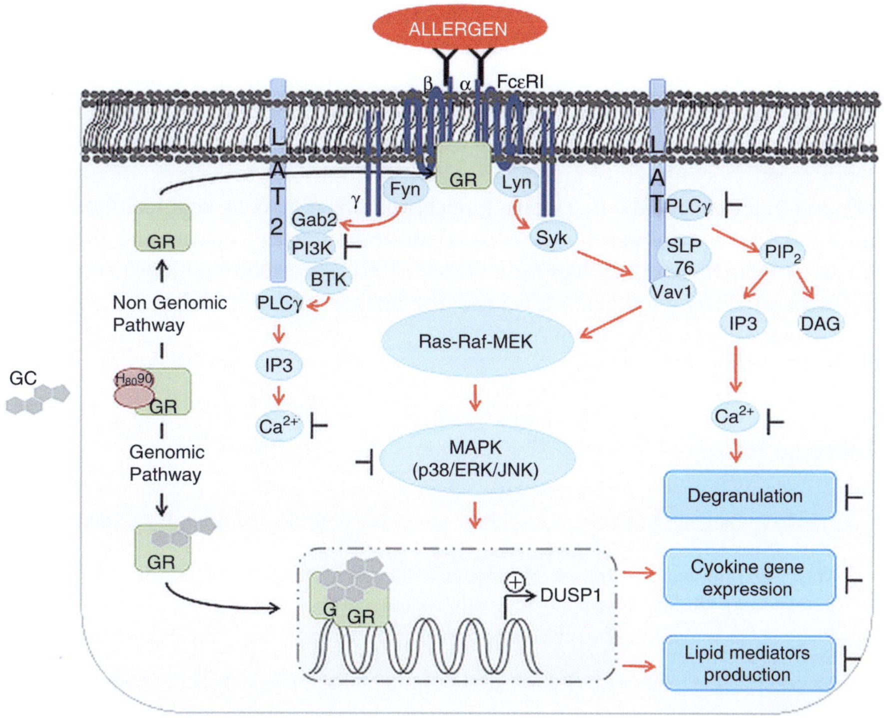

Fig. 9.1 Signal transduction in mast cells following crosslinking of FcεRI and activation of the glucocorticoid receptor (GR) in mast cells. Signaling pathways through the Lyn-Syk-LAT pathway following FcεRI crosslinking are indicated by *red arrows*. An alternative pathway involving Fyn-Gab2-PI3K initiated after aggregation of the FcεRI is also indicated by *red arrows*. Activation of the mast cell leads to three main downstream events: degranulation, cytokine gene expression and production of lipid mediators. The GR signaling pathway in mast cells is shown by *black arrows*. In the genomic effect of the GR, the receptor dissociates from heat shock proteins upon hormone binding and translocates to the nucleus to modulate gene expression such as the induction of expression of the DUSP1 gene, whose product inhibits MAPK activation. In the non-genomic action of glucocorticoids, allergen mediated activation of mast cells recruit the GR to the FcεRI complex. Components of the mast cell signaling pathway reported to be inhibited by glucocorticoids are indicated by the symbol

Conclusion

In the management of inflammatory diseases, both experimental and clinical data underscore the importance of glucocorticoids as the drugs of choice [107]. The potent anti-inflammatory effect of this hormone is due in part to its ability to modulate the behavior and function of immune cells. Over the years, much progress has been made in elucidating the effects of glucocorticoids in immune cells. However, the molecular details involved in the action of the hormone are still not completely understood since glucocorticoids exert diverse effects in these cells. Some effects of the hormone are apparently beneficial as they negatively regulate signaling pathways and downregulate pro-inflammatory cytokine gene expression. On the other hand, other actions such as hormone-increased survival and accumulation of neutrophils or enhanced Erk1/2 phosphorylation in mast cells are still not understood. However, positive regulatory effects of glucocorticoids may contribute to the anti-inflammatory action of glucocorticoids. In T-cells, glucocorticoids positively regulate the expression of IL-10, a cytokine that possesses anti-inflammatory effects and that has a valuable role in diseases such as asthma [108, 109]. These complexities in the action of glucocorticoids emphasize the need for further research to be carried out to clarify their effects in the immune system.

References

1. Grasso P, Gangolli S, Gaunt I. Essentials of pathology for toxicologist. Boca Raton: CRC; 2002. ISBN 978-0-415-25795-4.
2. Mayer G. Immunology: Innate (Non-Specific) Immunity (Chapter 1). In: Microbiology and Immunology On-line. University of South Carolina. 2009.
3. Janeway CPT, Walport M, Shlomchik M. Immunobiology. New York and London: Garland Science; 2001.
4. Medzhitov R. Origin and physiological roles of inflammation. Nature. 2008;454(7203): 428–35.
5. Okin D, Medzhitov R. Evolution of inflammatory diseases. Curr Biol. 2012;22(17):R733–40.
6. Medzhitov R. Inflammation 2010: new adventures of an old flame. Cell. 2010;140(6):771–6.
7. Baschant U, Tuckermann J. The role of the glucocorticoid receptor in inflammation and immunity. J Steroid Biochem Mol Biol. 2010;120(2–3):69–75.
8. Chinenov Y, Rogatsky I. Glucocorticoids and the innate immune system: crosstalk with the toll-like receptor signaling network. Mol Cell Endocrinol. 2007;275(1–2):30–42.
9. Hench P. Effects of cortisone in the rheumatic diseases. Lancet. 1950;2(6634):483–4.
10. Hench PS, Kendall EC, Slocumb CH, Polley HF. The antirheumatic effects of cortisone and pituitary ACTH. Trans Stud Coll Physicians Phila. 1950;18(3):95–102.
11. Hench PS, Kendall EC, Slocumb CH, Polley HF. Cortisone, its effects on rheumatoid arthritis, rheumatic fever, and certain other conditions. Merck Rep. 1950;59(4):9–14.
12. Alberts B, Johnson A, Lewis J, Raff M, Roberts K, Walter P. Molecular biology of the cell. 4th ed. New York: Garland Science; 2002.
13. Gartner LP, Hiatt JL. Color textbook of histology. 3rd ed. Philadelphia: Elsevier; 2007.
14. Borregaard N. Neutrophils, from marrow to microbes. Immunity. 2010;33(5):657–70.
15. Nathan C. Points of control in inflammation. Nature. 2002;420(6917):846–52.
16. Kobayashi SD, Voyich JM, Burlak C, DeLeo FR. Neutrophils in the innate immune response. Arch Immunol Ther Exp (Warsz). 2005;53(6):505–17.
17. Hogan SP, Foster PS, Rothenberg ME. Experimental analysis of eosinophil-associated gastrointestinal diseases. Curr Opin Allergy Clin Immunol. 2002;2(3):239–48.
18. Rothenberg ME, Hogan SP. The eosinophil. Annu Rev Immunol. 2006;24:147–74.
19. Turner H, Kinet JP. Signalling through the high-affinity IgE receptor Fc epsilonRI. Nature. 1999;402(6760 Suppl):B24–30.
20. Gilfillan AM, Rivera J. The tyrosine kinase network regulating mast cell activation. Immunol Rev. 2009;228(1):149–69.
21. Prussin C, Metcalfe DD. 4. IgE, mast cells, basophils, and eosinophils. J Allergy Clin Immunol. 2003;111(2 Suppl):S486–94.

22. Mills CD. M1 and M2 macrophages: oracles of health and disease. Crit Rev Immunol. 2012;32(6):463–88.
23. Banchereau J, Briere F, Caux C, et al. Immunobiology of dendritic cells. Annu Rev Immunol. 2000;18:767–811.
24. Banchereau J, Steinman RM. Dendritic cells and the control of immunity. Nature. 1998;392(6673):245–52.
25. Kumar R, Thompson EB. Gene regulation by the glucocorticoid receptor: structure:function relationship. J Steroid Biochem Mol Biol. 2005;94(5):383–94.
26. Grad I, Picard D. The glucocorticoid responses are shaped by molecular chaperones. Mol Cell Endocrinol. 2007;275(1–2):2–12.
27. Biddie SC, John S, Sabo PJ, et al. Transcription factor AP1 potentiates chromatin accessibility and glucocorticoid receptor binding. Mol Cell. 2011;43(1):145–55.
28. Revollo JR, Cidlowski JA. Mechanisms generating diversity in glucocorticoid receptor signaling. Ann N Y Acad Sci. 2009;1179:167–78.
29. Meijsing SH, Pufall MA, So AY, Bates DL, Chen L, Yamamoto KR. DNA binding site sequence directs glucocorticoid receptor structure and activity. Science. 2009;324(5925):407–10.
30. Lonard DM, Kumar R, O'Malley BW. Minireview: the SRC family of coactivators: an entree to understanding a subset of polygenic diseases? Mol Endocrinol. 2010;24(2):279–85.
31. Lonard DM, O'Malley BW. Nuclear receptor coregulators: judges, juries, and executioners of cellular regulation. Mol Cell. 2007;27(5):691–700.
32. Heck S, Kullmann M, Gast A, et al. A distinct modulating domain in glucocorticoid receptor monomers in the repression of activity of the transcription factor AP-1. EMBO J. 1994;13(17):4087–95.
33. Surjit M, Ganti KP, Mukherji A, et al. Widespread negative response elements mediate direct repression by agonist-liganded glucocorticoid receptor. Cell. 2011;145(2):224–41.
34. Hudson WH, Youn C, Ortlund EA. The structural basis of direct glucocorticoid-mediated transrepression. Nat Struct Mol Biol. 2013;20(1):53–8.
35. Kassel O, Herrlich P. Crosstalk between the glucocorticoid receptor and other transcription factors: molecular aspects. Mol Cell Endocrinol. 2007;275(1–2):13–29.
36. Croxtall JD, Choudhury Q, Flower RJ. Glucocorticoids act within minutes to inhibit recruitment of signalling factors to activated EGF receptors through a receptor-dependent, transcription-independent mechanism. Br J Pharmacol. 2000;130(2):289–98.
37. Oppong E, Flink N, Cato AC. Molecular mechanisms of glucocorticoid action in mast cells. Mol Cell Endocrinol. 2013;380(1–2):119–26.
38. Stahn C, Buttgereit F. Genomic and nongenomic effects of glucocorticoids. Nat Clin Pract Rheumatol. 2008;4(10):525–33.
39. Gametchu B. Glucocorticoid receptor-like antigen in lymphoma cell membranes: correlation to cell lysis. Science. 1987;236(4800):456–61.
40. Gametchu B, Watson CS, Pasko D. Size and steroid-binding characterization of membrane-associated glucocorticoid receptor in S-49 lymphoma cells. Steroids. 1991;56(8):402–10.
41. Gametchu B, Watson CS, Shih CC, Dashew B. Studies on the arrangement of glucocorticoid receptors in the plasma membrane of S-49 lymphoma cells. Steroids. 1991;56(8):411–9.
42. Gametchu B, Watson CS, Wu S. Use of receptor antibodies to demonstrate membrane glucocorticoid receptor in cells from human leukemic patients. FASEB J. 1993;7(13):1283–92.
43. Powell CE, Watson CS, Gametchu B. Immunoaffinity isolation of native membrane glucocorticoid receptor from S-49++ lymphoma cells: biochemical characterization and interaction with Hsp 70 and Hsp 90. Endocrine. 1999;10(3):271–80.
44. Gametchu B, Chen F, Sackey F, Powell C, Watson CS. Plasma membrane-resident glucocorticoid receptors in rodent lymphoma and human leukemia models. Steroids. 1999;64(1–2):107–19.
45. Buttgereit F, Scheffold A. Rapid glucocorticoid effects on immune cells. Steroids. 2002;67(6):529–34.

46. Scheffold A, Assenmacher M, Reiners-Schramm L, Lauster R, Radbruch A. High-sensitivity immunofluorescence for detection of the pro- and anti-inflammatory cytokines gamma interferon and interleukin-10 on the surface of cytokine-secreting cells. Nat Med. 2000;6(1):107–10.
47. Bartholome B, Spies CM, Gaber T, et al. Membrane glucocorticoid receptors (mGCR) are expressed in normal human peripheral blood mononuclear cells and up-regulated after in vitro stimulation and in patients with rheumatoid arthritis. FASEB J. 2004;18(1):70–80.
48. Spies CM, Bartholome B, Berki T, et al. Membrane glucocorticoid receptors (mGCR) on monocytes are up-regulated after vaccination. Rheumatology (Oxford). 2007;46(2):364–5.
49. Tryc AB, Spies CM, Schneider U, et al. Membrane glucocorticoid receptor expression on peripheral blood mononuclear cells in patients with ankylosing spondylitis. J Rheumatol. 2006;33(11):2249–53.
50. Strehl C, Gaber T, Jakstadt M, et al. High-sensitivity immunofluorescence staining: a comparison of the liposome procedure and the FASER technique on mGR detection. J Fluoresc. 2013;23(3):509–18.
51. Sekula-Neuner S, Maier J, Oppong E, Cato AC, Hirtz M, Fuchs H. Allergen arrays for antibody screening and immune cell activation profiling generated by parallel lipid dip-pen nanolithography. Small. 2012;8(4):585–91.
52. Oppong E, Hedde PN, Sekula-Neuner S, et al. Localization and dynamics of glucocorticoid receptor at the plasma membrane of activated mast cells. Small. 2014;10(10):1991–8.
53. Abraham SM, Lawrence T, Kleiman A, et al. Antiinflammatory effects of dexamethasone are partly dependent on induction of dual specificity phosphatase 1. J Exp Med. 2006;203(8):1883–9.
54. Abraham SM, Clark AR. Dual-specificity phosphatase 1: a critical regulator of innate immune responses. Biochem Soc Trans. 2006;34(Pt 6):1018–23.
55. Nissen RM, Yamamoto KR. The glucocorticoid receptor inhibits NFkappaB by interfering with serine-2 phosphorylation of the RNA polymerase II carboxy-terminal domain. Genes Dev. 2000;14(18):2314–29.
56. Rogatsky I, Zarember KA, Yamamoto KR. Factor recruitment and TIF2/GRIP1 corepressor activity at a collagenase-3 response element that mediates regulation by phorbol esters and hormones. EMBO J. 2001;20(21):6071–83.
57. Rogatsky I, Luecke HF, Leitman DC, Yamamoto KR. Alternate surfaces of transcriptional coregulator GRIP1 function in different glucocorticoid receptor activation and repression contexts. Proc Natl Acad Sci U S A. 2002;99(26):16701–6.
58. Chopra AR, Louet JF, Saha P, et al. Absence of the SRC-2 coactivator results in a glycogenopathy resembling Von Gierke's disease. Science. 2008;322(5906):1395–9.
59. Gehin M, Mark M, Dennefeld C, Dierich A, Gronemeyer H, Chambon P. The function of TIF2/GRIP1 in mouse reproduction is distinct from those of SRC-1 and p/CIP. Mol Cell Biol. 2002;22(16):5923–37.
60. Patchev AV, Fischer D, Wolf SS, et al. Insidious adrenocortical insufficiency underlies neuroendocrine dysregulation in TIF-2 deficient mice. FASEB J. 2007;21(1):231–8.
61. Chinenov Y, Gupte R, Dobrovolna J, et al. Role of transcriptional coregulator GRIP1 in the anti-inflammatory actions of glucocorticoids. Proc Natl Acad Sci U S A. 2012;109(29):11776–81.
62. Tuckermann JP, Kleiman A, Moriggl R, et al. Macrophages and neutrophils are the targets for immune suppression by glucocorticoids in contact allergy. J Clin Invest. 2007;117(5):1381–90.
63. Matasic R, Dietz AB, Vuk-Pavlovic S. Dexamethasone inhibits dendritic cell maturation by redirecting differentiation of a subset of cells. J Leukoc Biol. 1999;66(6):909–14.
64. Woltman AM, de Fijter JW, Kamerling SW, Paul LC, Daha MR, van Kooten C. The effect of calcineurin inhibitors and corticosteroids on the differentiation of human dendritic cells. Eur J Immunol. 2000;30(7):1807–12.
65. Schwarz BA, Bhandoola A. Trafficking from the bone marrow to the thymus: a prerequisite for thymopoiesis. Immunol Rev. 2006;209:47–57.

66. Broere F, Apasov SG, Sitkovsky MV, van Eden W. T cell subsets and T cell-mediated immunity. 3rd ed. New York: Springer; 2011.
67. Stemberger C, Neuenhahn M, Buchholz VR, Busch DH. Origin of CD8+ effector and memory T cell subsets. Cell Mol Immunol. 2007;4(6):399–405.
68. Hsieh CS, Macatonia SE, Tripp CS, Wolf SF, O'Garra A, Murphy KM. Development of TH1 CD4+ T cells through IL-12 produced by Listeria-induced macrophages. Science. 1993;260(5107):547–9.
69. Rogge L, D'Ambrosio D, Biffi M, et al. The role of Stat4 in species-specific regulation of Th cell development by type I IFNs. J Immunol. 1998;161(12):6567–74.
70. Ahmed R, Gray D. Immunological memory and protective immunity: understanding their relation. Science. 1996;272(5258):54–60.
71. Flammer JR, Rogatsky I. Minireview: glucocorticoids in autoimmunity: unexpected targets and mechanisms. Mol Endocrinol. 2011;25(7):1075–86.
72. Elenkov IJ. Glucocorticoids and the Th1/Th2 balance. Ann N Y Acad Sci. 2004;1024: 138–46.
73. Opferman JT, Korsmeyer SJ. Apoptosis in the development and maintenance of the immune system. Nat Immunol. 2003;4(5):410–5.
74. Gross A, McDonnell JM, Korsmeyer SJ. BCL-2 family members and the mitochondria in apoptosis. Genes Dev. 1999;13(15):1899–911.
75. Budd RC. Activation-induced cell death. Curr Opin Immunol. 2001;13(3):356–62.
76. Kluck RM, Bossy-Wetzel E, Green DR, Newmeyer DD. The release of cytochrome c from mitochondria: a primary site for Bcl-2 regulation of apoptosis. Science. 1997;275(5303): 1132–6.
77. Yang J, Liu X, Bhalla K, et al. Prevention of apoptosis by Bcl-2: release of cytochrome c from mitochondria blocked. Science. 1997;275(5303):1129–32.
78. Ashwell JD, Lu FW, Vacchio MS. Glucocorticoids in T cell development and function*. Annu Rev Immunol. 2000;18:309–45.
79. Herold MJ, McPherson KG, Reichardt HM. Glucocorticoids in T cell apoptosis and function. Cell Mol Life Sci. 2006;63(1):60–72.
80. Distelhorst CW. Recent insights into the mechanism of glucocorticosteroid-induced apoptosis. Cell Death Differ. 2002;9(1):6–19.
81. Zamoyska R, Basson A, Filby A, Legname G, Lovatt M, Seddon B. The influence of the src-family kinases, Lck and Fyn, on T cell differentiation, survival and activation. Immunol Rev. 2003;191:107–18.
82. Palacios EH, Weiss A. Function of the Src-family kinases, Lck and Fyn, in T-cell development and activation. Oncogene. 2004;23(48):7990–8000.
83. Lowenberg M, Tuynman J, Bilderbeek J, et al. Rapid immunosuppressive effects of glucocorticoids mediated through Lck and Fyn. Blood. 2005;106(5):1703–10.
84. Lowenberg M, Verhaar AP, Bilderbeek J, et al. Glucocorticoids cause rapid dissociation of a T-cell-receptor-associated protein complex containing LCK and FYN. EMBO Rep. 2006; 7(10):1023–9.
85. Bartis D, Boldizsar F, Szabo M, Palinkas L, Nemeth P, Berki T. Dexamethasone induces rapid tyrosine-phosphorylation of ZAP-70 in Jurkat cells. J Steroid Biochem Mol Biol. 2006;98(2–3):147–54.
86. Boldizsar F, Szabo M, Kvell K, et al. ZAP-70 tyrosines 315 and 492 transmit non-genomic glucocorticoid (GC) effects in T cells. Mol Immunol. 2013;53(1–2):111–7.
87. Fox S, Leitch AE, Duffin R, Haslett C, Rossi AG. Neutrophil apoptosis: relevance to the innate immune response and inflammatory disease. J Innate Immun. 2010;2(3):216–27.
88. Hallett JM, Leitch AE, Riley NA, Duffin R, Haslett C, Rossi AG. Novel pharmacological strategies for driving inflammatory cell apoptosis and enhancing the resolution of inflammation. Trends Pharmacol Sci. 2008;29(5):250–7.
89. Savill J. Apoptosis in resolution of inflammation. J Leukoc Biol. 1997;61(4):375–80.
90. Filep JG, El Kebir D. Neutrophil apoptosis: a target for enhancing the resolution of inflammation. J Cell Biochem. 2009;108(5):1039–46.

91. Lacy P. Mechanisms of degranulation in neutrophils. Allergy Asthma Clin Immunol. 2006; 2(3):98–108.
92. Cox G. Glucocorticoid treatment inhibits apoptosis in human neutrophils. Separation of survival and activation outcomes. J Immunol. 1995;154(9):4719–25.
93. Aoki K, Ishida Y, Kikuta N, Kawai H, Kuroiwa M, Sato H. Role of CXC chemokines in the enhancement of LPS-induced neutrophil accumulation in the lung of mice by dexamethasone. Biochem Biophys Res Commun. 2002;294(5):1101–8.
94. Saffar AS, Ashdown H, Gounni AS. The molecular mechanisms of glucocorticoids-mediated neutrophil survival. Curr Drug Targets. 2011;12(4):556–62.
95. Heasman SJ, Giles KM, Ward C, Rossi AG, Haslett C, Dransfield I. Glucocorticoid-mediated regulation of granulocyte apoptosis and macrophage phagocytosis of apoptotic cells: implications for the resolution of inflammation. J Endocrinol. 2003;178(1):29–36.
96. Barnes PJ. Inhaled corticosteroids are not beneficial in chronic obstructive pulmonary disease. Am J Respir Crit Care Med. 2000;161(2 Pt 1):342–4. discussion 344.
97. Nakagawa M, Terashima T, D'Yachkova Y, Bondy GP, Hogg JC, van Eeden SF. Glucocorticoid-induced granulocytosis: contribution of marrow release and demargination of intravascular granulocytes. Circulation. 1998;98(21):2307–13.
98. Kraft S, Kinet JP. New developments in FcepsilonRI regulation, function and inhibition. Nat Rev Immunol. 2007;7(5):365–78.
99. Rivera J, Gilfillan AM. Molecular regulation of mast cell activation. J Allergy Clin Immunol. 2006;117(6):1214–25. quiz 1226.
100. Scharenberg AM, Lin S, Cuenod B, Yamamura H, Kinet JP. Reconstitution of interactions between tyrosine kinases and the high affinity IgE receptor which are controlled by receptor clustering. EMBO J. 1995;14(14):3385–94.
101. Scharenberg AM, Kinet JP. Early events in mast cell signal transduction. Chem Immunol. 1995;61:72–87.
102. Metcalfe DD, Peavy RD, Gilfillan AM. Mechanisms of mast cell signaling in anaphylaxis. J Allergy Clin Immunol. 2009;124(4):639–46. quiz 647-638.
103. Kassel O, Sancono A, Kratzschmar J, Kreft B, Stassen M, Cato AC. Glucocorticoids inhibit MAP kinase via increased expression and decreased degradation of MKP-1. EMBO J. 2001;20(24):7108–16.
104. Andrade MV, Hiragun T, Beaven MA. Dexamethasone suppresses antigen-induced activation of phosphatidylinositol 3-kinase and downstream responses in mast cells. J Immunol. 2004;172(12):7254–62.
105. Zhou J, Liu D, Liu C, Kang ZM, Shen XH, Chen YZ, Xu T, Jiang CL. Glucocorticoids inhibit degranulation of mast cells in allergic asthma via nongenomic mechanism. Allergy. 2008;63:1177–85.
106. Liu C, Zhou J, Zhang LD, Wang YX, Kang ZM, Chen YZ, Jiang CL. Rapid inhibitory effect of corticosterone on histamine release from rat peritoneal mast cells. Horm Metab Res. 2007;39:273–7.
107. Barnes PJ. Anti-inflammatory actions of glucocorticoids: molecular mechanisms. Clin Sci (Lond). 1998;94(6):557–72.
108. Kui Wu YB. Kun Sun and Changzheng Wang IL-10-producing type 1 regulatory T cells and allergy. Cell Mol Immunol. 2007;4(4):269–75.
109. Hawrylowicz CM, O'Garra A. Potential role of interleukin-10-secreting regulatory T cells in allergy and asthma. Nat Rev Immunol. 2005;5(4):271–83.
110. Yamagata S, Tomita K, Sano H, et al. Non-genomic inhibitory effect of glucocorticoids on activated peripheral blood basophils through suppression of lipid raft formation. Clin Exp Immunol. 2012;170(1):86–93.
111. Yoshimura C, Miyamasu M, Nagase H, et al. Glucocorticoids induce basophil apoptosis. J Allergy Clin Immunol. 2001;108(2):215–20.
112. Moser M, De Smedt T, Sornasse T, et al. Glucocorticoids down-regulate dendritic cell function in vitro and in vivo. Eur J Immunol. 1995;25(10):2818–24.

113. Piemonti L, Monti P, Allavena P, et al. Glucocorticoids affect human dendritic cell differentiation and maturation. J Immunol. 1999;162(11):6473–81.
114. Piemonti L, Monti P, Allavena P, Leone BE, Caputo A, Di Carlo V. Glucocorticoid increase the endocytic activity of human dendritic cells. Int Immunol. 1999;11(9):1519–26.
115. Vizzardelli C, Pavelka N, Luchini A, et al. Effects of dexamethazone on LPS-induced activation and migration of mouse dendritic cells revealed by a genome-wide transcriptional analysis. Eur J Immunol. 2006;36(6):1504–15.
116. Gruver-Yates AL, Quinn MA, Cidlowski JA. Analysis of glucocorticoid receptors and their apoptotic response to dexamethasone in male murine B cells during development. Endocrinology. 2014;155(2):463–74.
117. Cupps TR, Edgar LC, Thomas CA, Fauci AS. Multiple mechanisms of B cell immunoregulation in man after administration of in vivo corticosteroids. J Immunol. 1984;132(1):170–5.
118. Cupps TR, Gerrard TL, Falkoff RJ, Whalen G, Fauci AS. Effects of in vitro corticosteroids on B cell activation, proliferation, and differentiation. J Clin Invest. 1985;75(2):754–61.

Chapter 10
Glucocorticoids and the Brain: Neural Mechanisms Regulating the Stress Response

Shawn N. Shirazi, Aaron R. Friedman, Daniela Kaufer, and Samuel A. Sakhai

Abstract In this chapter, we describe the central role of the brain in the glucocorticoid mediated stress response. We describe the mechanisms by which the brain gauges the severity of stress, mechanisms of hypothalamic-pituitary-adrenal axis (HPA) regulation, and how various sub-systems of the brain respond to glucocorticoid (GC) signaling to regulate stress behavior. In particular, we focus on the hippocampus, pre-frontal cortex, and amygdala, where GCs can induce a series of changes. Finally, we briefly discuss an apparent paradox in GC signaling: while exposure to glucocorticoids promotes the survival of an organism during acute stress, these same hormones in chronic excess can also cause damage and promote illness.

Keywords Stress • HPA axis • Negative feedback • Glucocorticoids • Behavior • Amygdala • Prefrontal cortex • Hippocampus

Introduction

Organisms face a wide variety of environmental conditions that can perturb homeostasis. To effectively respond to these "stressors," the organism must initiate a coordinated response across a variety of physiological systems. For example, the

---

S.N. Shirazi • A.R. Friedman
Department of Integrative Biology, University of California,
430 Li Ka Shing Center #3370, Berkeley, CA 94720, USA

D. Kaufer, Ph.D. (✉)
Department of Integrative Biology, Kaufer Lab, University of California,
400 Li Ka Shing Center, Berkeley, CA 94720, USA
e-mail: danielak@berkeley.edu

S.A. Sakhai, Ph.D. (✉)
Department of Psychology, University of California,
3210 Tolman Hall, Berkeley, CA 94720, USA
e-mail: ssakhai@berkeley.edu

J.-C. Wang, C. Harris (eds.), *Glucocorticoid Signaling*, Advances in Experimental Medicine and Biology 872, DOI 10.1007/978-1-4939-2895-8_10

organism must perceive the stressor and select appropriate behavioral strategies (brain) and optimize energy resources towards a "fight, flight or freeze" response (cardiac, respiratory, skeletal) in part by shutting down systems that are not immediately essential (digestive, reproductive, growth).[1] In the vertebrate stress response, the activation of these various systems is initiated by the release of glucocorticoid (GC) stress hormones from the adrenal glands, and also by catecholamine signaling. Importantly, the stressful situations that an organism encounters are diverse. Stressors may be acute and severe (e.g., predation), chronic and severe (e.g., drought), or mild (e.g., social interactions) and each type of stressor requires a unique adaptive response. On the other hand, some types of challenges are predictable, and in these cases GC secretion can allow the organism to prime its physiological response in anticipation of the pending challenge. For example, in diurnal animals GCs are secreted in a daily circadian cycle, with high GC secretion inducing arousal during the early morning and a GC trough promoting rest during the evening. To respond to these wide variety of environmental challenges, ranging from mild to severe and predictable to unpredictable, vertebrates have evolved a complex regulatory system, the hypothalamic-pituitary-adrenal (HPA) axis, to perceive the severity of environmental challenge and release an appropriate amount of GCs for a measured, homeostatic behavioral response.

In this chapter, we describe the central role of the brain in the GC-mediated stress response. We describe the mechanisms by which the brain gauges the severity of stress and initiates an appropriate systemic response—in other words, regulation via the HPA axis. Secondly, we describe how various sub-systems of the brain respond to GC signaling to regulate stress behavior. In particular, we focus on the hippocampus, pre-frontal cortex, and amygdala, where GCs can induce a series of changes (Fig. 10.1). These include alterations that underpin behavioral responses such as alertness and cognitive function, appetitive versus aversive thresholds to various threatening stimuli and rewards (i.e., motivation vs. avoidance), fear, and memory formation. On a cellular and molecular level, this entails modulations of neurotransmitter levels, alterations in dendritic morphology, receptor density, and changes in signal transduction. Thirdly, we briefly discuss an apparent paradox in GC signaling: while exposure to glucocorticoids promotes the survival of an organism during acute stress, these same hormones in chronic excess can also cause damage and promote illness. Chronic stress is a risk factor for multiple diseases, including diseases of central and peripheral nervous systems such as stroke, mental illness, and multiple sclerosis [122–127]. Within the CNS, chronic glucocorticoid exposure can suppress neurogenesis, bias cell fates of neural precursor cells, contribute to dendritic atrophy, and alter neuronal excitability in key regions of the brain involved in anxiety and depression [reference]. Therefore, an organism's best option is to mount as efficient a stress response as possible, limiting its exposure to high levels of catabolic and metabolically demanding glucocorticoids. Fine-tuning of the stress

[1] Hans Selye, the father of modern stress research defined stress as the "non-specific response of the body to any demand made upon it" [96].

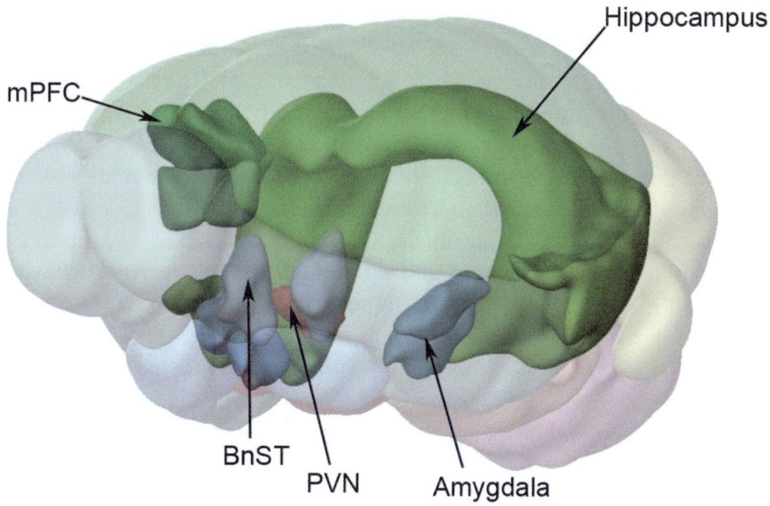

Fig. 10.1 Selected limbic structures involved with HPA axis regulation

response can have a dramatic influence on health. Importantly, the calibration and reactivity of the stress response is partly dependent upon early life environmental contexts and developmental programming, which help prepare organisms for future and current environmental challenges.

The Hypothalamic-Pituitary Adrenal Axis

As a first step in activating the HPA axis, the brain integrates external and internal sensory information pertaining to the immediate challenge, and this information is transduced into endocrine responses within the paraventricular nucleus of the hypothalamus (PVN) [45]. The hypophysiotropic neurons within the PVN secrete corticotropin-releasing hormone (CRH) and arginine vasopressin (AVP) into the hypophyseal portal system, a system of blood vessels that link the hypothalamus with the pituitary gland. Upon reaching the anterior pituitary, CRH stimulates the release of adrenocorticotropic hormone (ACTH) into circulation. Elevated ACTH levels, in turn, stimulate the synthesis and release of glucocorticoids via binding to melancortin-2 receptors within the cortex of the adrenal glands [1]. HPA activation results in a maximal rise in circulating GCs after 15–30 min, and returns to baseline levels at roughly one hour after the termination of a stressor [93]. The crucial ability to terminate the stress response, or inhibit the secretion of CRH and ACTH, is via glucocorticoid negative feedback on key neural regions, such as the PVN, anterior pituitary, medial prefrontal cortex (mPFC), and hippocampus.

The stress response, as a whole, does not solely depend on GCs to alter physiology and behavior—it also requires the concerted actions of several other neuropeptides. These include: urocortins, which interact with CRH; vasopressin, which is implicated in stress-related social memory and emotionality; and orexins, which are involved with stress-related energy and circadian homeostasis. Furthermore, CRH acts in many other brain regions outside of the PVN of the hypothalamus. For example, CRH is released in the bed nucleus of the stria terminalis (BnST) where it plays a role in stress-related anxiety. In the nucleus accumbens, CRH acts to suppress dopamine release in response to rewards, and shift appetitive and aversive behaviors [61, 113] while in the amygdala and in the hippocampus it is involved in stress-related emotional memories, anxiety, and learning processes [89, 92]. A review of the actions of GCs on the brain is incomplete without considering the coordinated influence of the aforementioned peptide mediators, however, its discussion exists outside the scope of this chapter. For an excellent review of CRH, see [53].

HPA Negative Feedback

The ability of the HPA axis to respond dynamically to stress or to tonic secretion of glucocorticoids via circadian rhythm is determined, in part, by the ability of glucocorticoids to adjust ACTH secretion. This negative feedback occurs when GCs penetrate the blood–brain barrier and exert rapid (non-genomic) and slower (genomic) effects on the various neural regions that regulate ACTH release [24, 38, 101]. Two classes of brain steroid receptors mediate negative feedback: the mineralocorticoid receptor (MR), and glucocorticoid receptor (GR). Both MR and GR belong to the nuclear receptor superfamily and function as transcription factors regulating gene expression [85].

MRs have a relatively limited distribution, exhibiting the highest expression within the subiculum/CA1 field and dentate gyrus of the hippocampus [85] (Fig. 10.2a). GRs are expressed nearly ubiquitously (Fig. 10.2b). There are, however, areas of greater GR density within the hippocampus, amygdala, cerebellum, hypothalamus (most notably the PVN), neurons of the ascending aminergic pathways of the brainstem, and to a lesser extent, the caudate nucleus and putamen [34, 78]. While both receptor subtypes bind corticosterone, MRs have a roughly tenfold greater affinity for GCs relative to GRs (Kd of ~0.5 nM for MR vs. Kd ~2.0–5.0 nM for GR) [85]. Consequently, MRs preferentially bind GCs over GRs and reach near-saturation levels during troughs of the circadian cycle (i.e., low basal levels), and are fully saturated during circadian peaks and stress. GRs are activated only when glucocorticoid levels reach a high concentration beyond the level that saturates MRs, such as during an acute stressor or during the zenith of the circadian rhythm. It is hypothesized that MRs are a critical component of the circadian regulation of baseline HPA tone (via fast-feedback, non-genomic actions), while GRs, occupied at higher corticosteroid concentrations, mediate feedback actions following stress [7]. Thus, the balance of MR and GR receptor types, their occupancy, and

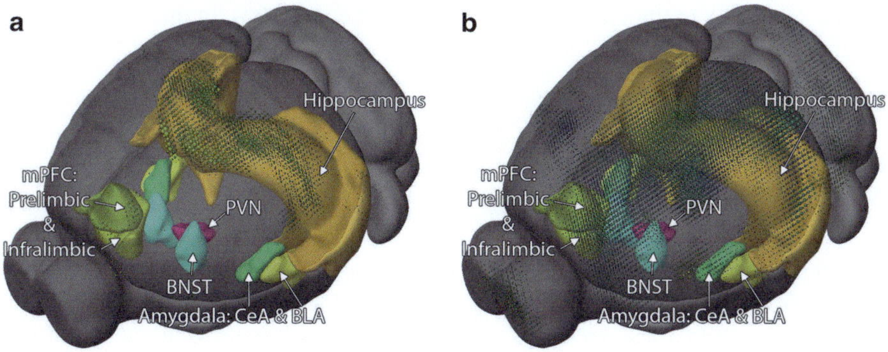

Fig. 10.2 (**a**) MR distribution in the mouse brain. (**b**) GR distribution in the mouse brain

their associated mechanism of action are intricately involved in HPA regulation. Might be worth citing the papers where they make tissue specific KO of GR in hypothalamus or hypothalamus + GR and get Cushings syndrome.

Major Brain Structures Involved in HPA Regulation

Four brain regions are strongly implicated as sites for HPA regulation and synthesis of CRH and AVP. These include the PVN of the hypothalamus, frontal cortex, amygdala and hippocampus.

Paraventricular Nucleus of the Hypothalamus

The PVN is the main gateway for initiating the hormonal stress response, and thus a primary target for regulating HPA negative feedback. It contains one of the densest populations of CRH neurons, which express GRs [17, 110]. Exogenous application of GCs in the PVN results in a rapid decrease in CRH mRNA expression [56] leading to a corollary decrease in HPA activation. Conversely, lesioning PVN afferents serves to increase expression of CRH and AVP mRNA demonstrating that neuronal inhibitory pathways are also necessary for the maintenance of HPA tone [44, 45].

Non-genomic, fast feedback inhibition of the HPA axis within the PVN is dependent on both endocannabinoid and GABAergic mechanisms [106]. GCs stimulate the synthesis and release of endocannabinoids within the PVN by binding to membrane-bound MRs. These endocannabinoids then bind to presynaptic CB1 receptors to suppress glutamatergic transmission, thus inhibiting the activation of PVN neurons and reducing secretion of CRH [24, 29, 48]. GCs also bind to receptors on inhibitory magnocellular neurons of the PVN to stimulate fast, G-protein-dependent release of GABA to inhibit downstream CRH secretion [106]. In this fashion, MRs act in a rapid, non-genomic pathway for negative feedback within the PVN of the hypothalamus.

Medial Prefrontal Cortex

In rodents, the medial prefrontal cortex (mPFC) is comprised of infra-limbic (IL), pre-limbic (PL), and anterior cingulate cortices (AC) (based on structural connectivity and function, these areas are thought to be homologous to human Brodman areas 25, 32 and 24b respectively) [111, 112]. The mPFC is a region of the brain that is involved in cognitive and executive functioning, including working memory, the ability to shift attention across perceptual dimensions, and rule-guided action to plan and guide behavioral sequences. Receiving diverse afferent inputs from the amygdala and ventral hippocampus, as well as providing direct efferent connections to hypothalamic and monoamine brain nuclei, the mPFC is well situated to regulate cognitive, emotional and physiological responses to stress [13].

The mPFC is highly involved in autonomic control and HPA inhibition [2, 6, 75]. Evidence for HPA suppression arises from lesion studies in which mPFC lesions lead to significantly increased plasma levels of ACTH and corticosterone following restraint stress and increased c-Fos activation in the PVN and medial amygdala [10, 27, 31, 102]. Furthermore, local injections of corticosterone into the mPFC are capable of dampening plasma levels of these same hormones [3]. However, exposure to chronic stressors, such as 4 weeks of daily restraint, leads to a down regulation of GR mRNA and protein levels in the PFC, resulting in attenuated PFC-mediated HPA negative feedback [74]. Of note, the primate brain expresses significantly different GR and MR distributions compared to rodents. In primates, GR levels are in greater abundance in the mPFC than the hippocampus, where there is a relative paucity in expression [91]. This suggests that the primate PFC may play a larger role in GR mediated feedback than the hippocampus.

The mPFC influence over the HPA axis is both intra-region specific and exhibits hemispheric functional lateralization [84, 102]. Pre-limbic and infra-limbic cortices exert opposing control over HPA tone. The pre-limbic cortex can be thought of as the 'brakes' whereas the infra-limbic cortex can be considered the 'gas pedal' of HPA regulation. Electrical stimulation of pre-limbic cortex activates parasympathetic systems, whereas infra-limbic stimulation results in robust HPA activation. More recent work has confirmed that this dual control of HPA regulation is GR dependent. GR knockdown in pre-limbic or infra-limbic cortices via short-hairpin RNA leads to differential regulation of HPA secretion, such that infra-limbic disruption leads to HPA hyper-reactivity, while pre-limbic GR knockdown contributes to stress hypo-reactivity in response to an acute psychogenic stressor [70].

HPA control also exhibits hemispheric lateralization. HPA activation is markedly lower after right mPFC lesions, but not left [102]. HPA axis down-regulation after right mPFC lesion was found to be greater in response to chronic stress than to acute stress, suggesting that the mPFC is associated with regulating HPA activity during highly stressful conditions [13].

Intra-region specificity of the mPFC (brake vs. gas) is stressor specific. It modulates its responses based on the nature of the environmental challenges presented, such as psychological stress vs. physical stress. The pre-limbic cortex (brake) is of particular note in its role in inhibiting the HPA axis, especially with respect to psychogenic stressors. This is evidenced by inhibition of CRH and AVP expression after GC infusion into mPFC during restraint stress, a psychological stressor, but not the anesthetic ether, a physical stressor. Conversely, the infra-limbic cortex can initiate HPA activity. It responds robustly to both physical and psychogenic stress [70, 84, 103]. For instance, repeated social stress, but not noise stress, significantly increases ΔFosB expression, an immediate early gene used as a marker for neuronal activation, within the infra-limbic mPFC [50]. This suggests that the mPFC plays a role in the ability to discriminate between psychogenic and physical stressors, thereby increasing the efficiency and specificity of HPA axis regulation [27, 31, 84].

Similarly to the PVN, mPFC-mediated HPA inhibition is dependent in part on GR-mediated endocannabinoid signaling. CB1 receptor antagonism within the mPFC up-regulates HPA activity and results in prolonged GC secretions. Furthermore, GC exposure results in endocannabinoid release, indicating homeostatic negative feedback. Mechanistically, increases in endocannabinoids lead to a decrease in GABA release. This results in a net gain in excitation on pre-limbic (brake) neurons [49]. The pre-limbic cortex has direct afferents onto GABAergic neurons within the BnST, which in turn send projections to the neurosecretory cells of the PVN. In this fashion, GC-induced CB1 activation of the pre-limbic mPFC results in an activation of inhibitory BnST-PVN circuitry. The net result is the suppression of HPA activity.

Responding to psychogenic and physiological stressors, the mPFC is a key component in the top-down regulation of the HPA axis. Part of this regulation is mediated by serotonergic mPFC-amygdala connectivity [32]. For example, decoupling serotonergic mPFC-amygdala circuitry leads to alterations in stress related behaviors [5, 116]. However, the amygdala also regulates the mPFC, thus is another critical component in the emotional guidance of behavior [26].

Amygdala

Receiving direct and indirect connections from limbic structures, including mPFC and hippocampus, the amygdala is thought to be a major integrating center for emotional and arousing stimuli. The amygdala is highly involved in the systemic stress response and is sensitive to both glucocorticoids and catecholamines. Direct stereotactic infusions of GC to the amygdala greatly increase CRH mRNA expression within the PVN during psychogenic stress, illustrating the amygdala's capacity to alter HPA activity during elevated GC exposure [97].

Like the mPFC, the amygdala's influence over HPA systems is region specific. The amygdala is a complex of many sub-nuclei, often segregated into three regions: corticomedial (MeA), central (CeA), and basolateral (BLA) nuclei groups [100]. The CeA is further divided into a lateral component (CeL), and a medial (CeM) component. The balance of excitation and inhibition in each of these sub-regions modulates HPA reactivity [12, 68, 98, 100]. The amygdala has both "anxiogenic" and "anxiolytic" pathways. Stimulation of the BLA itself has been shown to increase HPA activity (anxiogenic), while direct stimulation of the CeM, results in anxiolytic effects. Since both the CeA and the BLA sends projections to the PVN via the BNST, the net effect of amygdala activation on the HPA axis is contingent upon the circuitry that is invoked [80, 109].

Hippocampus

Involved in cognition and memory formation, the hippocampus is a critical locus in HPA regulation. It exhibits tremendous plasticity to stress and glucocorticoids. Within the rodent brain, the hippocampus expresses the highest level of GR and MR, hence, it is of little surprise that it serves as an important negative feedback center and regulator of the stress response [46]. Evidence for HPA negative feedback arises from early studies in which lesioning or blocking hippocampal GC receptors results in an up-regulation of CRH and AVP mRNA within the PVN. The consequence of this is hypersecretion of glucocorticoids [52, 93]. Conversely, activation of hippocampal GC receptors results in HPA axis inhibition [93].

However, the relationship between hippocampus and HPA regulation are specific to particular hippocampal sub-regions. Structurally and functionally heterogeneous, the hippocampus can be segregated across a septotemporal axis [8, 30]. In rodents, the dorsal hippocampus appears to be more involved in learning and memory, while the ventral hippocampus is implicated in the modulation of anxiety-like behavior and HPA regulation [8]. Lesions to dorsal hippocampus result in spatial memory deficits, whereas ventral hippocampal lesions result in anxiety-like behaviors alterations [8, 30, 79]. These differences in function can be explained in part by the connectivity of each region. For instance, regions of the ventral hippocampus project to areas involved in emotional regulation, most notably, the mPFC, amygdala, BnST, and the PVN [30].

The hippocampus is also one of two brain regions in which resident populations of neural stem cells (NSCs) produce new neurons in adult animals [33, 35]. These new neurons are thought to contribute to the plasticity of hippocampal networks and have roles in the classical hippocampal functions of learning and memory. However, several lines of recent evidence also suggest that NSCs play a role in HPA regulation. First, chronic, but not acute, activation of hippocampal GRs is associated with a general *increase* in HPA reactivity [86]. This failure in HPA axis negative feedback may be due, in part, to a reduction of the neurogenic pool resulting from chronic GC exposure. More direct evidence comes from ablation of hippocampal neurogenesis, through the use of techniques such as irradiation or by transgenic animal models, which results in impaired HPA negative feedback and elevated GC levels following recovery from restraint stress [99]. Animals with impaired neurogenesis also exhibit a depressed phenotype at baseline that can be reversed by antidepressant treatment [99]. Lastly, hippocampal neurogenesis appears to be required for antidepressants to restore HPA axis inhibition following chronic stress [104]. Taken as a whole, these and other findings have implicated hippocampal neurogenesis as a component of HPA axis regulation.

Effects of GCs on Brain and Behavior

As described above, many brain regions that integrate sensory information, such as the mPFC, amygdala, and hippocampus, can exert control of the PVN to fine tune the stress response according to the immediate experiences of the animal. In turn, once the stress response is initiated, the animal also has to enact appropriate behavioral strategies to cope with the stressor. Thus, beyond acting as a negative feedback signal, GCs also modulate brain function in these same regions to coordinate appropriate stress-response behaviors.

Medial Prefrontal Cortex

Stress, both mild and severe, can lead to functional and structural changes in the prefrontal cortex [6, 39, 51]. This includes alterations in dendritic arborization and spine density in all regions of the mPFC (IL, PL, and AC) and neighboring orbitofrontal cortex, which is driven in part, by GR signaling [11, 15, 62, 63, 82, 121]. For instance, 3 weeks of corticosterone administration [115] or daily restraint stress [20], is capable of reducing dendritic arborization and spine density in the mPFC and dorsomedial striatum [21, 25]. Functionally, both glucocorticoid administration and stress leads to deficits in working memory [11, 15, 76, 90], mPFC dependent set-shifting [63], as well as reversal learning [14, 15, 60].

Paradoxically, under some conditions, chronic stress can facilitate reversal learning [40, 41]. One hypothesis by Dias-Ferreira and colleagues as well as Schwab and Wolf, posits that stress leads to a disinhibition of PFC functions and towards striatal mediated learning [25]. The effect is bias in an organism's behavior towards habit formation [25, 41, 94, 95]. Indeed, severe, repeated stressors result in an increase in apical dendrite arborization in both the dorsolateral striatum and orbitofrontal cortex, regions involved in habitual strategies, reward valuation, and reversal learning [25, 63]. These studies suggest that the effects of stress and glucocorticoids may be beneficial to shift behaviors toward optimal behavioral adaption to environmental stress.

Amygdala

Its activation by GCs can lead to alterations in learning and memory [88]. However, it is important to note that amygdala contributions to autonomic functioning are with respect to emotional arousal, not circadian or homeostatic HPA regulation.

Both acute and chronic stress results in the remodeling of synapses and dendritic branching within the amygdala [72, 73]. Stress induced synaptic plasticity is modulated by GABAergic inputs [23]. These changes are correlated with an increase in anxiety-like behaviors and enhanced fear conditioning [19, 73, 118]. High levels of corticosterone reduce GABA transmission, which results in an increase in the firing rate of excitatory neurons in the basolateral amygdala [28]. This suggests that high levels of glucocorticoids can change the balance between excitation and inhibition, resulting in modifications in synaptic connectivity. These changes can influence neuronal plasticity even in distal brain regions. Recent findings demonstrate that the BLA can alter synaptic plasticity and long-term potentiation in the striatum and hippocampus. Therefore, it is becoming increasingly clear that glucocorticoids within the amygdala can be far-reaching and impactful [4, 81].

Finally, the amygdala also plays a central role in enhancing memory consolidation following emotionally arousing events. High levels of circulating GCs can improve the recall of a stressful event [9, 77, 89]. However, GCs effects may be mediated by ß-adrenergic activation; blockade of ß-ardrenergic receptors within the BLA prevents memory enhancements following GR activation. Furthermore, activation of BLA via emotional arousal is critical in GC-mediated memory enhancements [83]. Enhanced memory performance following a stressful event can be advantageous, as future encounters to similarly arousing stimuli would result in a feed-forward HPA activation to prime physiological systems in anticipation of a stressor. However, more investigations are needed to fully determine how stress, NE, and GCs influence different phases of fear learning and its expression in memory.

Hippocampus

The hippocampus responds dynamically to changes in GCs levels by modulating neuronal structure and function. GCs directly influence hippocampal function by acting as neuromodulators to influence neural excitability and signaling [16, 54, 55, 59, 64]. More broadly, GCs also affect the structural connectivity of the hippocampus by affecting dendritic arborization and formation of synapses [114, 119, 120]. Together, a model has emerged from these studies in which mild or acute stress increases hippocampal dendritic branching and long-term potentiation to boost hippocampal learning and memory, while chronic or high GC concentrations have opposite effects [19, 47, 58, 92]. However, the distribution of MR and GR receptors differs in dorsal versus ventral hippocampus, with ventral hippocampus having a much higher relative concentration of MR [87]. This suggests that the effect of stress on hippocampal function may be more nuanced and region-specific, such that high levels of GCs do not simply suppress hippocampal memory function in general, but rather specifically suppress the contextual memory functions of dorsal hippocampus while promoting the emotional cognitive functions of the ventral hippocampus [65, 66]. This model fits well with the overall paradigm of the stress response as an adaptive mechanism that manifests stress-specific behavioral strategies suited to overcoming stressful challenges [65, 66].

Stress effects on hippocampal-mediated behaviors may also be regulated through the contributions of hippocampal NSCs. NSCs express functional GRs (but not MRs) [18, 36], and their rate of proliferation and differentiation, as well as the survival of the new neurons that they produce, are altered by GCs [58, 117, 128–130]. The effects can be via direct activation of GR in the NSC [36] or indirectly, through activation of GR-dependent mechanisms in other cells in the hippocampal niche. For instance, acute corticosterone exposure elicits release of fibroblast growth factor-2 from astrocytes in the dorsal hippocampus, leading to increased proliferation of neural stem cells in the area [58].

The effects of stress on adult neurogenesis can be divided into the effects of acute stress and repeated, chronic stress. Chronic, repeated stressors inhibit NSC survival, proliferation, and neuronal differentiation within the dentate gyrus [57, 71, 117]. However, the effects of acute stress display a more mixed picture, ranging from a decrease, increase, or no change in NSC proliferation [22, 43, 107, 108]. One explanation for discrepancies in the literature may be that, like cognitive performance in response to stress, adult hippocampal neurogenesis follows an inverted U function—increasing in response to acute stressors and decreasing in response to high, chronic GC exposure. For example, high levels of transient GCs can inhibit NSC proliferation in the SGZ, and this effect can be blocked through adrenalectomy [105].

Beyond proliferation, high levels of GCs may also reduce the total number of new neurons by decreasing the survival of immature neurons as they begin to incorporate into the network [117]. Furthermore, GCs cause a shift in the cell fate of differentiation NSCs, causing them to more frequently differentiate into oligodendrocytes at the expense of neurogenesis [18]. Given that new neurons ultimately confer additional plasticity onto hippocampal networks, by forming new synaptic connections and showing enhanced capacity for LTP [37, 42, 67], the reductions in neurogenesis in response to elevated GCs may be one of the mechanisms underlying reduced memory capacity in stressed animals.

Conclusion

Glucocorticoids regulate the brain and behavior in multiple domains. They help adjust basal and peak HPA axis reactivity [93], as well as alter limbic structures (PVN, mPFC, amygdala, hippocampus), both structurally and functionally. This includes alterations in synaptic plasticity, long-term potentiation, and neurogenesis, which result in changes in appetitive and avoidant behaviors, and modifications in learning and memory. Additionally, the limbic system is responsible for the top-down and bottom-up regulation of the HPA axis through its complex micro-circuitry. In this sense, HPA regulation can be a recursive process since glucocorticoids modulate both initiators and terminators of the stress response.

Ultimately, stress is necessary for the optimization of behavior to environmental pressures. With respect to humans, it is HPA axis dysregulation that is implicated in the pathogenesis of many disease phenotypes such as anxiety and depression [69]. It is of paramount importance to efficiently initiate a stress response, as it promotes survival, while equally important to terminate the stress response, as glucocorticoids are metabolically demanding and can lead to disease.

References

1. Abdel-Malek ZA. Melanocortin receptors: their functions and regulation by physiological agonists and antagonists. Cell Mol Life Sci. 2001;58:434–41.
2. Ahima RS, Harlan RE. Charting of type II glucocorticoid receptor-like immunoreactivity in the rat central nervous system. Neuroscience. 1990;39:579–604.
3. Akana SF, Chu A, Soriano L, Dallman MF. Corticosterone exerts site-specific and state-dependent effects in prefrontal cortex and amygdala on regulation of adrenocorticotropic hormone, insulin and fat depots. J Neuroendocrinol. 2001;13:625–37.
4. Akirav I, Richter-Levin G. Mechanisms of amygdala modulation of hippocampal plasticity. J Neurosci. 2002;22:9912–21.
5. Andolina D, Maran D, Valzania A, Conversi D, Puglisi-Allegra S. Prefrontal/amygdalar system determines stress coping behavior through 5-HT/GABA connection. Neuropsychopharmacology. 2013;38:2057–67.
6. Arnsten AFT. Stress signalling pathways that impair prefrontal cortex structure and function. Nat Rev Neurosci. 2009;10:410–22.
7. Atkinson HC, Wood SA, Castrique ES, Kershaw YM, Wiles CC, Lightman SL. Corticosteroids mediate fast feedback of the rat hypothalamic-pituitary-adrenal axis via the mineralocorticoid receptor. Am J Physiol. 2008;294:E1011–22.
8. Bannerman DM, Rawlins JN, McHugh SB, Deacon RM, Yee BK, Bast T, Zhang WN, Pothuizen HH, Feldon J. Regional dissociations within the hippocampus–memory and anxiety. Neurosci Biobehav Rev. 2004;28:273–83.
9. Bohannon 3rd JN. Flashbulb memories for the space shuttle disaster: a tale of two theories. Cognition. 1988;29:179–96.
10. Brake WG, Flores G, Francis D, Meaney MJ, Srivastava LK, Gratton A. Enhanced nucleus accumbens dopamine and plasma corticosterone stress responses in adult rats with neonatal excitotoxic lesions to the medial prefrontal cortex. Neuroscience. 2000;96:687–95.
11. Butts KA, Weinberg J, Young AH, Phillips AG. Glucocorticoid receptors in the prefrontal cortex regulate stress-evoked dopamine efflux and aspects of executive function. Proc Natl Acad Sci. 2011;108:18459–64.
12. Carlsen J. Immunocytochemical localization of glutamate decarboxylase in the rat basolateral amygdaloid nucleus, with special reference to GABAergic innervation of amygdalostriatal projection neurons. J Comp Neurol. 1988;273:513–26.
13. Cerqueira JJ, Almeida OF, Sousa N. The stressed prefrontal cortex. Left? Right! Brain Behav Immun. 2008;22:630–8.

14. Cerqueira JJ, Mailliet F, Almeida OF, Jay TM, Sousa N. The prefrontal cortex as a key target of the maladaptive response to stress. J Neurosci. 2007;27:2781–7.
15. Cerqueira JJ, Pêgo JM, Taipa R, Bessa JM, Almeida OF, Sousa N. Morphological correlates of corticosteroid-induced changes in prefrontal cortex-dependent behaviors. J Neurosci. 2005;25:7792–800.
16. Chameau P, Qin Y, Spijker S, Smit G, Joëls M. Glucocorticoids specifically enhance L-type calcium current amplitude and affect calcium channel subunit expression in the mouse hippocampus. J Neurophysiol. 2007;97:5–14.
17. Chen J, Gomez-Sanchez CE, Penman A, May PJ, Gomez-Sanchez E. Expression of mineralocorticoid and glucocorticoid receptors in preautonomic neurons of the rat paraventricular nucleus. Am J Physiol. 2014;306:R328–40.
18. Chetty S, Friedman AR, Taravosh-Lahn K, Kirby ED, Mirescu C, Guo F, Krupik D, Nicholas A, Geraghty AC, Krishnamurthy A, Tsai MK, Covarrubias D, Wong AT, Francis DD, Sapolsky RM, Palmer TD, Pleasure D, Kaufer D. Stress and glucocorticoids promote oligodendrogenesis in the adult hippocampus. Mol Psychiatry. 2014;19(12):1275–83.
19. Conrad CD, LeDoux JE, Magarinos AM, McEwen BS. Repeated restraint stress facilitates fear conditioning independently of causing hippocampal CA3 dendritic atrophy. Behav Neurosci. 1999;113:902–13.
20. Cook SC, Wellman CL. Chronic stress alters dendritic morphology in rat medial prefrontal cortex. J Neurobiol. 2004;60:236–48.
21. Czeh B, Muller-Keuker JI, Rygula R, Abumaria N, Hiemke C, Domenici E, Fuchs E. Chronic social stress inhibits cell proliferation in the adult medial prefrontal cortex: hemispheric asymmetry and reversal by fluoxetine treatment. Neuropsychopharmacology. 2007;32: 1490–503.
22. Dagytė G, Van der Zee EA, Postema F, Luiten PGM, Den Boer JA, Trentani A, Meerlo P. Chronic but not acute foot-shock stress leads to temporary suppression of cell proliferation in rat hippocampus. Neuroscience. 2009;162:904–13.
23. Davis M, Rainnie D, Cassell M. Neurotransmission in the rat amygdala related to fear and anxiety. Trends Neurosci. 1994;17:208–14.
24. Di S, Malcher-Lopes R, Halmos KC, Tasker JG. Nongenomic glucocorticoid inhibition via endocannabinoid release in the hypothalamus: a fast feedback mechanism. J Neurosci. 2003;23:4850–7.
25. Dias-Ferreira E, Sousa JC, Melo I, Morgado P, Mesquita AR, Cerqueira JJ, Costa RM, Sousa N. Chronic stress causes frontostriatal reorganization and affects decision-making. Science. 2009;325:621–5.
26. Dilgen J, Tejeda HA, O'Donnell P. Amygdala inputs drive feedforward inhibition in the medial prefrontal cortex. J Neurophysiol. 2013;110:221–9.
27. Diorio D, Viau V, Meaney MJ. The role of the medial prefrontal cortex (cingulate gyrus) in the regulation of hypothalamic-pituitary-adrenal responses to stress. J Neurosci. 1993;13:3839–47.
28. Duvarci S, Pare D. Glucocorticoids enhance the excitability of principal basolateral amygdala neurons. J Neurosci. 2007;27:4482–91.
29. Evanson NK, Tasker JG, Hill MN, Hillard CJ, Herman JP. Fast feedback inhibition of the HPA axis by glucocorticoids is mediated by endocannabinoid signaling. Endocrinology. 2010;151:4811–9.
30. Fanselow MS, Dong HW. Are the dorsal and ventral hippocampus functionally distinct structures? Neuron. 2010;65:7–19.
31. Figueiredo HF, Bruestle A, Bodie B, Dolgas CM, Herman JP. The medial prefrontal cortex differentially regulates stress-induced c-fos expression in the forebrain depending on type of stressor. Eur J Neurosci. 2003;18:2357–64.
32. Fisher PM, Meltzer CC, Price JC, Coleman RL, Ziolko SK, Becker C, Moses-Kolko EL, Berga SL, Hariri AR. Medial prefrontal cortex 5-HT2A density is correlated with amygdala reactivity, response habituation, and functional coupling. Cereb Cortex. 2009;19:2499–507.

33. Friedman AR, Kaufer D. Emerging roles for hippocampal adult neural stem cells in memory. Princeton: Biota Publishing; 2013.
34. Fuxe K, Härfstrand A, Agnati LF, Yu ZY, Cintra A, Wikström AC, Okret S, Cantoni E, Gustafsson JÅ. Immunocytochemical studies on the localization of glucocorticoid receptor immunoreactive nerve cells in the lower brain stem and spinal cord of the male rat using a monoclonal antibody against rat liver glucocorticoid receptor. Neurosci Lett. 1985;60:1–6.
35. Gage FH. Mammalian neural stem cells. Science. 2000;287:1433–8.
36. Garcia A, Steiner B, Kronenberg G, Bick-Sander A, Kempermann G. Age-dependent expression of glucocorticoid- and mineralocorticoid receptors on neural precursor cell populations in the adult murine hippocampus. Aging Cell. 2004;3:363–71.
37. Ge S, Yang C-H, Hsu K-S, Ming G-L, Song H. A critical period for enhanced synaptic plasticity in newly generated neurons of the adult brain. Neuron. 2007;54:559–66.
38. Giguère V, Hollenberg SM, Rosenfeld MG, Evans RM. Functional domains of the human glucocorticoid receptor. Cell. 1986;46:645–52.
39. Goldwater DS, Pavlides C, Hunter RG, Bloss EB, Hof PR, McEwen BS, Morrison JH. Structural and functional alterations to rat medial prefrontal cortex following chronic restraint stress and recovery. Neuroscience. 2009;164:798–808.
40. Graybeal C, Feyder M, Schulman E, Saksida LM, Bussey TJ, Brigman JL, Holmes A. Paradoxical reversal learning enhancement by stress or prefrontal cortical damage: rescue with BDNF. Nat Neurosci. 2011;14:1507–9.
41. Graybeal C, Kiselycznyk C, Holmes A. Stress-induced impairments in prefrontal-mediated behaviors and the role of the N-methyl-D-aspartate receptor. Neuroscience. 2012;211: 28–38.
42. Gu Y, Arruda-Carvalho M, Wang J, Janoschka SR, Josselyn SA, Frankland PW, Ge S. Optical controlling reveals time-dependent roles for adult-born dentate granule cells. Nat Neurosci. 2012;15:1700–6.
43. Hanson ND, Owens MJ, Boss-Williams KA, Weiss JM, Nemeroff CB. Several stressors fail to reduce adult hippocampal neurogenesis. Psychoneuroendocrinology. 2011;36:1520–9.
44. Herman JP. In situ hybridization analysis of vasopressin gene transcription in the paraventricular and supraoptic nuclei of the rat: regulation by stress and glucocorticoids. J Comp Neurol. 1995;363:15–27.
45. Herman JP, Cullinan WE. Neurocircuitry of stress: central control of the hypothalamo-pituitary-adrenocortical axis. Trends Neurosci. 1997;20:78–84.
46. Herman JP, Cullinan WE, Young EA, Akil H, Watson SJ. Selective forebrain fiber tract lesions implicate ventral hippocampal structures in tonic regulation of paraventricular nucleus corticotropin-releasing hormone (CRH) and arginine vasopressin (AVP) mRNA expression. Brain Res. 1992;592:228–38.
47. Herman JP, Spencer R. Regulation of hippocampal glucocorticoid receptor gene transcription and protein expression in vivo. J Neurosci. 1998;18:7462–73.
48. Hill MN, McEwen BS. Involvement of the endocannabinoid system in the neurobehavioural effects of stress and glucocorticoids. Prog Neuropsychopharmacol Biol Psychiatry. 2010; 34:791–7.
49. Hill MN, McLaughlin RJ, Pan B, Fitzgerald ML, Roberts CJ, Lee TT, Karatsoreos IN, Mackie K, Viau V, Pickel VM, McEwen BS, Liu QS, Gorzalka BB, Hillard CJ. Recruitment of prefrontal cortical endocannabinoid signaling by glucocorticoids contributes to termination of the stress response. J Neurosci. 2011;31:10506–15.
50. Hinwood M, Tynan RJ, Day TA, Walker FR. Repeated social defeat selectively increases deltaFosB expression and histone H3 acetylation in the infralimbic medial prefrontal cortex. Cereb Cortex. 2011;21:262–71.
51. Holmes A, Wellman CL. Stress-induced prefrontal reorganization and executive dysfunction in rodents. Neurosci Biobehav Rev. 2009;33:773–83.
52. Jacobson L, Sapolsky R. The role of the hippocampus in feedback regulation of the hypothalamic-pituitary-adrenocortical axis. Endocr Rev. 1991;12:118–34.

53. Joels M, Baram TZ. The neuro-symphony of stress. Nat Rev Neurosci. 2009;10:459–66.
54. Karst H, Joëls M. Corticosterone slowly enhances miniature excitatory postsynaptic current amplitude in mice CA1 hippocampal cells. J Neurophysiol. 2005;94:3479–86.
55. Karst H, Karten Y, Reichardt H, De Kloet E, Schütz G, Joels M. Corticosteroid actions in hippocampus require DNA binding of glucocorticoid receptor homodimers. Nat Neurosci. 2000;3:977–8.
56. Keller-Wood ME, Dallman MF. Corticosteroid inhibition of ACTH secretion. Endocr Rev. 1984;5:1–24.
57. Kirby ED, Kaufer D. Stress and adult neurogenesis in the mammalian central nervous system. In: Soreq H, Friedman A, Kaufer D, editors. Stress—from molecules to behavior: a comprehensive analysis of the neurobiology of stress responses. Darmstadt: Wiley-VCH; 2009. p. 71–91.
58. Kirby ED, Muroy SE, Sun WG, Covarrubias D, Leong MJ, Barchas LA, Kaufer D. Acute stress enhances adult rat hippocampal neurogenesis and activation of newborn neurons via secreted astrocytic FGF2. eLife. 2013;2:e00362.
59. Krugers HJ, Alfarez DN, Karst H, Parashkouhi K, van Gemert N, Joëls M. Corticosterone shifts different forms of synaptic potentiation in opposite directions. Hippocampus. 2005; 15:697–703.
60. Lapiz-Bluhm MD, Soto-Pina AE, Hensler JG, Morilak DA. Chronic intermittent cold stress and serotonin depletion induce deficits of reversal learning in an attentional set-shifting test in rats. Psychopharmacology (Berl). 2009;202:329–41.
61. Lemos JC, Wanat MJ, Smith JS, Reyes BAS, Hollon NG, Van Bockstaele EJ, Chavkin C, Phillips PEM. Severe stress switches CRF action in the nucleus accumbens from appetitive to aversive. Nature. 2012;490:402–6.
62. Liston C, Gan W-B. Glucocorticoids are critical regulators of dendritic spine development and plasticity in vivo. Proc Natl Acad Sci. 2011;108:16074–9.
63. Liston C, Miller MM, Goldwater DS, Radley JJ, Rocher AB, Hof PR, Morrison JH, McEwen BS. Stress-induced alterations in prefrontal cortical dendritic morphology predict selective impairments in perceptual attentional set-shifting. J Neurosci. 2006;26:7870–4.
64. Maggio N, Segal M. Differential modulation of long-term depression by acute stress in the rat dorsal and ventral hippocampus. J Neurosci. 2009;29:8633–8.
65. Maggio N, Segal M. Corticosteroid regulation of synaptic plasticity in the hippocampus. Scientific World J. 2010;10:462–9.
66. Maggio N, Segal M. Steroid modulation of hippocampal plasticity: switching between cognitive and emotional memories. Front Cell Neurosci. 2012;6:12.
67. Marín-Burgin A, Mongiat LA, Pardi MB, Schinder AF. Unique processing during a period of high excitation/inhibition balance in adult-born neurons. Science. 2012;335:1238–42.
68. McDonald AJ. Cytoarchitecture of the central amygdaloid nucleus of the rat. J Comp Neurol. 1982;208:401–18.
69. McEwen BS. Stress, adaptation, and disease: allostasis and allostatic load. Ann N Y Acad Sci. 1998;840:33–44.
70. McKlveen JM, Myers B, Flak JN, Bundzikova J, Solomon MB, Seroogy KB, Herman JP. Role of prefrontal cortex glucocorticoid receptors in stress and emotion. Biol Psychiatry. 2013;74:672–9.
71. Mirescu C, Gould E. Stress and adult neurogenesis. Hippocampus. 2006;16:233–8.
72. Mitra R, Jadhav S, McEwen BS, Vyas A, Chattarji S. Stress duration modulates the spatiotemporal patterns of spine formation in the basolateral amygdala. Proc Natl Acad Sci U S A. 2005;102:9371–6.
73. Mitra R, Sapolsky RM. Acute corticosterone treatment is sufficient to induce anxiety and amygdaloid dendritic hypertrophy. Proc Natl Acad Sci U S A. 2008;105:5573–8.
74. Mizoguchi K, Ishige A, Aburada M, Tabira T. Chronic stress attenuates glucocorticoid negative feedback: involvement of the prefrontal cortex and hippocampus. Neuroscience. 2003;119:887–97.

75. Mizoguchi K, Ishige A, Takeda S, Aburada M, Tabira T. Endogenous glucocorticoids are essential for maintaining prefrontal cortical cognitive function. J Neurosci. 2004;24:5492–9.
76. Mizoguchi K, Yuzurihara M, Ishige A, Sasaki H, Chui D-H, Tabira T. Chronic stress induces impairment of spatial working memory because of prefrontal dopaminergic dysfunction. J Neurosci. 2000;20:1568–74.
77. Neisser U, Winograd E, Bergman ET, Schreiber CA, Palmer SE, Weldon MS. Remembering the earthquake: direct experience vs. hearing the news. Memory. 1996;4:337–57.
78. Patel PD, Lopez JF, Lyons DM, Burke S, Wallace M, Schatzberg AF. Glucocorticoid and mineralocorticoid receptor mRNA expression in squirrel monkey brain. J Psychiatr Res. 2000;34:383–92.
79. Pentkowski NS, Blanchard DC, Lever C, Litvin Y, Blanchard RJ. Effects of lesions to the dorsal and ventral hippocampus on defensive behaviors in rats. Eur J Neurosci. 2006;23: 2185–96.
80. Pitts MW, Takahashi LK. The central amygdala nucleus via corticotropin-releasing factor is necessary for time-limited consolidation processing but not storage of contextual fear memory. Neurobiol Learn Mem. 2011;95:86–91.
81. Popescu AT, Saghyan AA, Pare D. NMDA-dependent facilitation of corticostriatal plasticity by the amygdala. Proc Natl Acad Sci U S A. 2007;104:341–6.
82. Popoli M, Yan Z, McEwen BS, Sanacora G. The stressed synapse: the impact of stress and glucocorticoids on glutamate transmission. Nat Rev Neurosci. 2011;13:22–37.
83. Quirarte GL, Roozendaal B, McGaugh JL. Glucocorticoid enhancement of memory storage involves noradrenergic activation in the basolateral amygdala. Proc Natl Acad Sci U S A. 1997;94:14048–53.
84. Radley JJ, Arias CM, Sawchenko PE. Regional differentiation of the medial prefrontal cortex in regulating adaptive responses to acute emotional stress. J Neurosci. 2006;26:12967–76.
85. Reul JM, de Kloet ER. Two receptor systems for corticosterone in rat brain: microdistribution and differential occupation. Endocrinology. 1985;117:2505–11.
86. Ridder S, Chourbaji S, Hellweg R, Urani A, Zacher C, Schmid W, Zink M, Hortnagl H, Flor H, Henn FA, Schutz G, Gass P. Mice with genetically altered glucocorticoid receptor expression show altered sensitivity for stress-induced depressive reactions. J Neurosci. 2005;25: 6243–50.
87. Robertson DA, Beattie J, Reid I, Balfour D. Regulation of corticosteroid receptors in the rat brain: the role of serotonin and stress. Eur J Neurosci. 2005;21:1511–20.
88. Roozendaal B. Glucocorticoids and the regulation of memory consolidation. Psychoneuroendocrinology. 2000;25:213–38.
89. Roozendaal B, McEwen BS, Chattarji S. Stress, memory and the amygdala. Nat Rev Neurosci. 2009;10:423–33.
90. Roozendaal B, McReynolds JR, McGaugh JL. The basolateral amygdala interacts with the medial prefrontal cortex in regulating glucocorticoid effects on working memory impairment. J Neurosci. 2004;24:1385–92.
91. Sanchez MM, Young LJ, Plotsky PM, Insel TR. Distribution of coricosteriod receptors in the rhesus brain: relative absence of gluococorticoid receptors in the hippocampal formation. J Neurosci. 2000;20:4657–68.
92. Sapolsky R. Stress and plasticity in the limbic system. Neurochem Res. 2003;28:1735–42.
93. Sapolsky RM, Krey LC, McEwen BS. Glucocorticoid-sensitive hippocampal neurons are involved in terminating the adrenocortical stress response. Proc Natl Acad Sci. 1984; 81:6174–7.
94. Schwabe L, Wolf OT. Stress prompts habit behavior in humans. J Neurosci. 2009;29: 7191–8.
95. Schwabe L, Wolf OT. Stress-induced modulation of instrumental behavior: from goal-directed to habitual control of action. Behav Brain Res. 2011;219:321–8.
96. Selye H. Stress in health and disease. Boston: Butterworths; 1976.

97. Shepard JD, Barron KW, Myers DA. Stereotaxic localization of corticosterone to the amygdala enhances hypothalamo-pituitary–adrenal responses to behavioral stress. Brain Res. 2003;963:203–13.
98. Smith Y, Pare D. Intra-amygdaloid projections of the lateral nucleus in the cat: PHA-L anterograde labeling combined with postembedding GABA and glutamate immunocytochemistry. J Comp Neurol. 1994;342:232–48.
99. Snyder JS, Soumier A, Brewer M, Pickel J, Cameron HA. Adult hippocampal neurogenesis buffers stress responses and depressive behaviour. Nature. 2011;476:438–62.
100. Solano-Castiella E, Anwander A, Lohmann G, Weiss M, Docherty C, Geyer S, Reimer E, Friederici AD, Turner R. Diffusion tensor imaging segments the human amygdala in vivo. Neuroimage. 2010;49:2958–65.
101. Stahn C, Buttgereit F. Genomic and nongenomic effects of glucocorticoids. Nature clinical practice. Rheumatology. 2008;4:525–33.
102. Sullivan RM, Gratton A. Lateralized effects of medial prefrontal cortex lesions on neuroendocrine and autonomic stress responses in rats. J Neurosci. 1999;19:2834–40.
103. Sullivan RM, Gratton A. Behavioral effects of excitotoxic lesions of ventral medial prefrontal cortex in the rat are hemisphere-dependent. Brain Res. 2002;927:69–79.
104. Surget A, Tanti A, Leonardo ED, Laugeray A, Rainer Q, Touma C, Palme R, Griebel G, Ibarguen-Vargas Y, Hen R, Belzung C. Antidepressants recruit new neurons to improve stress response regulation. Mol Psychiatry. 2011;16:1177–88.
105. Tanapat P, Hastings NB, Rydel TA, Galea LA, Gould E. Exposure to fox odor inhibits cell proliferation in the hippocampus of adult rats via an adrenal hormone-dependent mechanism. J Comp Neurol. 2001;437:496–504.
106. Tasker JG, Di S, Malcher-Lopes R. Minireview: rapid glucocorticoid signaling via membrane-associated receptors. Endocrinology. 2006;147:5549–56.
107. Thomas RM, Hotsenpiller G, Peterson DA. Acute psychosocial stress reduces cell survival in adult hippocampal neurogenesis without altering proliferation. J Neurosci. 2007;27: 2734–43.
108. Thomas RM, Urban JH, Peterson DA. Acute exposure to predator odor elicits a robust increase in corticosterone and a decrease in activity without altering proliferation in the adult rat hippocampus. Exp Neurol. 2006;201:308–15.
109. Ulrich-Lai YM, Herman JP. Neural regulation of endocrine and autonomic stress responses. Nat Rev Neurosci. 2009;10:397–409.
110. Uth RM, McKelvy JF, Harrison RW, Bohn MC. Demonstration of glucocorticoid receptor-like immunoreactivity in glucocorticoid-sensitive vasopressin and corticotropin-releasing factor neurons in the hypothalamic paraventricular nucleus. J Neurosci Res. 1988;19: 405–11.
111. Uylings HBM, Groenewegen HJ, Kolb B. Do rats have a prefrontal cortex? Behav Brain Res. 2003;146:3–17.
112. Wallis JD. Cross-species studies of orbitofrontal cortex and value-based decision-making. Nat Neurosci. 2012;15:13–9.
113. Wanat MJ, Bonci A, Phillips PEM. CRF acts in the midbrain to attenuate accumbens dopamine release to rewards but not their predictors. Nat Neurosci. 2013;16:383–5.
114. Watanabe Y, Gould E, Daniels DC, Cameron H, McEwen BS. Tianeptine attenuates stress-induced morphological changes in the hippocampus. Eur J Pharmacol. 1992;222:157–62.
115. Wellman CL. Dendritic reorganization in pyramidal neurons in medial prefrontal cortex after chronic corticosterone administration. J Neurobiol. 2001;49:245–53.
116. Wellman CL, Izquierdo A, Garrett JE, Martin KP, Carroll J, Millstein R, Lesch K-P, Murphy DL, Holmes A. Impaired stress-coping and fear extinction and abnormal corticolimbic morphology in serotonin transporter knock-out mice. J Neurosci. 2007;27:684–91.
117. Wong EY, Herbert J. The corticoid environment: a determining factor for neural progenitors' survival in the adult hippocampus. Eur J Neurosci. 2004;20:2491–8.

118. Wood GE, Norris EH, Waters E, Stoldt JT, McEwen BS. Chronic immobilization stress alters aspects of emotionality and associative learning in the rat. Behav Neurosci. 2008;122: 282–92.
119. Woolley CS, Gould E, McEwen BS. Exposure to excess glucocorticoids alters dendritic morphology of adult hippocampal pyramidal neurons. Brain Res. 1990;531:225–31.
120. Yoshiya M, Komatsuzaki Y, Hojo Y, Ikeda M, Mukai H, Hatanaka Y, Murakami G, Kawata M, Kimoto T, Kawato S. Corticosterone rapidly increases thorns of CA3 neurons via synaptic/extranuclear glucocorticoid receptor in rat hippocampus. Front Neural Circuits. 2013;7:191.
121. Yuen EY, Liu W, Karatsoreos IN, Feng J, McEwen BS, Yan Z. Acute stress enhances glutamatergic transmission in prefrontal cortex and facilitates working memory. Proc Natl Acad Sci U S A. 2009;106:14075–9.
122. Bebbington P, Der G, MacCarthy B, Wykes T, Brugha T, Sturt P, Potter J. Stress incubation and the onset of affective disorders. Br J Psychiatry. 1993;162(3):358–62.
123. Brindley DN, Rolland Y. Possible connections between stress, diabetes, obesity, hypertension and altered lipoprotein metabolism that may result in atherosclerosis. Hypertension. 1989;23:33–351.
124. Cohen S, Herbert TB. Health psychology: psychological factors and physical disease from the perspective of human psychoneuroimmunology. Annu Rev Psychol. 1996;47(1):113–42.
125. Johnson EO, Kamilaris TC, Chrousos GP, Gold PW. Mechanisms of stress: a dynamic overview of hormonal and behavioral homeostasis. Neurosci Biobehav Rev. 1990;16:115–30.
126. Lupien SJ, McEwen BS, Gunnar MR, Heim C. Effects of stress throughout the lifespan on the brain, behavior and cognition. Nat Rev Neurosci. 2009;10:434–45.
127. Sternberg EM, Chrousos GP, Wilder RL, Gold PW. The stress response and the regulation of inflammatory disease. Ann Intern Med. 1992;117(10):854–66.
128. Dranovsky A, Hen R. Hippocampal neurogenesis: regulation by stress and antidepressants. Biol Psychiatry. 2006;59(12):1136–43. doi:10.1016/j.biopsych.2006.03.082.
129. Wong EYH, Herbert J. The corticoid environment: a determining factor for neural progenitors' survival in the adult hippocampus. Eur J Neurosci. 2006;20(10):2491–8. doi:10.1111/j.1460-9568.2004.03717.x.
130. Gould TD, Dao DT, Kovacsics CE. Mood and anxiety related phenotypes in mice, vol. 42. New York: Humana Press; 2009. p. 1–20.

Chapter 11
Glucocorticoid Regulation of Reproduction

Anna C. Geraghty and Daniela Kaufer

Abstract It is well accepted that stress, measured by increased glucocorticoid secretion, leads to profound reproductive dysfunction. In times of stress, glucocorticoids activate many parts of the fight or flight response, mobilizing energy and enhancing survival, while inhibiting metabolic processes that are not necessary for survival in the moment. This includes reproduction, an energetically costly procedure that is very finely regulated. In the short term, this is meant to be beneficial, so that the organism does not waste precious energy needed for survival. However, long-term inhibition can lead to persistent reproductive dysfunction, even if no longer stressed. This response is mediated by the increased levels of circulating glucocorticoids, which orchestrate complex inhibition of the entire reproductive axis. Stress and glucocorticoids exhibits both central and peripheral inhibition of the reproductive hormonal axis. While this has long been recognized as an issue, understanding the complex signaling mechanism behind this inhibition remains somewhat of a mystery. What makes this especially difficult is attempting to differentiate the many parts of both of these hormonal axes, and new neuropeptide discoveries in the last decade in the reproductive field have added even more complexity to an already complicated system. Glucocorticoids (GCs) and other hormones within the hypothalamic-pituitary-adrenal (HPA) axis (as well as contributors in the sympathetic system) can modulate the hypothalamic-pituitary-gonadal (HPG) axis at all levels—GCs can inhibit release of GnRH from the hypothalamus, inhibit gonadotropin synthesis and release in the pituitary, and inhibit testosterone synthesis and release from the gonads, while also influencing gametogenesis and sexual behavior. This chapter is not an exhaustive review of all the known literature, however is aimed at giving a brief look at both the central and peripheral effects of glucocorticoids on the reproductive function.

Keywords Reproduction • HPG axis • HPA axis • Glucocorticoids • Hypothalamus

A.C. Geraghty, B.A. • D. Kaufer, Ph.D. (✉)
Department of Integrative Biology, University of California, Berkeley, CA, USA
e-mail: danielak@berkeley.edu

J.-C. Wang, C. Harris (eds.), *Glucocorticoid Signaling*, Advances in Experimental Medicine and Biology 872, DOI 10.1007/978-1-4939-2895-8_11

Introduction

It is well accepted that stress, measured by increased glucocorticoid secretion, leads to profound reproductive dysfunction. In times of stress, glucocorticoids activate many parts of the fight or flight response, mobilizing energy and enhancing survival, while inhibiting metabolic processes that are not necessary for survival in the moment. This includes reproduction, an energetically costly procedure that is very finely regulated. In the short term, this is meant to be beneficial, so that the organism does not waste precious energy needed for survival. However, long-term inhibition can lead to persistent reproductive dysfunction, even after the stressor has been removed. This response is mediated by the increased levels of circulating glucocorticoids, which orchestrate complex inhibition of the entire reproductive axis. The reproductive and stress systems share a common architecture with central (hypothalamic stimulation of pituitary in both) and peripheral (gonads and adrenal respectively) components. Stress and glucocorticoids exhibits both central and peripheral inhibition of the reproductive hormonal axis. While this has long been recognized as an issue, understanding the complex signaling mechanism behind this inhibition remains somewhat of a mystery. In fact, the presenting complaint of Harvey Cushing's first patient with the syndrome that bears his name was secondary amenorrhea. She had normal menarche at age 14, but menses ceased two years later. What makes this especially difficult is attempting to differentiate the many parts of both of these hormonal axes, and new neuropeptide discoveries in the last decade in the reproductive field have added even more complexity to an already complicated system. Glucocorticoids (GCs) and other hormones within the hypothalamic-pituitary-adrenal (HPA) axis (as well as contributors in the sympathetic system) can modulate the hypothalamic-pituitary-gonadal (HPG) axis at all levels—GCs can inhibit release of GnRH from the hypothalamus, inhibit gonadotropin synthesis and release in the pituitary, and inhibit testosterone synthesis and release from the gonads, while also influencing gametogenesis and sexual behavior. This chapter is not an exhaustive review of all the known literature however is aimed at giving a brief look at both the central and peripheral effects of glucocorticoids on the reproductive function.

Reproductive Physiology: A Primer

Reproduction in mammals is a complex and precisely regulated hormonal process that requires the coordination of both the central nervous system and the peripheral reproductive organs for successful procreation. Negative and positive feedback signals tightly regulate the reproductive hormonal axis, also known as the hypothalamo-pituitary-gonadal (HPG) axis to maintain homeostasis. Perturbations in the axis, such as those caused by stress, can therefore have profound effects on reproductive ability, as even small changes can have large effects downstream, or even stop the axis in its tracks. Much of this chapter is focused on research conducted in rodents, however there are many species and sex differences in reproductive research indicating that stress can exert a myriad of effects to inhibit reproduction.

Studying the reproductive system of males offers a simpler "model" of HPG axis to begin examining reproductive physiology, due to its relative consistency over the lifetime post-puberty, as opposed to the cycling of female reproductive hormones during the post-pubertal and pre-menopausal years. Gonadotropin-releasing hormone (GnRH) is secreted from the hypothalamus in a pulsatile manner, crossing the hypophyseal portal system into the anterior pituitary. In the pituitary, GnRH stimulates the synthesis and release of luteinizing hormone (LH) and follicle-stimulating hormone (FSH). LH and FSH circulate systemically to trigger testosterone release from the testes and gametogenesis. In turn, testosterone (T), as well and LH and FSH, negatively feedback on the axis to maintain homeostasis [1–6] (Fig. 11.1).

Females, however, present a much more complicated picture. Unlike males, females experience hormonal surges to trigger ovulation, a phenomenon where estrogen switches from exerting negative feedback on the axis to positive feedback [7–12]. In rats, ovulation occurs once every 4 days. Similarly to males, GnRH is released from the hypothalamus, and LH and FSH circulate to the ovaries to trigger the release of estradiol (E2) and progesterone (P) as well as development of the ovum [6, 12–14]. Estradiol from the ovaries tightly regulates the HPG axis, exerting negative feedback onto the hypothalamus to inhibit GnRH release. However, when the developing ovum is near completion, and ovulation is due to occur, there is a

Fig. 11.1 General schematic of hypothalamic-pituitary-gonadal axis in the rodent model

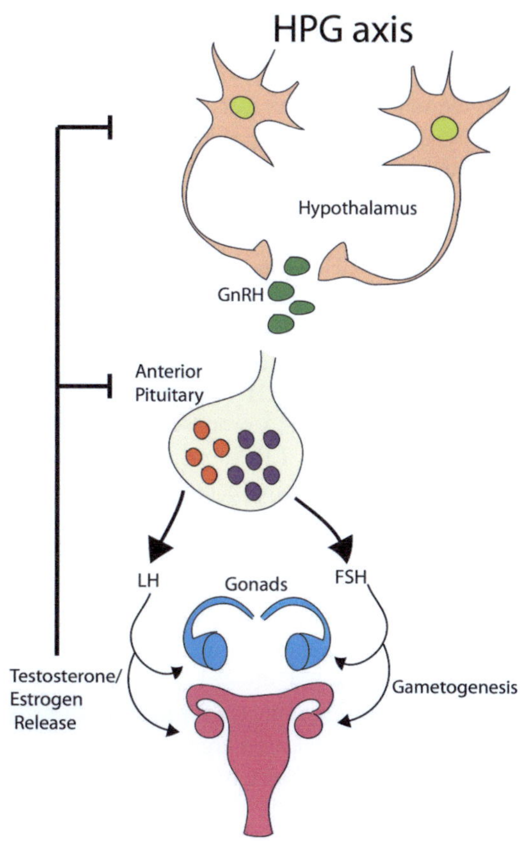

switch in the hormonal system. Increasing estradiol secretion from the developing follicle triggers a change from negative to positive feedback of estradiol on the hypothalamus, resulting in an increase of GnRH secretion and leading to a surge of LH secretion from the pituitary to trigger ovulation [15, 16]. However, until recently it was unknown what triggered the switch between negative and positive feedback.

Research in the last decade has shed more light on that switch during ovulation while also revealing that the axis is not nearly as simple as it appeared. There are currently two different hypotheses on what triggers the switch between negative and positive feedback in the HPG axis: The first one focuses on progesterone as the culprit, whereas the second one points to allopregnanolone. Supporting the first hypothesis, earlier research found that the estrogen surge initiates synthesis of progesterone receptors (PRs) in the hypothalamus, indicating progesterone is just as critical for ovulation as the estrogen surge [17–19]. However, studies on ovariectomized (OVX) and andrenalectomized (ADX) rats indicate that this progesterone does not come from the ovaries or adrenals, areas typical for steroid synthesis and systemic release. OVX rats given a normal dose of E2 can still exhibit LH surges [20]. Additionally, this surge can be blocked by administering trilostane, which inhibits the enzyme 3β-hydroxysteroid dehydrogenase (3β-HSD), an enzyme critical for synthesis of progesterone [20]. This data suggests that not only is pre-LH surge progesterone necessary for a successful LH surge, it is progesterone synthesized outside of the ovary and adrenals that regulates the surge. Hypothalamic progesterone synthesis, also known as neuroprogesterone, is a likely location, particularly hypothalamic astrocytes, which have been found to be the main source of neuroprogesterone in the hypothalamus, as they have all the steroidogenic enzyme machinery necessary for synthesis, regulate releasing factors and possess both types of estradiol receptors—Erα and ERβ receptors, which enable them to respond to the estradiol peak that precedes the LH surge [20–25].

While GnRH is the main coordinator of the HPG axis, secreted from the hypothalamus through the hypophyseal portal system to the anterior pituitary to stimulate release of LH, there are many factors upstream of GnRH that can impact its release, preventing successful ovulation. Two in particular, directly in the HPG axis, have been recently discovered: kisspeptin (KISS1) and gonadotropin-inhibitory hormone (GnIH) [26–34]. These two neuropeptides have opposing effects—KISS1 stimulates GnRH release from the hypothalamus [35–40], while GnIH inhibits it [30, 41–45]. A proposed mechanism of KISS1 action in the neuroprogesterone model of the LH surge states that the peak of estradiol also triggers an increase of progesterone receptors (PRs) on KISS1 neurons in the hypothalamus, which respond to the neuroprogesterone secreted from the astrocytes (activated also by estradiol to trigger the neuroprogesterone synthesis, likely through membrane estrogen receptors, specifically mERα). Neuroprogesterone, binding to the PRs on the KISS neurons, trigger the release of KISS, which then activates the GnRH neurons, leading to increased secretion of GnRH to trigger the LH surge from the pituitary [24, 46–48].

An alternative hypothesis for the switch between negative and positive feedback of E2 in the hypothalamus argues that progesterone is not the driving factor of ovulation, but it is a derivative of progesterone, allopregnanolone, that modulates GnRH release from the hypothalamus via increasing glutamate release. This in turn could act on NMDA receptors on the GnRH neurons, stimulating GnRH release [49–51]. While it is still not clear exactly what mechanism is truly responsible for positive feedback of E2 during ovulation, the many regulators of GnRH and GnRH itself are all impacted by glucocorticoids, and stress can lead to a disruption of homeostasis that results in both short-term reproductive dysfunction and long-term infertility.

Stress, whether psychological or physical, via the activation of the HPA axis, and subsequent increase in serum concentration of glucocorticoids, can inhibit the reproductive axis at every level, from the hypothalamus down to the ovaries or testes. Due to how closely regulated the HPG axis is, even small interferences in the hormonal milieu responsible for successful breeding can cause major dysfunction in the system (Fig. 11.2). The female ovulatory system discussed above, as well as

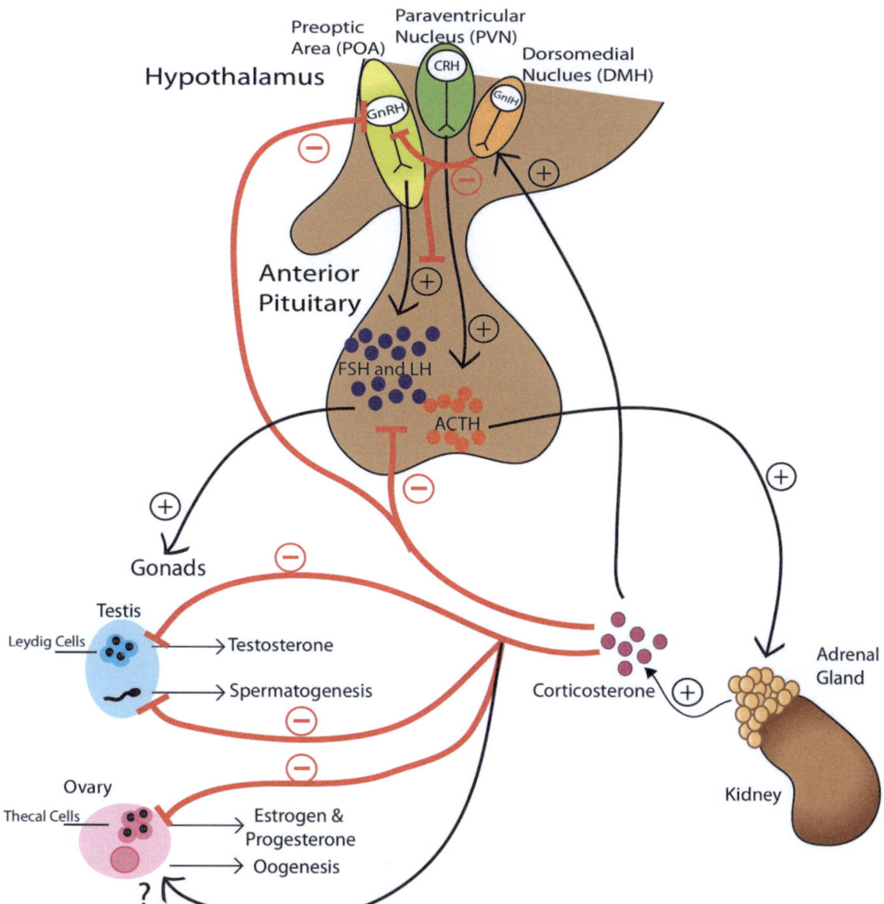

Fig. 11.2 Summary schematic of glucocorticoid and HPA axis interaction with the HPG axis and reproductive function

GnRH afferents such as GnIH and KISS, exhibit many points at which stress can disrupt the axis and cause fertility issues. While acute stress inhibiting the reproductive axis is found to be adaptive, preventing animals from breeding when times are not optimal for raising young, chronic stress and long-term shutdown of the reproductive axis can lead to prolonged dysfunction and infertility. In addition to causing infertility, this hypogonadism can contribute to other medical conditions such as osteoporosis. We will examine how stress impacts each part of the reproductive axis individually, and how this can add up to detrimental fertility issues.

Glucocorticoid Effects on the Hypothalamus

Stress causes the activation of the hypothalamic-pituitary-adrenal axis by a cascade of hormonal release: neural signals onto the hypothalamus cause the release of corticotropin-releasing hormone (CRH, previously known as CRF) to the hypophyseal portal system, via the median eminence. CRH in turn, activates cells in the pituitary to release adrenocorticotrophic hormone (ACTH) to the general circulation. ACTH stimulates cells in the adrenal cortex to synthesize and release the third hormone in this cascade of the glucocorticoid family, which in rodents is corticosterone (CORT). Stress and high glucocorticoids have a profound negative impact on the reproductive system, and within the hypothalamus both CRH and CORT affect GnRH and its afferents to inhibit reproductive success, as detailed in the next two sections. As discussed above, the timing of ovulation in female rodents is highly regulated by multiple factors controlling GnRH levels. Stress has been shown to impact each one of these factors, combining to effectively shut down the reproductive axis from the top. A difficulty in studying the effects of stress on reproduction is to accurately differentiate between the effects of CRH and the effects of CORT, both elevated during stress exposure, as well as determining specifically which level of the HPG axis is being directly affected. While it is well accepted that there are central effects of CRH directly on GnRH, completely separating those effects from downstream peripheral effects of CORT on GnRH is complex. Many manipulations of either CORT or CRH influence the other by feedback regulation, making the dissection of mechanisms even more difficult. In this section, we will try to differentiate the two to better understand how stress influences reproductive success.

Direct Regulation of GnRH by Glucocorticoids

All central influences on reproduction driven by stress converge on GnRH, and an inhibition of the GnRH signal from the hypothalamus can occur both directly on the GnRH neuron itself and indirectly via influence on GnRH afferents. While many peptides, steroids and neurotransmitters regulate the GnRH surge, stress and glucocorticoids can directly inhibit GnRH release from the hypothalamus. To regulate LH secretion from the pituitary, GnRH is released from the hypothalamus in pulses, and modulation of the pulsatile release of GnRH affects downstream gonadotropin release from the pituitary. Many types of stress have been found to affect the pulse generator of GnRH, and both CRH and CORT have been implicated in this mechanism of action [52–54]. GnRH crosses the hypophyseal portal system, a small circulatory system between the hypothalamus and the pituitary, to stimulate release of LH and FSH from the pituitary. Inhibiting the pulse generator of GnRH leads to a decrease in GnRH secretion from the hypothalamus. However, measuring GnRH release from the rodent hypothalamus is incredibly difficult, as the hypophyseal portal system is small and hard to sample from. To determine whether GC's are inhibiting GnRH release from the hypothalamus, researchers use downstream release of LH from the pituitary as a proxy for GnRH. If stress leads to a decrease in LH release, it is assumed that this is due to a decrease in GnRH release from the hypothalamus. Intracerebroventricular (ICV) administration of CRH has been found to inhibit LH pulses in rats [55] and this response can be blocked or reversed by administration of CRH antagonists [55, 56]. It appears that this suppression is mediated in part by both CRH receptor subtypes, CRH-R1 and CRH-R2, however which receptor is predominant depends on the type of stressor utilized [57–59]. CRH axon terminals directly interact with GnRH dendrites, and *in vitro* studies have shown that CRH can inhibit GnRH release. Other studies have also shown that infusing CRH into the medial preoptic area of females rats lead to a 60 % decrease in GnRH release from the hypothalamus [60, 61]. Though many studies have used *in vitro* systems to determine how GnRH can be inhibited by high CRH and CORT, recent studies using sheep, however, which have a much larger hypophyseal portal system than rodents, have allowed researchers to show *in vivo* that glucocorticoids inhibit GnRH release in the hypothalamus. Prolonged corticosterone administration IV caused a drop in GnRH release in the portal system of ewes [62] supporting the long-standing theory that glucocorticoids' central actions function to inhibit GnRH release from the hypothalamus. Release has also been inhibited in hypothalamic explants incubated with CORT and the glucocorticoid receptor agonist dexamethasone (DEX) in a dose-dependent manner [63].

Populations of GnRH neurons have been found to express both CRH receptors (CRH-R1, R2) [64] and glucocorticoid receptors (GRs) directly in both mice and rats [65, 66], indicating a method for which glucocorticoids (GCs) can act to directly inhibit GnRH synthesis and release. Glucocorticoids have also been found to inhibit GnRH transcription [67–69] and identified glucocorticoid response elements near the GnRH gene that regulate GnRH transcription. The GT1 cell line derived from GnRH-neurons, which synthesize and release GnRH [70, 71] express glucocorticoid receptors [72]. Further studies show that GCs can represses GnRH gene expression and release from these cells [73]. Studies *in vivo* have shown that chronic CORT treatment can inhibit GnRH expression, leading to a decrease in serum LH levels [74]. This however, had no effect on FSH levels, nor did it affect gonadotropin mRNA.

The effect of peripheral circulating glucocorticoids on GnRH and consequently the LH surge is highly variable and likely dependent on the severity and length of the stressor. In different acute stress studies glucocorticoids are shown to exert a full spectrum of effects on the LH surge that range from complete inhibition, to little or no effect to a positive activator of LH release. It emphasizes the importance of comparing stress paradigms and understanding how different stressors trigger the HPA axis. Acute stress tends to provide variable results on many different outcomes—life history, age and gender can influence this response greatly. In contrast, chronic stress has been consistently shown to inhibit the LH surge and ovulation in the literature, indicating that while short-term stressors may exert variable effects on reproduction, likely due to the type of stressor, long-term stress reliably causes reproductive dysfunction.

Indirect Regulation of GnRH by Glucocorticoids

Glucocorticoids may also exert an inhibitory influence on reproduction and GnRH output by influencing hormones upstream of GnRH, adding another level of control over reproduction. Two hormones in particular have been recently identified upstream of GnRH that respond directly to stress, affecting GnRH synthesis and secretion from the hypothalamus. Kisspeptin (KISS1) and gonadotropin-inhibitory hormone (GnIH), mentioned earlier, have opposing effects on GnRH. KISS1 is an activator of GnRH and plays a critical role in the maintenance of the GnRH pulse generator in the hypothalamus as well as the estrogen surge responsible for ovulation. Many different types of stressors, including exposure to the endotoxin lipopolysaccharide, acute hypoglycemia, immobilization (restraint) stress and social isolation, lead to downregulation of KISS1 and it's receptor KISS1R in the population of kisspeptin neurons in the median pre-optic area (mPOA) of the hypothalamus, as well as decreases in KISS1 expression in the arcuate nucleus, leading to downstream decreases in LH secretion [75–77]. Kisspeptin neurons have been shown to express both CRH-R and GR, implying that they could potentially respond directly to increase in either hormones following a stress stimulus [78]. These changes can transmit downstream into inhibition of the GnRH pulse generator, which has been shown to rely on kisspeptin input, and ultimately translate to downstream inhibition of LH secretion from the pituitary.

A subset of neurons in the hypothalamus expressing kisspeptin project directly to GnRH neurons [79–82], and has been implicated in modulation of the GnRH pulse generator. These neurons, known as the KNDy neurons, express two other neuropeptides, neurokinin B and dynorphin, and strongly respond to CORT. This subgroup offers yet another pathway in which stress via glucocorticoids can indirectly inhibit the reproductive axis. Dynorphin, an endogenous opioid in the brain signals through the kappa-subtype of the opioid receptor (KOR) in the hypothalamus, and administering dynorphin to female rats has been shown to inhibit LH pulses from the pituitary [83]. Stress has been shown to increase dynorphin release [84] which could lead to downstream inhibition of LH [85].

Gonadotropin-inhibitory hormone (GnIH) is another hormone upstream of GnRH that is regulated by stress to inhibit reproduction. GnIH was originally discovered in birds [30] and a mammalian orthologue, Rfamide-related peptide-3 (RFRP3) has since been identified in many species including rats, mice, hamsters and humans [26, 27, 86]. RFRP3 inhibits the GnRH pulse generator, decreasing GnRH release from the hypothalamus and leading to decreased gonadotropin secretion from the pituitary [41, 45, 87]. Both acute and chronic immobilization stress in male rats were found to increase RFRP3 mRNA and peptide levels in the hypothalamus and adrenalectomy blocked this effect, revealing that this is due to circulating glucocorticoid levels [88]. This increase in RFRP3 by glucocorticoids led to downstream inhibition of LH release from the pituitary. *In vitro* studies utilizing a cell line derived from RFRP3-ergic neurons has shown that RFRP3 neurons express the glucocorticoid receptor (GR) and also possess two glucocorticoid response elements (GREs) at the RFRP promoter region, further evidence pointing to a direct regulation of RFRP3 neurons by glucocorticoids [89, 90]. RFRP3 neurons in mammals can directly inhibit GnRH pulses from the hypothalamus, a response enhanced by high glucocorticoids after stress. In female rats, research has shown that high glucocorticoids after chronic stress can lead to an increase in RFRP3 levels in all stages of the estrous cycle and that this increase is sustained for at least one more estrous cycle after the stress has ceased. This increase leads to downstream reproductive dysfunction with fewer copulatory events in females exposed to chronic stress, lower pregnancy rates and decreased litter sizes [91]. This new research into the effect of glucocorticoids on RFRP3 levels and long-term reproductive dysfunction in female rodents could shed light on a more detailed mechanism of chronic inhibition of the HPG axis and reproductive success by glucocorticoids.

These are only a small sample of the possible indirect mechanisms of glucocorticoids on GnRH secretion. Much is still not well understood about the mechanism of action of glucocorticoids on GnRH levels, especially since there have been a slew of novel discoveries of peptides upstream of GnRH that increases the complexity of the regulation of reproductive neural systems by glucocorticoids. The difficulty in measuring GnRH levels from rodents has also made it difficult to come to generate conclusions about the mechanism of action of GCs on GnRH.

Glucocorticoid Effects in the Pituitary

While many of the stress effects on GnRH can lead to downstream pituitary dysfunction, glucocorticoids also directly influence reproduction at the level of the pituitary. This inhibition can happen via many different mechanisms, including modulating the sensitivity of the pituitary to changes in GnRH secretion, a decrease in synthesis of the gonadotropes LH and FSH, as well as decreasing the secretion of LH and FSH from the pituitary. However, the effects of glucocorticoids directly on the pituitary in secretion and synthesis are highly variable, indicating that the type of stressor and duration is very important when discussing this. This variability may stem from only looking at a part of the inhibition, such as focusing on just LH secretion as the output, rather than examining the mechanisms of synthesis and responsiveness of the pituitary itself.

Synthesis of LH and FSH is a highly regulated process involving not just GnRH levels, but steroid gonadal hormones as well [92, 93]. LH and FSH are glycoprotein heterodimers that consist of a common glycoprotein alpha-subunit (aGSU) and a specific β-subunit for either FSH or LH (LHβ and FSHβ) [94]. As rat and mice gonadotrope cells express GRs [95], it is likely that GR transcriptional regulation influences the synthesis of these subunits, both the specific β-subunits as well as the common alpha subunit to also regulate expression levels. Breen et.al. found that both daily immobilization stress and CORT administration led to reduced LH-β mRNA levels, as well as decreased LH release from the pituitary *in vivo*. They also showed *in vitro*, using a gonadotope cell line, Lβ T2 that CORT decreases GnRH-induced increase in LHβ mRNA levels as well as identified the LHβ promoter regions that are CORT and DEX responsive [96].

Similar to the effects seen in the hypothalamus, acute stress leads to variable results. Several studies utilizing multiple types of acute stressors on gonadotropin synthesis and release found that acute stress stimulated the HPG axis, leading to increased levels of LH, prolactin (PRL) and FSH in the plasma [97, 98]. Others have found though that acute administration of glucocorticoids can reduce the LH peak in females [99, 100] and injection of CRH peripherally has been shown to inhibit ovulation and block the LH surge completely [54]. Baldwin and Sawyer found that an acute injection of DEX early in the estrous cycle of rats can delay the onset of ovulation. Administration of LH though, in addition to DEX, can recover the ovulatory event, indicating that this delay of the estrous cycle is due to DEX inhibiting the LH surge necessary for ovulation, rather than a problem within the ovary [101]. There is some debate over timing of sample collection—some studies see an initial stimulation of LH but an inhibitory response later [102]. Collu et.al. found that after the first 15 min of an acute stressor, female rats exhibited an increase in LH plasma levels, however after 6 h of immobilization stress LH plasma levels in the females had dropped below control levels, supporting the theory that the acute stimulation of the HPG axis is only transitory [102]. Differences in the stressors, duration, sample collection and timing of the experiment may explain the differences found in LH release from the pituitary after acute stress and so it is hard to make concrete conclusions about acute stress and pituitary function in reproduction.

Chronic immobilization stress, on the other hand, has been shown to reliably reduce LH levels in both males and females [102–110]. One such study found that chronic immobilization stress reduced plasma levels of LH, and prolactin (PRL) with no change in FSH. However, pituitary levels of LH and FSH protein increased, showing that synthesis and secretion are not always matched up-it appears in this study, chronic stress did not influence synthesis, but inhibited release in some way [109]. This difference is critical when comparing studies—most studies do not look at both synthesis and secretion simultaneously, which may explain differences in findings.

Glucocorticoids can also change the responsiveness of the pituitary to GnRH secretion. This may occur via regulation of the GnRH receptor in the pituitary, however other studies have shown that this can also occur independently of changes in the GnRH receptor [111]. Baldwin (1979) found that stress affects estrogen feedback onto the pituitary, making it less responsive to GnRH secretion [112]. GnRH binding to its receptor on the gonadotropes is necessary to induce synthesis of aGSU, LHβ and FSHβ subunits, as well as stimulate the dimerization of the subunits for successful synthesis and release from the gonadotrope cells. GnRH receptors appear to be transcriptionally regulated by GR [113, 114].

This section focused predominately on the influence CORT and GR has on LH synthesis and secretion. CORT has been shown to also decrease FSH mRNA levels as well in some studies [110, 115], however, there is extreme variability in the FSH response in regards to stress. FSH secretion levels are rarely affected by stress, especially acute stress, showing little to no change in most studies looking at it [110, 116]. This likely is due to other regulatory signals on FSH beyond corticosterone. Many studies, in fact, show increases in FSHβ (beta) mRNA post-stress, both after immobilization stress and corticosterone or cortisol administration in many species in both acute and chronic studies, sometimes accompanied by decreases in LHβ or with no change in LH at all [96, 111, 117–119]. This also happens with little to no change in actual FSH secretion, so it is hard to draw conclusions on how stress regulates FSH levels.

In summary, based on current literature it seems impossible to draw clear answers about glucocorticoid inhibition of reproduction at the level of the pituitary. The range of studies utilizing different species, different sexes, a variety of stressors as well as different lengths of time of the stressors themselves is likely part of the cause of the confusion. Some studies show that acute stress increases the pituitary gonadotropins while some find decreases in gonadotropins. Some studies show changes in the synthesis of the gonadotropins, but no change in release and others find that release is altered with no change in synthesis. Much research is still to be done in this area to fully elucidate how glucocorticoids affect the pituitary directly in terms of reproduction.

Glucocorticoids and the Gonads

The final component of the HPG axis, the gonads, is yet another level in which GCs regulate the HPG axis. In the gonads, GCs can act to inhibit many critical steps to complete the reproductive process. Corticosterone can inhibit steroidogenesis, inhibiting the synthesis of testosterone (T), estrogen (E2) and progesterone (P), as well as directly inhibiting the release of these steroids from the gonads. GCs modulate the expression of the LH-receptor (LHR) on the gonads, changing how the gonads respond to LH and leading to downstream effects on steroids. GCs can also regulate gametogenesis, the development of mature sperm and ovum, to inhibit reproduction at the levels of the gamete. These effects can all be completed in the absence of influences from the hypothalamus and pituitary, emphasizing how profoundly stress can influence reproduction.

Glucocorticoid Effects in the Testes

Research has shown that GR is localized in several different cell populations within the testes, including importantly the Leydig cells, which is where steroidogenesis occurs within the testes, as well as in the Sertoli cells, primary spermatocytes and the epididymis [121–123]. This indicates that GRs can regulate not only steroidogenesis and the release of T, but spermatogenesis as well, either through affecting the primary population of cells or affecting the last steps of maturation in the epididymis. Both acute and chronic stress experiments have shown that high GCs inhibits testosterone secretion, spermatogenesis and libido [124–128] as expected. This effect is due specifically to circulating GC levels in the blood and action via GR because ACTH treatment in adrenalectomized animals fails to replicate these findings [129]. Some studies show that this decreased testosterone release can occur either via downregulation of the LH receptor in Leydig cells [130] or through inhibition of the enzymes necessary for testosterone biosynthesis [125, 126, 128, 129, 131–133]. Overall, these changes result in decreased testosterone synthesis and release from the gonads.

Glucocorticoids may also impact spermatogenesis, as GRs are present on the primary spermatocytes as well as within the epididymis. High glucocorticoids have been found to induce testicular germ cell apoptosis [134, 135] as well as Leydig cell apoptosis [136], which has a profound inhibitory influence on male reproductive abilities. Chronic stress has been shown to also decrease the number of spermatids within the testis [137], and in humans it has been shown that chronic stress leads to decreased sperm numbers, likely through a combination of the above responses [138]. Expression of GR in all these spermatogenic area indicates that glucocorticoids can act directly on the testes to regulate sperm production. Stress and high levels of GCs likely inhibit reproduction both indirectly and directly at the level of the gonads, with decreased secretion of LH from the pituitary decreasing testosterone release, and direct inhibition of testosterone synthesis and sperm production by GCs.

Glucocorticoid Effects in the Ovaries

The role of glucocorticoids within the ovary is somewhat more complicated than it is within the testes. Rather than a straight inhibitory role of GCs on ovarian function, some GC effects are actually beneficial to the ovaries and are necessary for maintenance of the follicular development pathway. During each cycle, many follicles are activated for development within the ovaries, however not all fully develop to maturity and it appears GCs are an active part in that selection process. This is necessary for normal ovarian function, but likely is very finely controlled, and high stress may tip the balance between a "good" level of GCs and a "maladaptive" level that leads to ovarian dysfunction.

A way in which the ovaries control levels of GCs through follicular development during the female estrous or menstrual cycle is via expression of the enzyme 11β-hydroxysteroid dehydrogenase (11β-HSD). 11β-HSD is a family of enzymes responsible for catalyzing the conversion of inactive cortisone to cortisol or vice versa to regulate glucocorticoid exposure. 11β-HSD1 activates cortisol predominately, however the reaction is bidirectional, and 11β-HSD1 can also inactivate cortisol. 11β-HSD2 on the other hand unidirectionally inactivates cortisol, converting it back to cortisone [139, 140]. Researchers have identified that many of the cells within the ovaries, including the follicles and corpus luteum, express 11β-HSD1, 11β-HSD2 and GR [141–145], indicating that there are possibly many regulatory effects of glucocorticoids on follicular development and ovarian function. Interestingly, the ovaries differentially regulate 11β-HSD1 and HSD2 throughout the cycle. 11β-HSD2, which inactivates GCs, is highly expressed in developing follicles in the ovary, while 11β-HDS1, which activates GCs, is highly expressed in follicles that have been luteinized, meaning they have been activated by an LH surge and ready for an ovulatory event [141, 146, 147]. This indicates that the ovaries upregulate 11β-HSD2 while developing in order to inactivate GCs present in the ovary while the follicles are maturing in order to enhance development, but choose to activate circulating GCs once the follicle is released for ovulation. These activated and functional GCs may act as an anti-inflammatory response triggered by the rupturing of the ovarian surface epithelium during ovulation [148, 149]. These two examples show how GCs are likely necessary for normal function of the ovaries, however their levels are tightly regulated via variability in expression of the 11β-HSD1 and 2. These enzymes are actually manipulated via gonadotropin signals from the pituitary, with LH controlling expression of 11β-HSD1 expression (thus activating GCs during ovulation). This regulation via gonadotropins provides a mechanism through which excess GCs could influence enzymatic regulation of GCs. As these two enzymes are so narrowly regulated during the ovarian cycle, stress and high GC secretion from the adrenals can easily dysregulate these signals and cause profound fertility problems in both ovarian function during ovulation and uterine function during fertilization, implantation and pregnancy.

In the ovaries, high amounts of GCs, surpassing the amount that is typically inactivated by 11β-HSD2, can suppress LH function and inhibit estrogen release and synthesis [131, 146, 150]. Studies both *in vivo* and *in vitro* have shown that GCs can influence not only LH response, but also inhibit transcription of the enzymes necessary for steroid biosynthesis, critically inhibiting p450 aromatase, necessary for conversion of testosterone to estrogen. In rat granulosa cell cultures, FSH triggers the increase of aromatase activity, promoting estrogen synthesis for ovulation. Administration of both CORT and DEX inhibited this FSH-induced increase, however stimulated progesterone synthesis and did not inhibit pre-existing aromatase function. This indicates that GCs act to inhibit induction of aromatase activity specifically, not necessarily affecting granulosa cell function as a whole [151]. Glucocorticoid treatment was also found to decrease LH receptor in cultured granulosa cells [152], indicating that GCs can act directly on the ovarian cells to decrease FSH-stimulated functions, including aromatase activity and LH receptor binding.

Interestingly, GC effects on oocyte maturation appear to be species dependent. Studies in humans and pig have shown that GCs can inhibit meiotic development in the oocytes [153, 154], however studies in sheep and mice have shown no effect of GCs on final oocyte maturation [155, 156]. However a recent study in mice showed that high levels of CRH in the cytoplasm of the ovaries due to restraint stress induced ovarian apoptosis, decreasing follicular development independent of GR. This increase of CRH was acting on thecal cells in the ovary, decreasing testosterone and estrogen levels and increasing progesterone, creating a hormonal imbalance between estrogen and progesterone that led to decreased oocyte success [157]. These differences are likely due to problems intrinsic to *in vitro* models that utilize only the ovarian granulosa cells. In addition, another *in vivo* study in mice utilizing predatory stress found that while high GCs did not affect oocyte maturation, blastocyst formation was significantly decreased in these mice, showing that GCs may have a stronger effect on embryo development or the oocyte potential for fertilization, rather than maturation of the oocytes in general [158]. The next section will explore GCs effects on pregnancy and fertilization more closely.

The role of glucocorticoid function in the ovary is incredibly complex and narrowly regulated. The actions of GCs are regulated via differential transcription of the two 11β-HSD enzymes, transcription of which is controlled through gonadotropin release form the pituitary. Glucocorticoids are critical for maintenance of ovarian function, involved in functional apoptosis of follicles to maintain normal follicular development, as well as its anti-inflammatory role necessary for ovulation to occur. However, high stress can tip the scales from functional to dysfunctional, overwhelming the ability of 11β-HSD to regulate GC levels and causing ovarian problems ranging from a decreased responsiveness to LH levels, decreased synthesis of estrogen due to inhibition of aromatase release, and potentially inhibiting the final step of oocyte maturation.

Glucocorticoids, Implantation and Pregnancy Success

Even if an ovum can be successfully developed in times of stress, and the HPG axis still functional enough to trigger ovulation, GCs can still act to influence the uterus to prevent successful implantation and completion of pregnancy. Glucocorticoids typically act in opposition to estrogenic actions, and this becomes increasingly critical in implantation. For successful implantation of a blastocyst, progesterone and estrogen regulate uterine cell proliferation, and are necessary for the changes in both the blastocyst and uterine epithelium for successful adhesion. Glucocorticoids inhibit estradiol-stimulated uterine growth and decreases estrogen receptor concentrations in the uterus [159–162].

Pregnancy itself requires a delicate immune balance and regulation of the maternal immune cells in order for survival of both the fetus and the mother. It is suggested that the high levels of progesterone (P) released from the corpus luteum of the ovary after ovulation and sustained by the placenta throughout pregnancy help regulate the mothers immune system. There is some research indicating that membrane-bound progesterone receptors act to inhibit maternal T-cell during pregnancy [163, 164]. This combines with a series of other downstream immune events that allows the maternal immune system to accept the foreign fetus and expression of progesterone and related progesterone factors such as progesterone-induced blocking factor (PIBF) continues to increase through pregnancy. High levels of Th1 cytokines in mice have been shown to be abortogenic, and progesterone during pregnancy binding to progesterone receptors have been shown to release PIBF, which in turn decreases natural killer (NK) cells in the uterus and induces Th2 cytokine development, changing the balance towards an anti-abortive immune response [163–167]. However studies using restraint stress in rodents has found that stress early in pregnancy leads to decreased embryo success, showing higher abortion rates in the mice and smaller litter sizes [168–170]. Wiebold et.al. found that this was due to decreases in corpora lutea, lower levels of serum progesterone and fewer implantation sites [168]. In humans, circadian cortisol levels are suppressed in early pregnancy, and women who have been found to have high morning cortisol levels in the first weeks of pregnancy were more likely to experience spontaneous abortions [130]. This however appears to be specific to the peri-implantation time, as studies have not found that circadian levels of cortisol are indicative of likelihood of a miscarriage later in the first trimester [131]. Human data on the subject is inconclusive, but the immunosuppressive effects of high cortisol, as well GC's influence on decreasing progesterone and PIBF release support the idea that stress can have a profound influence on early miscarriage rates.

Glucocorticoids play a significant role in pregnancy maintenance, opposing estrogen's ability to ready the uterus for implantation and inhibiting progesterone's anti-abortive immune response. While much of research into this focuses on stress during the pregnancy itself, there could be long-term effects of stress prior to the pregnancy that could affect pregnancy success as well, maybe via long-term inhibition of progesterone.

Conclusion: Stress and Its Many Effects on Reproductive Ability

Physiologically, glucocorticoids exert many effects on surrounding cells and are necessary for life. Within normal ranges, GCs regulate homeostasis and are critical for our stress response. In times of stress, high levels of GCs shut down physiological processes not relevant for survival in that time, including reproduction. GCs and the

HPA axis can act upon every level of the HPG axis, both directly and indirectly inhibiting gonadotropin release from the pituitary and exerting direct effects on the gonads. Stress and high GCs decrease the release of GnRH from the hypothalamus, either by directly inhibiting GnRH pulses or inhibiting upstream regulators of GnRH release. This can lead to downstream decreases in LH release form the pituitary, however GCs can also directly inhibit the synthesis and release of gonadotropins from the anterior pituitary. The decrease of LH and sometimes FSH from the pituitary can decrease steroid release form the gonads, and circulating GCs can also act directly on the gonads to inhibit the transcription of enzymes necessary for gonadal steroid biosynthesis. There are sex and species differences in all these responses. It is a complex and confusing field, however new techniques utilizing cell-specific knockdowns of GR and/or other peptides involved in this response can help clarify the more specific roles of GCs and reproductive dysfunction. This becomes increasingly important as we find that infertility rates continue to increase in humans, likely due to high stress exposure in day-to-day lives. Understanding the molecular mechanisms behind how stress impairs fecundity and reproductive success, especially in females, is critical to helping improve fertility rates.

References

1. Handa RJ, Weiser MJ. Gonadal steroid hormones and the hypothalamo-pituitary-adrenal axis. Front Neuroendocrinol. 2014;35(2):197–220. doi:10.1016/j.yfrne.2013.11.001.
2. Jennes L, Conn PM. Gonadotropin-releasing hormone and its receptors in rat brain. Front Neuroendocrinol. 1994;15(1):51–77. doi:10.1006/frne.1994.1003.
3. King JC, Tobet SA, Snavely FL, Arimura AA. LHRH immunopositive cells and their projections to the median eminence and organum vasculosum of the lamina terminalis. J Comp Neurol. 1982;209(3):287–300. doi:10.1002/cne.902090307.
4. Levine JE, Bauer-Dantoin AC. Neuroendocrine regulation of the luteinizing hormone-releasing hormone pulse generator in the rat. Recent Prog Horm Res. 1991;47:97–151.
5. Moenter SM, Anthony DeFazio R, Pitts GR, Nunemaker CS. Mechanisms underlying episodic gonadotropin-releasing hormone secretion. Front Neuroendocrinol. 2003;24(2):79–93. doi:10.1016/S0091-3022(03)00013-X.
6. Haisenleder DJ, Dalkin AC, Ortolano GA, Marshall JC, Shupnik MA. A pulsatile gonadotropin-releasing hormone stimulus is required to increase transcription of the gonadotropin subunit genes: evidence for differential regulation of transcription by pulse frequency in vivo. Endocrinology. 1991;128(1):509–17. doi:10.1210/endo-128-1-509.
7. Sarkar DK, Chiappa SA, Fink G, Sherwood NM. Gonadotropin-releasing hormone surge in pro-oestrous rats. Nature. 1976;264(5585):461–3. doi:10.1038/264461a0.
8. Park O-K, Ramirez VD. Spontaneous changes in LHRH release during the rat estrous cycle, as measured with repetitive push-pull perfusions of the pituitary gland in the same female rats. Neuroendocrinology. 1989;50(1):66–72. doi:10.1159/10.1159/000125203.
9. Marshall JC, Griffin ML. The role of changing pulse frequency in the regulation of ovulation. Hum Reprod. 1993;8 Suppl 2:57–61. http://www.ncbi.nlm.nih.gov/pubmed/8276970. Accessed 27 May 2014.
10. Marshall JC, Dalkin AC, Haisenleder DJ, Griffin ML, Kelch RP. GnRH pulses—the regulators of human reproduction. Trans Am Clin Climatol Assoc. 1993;104:31–46. http://www.pubmedcentral.nih.gov/articlerender.fcgi?artid=2376610&tool=pmcentrez&rendertype=abstract. Accessed 27 May 2014.

11. Herbison AE. Estrogen positive feedback to gonadotropin-releasing hormone (GnRH) neurons in the rodent: the case for the rostral periventricular area of the third ventricle (RP3V). Brain Res Rev. 2008;57(2):277–87. doi:10.1016/j.brainresrev.2007.05.006.
12. Chazal G, Faudon M, Gogan F, Laplante E. Negative and positive effects of oestradiol upon luteinizing hormone secretion in the female rat. J Endocrinol. 1974;61(3):511–2. doi:10.1677/joe.0.0610511.
13. Shupnik MA. Gonadotropin gene modulation by steroids and gonadotropin-releasing hormone. Biol Reprod. 1996;54(2):279–86. doi:10.1095/biolreprod54.2.279.
14. Knobil E. The neuroendocrine control of the menstrual cycle. Recent Prog Horm Res. 1980;36:53–88.
15. Baird DT, McNeilly AS. Gonadotrophic control of follicular development and function during the oestrous cycle of the ewe. J Reprod Fertil Suppl. 1981;30:119–33. http://europepmc.org/abstract/MED/6300383. Accessed 2 Sept 2014.
16. Legan SJ, Karsch FJ. A daily signal for the LH surge in the rat. Endocrinology. 1975;96(1):57–62. doi:10.1210/endo-96-1-57.
17. Ferin M, Tempone A, Zimmering PE, Van de Wiele RL. Effect of antibodies to 17beta-estradiol and progesterone on the estrous cycle of the rat. Endocrinology. 1969;85(6):1070–8. doi:10.1210/endo-85-6-1070.
18. Labhsetwar AP. Role of estrogens in ovulation: a study using the estrogen-antagonist, I.C.I. 46,474. Endocrinology. 1970;87(3):542–51. doi:10.1210/endo-87-3-542.
19. Chappell PE, Levine JE. Stimulation of gonadotropin-releasing hormone surges by estrogen. I. Role of hypothalamic progesterone receptors. Endocrinology. 2000;141(4):1477–85. doi:10.1210/endo.141.4.7428.
20. Micevych P, Sinchak K, Mills RH, Tao L, LaPolt P, Lu JKH. The luteinizing hormone surge is preceded by an estrogen-induced increase of hypothalamic progesterone in ovariectomized and adrenalectomized rats. Neuroendocrinology. 2003;78(1):29–35. doi:10.1159/000071703.
21. Kuo J, Hamid N, Bondar G, Prossnitz ER, Micevych P. Membrane estrogen receptors stimulate intracellular calcium release and progesterone synthesis in hypothalamic astrocytes. J Neurosci. 2010;30(39):12950–7. doi:10.1523/JNEUROSCI.1158-10.2010.
22. Micevych P, Soma KK, Sinchak K. Neuroprogesterone: key to estrogen positive feedback? Brain Res Rev. 2008;57(2):470–80. doi:10.1016/j.brainresrev.2007.06.009.
23. Micevych PE, Chaban V, Ogi J, Dewing P, Lu JKH, Sinchak K. Estradiol stimulates progesterone synthesis in hypothalamic astrocyte cultures. Endocrinology. 2007;148(2):782–9. doi:10.1210/en.2006-0774.
24. Micevych P, Sinchak K. The neurosteroid progesterone underlies estrogen positive feedback of the LH surge. Front Endocrinol (Lausanne). 2011;2:90. doi:10.3389/fendo.2011.00090.
25. Chaban VV, Lakhter AJ, Micevych P. A membrane estrogen receptor mediates intracellular calcium release in astrocytes. Endocrinology. 2004;145(8):3788–95. doi:10.1210/en.2004-0149.
26. Ubuka T, Inoue K, Fukuda Y, et al. Identification, expression, and physiological functions of Siberian hamster gonadotropin-inhibitory hormone. Endocrinology. 2012;153(1):373–85. papers://cf7c60b8-94a1-4c79-88e4-6c57345fd583/Paper/p1434.
27. Ubuka T, Morgan K, Pawson A, et al. Identification of human GnIH homologs, RFRP-1 and RFRP-3, and the cognate receptor, GPR147 in the human hypothalamic pituitary axis. PLoS One. 2009;4(12):1334–9. papers://cf7c60b8-94a1-4c79-88e4-6c57345fd583/Paper/p1233.
28. Ubuka T, Lai H, Kitani M, et al. Gonadotropin-inhibitory hormone identification, cDNA cloning, and distribution in rhesus macaque brain. J Comp Neurol. 2009;517:841–55. doi:10.1002/cne.22191.
29. Ukena K, Iwakoshi E, Minakata H, Tsutsui K. A novel rat hypothalamic RFamide-related peptide identified by immunoaffinity chromatography and mass spectrometry. FEBS Lett. 2002;512(1–3):255–8. papers://cf7c60b8-94a1-4c79-88e4-6c57345fd583/Paper/p1299.
30. Tsutsui K, Saigoh E, Ukena K, et al. A novel avian hypothalamic peptide inhibiting gonadotropin release. Biochem Biophys Res Commun. 2000;275(2):661–7. doi:10.1006/bbrc.2000.3350.

31. De Roux N, Genin E, Carel J-C, Matsuda F, Chaussain J-L, Milgrom E. Hypogonadotropic hypogonadism due to loss of function of the KiSS1-derived peptide receptor GPR54. Proc Natl Acad Sci U S A. 2003;100(19):10972–6. doi:10.1073/pnas.1834399100.
32. Thompson EL, Patterson M, Murphy KG, et al. Central and peripheral administration of kisspeptin-10 stimulates the hypothalamic-pituitary-gonadal axis. J Neuroendocrinol. 2004;16(10):850–8. doi:10.1111/j.1365-2826.2004.01240.x.
33. Gottsch ML, Cunningham MJ, Smith JT, et al. A role for kisspeptins in the regulation of gonadotropin secretion in the mouse. Endocrinology. 2004;145(9):4073–7. doi:10.1210/en.2004-0431.
34. Navarro VM, Castellano JM, Fernández-Fernández R, et al. Effects of KiSS-1 peptide, the natural ligand of GPR54, on follicle-stimulating hormone secretion in the rat. Endocrinology. 2005;146(4):1689–97. doi:10.1210/en.2004-1353.
35. Dhillo WS, Chaudhri OB, Patterson M, et al. Kisspeptin-54 stimulates the hypothalamic-pituitary gonadal axis in human males. J Clin Endocrinol Metab. 2005;90(12):6609–15. doi:10.1210/jc.2005-1468.
36. Li X-F, Kinsey-Jones JS, Cheng Y, et al. Kisspeptin signalling in the hypothalamic arcuate nucleus regulates GnRH pulse generator frequency in the rat. Tena-Sempere M, ed. PLoS One. 2009;4(12):e8334. doi:10.1371/journal.pone.0008334.
37. Maeda K-I, Ohkura S, Uenoyama Y, et al. Neurobiological mechanisms underlying GnRH pulse generation by the hypothalamus. Brain Res. 2010;1364:103–15. doi:10.1016/j.brainres.2010.10.026.
38. Roseweir AK, Kauffman AS, Smith JT, et al. Discovery of potent kisspeptin antagonists delineate physiological mechanisms of gonadotropin regulation. J Neurosci. 2009;29(12):3920–9. doi:10.1523/JNEUROSCI.5740-08.2009.
39. Messager S, Chatzidaki EE, Ma D, et al. Kisspeptin directly stimulates gonadotropin-releasing hormone release via G protein-coupled receptor 54. Proc Natl Acad Sci U S A. 2005;102(5):1761–6. papers://cf7c60b8-94a1-4c79-88e4-6c57345fd583/Paper/p1232.
40. Millar RP, Roseweir AK, Tello JA, et al. Kisspeptin antagonists: unraveling the role of kisspeptin in reproductive physiology. Brain Res. 2010;1364:81–9. doi:10.1016/j.brainres.2010.09.044.
41. Pineda R, Garcia-Galiano D, Sanchez-Garrido MA, et al. Characterization of the inhibitory roles of RFRP3, the mammalian ortholog of GnIH, in the control of gonadotropin secretion in the rat: in vivo and in vitro studies. Am J Physiol Endocrinol Metab. 2010;299(1):E39–46. papers://cf7c60b8-94a1-4c79-88e4-6c57345fd583/Paper/p1269.
42. Khan AR, Kauffman AS. The role of kisspeptin and RFamide-related peptide-3 neurones in the circadian-timed preovulatory luteinising hormone surge. J Neuroendocrinol. 2012;24(1):131–43. papers://cf7c60b8-94a1-4c79-88e4-6c57345fd583/Paper/p1435.
43. Clarke I, Smith J, Henry B, et al. Gonadotropin-inhibitory hormone is a hypothalamic peptide that provides a molecular switch between reproduction and feeding. Neuroendocrinology. 2012;95(4):305–16. papers://cf7c60b8-94a1-4c79-88e4-6c57345fd583/Paper/p1596.
44. Wu M, Dumalska I, Morozova E, van den Pol AN, Alreja M. Gonadotropin inhibitory hormone inhibits basal forebrain vGluT2-gonadotropin-releasing hormone neurons via a direct postsynaptic mechanism. J Physiol. 2009;587(7):1401. papers://cf7c60b8-94a1-4c79-88e4-6c57345fd583/Paper/p1234.
45. Kriegsfeld LJ, Gibson EM, Williams WP, et al. The roles of RFamide-related peptide-3 in mammalian reproductive function and behaviour. J Neuroendocrinol. 2010;22(7):692–700. doi:10.1111/j.1365-2826.2010.02031.x.
46. Clarkson J, d'Anglemont de Tassigny X, Moreno AS, Colledge WH, Herbison AE. Kisspeptin-GPR54 signaling is essential for preovulatory gonadotropin-releasing hormone neuron activation and the luteinizing hormone surge. J Neurosci. 2008;28(35):8691–7. doi:10.1523/JNEUROSCI.1775-08.2008.
47. Sinchak K, Wagner EJ. Estradiol signaling in the regulation of reproduction and energy balance. Front Neuroendocrinol. 2012;33(4):342–63. doi:10.1016/j.yfrne.2012.08.004.

48. Christensen A, Bentley GE, Cabrera R, et al. Hormonal regulation of female reproduction. Horm Metab Res. 2012;44(8):587–91. doi:10.1055/s-0032-1306301.
49. Giuliani FA, Yunes R, Mohn CE, Laconi M, Rettori V, Cabrera R. Allopregnanolone induces LHRH and glutamate release through NMDA receptor modulation. Endocrine. 2011;40(1):21–6. doi:10.1007/s12020-011-9451-8.
50. Sim JA, Skynner MJ, Herbison AE. Direct regulation of postnatal GnRH neurons by the progesterone derivative allopregnanolone in the mouse. Endocrinology. 2001;142(10):4448–53. doi:10.1210/endo.142.10.8451.
51. el-Etr M, Akwa Y, Fiddes RJ, Robel P, Baulieu EE. A progesterone metabolite stimulates the release of gonadotropin-releasing hormone from GT1-1 hypothalamic neurons via the gamma-aminobutyric acid type A receptor. Proc Natl Acad Sci U S A. 1995;92(9):3769–73. http://www.pubmedcentral.nih.gov/articlerender.fcgi?artid=42043&tool=pmcentrez&rendertype=abstract. Accessed 27 May 2014.
52. Breen KM, Karsch FJ. New insights regarding glucocorticoids, stress and gonadotropin suppression. Front Neuroendocrinol. 2006;27(2):233–45. doi:10.1016/j.yfrne.2006.03.335.
53. Li XF, Knox AMI, O'Byrne KT. Corticotrophin-releasing factor and stress-induced inhibition of the gonadotrophin-releasing hormone pulse generator in the female. Brain Res. 2010;1364:153–63. doi:10.1016/j.brainres.2010.08.036.
54. Rivier C, Vale W. Influence of corticotropin-releasing factor on reproductive functions in the rat. Endocrinology. 1984;114(3):914–21. doi:10.1210/endo-114-3-914.
55. Cates PS, Li XF, O'Byrne KT. The influence of 17beta-oestradiol on corticotrophin-releasing hormone induced suppression of luteinising hormone pulses and the role of CRH in hypoglycaemic stress-induced suppression of pulsatile LH secretion in the female rat. Stress. 2004;7(2):113–8. doi:10.1080/1025389042000218988.
56. Bowe JE, Li XF, Kinsey-Jones JS, Brain SD, Lightman SL, O'Byrne KT. The role of corticotrophin-releasing hormone receptors in the calcitonin gene-related peptide-induced suppression of pulsatile luteinising hormone secretion in the female rat. Stress. 2008;11(4):312–9. doi:10.1080/10253890701801448.
57. Li XF, Bowe JE, Kinsey-Jones JS, Brain SD, Lightman SL, O'Byrne KT. Differential role of corticotrophin-releasing factor receptor types 1 and 2 in stress-induced suppression of pulsatile luteinising hormone secretion in the female rat. J Neuroendocrinol. 2006;18(8):602–10. papers://cf7c60b8-94a1-4c79-88e4-6c57345fd583/Paper/p701.
58. Li XF, Bowe JE, Lightman SL, O'Byrne KT. Role of corticotropin-releasing factor receptor-2 in stress-induced suppression of pulsatile luteinizing hormone secretion in the rat. Endocrinology. 2005;146(1):318–22. papers://cf7c60b8-94a1-4c79-88e4-6c57345fd583/Paper/p710.
59. Kinsey-Jones J, Li X, Knox A, et al. Corticotrophin-releasing factor alters the timing of puberty in the female rat. J Neuroendocrinol. 2010;22(2):102–109. file:///Users/annageraghty/Documents/Papers/2010/Kinsey-Jones/Journal of Neuroendocrinology 2010 Kinsey-Jones.pdf.
60. Rivier C, Rivest S. Effect of stress on the activity of the hypothalamic-pituitary-gonadal axis: peripheral and central mechanisms. Biol Reprod. 1991;45(4):523–32. http://www.ncbi.nlm.nih.gov/pubmed/1661182. Accessed 9 Sept 2013.
61. Rivest S, Rivier C. Central mechanisms and sites of action involved in the inhibitory effects of CRF and cytokines on LHRH neuronal activity. Ann N Y Acad Sci. 1993;697(1 Corticotropin):117–41. doi:10.1111/j.1749-6632.1993.tb49928.x.
62. Oakley AE, Breen KM, Clarke IJ, Karsch FJ, Wagenmaker ER, Tilbrook AJ. Cortisol reduces gonadotropin-releasing hormone pulse frequency in follicular phase ewes: influence of ovarian steroids. Endocrinology. 2009;150(1):341–9. doi:10.1210/en.2008-0587.
63. Calogero AE, Burrello N, Bosboom AM, Garofalo MR, Weber RF, D'Agata R. Glucocorticoids inhibit gonadotropin-releasing hormone by acting directly at the hypothalamic level. J Endocrinol Invest. 1999;22(9):666–70. http://www.ncbi.nlm.nih.gov/pubmed/10595829. Accessed 28 May 2014.

64. Jasoni CL, Todman MG, Han S-K, Herbison AE. Expression of mRNAs encoding receptors that mediate stress signals in gonadotropin-releasing hormone neurons of the mouse. Neuroendocrinology. 2005;82(5–6):320–8. doi:10.1159/000093155.
65. Ahima RS, Harlan RE. Glucocorticoid receptors in LHRH neurons. Neuroendocrinology. 1992;56(6):845–50. http://www.ncbi.nlm.nih.gov/pubmed/1369593. Accessed 27 May 2014.
66. Dondi D, Piccolella M, Messi E, et al. Expression and differential effects of the activation of glucocorticoid receptors in mouse gonadotropin-releasing hormone neurons. Neuroendocrinology. 2005;82(3–4):151–63. doi:10.1159/000091693.
67. DeFranco DB, Attardi B, Chandran UR. Glucocorticoid receptor-mediated repression of GnRH gene expression in a hypothalamic GnRH-secreting neuronal cell line. Ann N Y Acad Sci. 1994;746:473–5. http://www.ncbi.nlm.nih.gov/pubmed/7825918. Accessed 28 May 2014.
68. Tellam DJ, Perone MJ, Dunn IC, et al. Direct regulation of GnRH transcription by CRF-like peptides in an immortalized neuronal cell line. Neuroreport. 1998;9(14):3135–40. doi:10.1097/00001756-199810050-00003.
69. Tellam DJ, Mohammad YN, Lovejoy DA. Molecular integration of hypothalamo-pituitary-adrenal axis-related neurohormones on the GnRH neuron. 2011. http://www.nrcresearchpress.com/doi/abs/10.1139/o00-060#.U4Z5rlhdX-Y. Accessed 29 May 2014.
70. Mellon PL, Windle JJ, Goldsmith PC, Padula CA, Roberts JL, Weiner RI. Immortalization of hypothalamic GnRH by genetically targeted tumorigenesis. Neuron. 1990;5(1):1–10. doi:10.1016/0896-6273(90)90028-E.
71. Wetsel WC, Mellon PL, Weiner RI, Negro-Vilar A. Metabolism of pro-luteinizing hormone-releasing hormone in immortalized hypothalamic neurons. Endocrinology. 1991;129(3):1584–95. doi:10.1210/endo-129-3-1584.
72. Chandran UR, Attardi B, Friedman R, Dong KW, Roberts JL, DeFranco DB. Glucocorticoid receptor-mediated repression of gonadotropin-releasing hormone promoter activity in GT1 hypothalamic cell lines. Endocrinology. 1994;134(3):1467–74. doi:10.1210/endo.134.3.8119188.
73. Attardi B, Tsujii T, Friedman R, et al. Glucocorticoid repression of gonadotropin-releasing hormone gene expression and secretion in morphologically distinct subpopulations of GT1-7 cells. Mol Cell Endocrinol. 1997;131(2):241–55. http://www.ncbi.nlm.nih.gov/pubmed/9296383. Accessed 28 May 2014.
74. Gore A, Attardi B, DeFranco D. Glucocorticoid repression of the reproductive axis: effects on GnRH and gonadotropin subunit mRNA levels. Mol Cell Endocrinol. 2006;256(1-2):40–8. Papers
75. Kinsey-Jones J, Li X, Knox A, et al. Down-regulation of hypothalamic kisspeptin and its receptor, Kiss1r, mRNA expression is associated with stress-induced suppression of luteinising hormone secretion in the female rat. J Neuroendocrinol. 2009;21(1):20–29. file:///Users/annageraghty/Documents/Papers/2009/Kinsey-Jones/Journal of Neuroendocrinology 2009 Kinsey-Jones-1.pdf.
76. Iwasa T, Matsuzaki T, Murakami M, et al. Decreased expression of kisspeptin mediates acute immune/inflammatory stress-induced suppression of gonadotropin secretion in female rat. J Endocrinol Invest. 2008;31(7):656–9. http://europepmc.org/abstract/MED/18787387. Accessed 28 May 2014.
77. Grachev P, Li XF, O'Byrne K. Stress regulation of kisspeptin in the modulation of reproductive function. Adv Exp Med Biol. 2013;784:431–54. doi:10.1007/978-1-4614-6199-9_20.
78. Takumi K, Iijima N, Higo S, Ozawa H. Immunohistochemical analysis of the colocalization of corticotropin-releasing hormone receptor and glucocorticoid receptor in kisspeptin neurons in the hypothalamus of female rats. Neurosci Lett. 2012;531(1):40–5. doi:10.1016/j.neulet.2012.10.010.
79. Grachev P, Li XF, Hu MH, et al. Neurokinin B signaling in the female rat: a novel link between stress and reproduction. Endocrinology. 2014;155(7):2589–601. doi:10.1210/en.2013-2038.
80. Goodman RL, Hileman SM, Nestor CC, et al. Kisspeptin, neurokinin B, and dynorphin act in the arcuate nucleus to control activity of the GnRH pulse generator in ewes. Endocrinology. 2013;154(11):4259–69. doi:10.1210/en.2013-1331.

81. Okamura H, Tsukamura H, Ohkura S, Uenoyama Y, Wakabayashi Y, Maeda K. Kisspeptin and GnRH pulse generation. Adv Exp Med Biol. 2013;784:297–323. doi:10.1007/978-1-4614-6199-9_14.
82. Wakabayashi Y, Yamamura T, Sakamoto K, Mori Y, Okamura H. Electrophysiological and morphological evidence for synchronized GnRH pulse generator activity among Kisspeptin/neurokinin B/dynorphin A (KNDy) neurons in goats. J Reprod Dev. 2013;59(1):40–8. http://www.pubmedcentral.nih.gov/articlerender.fcgi?artid=3943231&tool=pmcentrez&rendertype=abstract. Accessed 28 May 2014.
83. Bowe JE, Li XF, Kinsey-Jones JS, et al. Calcitonin gene-related peptide-induced suppression of luteinizing hormone pulses in the rat: the role of endogenous opioid peptides. J Physiol. 2005;566(Pt 3):921–8. doi:10.1113/jphysiol.2005.085662.
84. Nabeshima T, Katoh A, Wada M, Kameyama T. Stress-induced changes in brain Met-enkephalin, Leu-enkephalin and dynorphin concentrations. Life Sci. 1992;51(3):211–7. http://www.ncbi.nlm.nih.gov/pubmed/1352028. Accessed 28 May 2014.
85. Petraglia F, Vale W, Rivier C. Opioids act centrally to modulate stress-induced decrease in luteinizing hormone in the rat. Endocrinology. 1986;119(6):2445–50. doi:10.1210/endo-119-6-2445.
86. Ukena K, Iwakoshi E, Minakata H, Tsutsui K. A novel rat hypothalamic RFamide-related peptide identified by immunoaffinity chromatography and mass spectrometry. FEBS Lett. 2002;512(1–3):255–8. http://www.ncbi.nlm.nih.gov/pubmed/11852091. Accessed 9 Sept 2013.
87. Sari IP, Rao A, Smith JT, Tilbrook AJ, Clarke IJ. Effect of RF-amide-related peptide-3 on luteinizing hormone and follicle-stimulating hormone synthesis and secretion in ovine pituitary gonadotropes. Endocrinology. 2009;150(12):5549–56. papers://cf7c60b8-94a1-4c79-88e4-6c57345fd583/Paper/p1243.
88. Kirby ED, Geraghty AC, Ubuka T, Bentley GE, Kaufer D. Stress increases putative gonadotropin inhibitory hormone and decreases luteinizing hormone in male rats. Proc Natl Acad Sci U S A. 2009;106(27):11324–9. papers://cf7c60b8-94a1-4c79-88e4-6c57345fd583/Paper/p1273.
89. Lee Son Y, Ubuka T, Narihiro M, et al. Molecular basis for the activation of gonadotropin-inhibitory hormone gene transcription by corticosterone. Endocrinology. 2014;155(5):1817–26. doi:10.1210/en.2013-2076.
90. Gojska NM, Belsham DD. Glucocorticoid receptor-mediated regulation of Rfrp (GnIH) and Gpr147 (GnIH-R) synthesis in immortalized hypothalamic neurons. Mol Cell Endocrinol. 2014;384(1–2):23–31. doi:10.1016/j.mce.2013.12.015.
91. Geraghty A, Muroy S, Zhao S, Bentley G, Kriegsfeld L, Kaufer D. Chronic stress causes an increase in RFRP expression and leads to reproductive dysfunction in the adult female rat. [abstract]. In: 2013 Neuroscience Meet Plan, Society for Neuroscience; 2013.
92. Kaiser UB, Jakubowiak A, Steinberger A, Chin WW. Differential effects of gonadotropin-releasing hormone (GnRH) pulse frequency on gonadotropin subunit and GnRH receptor messenger ribonucleic acid levels in vitro. Endocrinology. 1997;138(3):1224–31. doi:10.1210/endo.138.3.4968.
93. Vale W, Rivier C, Brown M. Regulatory peptides of the hypothalamus. Annu Rev Physiol. 1977;39:473–527. doi:10.1146/annurev.ph.39.030177.002353.
94. Pierce JG, Parsons TF. Glycoprotein hormones: structure and function. Annu Rev Biochem. 1981;50:465–95. doi:10.1146/annurev.bi.50.070181.002341.
95. Kononen J, Honkaniemi J, Gustafsson JA, Pelto-Huikko M. Glucocorticoid receptor colocalization with pituitary hormones in the rat pituitary gland. Mol Cell Endocrinol. 1993;93(1):97–103. http://www.ncbi.nlm.nih.gov/pubmed/8319836. Accessed 28 May 2014.
96. Breen KM, Thackray VG, Hsu T, Mak-McCully RA, Coss D, Mellon PL. Stress levels of glucocorticoids inhibit LHβ-subunit gene expression in gonadotrope cells. Mol Endocrinol. 2012;26(10):1716–31. doi:10.1210/me.2011-1327.

97. Armario A, Lopez-Calderon A, Jolin T, Balasch J. Response of anterior pituitary hormones to chronic stress. The specificity of adaptation. Neurosci Biobehav Rev. 1986;10(3):245–50. doi:10.1016/0149-7634(86)90011-4.
98. Lopez-Calderon A, Gonzalez-Quijano MI, Tresguerres JAF, Ariznavarreta C. Role of LHRH in the gonadotrophin response to restraint stress in intact male rats. J Endocrinol. 1990; 124(2):241–6. doi:10.1677/joe.0.1240241.
99. Blake CA. Effects of "stress" on pulsatile luteinizing hormone release in ovariectomized rats. Proc Soc Exp Biol Med. 1975;148(3):813–5. http://www.ncbi.nlm.nih.gov/pubmed/165534. Accessed 29 May 2014.
100. Kamel F, Kubajak CL. Modulation of gonadotropin secretion by corticosterone: interaction with gonadal steroids and mechanism of action. Endocrinology. 1987;121(2):561–8. doi:10.1210/endo-121-2-561.
101. Baldwin DM, Sawyer CH. Effects of dexamethasone on LH release and ovulation in the cyclic rat. Endocrinology. 1974;94(5):1397–403. doi:10.1210/endo-94-5-1397.
102. Collu R, Taché Y, Ducharme J. Hormonal modifications induced by chronic stress in rats. J Steroid Biochem. 1979;11(1):989–1000. doi:10.1016/0022-4731(79)90042-6.
103. Vreeburg JT, de Greef WJ, Ooms MP, van Wouw P, Weber RF. Effects of adrenocorticotropin and corticosterone on the negative feedback action of testosterone in the adult male rat. Endocrinology. 1984;115(3):977–83. doi:10.1210/endo-115-3-977.
104. Taché Y, Ducharme JR, Charpenet G, Haour F, Saez J, Collu R. Effect of chronic intermittent immobilization stress on hypophyso-gonadal function of rats. Acta Endocrinol (Copenh). 1980;93(2):168–74. http://www.ncbi.nlm.nih.gov/pubmed/7376788. Accessed 28 May 2014.
105. Briski KP, Sylvester PW. Differential impact of naltrexone on luteinizing hormone release during single versus repetitive exposure to restraint stress. Psychoneuroendocrinology. 1992;17(2–3):125–33. http://www.ncbi.nlm.nih.gov/pubmed/1332097. Accessed 28 May 2014.
106. Li XF, Edward J, Mitchell JC, et al. Differential effects of repeated restraint stress on pulsatile lutenizing hormone secretion in female Fischer, Lewis and Wistar rats. J Neuroendocrinol. 2004;16(7):620–7. doi:10.1111/j.1365-2826.2004.01209.x.
107. Briski KP, Sylvester PW. Effects of repetitive daily acute stress on pituitary LH and prolactin release during exposure to the same stressor or a second novel stress. Psychoneuroendocrinology. 1987;12(6):429–37. http://www.ncbi.nlm.nih.gov/pubmed/3441582. Accessed 28 May 2014.
108. Sakakura M, Takebe K, Nakagawa S. Inhibition of luteinizing hormone secretion induced by synthetic LRH by long-term treatment with glucocorticoids in human subjects. J Clin Endocrinol Metab. 1975;40(5):774–9. doi:10.1210/jcem-40-5-774.
109. Du Ruisseau P, Taché Y, Brazeau P, Collu R. Effects of chronic immobilization stress on pituitary hormone secretion, on hypothalamic factor levels, and on pituitary responsiveness to LHRH and TRH in female rats. Neuroendocrinology. 1979;29(2):90–9. http://www.ncbi.nlm.nih.gov/pubmed/116141. Accessed 9 Sept 2013.
110. Rivier C, Vale W. Effect of the long-term administration of corticotropin-releasing factor on the pituitary-adrenal and pituitary-gonadal axis in the male rat. J Clin Invest. 1985;75(2):689–94. doi:10.1172/JCI111748.
111. Suter DE, Schwartz NB, Ringstrom SJ. Dual role of glucocorticoids in regulation of pituitary content and secretion of gonadotropins. Am J Physiol. 1988;254(5 Pt 1):E595–600. http://ajpendo.physiology.org/content/254/5/E595.abstract. Accessed 20 May 2014.
112. Baldwin DM. The effect of glucocorticoids on estrogen-dependent luteinizing hormone release in the ovariectomized rat and on gonadotropin secretin in the intact female rat. Endocrinology. 1979;105(1):120–8. doi:10.1210/endo-105-1-120.
113. Maya-Núñez G, Conn PM. Transcriptional regulation of the GnRH receptor gene by glucocorticoids. Mol Cell Endocrinol. 2003;200(1–2):89–98. doi:10.1016/S0303-7207(02)00419-7.
114. Kotitschke A, Sadie-Van Gijsen H, Avenant C, Fernandes S, Hapgood JP. Genomic and non-genomic cross talk between the gonadotropin-releasing hormone receptor and glucocorticoid receptor signaling pathways. Mol Endocrinol. 2009;23(11):1726–45. doi:10.1210/me.2008-0462.

115. Bronson FH. Establishment of social rank among grouped male mice: relative effects on circulatingFSH,LH,andcorticosterone.PhysiolBehav.1973;10(5):947–51.doi:10.1016/0031-9384(73)90065-6.
116. Du Ruisseau P, Taché Y, Brazeau P, Colin R. Effects of chronic immobilization stress on pituitary hormone secretion, on hypothalamic factor levels, and on pituitary responsiveness to LHRH and TRH in female rats. Neuroendocrinology. 1979;29(2):90–9. doi:10.1159/000122910.
117. Ringstrom SJ, Schwartz NB. Differential effect of glucocorticoids on synthesis and secretion of luteinizing hormone (LH) and follicle stimulating hormone (FSH). J Steroid Biochem. 1987;27(1–3):625–30. doi:10.1016/0022-4731(87)90362-1.
118. Ringstrom SJ, McAndrews JM, Rahal JO, Schwartz NB. Cortisol in vivo increases FSH beta mRNA selectively in pituitaries of male rats. Endocrinology. 1991;129(5):2793–5. doi:10.1210/endo-129-5-2793.
119. Thackray VG, McGillivray SM, Mellon PL. Androgens, progestins, and glucocorticoids induce follicle-stimulating hormone beta-subunit gene expression at the level of the gonadotrope. Mol Endocrinol. 2006;20(9):2062–79. doi:10.1210/me.2005-0316.
120. Smals AG, Kloppenborg PW, Benraad TJ. Plasma testosterone profiles in Cushing's syndrome. J Clin Endocrinol Metab. 1977;45(2):240–5. doi:10.1210/jcem-45-2-240.
121. Whirledge S, Cidlowski JA. Glucocorticoids, stress, and fertility. Minerva Endocrinol. 2010;35(2):109–25. http://www.pubmedcentral.nih.gov/articlerender.fcgi?artid=3547681&tool=pmcentrez&rendertype=abstract. Accessed 20 May 2014.
122. Schultz R, Isola J, Parvinen M, et al. Localization of the glucocorticoid receptor in testis and accessory sexual organs of male rat. Mol Cell Endocrinol. 1993;95(1–2):115–20. doi:10.1016/0303-7207(93)90036-J.
123. Silva EJR, Queiróz DBC, Honda L, Avellar MCW. Glucocorticoid receptor in the rat epididymis: expression, cellular distribution and regulation by steroid hormones. Mol Cell Endocrinol. 2010;325(1–2):64–77. doi:10.1016/j.mce.2010.05.013.
124. Bernier M, Gibb W, Collu R, Ducharme JR. Effect of glucocorticoids on testosterone production by porcine Leydig cells in primary culture. Can J Physiol Pharmacol. 1984;62(9):1166–9. doi:10.1139/y84-195.
125. Orr T. Effects of restraint stress on plasma LH and testosterone concentrations, Leydig cell LH/HCG receptors, and in vitro testicular steroidogenesis in adult rats. Horm Behav. 1990;24(3):324–41. doi:10.1016/0018-506X(90)90013-N.
126. Orr T. Role of glucocorticoids in the stress-induced suppression of testicular steroidogenesis in adult male rats. Horm Behav. 1992;26(3):350–63. doi:10.1016/0018-506X(92)90005-G.
127. Cumming DC, Quigley ME, Yen SS. Acute suppression of circulating testosterone levels by cortisol in men. J Clin Endocrinol Metab. 1983;57(3):671–3. doi:10.1210/jcem-57-3-671.
128. Marić D, Kostić T, Kovačević R. Effects of acute and chronic immobilization stress on rat Leydig cell steroidogenesis. J Steroid Biochem Mol Biol. 1996;58(3):351–5.
129. Saez JM, Morera AM, Haour F, Evain D. Effects of in vivo administration of dexamethasone, corticotropin and human chorionic gonadotropin on steroidogenesis and protein and DNA synthesis of testicular interstitial cells in prepuberal rats. Endocrinology. 1977;101(4):1256–63. doi:10.1210/endo-101-4-1256.
130. Bambino TH, Hsueh AJ. Direct inhibitory effect of glucocorticoids upon testicular luteinizing hormone receptor and steroidogenesis in vivo and in vitro. Endocrinology. 1981;108(6):2142–8. doi:10.1210/endo-108-6-2142.
131. Hales DB, Payne AH. Glucocorticoid-mediated repression of P450scc mRNA and de novo synthesis in cultured Leydig cells. Endocrinology. 1989;124(5):2099–104. doi:10.1210/endo-124-5-2099.
132. Payne AH, Sha LL. Multiple mechanisms for regulation of 3 beta-hydroxysteroid dehydrogenase/delta 5—delta 4-isomerase, 17 alpha-hydroxylase/C17-20 lyase cytochrome P450, and cholesterol side-chain cleavage cytochrome P450 messenger ribonucleic acid levels in primary cultures of mouse Leydig cells. Endocrinology. 1991;129(3):1429–35. doi:10.1210/endo-129-3-1429.

133. Martin LJ, Tremblay JJ. Glucocorticoids antagonize cAMP-induced Star transcription in Leydig cells through the orphan nuclear receptor NR4A1. J Mol Endocrinol. 2008;41(3): 165–75. doi:10.1677/JME-07-0145.
134. Sasagawa I, Yazawa H, Suzuki Y, Nakada T. Stress and testicular germ cell apoptosis. 2009. http://informahealthcare.com/doi/abs/10.1080/014850101753145924. Accessed 20 May 2014.
135. Yazawa H. Apoptosis of testicular germ cells induced by exogenous glucocorticoid in rats. Hum Reprod. 2000;15(9):1917–20. doi:10.1093/humrep/15.9.1917.
136. Gao H-B, Tong M-H, Hu Y-Q, Guo Q-S, Ge R, Hardy MP. Glucocorticoid induces apoptosis in rat Leydig cells. 2013. http://press.endocrine.org/doi/abs/10.1210/endo.143.1.8604?url_ver=Z39.88-2003&rfr_id=ori:rid:crossref.org&rfr_dat=cr_pub=pubmed. Accessed 20 May 2014.
137. Almeida SA, Petenusci SO, Anselmo-Franci JA, Rosa-e-Silva AAM, Lamano-Carvalho TL. Decreased spermatogenic and androgenic testicular functions in adult rats submitted to immobilization-induced stress from prepuberty. Braz J Med Biol Res. 1998;31(11):1443–8. doi:10.1590/S0100-879X1998001100013.
138. Zorn B, Auger J, Velikonja V, Kolbezen M, Meden-Vrtovec H. Psychological factors in male partners of infertile couples: relationship with semen quality and early miscarriage. Int J Androl. 2008;31(6):557–64. doi:10.1111/j.1365-2605.2007.00806.x.
139. Seckl JR. 11Beta-hydroxysteroid dehydrogenase in the brain: a novel regulator of glucocorticoid action? Front Neuroendocrinol. 1997;18(1):49–99. doi:10.1006/frne.1996.0143.
140. Seckl JR, Walker BR. Minireview: 11beta-hydroxysteroid dehydrogenase type 1—a tissue-specific amplifier of glucocorticoid action. Endocrinology. 2001;142(4):1371–6. doi:10.1210/endo.142.4.8114.
141. Tetsuka M. Expression of 11 beta-hydroxysteroid dehydrogenase, glucocorticoid receptor, and mineralocorticoid receptor genes in rat ovary. Biol Reprod. 1999;60(2):330–5. doi:10.1095/biolreprod60.2.330.
142. Benediktsson R, Yau JLW, Brett LP, Cooke BE, Edwards CRW, Seckl JR. 11β-Hydroxysteroid dehydrogenase in the rat ovary: high expression in the oocyte. J Endocrinol. 1992;135(1):53–58. http://www.scopus.com/inward/record.url?eid=2-s2.0-0026757114&partnerID=tZOtx3y1.
143. McDonald SE, Henderson TA, Gomez-Sanchez CE, Critchley HOD, Mason JI. 11Beta-hydroxysteroid dehydrogenases in human endometrium. Mol Cell Endocrinol. 2006;248 (1–2):72–8. doi:10.1016/j.mce.2005.12.010.
144. Michael AE, Evagelatou M, Norgate DP, et al. Isoforms of 11β-hydroxysteroid dehydrogenase in human granulosa-lutein cells. Mol Cell Endocrinol. 1997;132(1–2):43–52. doi:10.1016/S0303-7207(97)00118-4.
145. Schreiber JR, Nakamura K, Erickson GF. Rat ovary glucocorticoid receptor: identification and characterization. Steroids. 1982;39(5):569–84. doi:10.1016/0039-128X(82)90057-5.
146. Michael A, Cooke B. A working hypothesis for the regulation of steroidogenesis and germ cell development in the gonads by glucocorticoids and 11β-hydroxysteroid dehydrogenase (11βHSD). Mol Cell Endocrinol. 1994;100(1–2):55–63. doi:10.1016/0303-7207(94) 90279-8.
147. Albiston AL, Smith RE, Krozowski ZS. Changes in the levels of 11β-hydroxysteroid dehydrogenase mRNA over the oestrous cycle in the rat. J Steroid Biochem Mol Biol. 1995;52(1):45–8. doi:10.1016/0960-0760(94)00154-E.
148. Hillier SG. Molecular biology of the female reproductive system. Amsterdam: Elsevier; 1994. p. 1–37. doi:10.1016/B978-0-08-091819-8.50005-9.
149. Hillier S, Tetsuka M. An anti-inflammatory role for glucocorticoids in the ovaries? J Reprod Immunol. 1998;39(1–2):21–7. doi:10.1016/S0165-0378(98)00011-4.
150. Michael AE, Pester LA, Curtis P, Shaw RW, Edwards CR, Cooke BA. Direct inhibition of ovarian steroidogenesis by cortisol and the modulatory role of 11 beta-hydroxysteroid dehydrogenase. Clin Endocrinol (Oxf). 1993;38(6):641–4. http://www.ncbi.nlm.nih.gov/pubmed/8334750. Accessed 29 May 2014.

151. Hsueh A. Glucocorticoid inhibition of FSH-induced estrogen production in cultured rat granulosa cells. Steroids. 1978;32(5):639–48. doi:10.1016/0039-128X(78)90074-0.
152. Schoonmaker JN, Erickson GF. Glucocorticoid modulation of follicle-stimulating hormone-mediated granulosa cell differentiation. Endocrinology. 1983;113(4):1356–63. doi:10.1210/endo-113-4-1356.
153. Yang J-G. Effects of glucocorticoids on maturation of pig oocytes and their subsequent fertilizing capacity in vitro. Biol Reprod. 1999;60(4):929–36. doi:10.1095/biolreprod60.4.929.
154. Jimena P, Castilla JA, Peran F, et al. Adrenal hormones in human follicular fluid. Eur J Endocrinol. 1992;127(5):403–6. doi:10.1530/acta.0.1270403.
155. González R, Ruiz-León Y, Gomendio M, Roldan ERS. The effect of glucocorticoids on ERK-1/2 phosphorylation during maturation of lamb oocytes and their subsequent fertilization and cleavage ability in vitro. Reprod Toxicol. 2010;29(2):198–205. doi:10.1016/j.reprotox.2009.10.009.
156. Andersen CY. Effect of glucocorticoids on spontaneous and follicle-stimulating hormone induced oocyte maturation in mouse oocytes during culture. J Steroid Biochem Mol Biol. 2003;85(2–5):423–7. doi:10.1016/S0960-0760(03)00190-0.
157. Liang B, Wei D-L, Cheng Y-N, et al. Restraint stress impairs oocyte developmental potential in mice: role of CRH-induced apoptosis of ovarian cells. Biol Reprod. 2013;89(3):64. doi:10.1095/biolreprod.113.110619.
158. Liu Y-X, Cheng Y-N, Miao Y-L, et al. Psychological stress on female mice diminishes the developmental potential of oocytes: a study using the predatory stress model. PLoS One. 2012;7(10), e48083. doi:10.1371/journal.pone.0048083.
159. Rabin DS. Glucocorticoids inhibit estradiol-mediated uterine growth: possible role of the uterine estradiol receptor. Biol Reprod. 1990;42(1):74–80. doi:10.1095/biolreprod42.1.74.
160. Bever AT, Hisaw FL, Velardo JT. Inhibitory action of desoxycorticosterone acetate, cortisone acetate, and testosterone on uterine growth induced by estradiol-17beta. Endocrinology. 1956;59(2):165–9. doi:10.1210/endo-59-2-165.
161. Johnson DC, Dey SK. Role of Histamine in implantation: dexamethasone inhibits estradiol-induced implantation in the rat. Biol Reprod. 1980;22(5):1136–1141. http://www.biolreprod.org/content/22/5/1136.abstract. Accessed 26 May 2014.
162. Zhao L-H, Cui X-Z, Yuan H-J, et al. Restraint stress inhibits mouse implantation: temporal window and the involvement of HB-EGF, estrogen and progesterone. PLoS One. 2013;8(11), e80472. doi:10.1371/journal.pone.0080472.
163. Chien EJ, Liao C-F, Chang C-P, et al. The non-genomic effects on Na+/H+-exchange 1 by progesterone and 20alpha-hydroxyprogesterone in human T cells. J Cell Physiol. 2007;211(2):544–50. doi:10.1002/jcp.20962.
164. Szekeres-Bartho J, Barakonyi A, Miko E, Polgar B, Palkovics T. The role of gamma/delta T cells in the feto-maternal relationship. Semin Immunol. 2001;13(4):229–33. doi:10.1006/smim.2000.0318.
165. Arck P, Hansen PJ, Mulac Jericevic B, Piccinni MP, Szekeres-Bartho J. Progesterone during pregnancy: endocrine-immune cross talk in mammalian species and the role of stress. Am J Reprod Immunol (New York, NY 1989). 2007;58(3):268–279. papers://cf7c60b8-94a1-4c79-88e4-6c57345fd583/Paper/p1437.
166. Szekeres-Bartho J, Wegmann TG. A progesterone-dependent immunomodulatory protein alters the Th1/Th2 balance. J Reprod Immunol. 1996;31(1–2):81–95. http://www.ncbi.nlm.nih.gov/pubmed/8887124. Accessed 29 May 2014.
167. Polgar B, Kispal G, Lachmann M, et al. Molecular cloning and immunologic characterization of a novel cDNA coding for progesterone-induced blocking factor. J Immunol. 2003;171(11):5956–63. http://www.ncbi.nlm.nih.gov/pubmed/14634107. Accessed 29 May 2014.
168. Wiebold JL, Stanfield PH, Becker WC, Hillers JK. The effect of restraint stress in early pregnancy in mice. Reproduction. 1986;78(1):185–92. doi:10.1530/jrf.0.0780185.

169. Joachim R, Zenclussen AC, Polgar B, et al. The progesterone derivative dydrogesterone abrogates murine stress-triggered abortion by inducing a Th2 biased local immune response. Steroids. 2003;68(10–13):931–40. doi:10.1016/j.steroids.2003.08.010.
170. Blois SM, Joachim R, Kandil J, et al. Depletion of CD8+ cells abolishes the pregnancy protective effect of progesterone substitution with dydrogesterone in mice by altering the Th1/Th2 cytokine profile. J Immunol. 2004;172(10):5893–9. doi:10.4049/jimmunol.172.10.5893.
171. Kalinka J, Szekeres-Bartho J. The impact of dydrogesterone supplementation on hormonal profile and progesterone-induced blocking factor concentrations in women with threatened abortion. Am J Reprod Immunol. 2005;53(4):166–71. doi:10.1111/j.1600-0897.2005.00261.x.
172. Raghupathy R, Al Mutawa E, Makhseed M, Azizieh F, Szekeres-Bartho J. Modulation of cytokine production by dydrogesterone in lymphocytes from women with recurrent miscarriage. BJOG. 2005;112(8):1096–101. doi:10.1111/j.1471-0528.2005.00633.x.

Chapter 12
Glucocorticoids and the Lung

Anthony N. Gerber

Abstract The lung is a major clinical target of glucocorticoid-based therapeutics, and GR signaling has broad effects on respiratory physiology and inflammation. During lung development, expression of GR in the mesenchyme is required for normal terminal alveolar epithelial differentiation. Prenatal administration of exogenous glucocorticoids (GCs) to prevent neonatal respiratory distress syndrome, however, promotes alveolar maturation and accelerates surfactant expression in a manner consistent with direct effects on the developing alveolar epithelium. Likewise, cell autonomous effects of GCs in regulating gene expression and phenotype of the airway epithelium and airway smooth muscle have been demonstrated to control important therapeutic effects of GCs in treating asthma and chronic obstructive pulmonary disease. Here, mechanisms and consequences of GR signaling in the developing lung and in treating obstructive lung disease are reviewed, with a focus on direct effects of GR signaling on alveolar differentiation, surfactant expression, and airway epithelial and smooth muscle pathophysiology.

Keywords Glucocorticoid receptor • Surfactant • Asthma • Airway smooth muscle • Airway epithelium • Feed-forward loop

Introduction: Glucocorticoids and the Lung

Glucocorticoids are used to treat a wide variety of lung disease, ranging from maternal administration to prevent newborn respiratory distress syndrome of prematurity [1], to their dominant role in treating inflammatory lung diseases, such as asthma, in both children and adults [2]. A great deal of research has therefore focused on the effects of exogenous glucocorticoids on various aspects of lung biology, and our overall understanding of glucocorticoid signaling in the lung is largely informed by *in vitro* and *in vivo* studies on the effects of supplemental, generally synthetic, GR agonists on a subset of lung cell and tissue types, and pulmonary developmental

A.N. Gerber, M.D., Ph.D. (✉)
Department of Medicine, National Jewish Health, University of Colorado, Denver, 1400 Jackson Street, Room K621b, Denver, CO 80206, USA
e-mail: gerbera@njhealth.org

J.-C. Wang, C. Harris (eds.), *Glucocorticoid Signaling*, Advances in Experimental Medicine and Biology 872, DOI 10.1007/978-1-4939-2895-8_12

processes. Reflecting clinical use, particular areas of focus have been aimed at understanding how glucocorticoids regulate lung maturation and surfactant expression, and also how glucocorticoids exert therapeutic effects on airway inflammation, notably in airway epithelial and smooth muscle cells. In this chapter, I will review the literature that has established a key role for supplemental glucocorticoids in promoting airway maturation and surfactant expression, and I will provide an overview of our understanding of the molecular mechanisms underpinning these effects. I will also summarize recent loss of function studies that have indicated a seemingly contradictory role for endogenous GR signaling in the developing lung. In addition, I will review the use of glucocorticoids in the treatment of obstructive lung disease, and summarize the large and growing body of literature exploring the mechanisms responsible for therapeutic effects of glucocorticoids in airway inflammation. Important areas for future research will also be discussed.

Glucocorticoids and Neonatal Respiratory Distress Syndrome

In 1969, Dr. Graham Liggins published a landmark paper entitled Premature Delivery of Foetal Lambs Infused With Glucocorticoid [3]. In this study, Liggins found that injection of pre-term pregnant ewes with dexamethasone resulted in the delivery of viable premature lambs. Lungs from these animals were inflated after birth, in contrast to a complete lack of aeration in control animals. This observation led Liggins to speculate that glucocorticoid (GC) signaling promoted surfactant expression, thus allowing postnatal maintenance of alveolar integrity. Based on this hypothesis, Liggins and his colleague, Dr. Ross Howie, conducted a remarkable randomized clinical trial in which betamethasone or control was infused antepartum to 282 women with threatened premature labor or planned early delivery for medical reasons (e.g. pre-eclampsia). The primary endpoint of the trial was the occurrence of neonatal respiratory distress syndrome (RDS), which is a major complication of preterm birth that has high associated morbidity and mortality. In infants born to the subset of mothers with premature labor who were treated with betamethasone, the rate of RDS was 9 %, while the rate of RDS in babies born to mothers with premature labor that were in the control arm was 25.8 % ($P<0.03$) [4]. The beneficial effect was confined to babies of less than 32 weeks gestational age. This extraordinary finding, which has been replicated in numerous subsequent studies [5], led to the prevention of hundreds of thousand of potentially fatal cases of neonatal respiratory distress syndrome and transformed the standard of care for pre-term labor.

Glucocorticoid Signaling, Surfactogenesis, and Lung Maturation

The potent clinical effect of maternal GC administration in reducing the incidence of newborn RDS following premature birth has spawned substantial research into the mechanisms responsible for the effects of GCs on lung maturation and function.

GR is highly expressed in the developing lung. Administration of exogenous GCs to pregnant rats, rabbits and monkeys, among other animals, leads to accelerated transition through the pseudoglandular phase of lung development, in association with premature terminal differentiation [6–8], precocious secretion of surfactant, and the initiation of structural changes required for gas exchange, most notably alveolar thinning [9–11]. These effects of GCs on terminal lung maturation *in vivo* are reflected *in vitro* in both explants from developing lungs and in cultured cell models. In human lung explants, glucocorticoids induce expression of surfactant B (*SFTPB*) and surfactant C (*SFTPB*) mRNA, and have a complex biphasic effect on surfactant A (*SFTPA*) expression [12–14]. In cultured Type II alveolar cells, exogenous GCs induce the expression of numerous markers terminal differentiation of the respiratory epithelium [15], including TTF-1 (also known as *NKX2-1*), a transcription factor that is implicated in directly regulating surfactant C expression [16], and the expression of other genes associated with mature alveolar epithelium. The effects of GCs on isolated alveolar epithelial cells thus appear to recapitulate many of the important effects of exogenous GCs *in vivo*, supporting an epithelial cell autonomous model for the effects of GCs on surfactogenesis in which ligand-activated GR in the developing epithelium drives terminal differentiation.

Glucocorticoids and Surfactant Expression: Molecular Mechanisms

A number of studies have addressed putative epithelial-autonomous mechanisms through which GR signaling induces surfactant gene expression and promotes differentiation. Similar to explants, which include both epithelial and mesenchymal components, exposure of isolated cultured fetal alveolar type II cells to a differentiation cocktail containing dexamethasone led to *SFTPB* and *SFTPC* mRNA levels increasing over time [17]. In contrast, induction of *SFTPA* expression by treatment with cAMP and IL1B was inhibited by GCs such as dexamethasone [18]; these dichotomous effects are concordant with complex biphasic changes in *SFTPA* mRNA levels that have been noted in association with GC treatment in explants [13, 19]. Repression of *SFTPA* by GCs occurs both through down-regulation of *SFTPA* transcription and through post-transcriptional destabilization of *SFTPA* mRNA [18, 20, 21]. Work from Mendelson and colleagues, encompassed in several publications, indicates that direct repression of *SFTPA* transcription by GCs results from agonist-bound GR interacting with a composite *TTF-1/NF-kB* regulatory element upstream of the *SFTPA* start site [20, 22, 23]. This TTF binding element has been implicated in the induction of *SFTPA* by both TTF-1 and NF-kB in association with histone modifications indicative of active transcription. Activation of GR signaling modified the histone acetylation pattern at the *SFTPA* locus to marks reflective of transcriptional repression [18]. This suggests that antagonism of TTF-1 and NF-kB activity at the *SFTPA* promoter by GR occurs through canonical trans-repression mechanisms in which GR tethering to a TTF-1/NF-kB complex results in altered co-regulator recruitment and assembly of a repressive complex [24, 25].

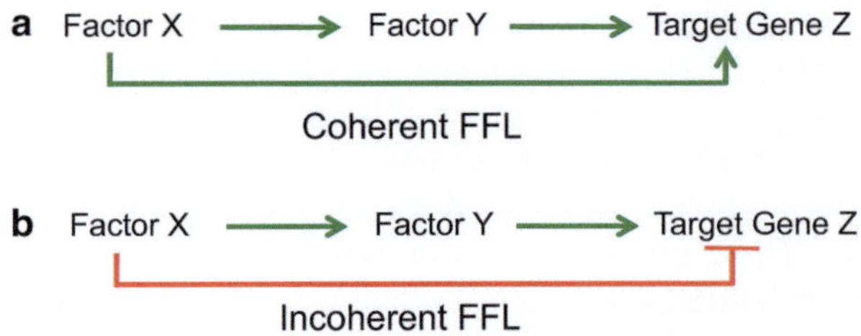

Fig. 12.1 The structures of the Type 1 coherent and incoherent feed-forward loop (FFL) as indicated. (**a**) In a coherent FFL, factor A induces factor B, and both factors exert the same effects on expression of a third downstream gene, referred to here as target gene Z. (**b**) In an incoherent FFL, the two factors exert opposing effects on target gene Z expression

The transrepression model for the effects of GR on *SFTPA* expression is, however, complicated by data indicating that treatment of Type II cells with GCs results in increased *TTF-1* expression [26]. Thus, GR appears to exert an inductive effect on *SFTPA* expression through promoting *TTF-1* expression, while also directly antagonizing TTF-1 function at the *SFTPA* promoter. These opposing effects provide a possible mechanistic basis for the complex biphasic expression response of *SFTPA* to GCs. The pattern of seemingly antagonistic regulation of *SFTPA* expression by GR is consistent with GR and TTF-1 comprising an incoherent feed-forward loop that controls *SFTPA* expression. The general circuit design of coherent and incoherent feed-forward loops are depicted in Fig. 12.1. Feed-forward loops are widely utilized in bacteria and yeast to confer specific properties to downstream gene expression, including delayed induction and biphasic expression patterns [27, 28]. Two recent studies have indicated that feed-forward logic may result in similar expression patterns in response to GCs [29, 30], with incoherent feed-forward regulatory logic implicated as the mechanism underlying biphasic expression responses of several genes to GCs in U2OS and A549 cells. A theoretical GR:TTF-1 incoherent feed-forward loop controlling *SFTPA* expression is shown in Fig. 12.2. Whether GR directly induces *TTF-1*, thus fulfilling the formal requirements for GR:TTF1 feed-forward control of *SFTPA*, remains to be determined.

Similar to *SFTPA*, *SFTPB* and *SFTPC* are both transcriptional targets of TTF-1 [31–33]. This suggests that the inductive effect of GCs on *TTF-1* expression is at least one mechanism through which GR signaling promotes *SFTPB* and *SFTPC* expression in a cell-autonomous manner during alveolar epithelial differentiation. It is likely, however, that GCs exert additional effects on *SFTPB* and *SFTPC* expression beyond a simple linear cascade in which GR induces *TTF-1*, with TTF-1 in turn regulating *SFTPB* and *SFTPC*. In elegant work by Kolla *et al*, the effects of viral-driven TTF-1 over-expression on various markers of differentiated alveolar epithelium were assessed [17]. In this study, *TTF-1* over-expression increased *SFTPB* and *SFTPC* expression. However, expression of both *SFTPB* and *SFTPC* was further increased when a dexamethasone containing cocktail was added to the culture medium. The mechanisms underlying the additional induction of *SFTPB* and *SFTPC* expression have yet to be established. One possibility is that GR, in addition

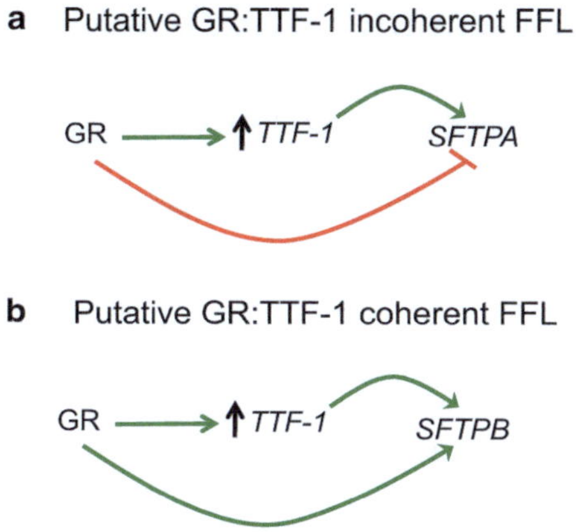

Fig. 12.2 Putative feed-forward regulation of *SFTPA* and *SFTPB* by GR and TTF-1. (**a**) In this model, GR induces *TTF-1* expression, which induces *SFTPA*. However, GR also negatively regulates *SFTPA* expression, thus creating a feed-forward loop that may be responsible for biphasic expression patterns of SFTPA. (**b**) In this case, GR induces *TTF-1* and both GR and TTF-1 induce *SFTPB* expression

to inducing *TTF-1*, also enhances the expression of a second transcriptional driver of surfactant expression. An alternative, but not mutually exclusive possibility, is that GR itself directly regulates *SFTPB* and *SFTPC* expression, possibly in cooperation with TTF-1. ChIP-seq data from Reddy et al. suggest that this may be the case [34]. In this study, A549 cells, an alveolar adenocarcinoma-derived cell line, were treated with dexamethasone, and genome wide GR binding events were assayed using chromatin immunoprecipitation (ChIP) followed by deep sequencing. Identified binding sites were subsequently annotated in the public domain through ENCODE and can be visualized on the UCSC genome browser [35]. Scanning of the *SFTPB* and *SFTPC* loci in the UCSC genome browser revealed GR binding in regions upstream of both the *SFTPB* and *SFTPC* promoters. These putative binding sites have not been validated by ChIP-PCR, however, the original publication showed that a high percentage of randomly selected GR binding sites identified by sequencing could also be detected through standard PCR-based analysis [34]. A second limitation of this study in the context of understanding surfactant regulation is that the interactions between GR and the *SFTPB* and *SFTPC* loci were detected in A549 cells, and thus may not reflect events in the immature respiratory epithelium. In addition, GR binding to these areas has not been functionally implicated in al regulation of *SFTPB* and *SFTPC*. Nevertheless, despite these caveats, the A549-based GR binding data support a model in which GR and TTF1 regulate *SFTPB* and *SFTPC* expression though a coherent feed-forward loop system (see Figs. 12.1 and 12.2b). Establishing whether GR:TTF1 coherent feed-forward loops regulate *SFTPB* and *SFTPC* expression could be accomplished through temporal analysis of genome wide binding of TTF1 and GR during alveolar epithelial cell differentiation, along with functional characterization of any binding sites identified within the *SFTPB* and *SFTPC* loci.

Function of Endogenous GR in Lung Development

The data reviewed above strongly suggest that exogenous administration of glucocorticoids promotes surfactant expression and epithelial maturation through epithelial cell-autonomous mechanisms. In further support of this notion, deletion of GR in the lung epithelium during murine development reduced, but did not eliminate surfactant expression [36], and also reduced viability. However, additional experiments comparing systemic and mesenchymal restricted knockout of GR in mice indicate that the dominant physiological effects of endogenous GR signaling during lung development likely occur through mesenchymally expressed GR. Systemic knockout of the murine GR resulted in highly penetrant perinatal lethality secondary to defective lung development. In particular, there was a marked reduction in alveolar Type I cell number and surfactant expression, indicating a defect in terminal alveolar differentiation [37]. In an initially surprising finding, given the important effects of exogenous GCs on epithelial maturation, the pulmonary phenotype of $GR^{-/-}$ mice was largely recapitulated with tissue-specific removal of GR from the mesenchyme [38, 39]. For example, in one recent study, germline deletion of GR was compared with conditional removal of GR in the endothelium, mesenchyme and epithelium. Mice lacking GR in the mesenchyme were found to display lethal effects similar to germline GR knockout, while the epithelial and endothelial KO strains did not exhibit perinatal respiratory failure [40]. Although there was reduced viability in a third publication that reported on epithelial deletion of GR [36], and hyper-cellularity has been observed consistently with epithelial GR deletion, in aggregate, the data indicate that normal lung development and surfactogenesis has a greater requirement for GR activity in the mesenchyme than within the developing airway epithelium. Thus, the effects of exogenous GCs, which can drive cell autonomous maturation of the epithelium, and the physiologic role of endogenous GR signaling in murine lung development, which appears to predominantly depend on GR activity in the mesenchyme, are somewhat divergent.

Efforts to explain this non-cell autonomous role of GR signaling on epithelial development have focused primarily on well-known, obligate signaling between the epithelial and mesenchyme layers that mediates branching morphogenesis and later stages of lung development [41]. One model that is consistent with mesenchymal GR regulating epithelial proliferation and differentiation is that GR induces a secreted mesenchymal factor (or factors) that is required for normal alveolar differentiation [42]. In support of this general notion, Sweezey and colleagues have shown that *LGL1* (also known as *CRIPSDL2*) is a mesenchymal target of GR that signals to the epithelium [42]. Specifically, in rat lung, immunohistochemistry revealed mesenchymal *LGL1* expression and co-localization of LGL1 with markers of the Golgi and ER, consistent with it functioning as a secreted protein. Moreover epithelial cells were shown to import extracellular LGL1. Antisense inhibition of *LGL1* in developing rats resulted in decreased branching morphogenesis [43]. Thus, LGL1, although unlikely to be the primary mesenchymal factor that drives the effects of GR on the epithelium, exemplifies the notion that secreted GR-regulated proteins in the mesenchyme can control epithelial development. Intriguingly, glucocorticoids induce the expression of *LGL1* in Beas-2B airway epithelial cells [44].

This suggests that high doses of exogenous GCs may promote ectopic expression of mesenchymal factors such as *LGL1*, thus driving cell-autonomous effects of GR signaling on epithelial cells, as is observed in cultured systems. In that regard, it would be interesting to determine whether prenatal administration of exogenous GCs to fetal mice would reverse, at least partially, the severe basal phenotype associated with mesenchymal GR deletion.

In summary, GR signaling is required for normal lung development and treatment of prenatal mammals with exogenous GCs accelerates alveolar maturation and surfactant expression. The latter effect is the basis for widespread and effective use of GCs during threatened pre-term labor to prevent neonatal respiratory distress. However, the mechanisms underpinning the discordance between the effects of exogenous GCs on the epithelium, and the requirement for mesenchymal GR in mediating normal lung development, have yet to be fully elucidated. Future mechanistic studies of GR action in the pulmonary mesenchyme, epithelium, and representative cultured cells, using both loss of function and exogenous GC administration, will undoubtedly further enhance our understanding of the molecular basis of the clinical effects of GCs on alveolar maturation and surfactogenesis.

Glucocorticoids and Obstructive Lung Disease

In addition to the use of GCs to prevent RDS of prematurity, GCs are used to treat a spectrum of lung disease in children and adults, including autoimmune associated interstitial lung disease [45], hypersensitivity pneumonitis [46], cryptogenic organizing pneumonia [47], eosinsophilic pneumonia [48], asthma [49], and chronic obstructive pulmonary disease (COPD) [50]. Of these, the most extensive literature involves the use of GCs in two of the most prevalent chronic diseases of the lung—asthma and COPD. Moreover, therapeutic responses in these diseases are at least in part due to direct effects of GCs on pulmonary structural cells, notably the airway epithelium and airway smooth muscle, which are the primary tissue types found in the conducting airways (see Fig. 12.3). Although tissue resident immune cells certainly contribute to airway pathology and corticosteroid effects [51], the role of GCs in modulating immune cell function is covered elsewhere in this book. In the remainder of this chapter, therefore, I will predominantly focus on mechanisms and consequences of GR signaling in the airway epithelium and smooth muscle.

Introduction to Asthma and COPD: Similarities and Differences

Asthma and COPD both fall within the broad category of obstructive lung disease and there is considerable overlap in the symptoms associated with both diseases [52], and to a lesser extent, their associated physiologic perturbations and

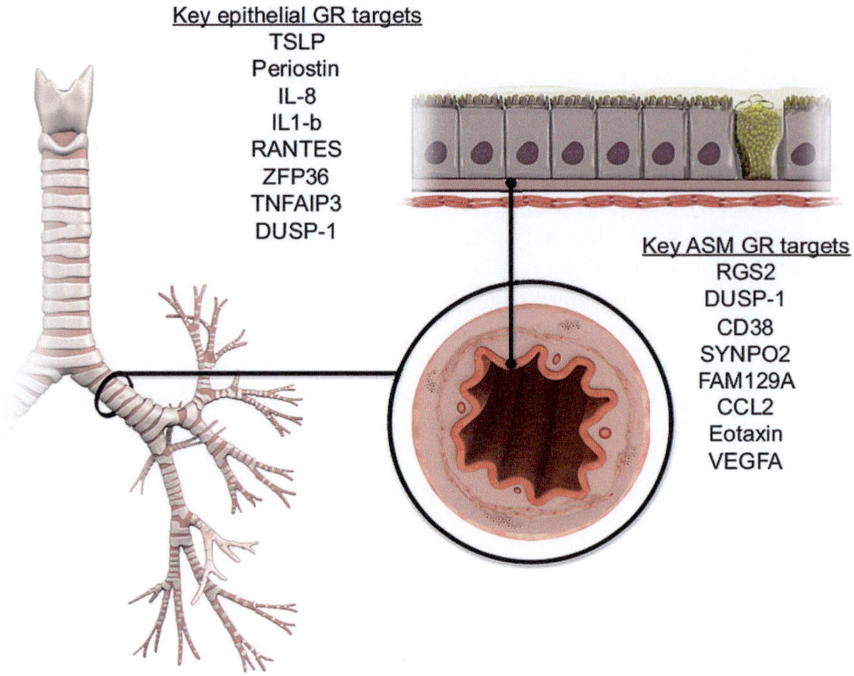

Fig. 12.3 A schematic of airway structure and a partial list of therapeutically relevant targets of glucocorticoid signaling in the airway epithelium and in airway smooth muscle

molecular pathology [53, 54]. Obstructive lung disease is defined as an abnormality in pulmonary air movement that limits exhalation; the primary symptom associated with obstructive lung disease is shortness of breath [55]. In asthma, airway obstruction and symptoms are generally intermittent and are caused, in part, by exuberant contraction of the smooth muscle layer that surrounds conducting airways [56]. Alterations in the epithelium also occur in asthma, including excessive production of mucus, which can also serve to limit airflow and to precipitate coughing. Additionally, allergic inflammation is frequently, but not always, present in asthmatic airways. In partial contrast, while airway obstruction in COPD often has a reversible component in which the airway smooth muscle response to beta agonists results in improved lung function, COPD also consistently features a component of "fixed" airway obstruction that is essentially irreversible and is associated with persistent shortness of breath. Fixed airway obstruction in COPD can result from several distinct pathologic processes including airway thickening, excess mucus production, and emphysematous destruction of the integrity of distal bronchioles and alveoli [53]. Although as we will review below, both diseases are treated with inhaled GCs, the distinct pathophysiologic features of COPD and asthma indicate that therapeutic responses to GCs likely involve both overlapping and non-overlapping molecular actions of GR on the airway epithelium and airway smooth muscle.

GCs in Asthma and COPD: Clinical Perspective

In the late 1940s and 1950s it was recognized that administration of cortisone, and various derivative formulations, provides substantial clinical benefit to asthma patients [57, 58]. In 1951, a paper in the New England Journal of Medicine from Gelfand reported on the use of an aerosolized cortisone preparation to treat asthma [59]; this intervention improved symptoms in a small cohort of patients and foreshadowed the eventual predominant role of inhaled corticosteroids (ICS) in asthma control. By the early 1970s, with the development of more modern ICS formulations, the promise of the pilot data from the Gelfand study was realized and expanded on by numerous other investigators [60, 61]. For example, Lal *et al.* reported on an inhaled preparation of beclomethasone that exhibited similar clinical efficacy to systemic prednisolone [62]. Increasingly potent ICS formulations have led to dramatically improved control of asthma symptoms, with the localized dosing afforded through inhalation resulting in substantially fewer side effects and incidence of adrenal suppression in comparison to systemic GCs [24, 63]. Consequently, ICSs are now the standard of care for treating the majority of asthma patients [2]. Similarly, secondary to reducing the frequency of disease exacerbations, ICSs are also frequently prescribed in COPD [50], whereas systemic use of GCs has been associated with a number of debilitating morbidities.

In addition to these important clinical benefits of ICSs over systemic GCs, the efficacy of ICS formulations in limiting asthma symptoms suggests that airway structural cells, which receive the majority of GCs dosed through inhaled preparations, are likely crucial in mediating therapeutic effects of GCs. The primary constituents of the airway are the airway epithelium and smooth muscle. A variety of studies indicate that both tissues harbor an active GR signaling apparatus and that both tissues respond to inhaled steroids in human subjects at the transcriptional level. Mechanisms and consequences of GCs on both tissue types are reviewed below.

Effects of GCs on Airway Epithelial Cells

The airway contains an internal epithelial layer that serves as the conducting surface through which air passes to and from the alveoli during the respiratory cycle (Fig. 12.3). Although this surface contains a number of different cell types, including goblet and club cells, few studies have discriminated between epithelial cell subtype and GC effect, nor shall distinct effects of GCs on cell types within the airway epithelium be addressed here. With that general caveat, beginning in the early 1990s, a large number of studies have employed various airway epithelial cell models to examine the effects of GCs on the expression of individual cytokines and other factors implicated in asthma and/or COPD pathogenesis. For example, in a representative publication, van de Stolpe and colleagues showed that *ICAM* expression was repressed by GCs in H292 cells [64], a cancer cell model of mucus producing airway epithelium. Similarly a study by Levine and colleagues in 1993 established that corticosteroids could block the secretion of IL-6, IL-8 and G-CSF from Beas2B cells [65], a standard model of cultured human airway epithelial cells. Another study showed that *TSLP*, whose misregulation is implicated as a genetic risk factor in asthma [66], is induced by dsRNA in normal human bronchial epithelial cells; this induction is repressed by GCs. Several groups have also shown that GCs reduce mucin gene expression, notably *MUC5AC*, both in cultured human airway epithelial cells and in *ex vivo* murine airway epithelial progenitor cells [67]. As increased mucus is strongly implicated as driving symptoms in asthma and in COPD, down-regulation of *MUC5AC* by GCs in the airway epithelium is thus implicated as at least part of the underlying therapeutic effects of GCs in both diseases. Taken together, these and other *in vitro* studies, as reviewed in greater detail by Stellato [68], clearly establish that GCs modulate transcriptional programming in a cell autonomous manner in airway epithelial cells, including exerting repressive effects on cytokine expression and other pathways implicated in asthma and COPD pathogenesis.

In Vivo Activity of GCs in the Airway Epithelium

These *in vitro* studies have been mirrored by various clinical studies in which inhaled corticosteroids have been shown to reduce the expression of pro-inflammatory cytokines [69] and other aspects of airway pathology in airway epithelial cells in patients. For example in a study by Sousa et al. [69], inhaled beclomethasone led to a significant decrease in the expression of GM-CSF and IL-8 in the airway epithelium. Likewise *RANTES* expression was reduced in patients after treatment with corticosteroids [70]. Thus, the clinical efficacy of GCs in treating asthma is associated with both *in vivo* and *in vitro* effects on airway epithelial cells, establishing the airway epithelium as a steroid responsive tissue that mediates important therapeutic effects of GCs in the lung.

More recent investigations have expanded on these studies through applying genomics technology to study *in vivo* effects of glucocorticoids on airway epithelial gene expression. One particularly well–designed study assessed airway gene expression on a genome-wide basis in normal volunteers, smokers and asthmatics, both before and after treatment with an ICS for 2 weeks [71]. Microarray-based expression profiling of epithelial cells from airway brushings identified a set of genes whose expression changed significantly after 1 week of inhaled fluticasone. The investigators also found a set of genes whose expression levels differed in the airway epithelium between asthmatics and healthy controls. The overlap of these sets was comprised of three genes, *periostin*, *Clca1*, and *serpinB2*. Each of these three genes exhibited elevated expression in asthmatic epithelium in comparison to control, and expression of each gene was reduced after treatment with fluticasone. Moreover, the decrease in *serpinB2* expression caused by fluticasone was correlated with treatment-associated improvements in lung function. The authors went on to use cultured normal human bronchial epithelial cells to show that IL-13, a major contributor to the Th2-dominant asthma phenotype, increased the expression of *periostin*, *Clca1*, and *serpinB2*. Conversely, expression of these three genes was repressed by dexamethasone. Taken together, this study, along with several other studies, support a model in which Th2-dominant asthma responds to GC treatment, at least in part through direct repression of specific IL-4/IL-13 targets in the airway epithelium.

A recent study also applied microarray technology to examine the effects of fluticasone delivered in combination with a long acting beta agonist, salmeterol, on airway gene expression in COPD [72]. They found a set of genes whose expression is increased COPD that tended to return to normal expression levels with treatment; conversely, a second set of genes exhibited decreased expression in COPD that normalized after fluticasone/salmeterol therapy. Gene set enrichment analysis indicated that epithelial barrier function genes were overrepresented amongst genes with COPD specific expression patterns that normalized with fluticasone/salmeterol treatment, suggesting that beneficial effects of GCs in COPD may occur through non-inflammatory mechanisms. Moreover, these data, similar to data from the asthma study of Woodruff *et al.* described previously above, support an epithelial cell-autonomous mode of GC action contributing to therapeutic effects in COPD.

Molecular Mechanisms of GR Signaling in the Airway Epithelium

The wide range of studies, which indicate that GCs exert beneficial effects in obstructive airway disease through induction of GR signaling in the epithelium, has led to a growing body of research focused on the molecular mechanisms underpinning GR-mediated gene regulation in the airway epithelium. Although such work is complicated by intricacies inherent to airway disease, including multiple phenotypes, and environmental and infectious contributions to airway pathology, a number of underlying mechanistic principals governing GR activity in airway epithelial cells have emerged. First, and similar to results from a range of non-airway cell types and disease models, repression of inflammatory transcription factor activity, exemplified by NF-kB, appears to be an important mechanism through which GCs repress cytokine expression in the airway epithelium. This appears to occur through a combination of direct effects of GR on NF-kB (or on other inflammatory transcription factors, e.g. AP-1) activity, and through two-step processes in which GR induces an inhibitor of inflammatory gene expression, such as DUSP1 [73]. Evidence for the first mode of repression (generally referred to as transrepression) occurring specifically in normal airway epithelial cells (i.e. not tumor-derived cell) is somewhat limited, with support garnered from studies that applied protein synthesis and or transcriptional inhibitors to infer both direct and indirect repression of cytokine transcription by ligand-activated GR [73, 74]. Corroborating these findings, repression of NF-kB activity in airway epithelial cells has been associated with changes in histone acetylation to marks consistent with gene repression at several loci [75]. For example, work from Rose and colleagues encompassed in several publications established that GR directly represses *MUC5AC* expression through recruiting HDAC2 [76]. Moreover, stimulation of *MUC5AC* expression by IL-1b, which occurs through induction of NF-kB activity, is antagonized by dexamethasone [77], implicating GR-mediated recruitment of HDAC2 as a mechanism through which NF-kB activity is repressed by GCs in the airway epithelium.

Despite these insights into GR-mediated repression of *MUC5AC*, and several other studies of the molecular events involved in repression of NF-kB activity at specific loci in airway epithelial cells [78], genome-wide analysis of GR cross-talk with inflammatory transcriptional regulators implicated in airway disease, such as NF-kB, in relationship to RNA Polymerase II occupancy and histone modification has yet to be reported on in airway epithelial models. Such studies would dramatically enhance our understanding of mechanisms through which GR represses cytokine expression in the airway and would also serve as a foundation for studying the mechanisms through which specific airway disease phenotypes and environmental factors reduce the efficacy of GCs, which is a major clinical problem in asthma and COPD [79]. In that regard, although a number of mechanisms through which GR activity is potentially reduced in airway disease have been proposed [80, 81], without a more detailed understanding of GR activity in specific disease-relevant models, e.g. mechanisms whereby GR reduces symptoms in IL-4/IL-13 dominant asthma, it will be difficult to fully characterize resistance mechanisms, and to rationally improve GC-based therapies.

In addition to the direct effects of GR on NF-kB and other inflammatory transcription factors, it is also clear the GR induces the expression of various factors that negatively regulate inflammatory gene expression. For example, the induction of *DUSP1* by GR, which occurs through GR binding to several canonical GC response elements within the *DUSP1* locus [82], is strongly implicated as contributing to repression of cytokine expression in airway epithelial cells by GCs [73]. Likewise, we have recently shown that *TNFAIP3*, a potent negative feedback regulator of NF-kB, is induced by ligand-activated GR [83]. Intriguingly, induction of *TNFAIP3* by GR involves cooperation with NF-kB, indicating that crosstalk between GR and inflammatory transcription factors is likely to be combinatorially complex and result in divergent regulatory outcomes depending on the specific target gene. Yet a third GR-induced gene that mediates inflammatory repression is *ZFP36*, which decreases the stability of various cytokine mRNAs, thus implicating post-transcriptional regulation as another important element of GR-mediated inflammatory repression [84]. Thus, genes that are induced by GCs are implicated in mediating a variety of effects on airway epithelial inflammatory gene expression, and likely contribute to therapeutic effects.

In summary, agonist-induced GR directly regulates gene expression in the airway epithelium, which contributes to the effectiveness of GC-based therapies in treating airway disease. A number of specific genes and mechanisms have been identified as important in this process, with notable findings including profound repressive effects of GCs on cytokine expression, a reduction in the expression of specific IL-13 targets in TH2 predominant asthma, and an impact of GCs on airway epithelial barrier function in COPD. Both the targets and mechanisms of GR action in the airway remain under active investigation, with an expectation that increasingly sophisticated genomics technology will greatly expand our understanding of GR signaling in the airway epithelium.

GCs and Airway Smooth Muscle: Another GR-Responsive Tissue Implicated in Mediating Therapeutic Effects

The other major tissue type that is targeted by glucocorticoid-based therapies in the treatment of obstructive airway diseases is airway smooth muscle (ASM), which forms the outside layer of the conducting airways (Fig. 12.3). ASM dysfunction, including hyperresponsiveness to methacholine, a specific constrictor of ASM, is one of the major pathophysiologic hallmarks of asthma [85], and ASM also exhibits significant pathology in COPD [86]. In addition to hyperresponsiveness and exuberant contractility, ASM can also show both hypertrophy and hyperplasia in asthma [87, 88], which is believed to contribute to chronic airway thickening and fixed airway obstruction [89]. Given these central roles for ASM pathology in obstructive airway disease, a growing body of research has been focused on the potential that GC signaling in ASM may result in therapeutic benefit.

Investigations of GC action in ASM have been enabled by the development of techniques that allow for culturing ASM from explanted and post-mortem lungs [90], including following fatal asthma attacks, and also from material obtained with airway biopsies [91]. Culture cells derived from these sources have been shown to have many characteristics that are associated with airway smooth in vivo. These include response to various asthma associated mitogenic factors, expression of smooth muscle contractile markers, and hypertrophy in response to stimuli such as TGF beta, which is also implicated in hypertrophy in vivo [92–94]. With physiological relevance of the cultured ASM cell model established, investigators have gone on to study specific effects of GCs on ASM that may be responsible for limiting inflammation and ASM dysfunction. Similar to analogous work on cultured airway epithelial cells described above, in the pre-genomics era, a number of studies analyzed the effects of glucocorticoids on gene and protein expression to establish that glucocorticoids suppress ASM cytokine expression [95]. This led to the notion that ASM, in addition to serving a physiologic role in asthma pathology, likely also serves an immunomodulatory role in which cytokines released from ASM promote epithelial dysfunction [96]. Again, similar to epithelial cells, mechanisms underlying the effects of GCs on ASM include trans-repression of inflammatory transcription factor function by GR, with co-regulators such as GRIP1 and IRF1 implicated in this process [97]. Induction of negative regulators by GR also appears to be important for cytokine repression, with DUSP1 well characterized as a mediator of IL6 repression by GCs in ASM; GC-mediated repression of CD38, which is implicated in regulating aberrant ASM contraction in asthma and is the target of several inflammatory pathways [98, 99], also appears to involve DUSP1 [100, 101].

The genomics era ushered in a number of genome-wide studies on the effects of GCs on ASM that highlighted both a role for GR signaling in suppressing ASM-mediated cytokine secretion, and also in modulating other aspects of ASM gene expression, potentially including pathways associated with hypertrophy and hyperplasia. For example, studies by Hakonarson *et al*, Misior *et al*, and Masuno *et al* showed robust changes in ASM gene expression caused by GC treatment [102–104]; these effects were modulated by co-treatment with cytokines, growth factors, PKA activation, and also depended on treatment time. A single study has reported on *in vivo* effects of GCs on human ASM [105]; this work defined two genes *SYNPO2* and *FAM129A*, whose expression changed in ASM after systemic GC treatment and that correlated with lung function. Additional studies are needed to validate these findings and also to test whether genes such as *RGS2*, implicated in animal and cell culture models as mediating therapeutic effects of GCs via modulating calcium signaling, are also regulated in patients by GCs.

Another area of intense investigation in ASM and the glucocorticoid response in asthma has been cross-talk between GR and beta agonists [106]. Indeed these two drugs are frequently packaged in combination, with beta agonists primarily implicated as promoting ASM relaxation. However a variety of studies have suggested that GCs and beta-*agonists* have synergistic effects on gene regulation in ASM [107]. For example, RGS2 appears to be co-regulated by both agents [108]. Paradoxically, there is evidence that GCs may promote ASM proliferation in a beta-agonist dependent fashion [103], potentially indicating that long term GC/beta-agonist treatment may have deleterious effects in a subset of patients. The underlying molecular control of transcriptional interactions between GC and beta-agonist signaling remains incompletely understood and requires further investigation.

In closing, in addition to effects of GR on the developing lung and the airway epithelium, GR signaling is also active in ASM where it appears to mediate therapeutic effects on ASM pathology in asthma and COPD through a range of mechanisms. These include down-regulation of cytokine expression, and induction of genes such as *RGS2*, which promote ASM relaxation. GR also cooperates with beta-agonists at the transcriptional level to modulate ASM gene expression and phenotype. Additional *in vivo* determination of ASM targets of GR may aid in developing better therapies to treat ASM pathology in obstructive airway disease.

References

1. Hallman M, Peltoniemi O, Kari MA. Enhancing functional maturity before preterm birth. Neonatology. 2010;97(4):373–8.
2. Stoloff SW, Kelly HW. Updates on the use of inhaled corticosteroids in asthma. Curr Opin Allergy Clin Immunol. 2011;11(4):337–44.

3. Liggins GC. Premature delivery of foetal lambs infused with glucocorticoids. J Endocrinol. 1969;45(4):515–23.
4. Liggins GC, Howie RN. A controlled trial of antepartum glucocorticoid treatment for prevention of the respiratory distress syndrome in premature infants. Pediatrics. 1972;50(4): 515–25.
5. Brownfoot FC, Gagliardi DI, Bain E, Middleton P, Crowther CA. Different corticosteroids and regimens for accelerating fetal lung maturation for women at risk of preterm birth. Cochrane Database Syst Rev. 2013;(8):CD006764.
6. Oshika E, Liu S, Ung LP, et al. Glucocorticoid-induced effects on pattern formation and epithelial cell differentiation in early embryonic rat lungs. Pediatr Res. 1998;43(3):305–14.
7. Mendelson CR, Gao E, Young PP, Michael LF, Alcorn JL. Transcriptional regulation of the surfactant protein-A gene in fetal lung. Chest. 1997;111(6 Suppl):96S–104.
8. Vyas J, Kotecha S. Effects of antenatal and postnatal corticosteroids on the preterm lung. Arch Dis Child Fetal Neonatal Ed. 1997;77(2):F147–50.
9. Mendelson CR. Role of transcription factors in fetal lung development and surfactant protein gene expression. Annu Rev Physiol. 2000;62:875–915.
10. Blanco LN, Massaro GD, Massaro D. Alveolar dimensions and number: developmental and hormonal regulation. Am J Physiol. 1989;257(4 Pt 1):L240–7.
11. Whitsett JA, Matsuzaki Y. Transcriptional regulation of perinatal lung maturation. Pediatr Clin North Am. 2006;53(5):873–87, viii.
12. Venkatesh VC, Iannuzzi DM, Ertsey R, Ballard PL. Differential glucocorticoid regulation of the pulmonary hydrophobic surfactant proteins SP-B and SP-C. Am J Respir Cell Mol Biol. 1993;8(2):222–8.
13. Boggaram V, Smith ME, Mendelson CR. Regulation of expression of the gene encoding the major surfactant protein (SP-A) in human fetal lung in vitro. Disparate effects of glucocorticoids on transcription and on mRNA stability. J Biol Chem. 1989;264(19):11421–7.
14. Odom MJ, Snyder JM, Boggaram V, Mendelson CR. Glucocorticoid regulation of the major surfactant associated protein (SP-A) and its messenger ribonucleic acid and of morphological development of human fetal lung in vitro. Endocrinology. 1988;123(4):1712–20.
15. Gonzales LW, Guttentag SH, Wade KC, Postle AD, Ballard PL. Differentiation of human pulmonary type II cells in vitro by glucocorticoid plus cAMP. Am J Physiol Lung Cell Mol Physiol. 2002;283(5):L940–51.
16. Maeda Y, Hunter TC, Loudy DE, Dave V, Schreiber V, Whitsett JA. PARP-2 interacts with TTF-1 and regulates expression of surfactant protein-B. J Biol Chem. 2006;281(14): 9600–6.
17. Kolla V, Gonzales LW, Gonzales J, et al. Thyroid transcription factor in differentiating type II cells: regulation, isoforms, and target genes. Am J Respir Cell Mol Biol. 2007;36(2): 213–25.
18. Islam KN, Mendelson CR. Glucocorticoid/glucocorticoid receptor inhibition of surfactant protein-A (SP-A) gene expression in lung type II cells is mediated by repressive changes in histone modification at the SP-A promoter. Mol Endocrinol. 2008;22(3):585–96.
19. Boggaram V, Smith ME, Mendelson CR. Posttranscriptional regulation of surfactant protein-A messenger RNA in human fetal lung in vitro by glucocorticoids. Mol Endocrinol. 1991;5(3):414–23.
20. Alcorn JL, Islam KN, Young PP, Mendelson CR. Glucocorticoid inhibition of SP-A gene expression in lung type II cells is mediated via the TTF-1-binding element. Am J Physiol Lung Cell Mol Physiol. 2004;286(4):L767–76.
21. Huang HW, Bi W, Jenkins GN, Alcorn JL. Glucocorticoid regulation of human pulmonary surfactant protein-B mRNA stability involves the 3′-untranslated region. Am J Respir Cell Mol Biol. 2008;38(4):473–82.
22. Liu D, Yi M, Smith M, Mendelson CR. TTF-1 response element is critical for temporal and spatial regulation and necessary for hormonal regulation of human surfactant protein-A2 promoter activity. Am J Physiol Lung Cell Mol Physiol. 2008;295(2):L264–71.

23. Islam KN, Mendelson CR. Potential role of nuclear factor kappaB and reactive oxygen species in cAMP and cytokine regulation of surfactant protein-A gene expression in lung type II cells. Mol Endocrinol. 2002;16(6):1428–40.
24. Barnes PJ. Glucocorticosteroids: current and future directions. Br J Pharmacol. 2011;163(1): 29–43.
25. De Bosscher K, Van Craenenbroeck K, Meijer OC, Haegeman G. Selective transrepression versus transactivation mechanisms by glucocorticoid receptor modulators in stress and immune systems. Eur J Pharmacol. 2008;583(2–3):290–302.
26. Wade KC, Guttentag SH, Gonzales LW, et al. Gene induction during differentiation of human pulmonary type II cells in vitro. Am J Respir Cell Mol Biol. 2006;34(6):727–37.
27. Kim D, Kwon YK, Cho KH. The biphasic behavior of incoherent feed-forward loops in biomolecular regulatory networks. Bioessays. 2008;30(11–12):1204–11.
28. Eichenberger P, Fujita M, Jensen ST, et al. The program of gene transcription for a single differentiating cell type during sporulation in Bacillus subtilis. PLoS Biol. 2004;2(10):e328.
29. Sasse SK, Mailloux CM, Barczak AJ, et al. The glucocorticoid receptor and KLF15 regulate gene expression dynamics and integrate signals through feed-forward circuitry. Mol Cell Biol. 2013;33(11):2104–15.
30. Chen SH, Masuno K, Cooper SB, Yamamoto KR. Incoherent feed-forward regulatory logic underpinning glucocorticoid receptor action. Proc Natl Acad Sci U S A. 2013;110(5): 1964–9.
31. Yang MC, Guo Y, Liu CC, Weissler JC, Yang YS. The TTF-1/TAP26 complex differentially modulates surfactant protein-B (SP-B) and -C (SP-C) promoters in lung cells. Biochem Biophys Res Commun. 2006;344(2):484–90.
32. Park KS, Whitsett JA, Di Palma T, Hong JH, Yaffe MB, Zannini M. TAZ interacts with TTF-1 and regulates expression of surfactant protein-C. J Biol Chem. 2004;279(17): 17384–90.
33. Margana RK, Boggaram V. Functional analysis of surfactant protein B (SP-B) promoter. Sp1, Sp3, TTF-1, and HNF-3alpha transcription factors are necessary for lung cell-specific activation of SP-B gene transcription. J Biol Chem. 1997;272(5):3083–90.
34. Reddy TE, Pauli F, Sprouse RO, et al. Genomic determination of the glucocorticoid response reveals unexpected mechanisms of gene regulation. Genome Res. 2009;19(12):2163–71.
35. ENCODE Project Consortium. A user's guide to the encyclopedia of DNA elements (ENCODE). PLoS Biol. 2011;9(4):e1001046.
36. Manwani N, Gagnon S, Post M, et al. Reduced viability of mice with lung epithelial-specific knockout of glucocorticoid receptor. Am J Respir Cell Mol Biol. 2010;43(5):599–606.
37. Cole TJ, Solomon NM, Van Driel R, et al. Altered epithelial cell proportions in the fetal lung of glucocorticoid receptor null mice. Am J Respir Cell Mol Biol. 2004;30(5):613–9.
38. Habermehl D, Parkitna JR, Kaden S, et al. Glucocorticoid activity during lung maturation is essential in mesenchymal and less in alveolar epithelial cells. Mol Endocrinol. 2011; 25(8):1280–8.
39. Li A, Hardy R, Stoner S, Tuckermann J, Seibel M, Zhou H. Deletion of mesenchymal glucocorticoid receptor attenuates embryonic lung development and abdominal wall closure. PLoS One. 2013;8(5):e63578.
40. Bird AD, Choo YL, Hooper SB, McDougall AR, Cole TJ. Mesenchymal glucocorticoid receptor regulates the development of multiple cell layers of the mouse lung. Am J Respir Cell Mol Biol. 2014;50(2):419–28.
41. Morrisey EE, Hogan BL. Preparing for the first breath: genetic and cellular mechanisms in lung development. Dev Cell. 2010;18(1):8–23.
42. Oyewumi L, Kaplan F, Sweezey NB. Lgl1, a mesenchymal modulator of early lung branching morphogenesis, is a secreted glycoprotein imported by late gestation lung epithelial cells. Biochem J. 2003;376(Pt 1):61–9.
43. Oyewumi L, Kaplan F, Gagnon S, Sweezey NB. Antisense oligodeoxynucleotides decrease LGL1 mRNA and protein levels and inhibit branching morphogenesis in fetal rat lung. Am J Respir Cell Mol Biol. 2003;28(2):232–40.

44. Greer S, Page CW, Joshi T, Yan D, Newton R, Giembycz MA. Concurrent agonism of adenosine A2B and glucocorticoid receptors in human airway epithelial cells cooperatively induces genes with anti-inflammatory potential: a novel approach to treat chronic obstructive pulmonary disease. J Pharmacol Exp Ther. 2013;346(3):473–85.
45. Kim R, Meyer KC. Therapies for interstitial lung disease: past, present and future. Ther Adv Respir Dis. 2008;2(5):319–38.
46. Selman M, Buendia-Roldan I. Immunopathology, diagnosis, and management of hypersensitivity pneumonitis. Semin Respir Crit Care Med. 2012;33(5):543–54.
47. Epler GR. Bronchiolitis obliterans organizing pneumonia, 25 years: a variety of causes, but what are the treatment options? Expert Rev Respir Med. 2011;5(3):353–61.
48. Rhee CK, Min KH, Yim NY, et al. Clinical characteristics and corticosteroid treatment of acute eosinophilic pneumonia. Eur Respir J. 2013;41(2):402–9.
49. Calhoun WJ, Ameredes BT, King TS, et al. Comparison of physician-, biomarker-, and symptom-based strategies for adjustment of inhaled corticosteroid therapy in adults with asthma: the BASALT randomized controlled trial. JAMA. 2012;308(10):987–97.
50. Calverley PM, Anderson JA, Celli B, et al. Salmeterol and fluticasone propionate and survival in chronic obstructive pulmonary disease. N Engl J Med. 2007;356(8):775–89.
51. Wang JH, Trigg CJ, Devalia JL, Jordan S, Davies RJ. Effect of inhaled beclomethasone dipropionate on expression of proinflammatory cytokines and activated eosinophils in the bronchial epithelium of patients with mild asthma. J Allergy Clin Immunol. 1994;94(6 Pt 1):1025–34.
52. Mosenifar Z. Differentiating COPD from asthma in clinical practice. Postgrad Med. 2009;121(3):105–12.
53. Carolan BJ, Sutherland ER. Clinical phenotypes of chronic obstructive pulmonary disease and asthma: recent advances. J Allergy Clin Immunol. 2013;131(3):627–34, quiz 635.
54. Postma DS, Reddel HK, ten Hacken NH, van den Berge M. Asthma and chronic obstructive pulmonary disease: similarities and differences. Clin Chest Med. 2014;35(1):143–56.
55. Scano G, Stendardi L. Dyspnea and asthma. Curr Opin Pulm Med. 2006;12(1):18–22.
56. Hershenson MB, Brown M, Camoretti-Mercado B, Solway J. Airway smooth muscle in asthma. Annu Rev Pathol. 2008;3:523–55.
57. Randolph TG, Rollins JP. The effect of cortisone on bronchial asthma. J Allergy. 1950;21(4):288–95.
58. Friedlaender S, Friedlaender AS. Effect of cortisone administered orally in bronchial asthma. J Am Med Assoc. 1951;146(15):1381–2.
59. Gelfand ML. Administration of cortisone by the aerosol method in the treatment of bronchial asthma. N Engl J Med. 1951;245(8):293–4.
60. Clark TJ. Effect of beclomethasone dipropionate delivered by aerosol in patients with asthma. Lancet. 1972;1(7765):1361–4.
61. Brown HM, Storey G, George WH. Beclomethasone dipropionate: a new steroid aerosol for the treatment of allergic asthma. Br Med J. 1972;1(5800):585–90.
62. Lal S, Harris DM, Bhalla KK, Singhal SN, Butler AG. Comparison of beclomethasone dipropionate aerosol and prednisolone in reversible airways obstruction. Br Med J. 1972;3(5822):314–7.
63. Inhaled corticosteroids compared with oral prednisone in patients starting long-term corticosteroid therapy for asthma. A controlled trial by the British Thoracic and Tuberculosis Association. Lancet. 1975;2(7933):469–73.
64. van de Stolpe A, Caldenhoven E, Raaijmakers JA, van der Saag PT, Koenderman L. Glucocorticoid-mediated repression of intercellular adhesion molecule-1 expression in human monocytic and bronchial epithelial cell lines. Am J Respir Cell Mol Biol. 1993;8(3):340–7.
65. Levine SJ, Larivee P, Logun C, Angus CW, Shelhamer JH. Corticosteroids differentially regulate secretion of IL-6, IL-8, and G-CSF by a human bronchial epithelial cell line. Am J Physiol. 1993;265(4 Pt 1):L360–8.

66. Kato A, Favoreto Jr S, Avila PC, Schleimer RP. TLR3- and Th2 cytokine-dependent production of thymic stromal lymphopoietin in human airway epithelial cells. J Immunol. 2007;179(2):1080–7.
67. Chen H, Sun X, Chi R, et al. Glucocorticoid dexamethasone regulates the differentiation of mouse conducting airway epithelial progenitor cells. Steroids. 2014;80:44–50.
68. Stellato C. Glucocorticoid actions on airway epithelial responses in immunity: functional outcomes and molecular targets. J Allergy Clin Immunol. 2007;120(6):1247–63. quiz 1264–5.
69. Sousa AR, Poston RN, Lane SJ, Nakhosteen JA, Lee TH. Detection of GM-CSF in asthmatic bronchial epithelium and decrease by inhaled corticosteroids. Am Rev Respir Dis. 1993;147(6 Pt 1):1557–61.
70. Wang JH, Devalia JL, Xia C, Sapsford RJ, Davies RJ. Expression of RANTES by human bronchial epithelial cells in vitro and in vivo and the effect of corticosteroids. Am J Respir Cell Mol Biol. 1996;14(1):27–35.
71. Woodruff PG, Boushey HA, Dolganov GM, et al. Genome-wide profiling identifies epithelial cell genes associated with asthma and with treatment response to corticosteroids. Proc Natl Acad Sci U S A. 2007;104(40):15858–63.
72. van den Berge M, Steiling K, Timens W, et al. Airway gene expression in COPD is dynamic with inhaled corticosteroid treatment and reflects biological pathways associated with disease activity. Thorax. 2014;69(1):14–23.
73. King EM, Holden NS, Gong W, Rider CF, Newton R. Inhibition of NF-kappaB-dependent transcription by MKP-1: transcriptional repression by glucocorticoids occurring via p38 MAPK. J Biol Chem. 2009;284(39):26803–15.
74. Newton R, King EM, Gong W, et al. Glucocorticoids inhibit IL-1beta-induced GM-CSF expression at multiple levels: roles for the ERK pathway and repression by MKP-1. Biochem J. 2010;427(1):113–24.
75. Kim SH, Kim DH, Lavender P, et al. Repression of TNF-alpha-induced IL-8 expression by the glucocorticoid receptor-beta involves inhibition of histone H4 acetylation. Exp Mol Med. 2009;41(5):297–306.
76. Chen Y, Watson AM, Williamson CD, et al. Glucocorticoid receptor and histone deacetylase-2 mediate dexamethasone-induced repression of MUC5AC gene expression. Am J Respir Cell Mol Biol. 2012;47(5):637–44.
77. Chen Y, Garvin LM, Nickola TJ, Watson AM, Colberg-Poley AM, Rose MC. IL-1beta induction of MUC5AC gene expression is mediated by CREB and NF-kappaB and repressed by dexamethasone. Am J Physiol Lung Cell Mol Physiol. 2014;306(8):L797–807.
78. Matsukura S, Kokubu F, Kurokawa M, et al. Molecular mechanisms of repression of eotaxin expression with fluticasone propionate in airway epithelial cells. Int Arch Allergy Immunol. 2004;134 Suppl 1:12–20.
79. Barnes PJ. Corticosteroid resistance in patients with asthma and chronic obstructive pulmonary disease. J Allergy Clin Immunol. 2013;131(3):636–45.
80. Papi A, Contoli M, Adcock IM, et al. Rhinovirus infection causes steroid resistance in airway epithelium through nuclear factor kappaB and c-Jun N-terminal kinase activation. J Allergy Clin Immunol. 2013;132(5):1075–85.e6.
81. Tliba O, Cidlowski JA, Amrani Y. CD38 expression is insensitive to steroid action in cells treated with tumor necrosis factor-alpha and interferon-gamma by a mechanism involving the up-regulation of the glucocorticoid receptor beta isoform. Mol Pharmacol. 2006;69(2):588–96.
82. Shipp LE, Lee JV, Yu CY, et al. Transcriptional regulation of human dual specificity protein phosphatase 1 (DUSP1) gene by glucocorticoids. PLoS One. 2010;5(10):e13754.
83. Altonsy MO, Sasse SK, Phang TL, Gerber AN. Context-dependent cooperation between nuclear factor kappaB (NF-kappaB) and the glucocorticoid receptor at a TNFAIP3 intronic enhancer: a mechanism to maintain negative feedback control of inflammation. J Biol Chem. 2014;289(12):8231–9.

84. Fan J, Ishmael FT, Fang X, et al. Chemokine transcripts as targets of the RNA-binding protein HuR in human airway epithelium. J Immunol. 2011;186(4):2482–94.
85. Prakash YS. Airway smooth muscle in airway reactivity and remodeling: what have we learned? Am J Physiol Lung Cell Mol Physiol. 2013;305(12):L912–33.
86. Mauad T, Dolhnikoff M. Pathologic similarities and differences between asthma and chronic obstructive pulmonary disease. Curr Opin Pulm Med. 2008;14(1):31–8.
87. Woodruff PG, Fahy JV. Airway remodeling in asthma. Semin Respir Crit Care Med. 2002;23(4):361–7.
88. Woodruff PG, Dolganov GM, Ferrando RE, et al. Hyperplasia of smooth muscle in mild to moderate asthma without changes in cell size or gene expression. Am J Respir Crit Care Med. 2004;169(9):1001–6.
89. Tliba O, Panettieri Jr RA. Noncontractile functions of airway smooth muscle cells in asthma. Annu Rev Physiol. 2009;71:509–35.
90. Panettieri RA, Murray RK, DePalo LR, Yadvish PA, Kotlikoff MI. A human airway smooth muscle cell line that retains physiological responsiveness. Am J Physiol. 1989;256(2 Pt 1):C329–35.
91. Chang PJ, Bhavsar PK, Michaeloudes C, Khorasani N, Chung KF. Corticosteroid insensitivity of chemokine expression in airway smooth muscle of patients with severe asthma. J Allergy Clin Immunol. 2012;130(4):877–85.e5.
92. Zhou L, Hershenson MB. Mitogenic signaling pathways in airway smooth muscle. Respir Physiol Neurobiol. 2003;137(2-3):295–308.
93. Bentley JK, Hershenson MB. Airway smooth muscle growth in asthma: proliferation, hypertrophy, and migration. Proc Am Thorac Soc. 2008;5(1):89–96.
94. Ma L, Brown M, Kogut P, et al. Akt activation induces hypertrophy without contractile phenotypic maturation in airway smooth muscle. Am J Physiol Lung Cell Mol Physiol. 2011; 300(5):L701–9.
95. Ammit AJ, Lazaar AL, Irani C, et al. Tumor necrosis factor-alpha-induced secretion of RANTES and interleukin-6 from human airway smooth muscle cells: modulation by glucocorticoids and beta-agonists. Am J Respir Cell Mol Biol. 2002;26(4):465–74.
96. Damera G, Tliba O, Panettieri Jr RA. Airway smooth muscle as an immunomodulatory cell. Pulm Pharmacol Ther. 2009;22(5):353–9.
97. Bhandare R, Damera G, Banerjee A, et al. Glucocorticoid receptor interacting protein-1 restores glucocorticoid responsiveness in steroid-resistant airway structural cells. Am J Respir Cell Mol Biol. 2010;42(1):9–15.
98. Tirumurugaan KG, Kang BN, Panettieri RA, Foster DN, Walseth TF, Kannan MS. Regulation of the cd38 promoter in human airway smooth muscle cells by TNF-alpha and dexamethasone. Respir Res. 2008;9:26.
99. Abcejo AJ, Sathish V, Smelter DF, et al. Brain-derived neurotrophic factor enhances calcium regulatory mechanisms in human airway smooth muscle. PLoS One. 2012;7(8):e44343.
100. Kang BN, Jude JA, Panettieri Jr RA, Walseth TF, Kannan MS. Glucocorticoid regulation of CD38 expression in human airway smooth muscle cells: role of dual specificity phosphatase 1. Am J Physiol Lung Cell Mol Physiol. 2008;295(1):L186–93.
101. Jude JA, Panettieri Jr RA, Walseth TF, Kannan MS. TNF-alpha regulation of CD38 expression in human airway smooth muscle: role of MAP kinases and NF-kappaB. Adv Exp Med Biol. 2011;691:449–59.
102. Hakonarson H, Halapi E, Whelan R, Gulcher J, Stefansson K, Grunstein MM. Association between IL-1beta/TNF-alpha-induced glucocorticoid-sensitive changes in multiple gene expression and altered responsiveness in airway smooth muscle. Am J Respir Cell Mol Biol. 2001;25(6):761–71.
103. Misior AM, Deshpande DA, Loza MJ, Pascual RM, Hipp JD, Penn RB. Glucocorticoid- and protein kinase A-dependent transcriptome regulation in airway smooth muscle. Am J Respir Cell Mol Biol. 2009;41(1):24–39.
104. Masuno K, Haldar SM, Jeyaraj D, et al. Expression profiling identifies Klf15 as a glucocorticoid target that regulates airway hyperresponsiveness. Am J Respir Cell Mol Biol. 2011;45(3):642–9.

105. Yick CY, Zwinderman AH, Kunst PW, et al. Glucocorticoid-induced changes in gene expression of airway smooth muscle in patients with asthma. Am J Respir Crit Care Med. 2013;187(10):1076–84.
106. Kaur M, Chivers JE, Giembycz MA, Newton R. Long-acting beta2-adrenoceptor agonists synergistically enhance glucocorticoid-dependent transcription in human airway epithelial and smooth muscle cells. Mol Pharmacol. 2008;73(1):203–14.
107. Manetsch M, Rahman MM, Patel BS, et al. Long-acting beta2-agonists increase fluticasone propionate-induced mitogen-activated protein kinase phosphatase 1 (MKP-1) in airway smooth muscle cells. PLoS One. 2013;8(3):e59635.
108. Holden NS, Bell MJ, Rider CF, et al. Beta2-adrenoceptor agonist-induced RGS2 expression is a genomic mechanism of bronchoprotection that is enhanced by glucocorticoids. Proc Natl Acad Sci U S A. 2011;108(49):19713–8.

Chapter 13
Glucocorticoids and the Cardiovascular System

Julie E. Goodwin

Abstract Glucocorticoids affect the developing and mature cardiovascular system in profound and, at times, contradictory ways. The glucocorticoid receptor is ubiquitous in most cell types and conserved across species, highlighting its importance in development and homeostasis. Despite the fact that the glucocorticoid receptor is widely expressed, tissue-specific effects of glucocorticoids may have pronounced effects on whole organism phenotypes. Here we will review the interactions between glucocorticoids and the cardiovascular system.

Keywords Glucocorticoids • Hypertension • Vascular smooth muscle • Endothelium • Dexamethasone • Nitric oxide • Cardiomyocyte • Nephron • Stent

Introduction

Glucocorticoids affect the developing and mature cardiovascular system in profound and, at times, contradictory ways. The glucocorticoid receptor is ubiquitous in most cell types and conserved across species, highlighting its importance in development and homeostasis. Despite the fact that the glucocorticoid receptor is widely expressed, tissue-specific effects of glucocorticoids may have pronounced effects on whole organism phenotypes. Here we will review the interactions between glucocorticoids and the cardiovascular system.

J.E. Goodwin, M.D. (✉)
Department of Pediatrics, Yale University School of Medicine,
333 Cedar Street, PO Box 208064, New Haven, CT 06520-8064, USA
e-mail: Julie.goodwin@yale.edu

J.-C. Wang, C. Harris (eds.), *Glucocorticoid Signaling*, Advances in Experimental Medicine and Biology 872, DOI 10.1007/978-1-4939-2895-8_13

Blood Pressure Regulation

Excess glucocorticoid, whether endogenous as in Cushing's syndrome, or exogenous, via pharmacologic administration of glucocorticoids, induces hypertension. Approximately 80 % of patients with Cushing's syndrome have hypertension, while 20 % of patients receiving chronic pharmacologic treatment with glucocorticoids have hypertension [1]. In the general population, hypertension increases the risk for cardiovascular disease with a doubling in mortality from ischemic heart disease and stroke for every 20 mmHg systolic or 10 mmHg diastolic increase in blood pressure [2]. This likely contributes to increased mortality; patients with Cushing's disease have a fivefold increased standard mortality ratio for vascular disease [3].

Given the widespread usage of glucocorticoids in the treatment of a variety of diseases the significance of glucocorticoid-induced hypertension is clear. Despite this, however, the mechanisms underlying this association are unclear. This form of hypertension has traditionally been believed to be mediated through promiscuous activation of the mineralocorticoid receptor by excess glucocorticoid [2, 4]. Cortisol is able to bind the mineralocorticoid receptor in the kidney, but is normally prevented by local inactivation of glucocorticoids in the kidney by the enzyme 11-b-HSD2. In cases of endogenous hypercortisolism, the levels of cortisol can overcome inhibition by 11-b-HSD to activate the mineralocorticoid receptor resulting in hypertension and hypokalemia. However, increasing evidence argues against this being the only, or perhaps even the most significant, factor. The hypertensive response to glucocorticoids in patients suffering from Addisonian shock occurs much too rapidly to be explained by renal mechanisms alone and suggests a crucial role of vascular tissues in *acute* blood pressure responses [5]. More significantly, however, many lines of evidence indicate that glucocorticoids are capable of inducing *sustained* elevations of blood pressure in both humans and in animal models via effects entirely independent of the mineralocorticoid receptor. For example, budesonide, a potent glucocorticoid with essentially no activity at the mineralocorticoid receptor, produces sustained hypertension in humans [6], while RU486, an anti-glucocorticoid agent without anti-mineralocorticoid effects, reduces blood pressure in both humans with Cushing's syndrome [7–9] as well as in rat models of glucocorticoid-induced hypertension [10, 11]. Similarly, spironolactone does not inhibit glucocorticoid-mediated hypertension in humans [12] or in rat models [13, 14] despite blocking salt and water retention.

In fact, in vivo studies consistently demonstrate that glucocorticoids acutely increase renal sodium excretion in adrenalectomized rats [15–18], although sustained effects of glucocorticoids on renal sodium handling are more challenging to assess.

Glucocorticoids and the Kidney

The glucocorticoid receptor is widely expressed in the kidney, and yet, its role there is poorly understood. While glucocorticoid receptor mRNA has been identified in most cells of the (rat and rabbit) nephron, it is most abundant in the glomerulus, proximal tubule, and thick ascending limb segments [19]. Conversely, immunoreactive (human) glucocorticoid receptor has been demonstrated by light microscopy primarily in the distal convoluted tubules and collecting ducts, with moderate detection in the glomerulus, and a faint presence in the proximal tubule. Within the glomerulus, light microscopy and electron microscopy identified the glucocorticoid receptor in all cell types, including visceral epithelial cells, parietal epithelial cells, mesangial cells, and endothelial cells [20].

Despite its wide expression within the kidney, the role of glucocorticoids within any cell of the nephron is not clear, as data on functional effects of glucocorticoids in the kidney have been contradictory [19]. Glucocorticoids have been ascribed a number of roles within the kidney via in vitro studies, including a role in increasing glomerular filtration rate [21], regulation of renal ammoniagenesis [22], gluconeogenesis [23], sodium-hydrogen ion exchange [24, 25], and sodium-phosphate cotransport [24, 26]. Micropuncture studies suggest that dexamethasone increases the rate of sodium transport in the loop of Henle [27], as well as Na-K-ATPase activity in the medullary thick ascending limb [28], but in situ microperfusion showed no effect of glucocorticoid hormones on loop sodium transport [29]. In the distal nephron, there is increasing evidence that glucocorticoids serve to regulate the sodium reabsorption machinery. Dexamethasone increases mRNA abundance of a number of genes central to renal sodium reabsorption, including SGK1, α-ENaC, and GILZ [17]. RU28362, a pure glucocorticoid agonist devoid of mineralocorticoid effects, induces mineralocorticoid like effects in cultured collecting duct cells; this activity is blocked by glucocorticoid but not mineralocorticoid antagonists [30]. Finally, dexamethasone induces a kidney-specific WNK1 isoform in cultured cells [31] while decreasing expression of WNK4 as well [32]; elevated expression of WNK1 and decreased activity of WNK4 have both been associated with a hereditary form of human hypertension and hyperkalemia via up regulation of thiazide sensitive cotransporter activity [33–35]. It should be noted, of course, that this potential activation of the distal nephron glucocorticoid receptor by glucocorticoids, as with activation of the mineralocorticoid receptor, would require glucocorticoids to evade 11β-HSD2 machinery metabolizing glucocorticoids, at least in aldosterone-sensitive epithelia.

Although the sum of these in vitro observations would seem to point to an anti-natriuretic role of glucocorticoids in the distal nephron, studies performed in live animals have consistently demonstrated that glucocorticoids induce natriuresis in conjunction with a more prominent kaliuresis [16, 17]. As such, the contribution of glucocorticoids to renal sodium handling has remained difficult to comprehend. We believe glucocorticoid infusion acutely increases blood pressure in these rats via vascular mechanisms, as noted above, thus increasing renal perfusion and triggering a pressure natriuresis independent of any renal effects of glucocorticoids.

Investigation of the role of the glucocorticoid receptor in the distal nephron has been accomplished via a mouse model yielding a tissue-specific deletion of the glucocorticoid receptor, using the Ksp-cadherin Cre model. Ksp-Cre mice express Cre under the control of the cadherin 16 promoter which is active in the renal tubules and collecting ducts. The only extrarenal site is the Mullerian and Wolffian ducts. Though the initial prediction was that the mice lacking the glucocorticoid receptor in the distal nephron would become acutely hypertensive but would be unable to sustain this elevated blood pressure chronically, in vivo studies have shown that these mutant mice develop the same degree of acute and chronic hypertension as controls when treated with dexamethasone [36]. These mice also demonstrated similar nephron number and aldosterone levels as controls though they had a small but significant elevation in mean blood pressure (7 mmHg) at baseline. Since the excision of the glucocorticoid receptor using the Cre-lox system is not 100 % efficient the authors could not exclude the possibility that only a small percentage of existing glucocorticoid receptor need be present to perpetuate steroid-mediated hypertension, nor could they rule out the possibility that proximal up regulation of the glucocorticoid receptor was playing a role.

Glucocorticoids and Vascular Smooth Muscle

The rapidity with which glucocorticoids induce a rise in blood pressure is inconsistent with glucocorticoids inducing hypertension via augmentation of renal sodium reabsorption and suggests, perhaps, a profound effect of glucocorticoids on vascular tone. Perhaps the most concentrated efforts to date have been focused on trying to understand the vascular effects of glucocorticoids and how they may contribute to clinically observed hypertension. There is evidence that the glucocorticoid receptor is present both in vascular smooth muscle [37–39] and in the vascular endothelium [38, 40, 41]. The majority of studies that have investigated glucocorticoid activity at either of these sites have employed in vitro models.

Studies in smooth muscle cell culture have observed an up regulation in angiotensin II type I (AT I) receptors induced by glucocorticoids, which are hypothesized, to result in alterations in blood pressure [42, 43]. Other in vitro studies using vascular smooth muscle cells have indicated that the glucocorticoid receptor as well as the mineralocorticoid receptor may mediate the effect of glucocorticoids on the influx of Na^+ and/or Ca^{2+} into cells [38, 44]. Molnar et al. previously demonstrated that corticosterone-induced signaling pathways in rat aortic vascular smooth muscle cells involved phosphorylation of the mitogen-activated protein kinases, Src and Akt which is believed to be mediated through the classical mineralocorticoid receptor. In the same study, the authors showed that corticosterone enhanced phenylephrine-induced vasoconstriction in isolated aortic rings in a dose-dependent manner, and that this effect was blocked by pre-treatment of the rings with L-NAME or indomethacin, suggesting that the effect of corticosterone is endothelial-dependent. Pre-incubation with RU486, a glucocorticoid receptor blocker lessened the observed corticosterone-dependent effects, while pre-incubation with spironolactone, a mineralocorticoid blocker, did not. Ultimately, the authors suggest that glucocorticoids may activate signaling of the mineralocorticoid receptor in non-epithelial tissues resulting in the formation of a protein complex that may also include the glucocorticoid receptor, which then transmits the signal [45].

A recent study in rats by Ong et al. is one of the only attempts in vivo to try to dissect the underlying mechanism of hypertension by examining regional blood flow [46]. In this study, central and regional blow flows of dexamethasone-treated rats were compared to those of rats treated with minoxidil and dexamethasone, saline alone and minoxidil alone. The authors were able to measure cardiac output and regional blood flow directly by placing a perivascular probe around the aorta, renal artery, superior mesenteric artery and common iliac artery of anesthetized animals. Dexamethasone increased blood pressure in association with an increase in total peripheral resistance. Interestingly, there was no detectable increase in the peripheral resistance in any of the regional vascular beds assessed. The authors speculated that this discrepancy could be explained by the possibility that the increase in total peripheral resistance might be a reflection of a cumulative increase in many different peripheral vascular beds, only a few of which were measured in this study. Furthermore, they acknowledged that their technique may not be sensitive enough to measure small changes in resistance in individual vascular beds. In the group treated with both minoxidil and dexamethasone, the authors show that these rats have significantly lower total peripheral resistance and significantly higher cardiac output than the rats treated with dexamethasone alone, though there is no significant decrease in mean arterial pressure. Since the increase in total peripheral resistance was prevented while the increase in blood pressure was not, the authors suggest that increased total peripheral resistance is not necessary for the development of glucocorticoid-induced hypertension. They speculate that changes in cardiac output and specifically stroke volume and resulting plasma expansion may have offset the vasodilator effects of minoxidil though there were no observed differences in hematocrit between the dexamethasone only treated rats and the minoxidil and dexamethasone treated rats in this study [46].

A second study has used advances in mouse genetics to enable a more targeted in vivo approach. In this study, a mouse model with a tissue-specific deletion of the glucocorticoid receptor in the vascular smooth muscle tissues via the use of SM22-Cre was characterized. While knockout and control mice had similar baseline blood pressures with normal diurnal variation, knockout animals demonstrated a statistically significant attenuation of the acute hypertensive response to exogenous dexamethasone [47]. This persisted through 1 week of dexamethasone administration, but the difference in blood pressure was not statistically significant after 2 weeks. Interestingly, in this study, the acute natriuretic response to dexamethasone was preserved in the knockout animals even though the acute hypertensive response was not.

This finding raises the possibility that the preserved "pressure natriuresis" is responsible for the absent hypertensive response and that the glucocorticoid receptor's presence in the vascular smooth muscle serves as an obstacle to pressure natriuresis. This proposal is speculative, but it seems clear that the preserved "pressure natriuresis" in the absence of elevated blood pressure suggests a complex relationship between glucocorticoids and the systemic and renal vasculature, which requires further investigation.

Glucocorticoids and the Endothelium

There is evidence to confirm the presence of the glucocorticoid in the vascular endothelium [48–50], but little is known about its role here. In vitro experiments with endothelial cell culture models have suggested that glucocorticoids regulate vascular reactivity via suppression of the production of vasodilators such as prostacyclin and nitric oxide, but conflicting evidence exists in this regard. Provencher et al. demonstrated an increase in angiotensin II receptor levels, yet a decrease in endothelin-1 levels in response to synthetic glucocorticoids in a cell culture model of vascular smooth muscle cells [37]. Endothelin-1, secreted only by the endothelium, is the strongest vasoconstrictor known. These contradictory results led the authors to speculate that glucocorticoids function as modulators of vascular inflammation and not solely as vasoconstrictive agents.

Experiments performed using vascular endothelial models have generally employed vessels in which the endothelium has been mechanically stripped, and hence is devoid of all endothelial function. Wallerath et al. suggest that it is the ability of glucocorticoids to destabilize endothelial nitric oxide synthase (eNOS) mRNA and reduce eNOS protein expression that is responsible for the ensuing hypertension observed in rats treated with dexamethasone [51]. More recent studies in isolated rat aortas showed that, through the glucocorticoid receptor, glucocorticoids could decrease expression of guanosine triphosphate cyclohydrolase 1 (GTPCH1) mRNA, the rate-limiting enzyme in the production of tetrahydrobiopterin (BH4), a cofactor for nitric oxide synthase [52]. However, in this series, eNOS mRNA levels were not significantly different in control vessels than in those treated with dexamethasone.

Using a mouse model used to investigate tissue-specific contributions of the glucocorticoid receptor, this receptor was removed from the vascular endothelium by using Tie-1 Cre. Interestingly, these knockout animals had a small but statistically significant increase in their baseline blood pressure, the source of which could not be entirely explained. In addition, the knockout animals were almost completely resistant to glucocorticoid-induced hypertension and conspicuously lacked the pressure natriuresis observed in the vascular smooth muscle glucocorticoid receptor knockout model. Intravital microscopy studies done in real-time on resistance vessels from these knockout animals revealed a statistically significant decrease in vessel contractility to the glucocorticoid receptor-specific ligand dexamethasone when compared to wild-type animals. The contractility to phenylephrine was similar between the two groups suggesting that loss of the endothelial glucocorticoid receptor confers a specific contractile defect in these animals, presumably preventing them from mounting a hypertensive response to systemic dexamethasone. In this study whole blood nitric oxide levels were not different between the two groups though the possibility of differences in nitric oxide in local vascular beds could not be ruled out [53].

Other laboratories have examined nitric oxide derangements via *ex vivo* investigations in rats. In 2009, Aras-Lopez et al. evaluated electrical-field stimulation induced neuronal nitric oxide release in mesenteric arteries of Wistar-Kyoto (WKY) rats and spontaneously hypertensive rats (SHRs) and the role of protein kinase C (PKC) in these responses [54]. Through a series of manipulations involving various PKC inhibitors, the authors demonstrated that dexamethasone was able to reduce neuronal nitric oxide release in arteries from SHRs but not WKY rats, and that this effect was mediated though activation of glucocorticoid receptors. This is the first study to address the possibility that glucocorticoids exert their hypertensive effects, at least in part, by altering the neuronal nitric oxide release of the perivascular innervation of these tissues [54]. This study raises important questions with regard to differential effects of glucocorticoids on the various nitric oxide isoforms, which appear to be affected by the overall milieu.

To date, there have been few reliable ways to assess the role of glucocorticoids in the vascular endothelium *in vivo*. In addition, there is increasing focus on the purported non-genomic effects of glucocorticoids which would allow very rapid signaling responses as opposed to the traditional notion of nuclear translocation and transcription which theoretically take much longer. The mystery as to how a receptor with only one known gene can effect such a vast array of physiologic and pharmacologic consequences suggests that there must be multiple promoters, mRNAs, translational isoforms and/or post-translational modifications to account for such diversity. The discovery of profound species diversity in glucocorticoid sensitivity also suggests that there may be specific environmental or epigenetic influences that impact cellular responses to glucocorticoids. Given all of these variables, it is conceivable that a dozen or more GR isoforms may be able to be generated in a single cell [55]. Therefore, it is not particularly surprising that there is much that is poorly understood about vascular responses to glucocorticoids.

Glucocorticoids and Adipose Tissue

Transgenic expression of 11-b-HSD1 in adipose tissue increases local glucocorticoid levels in adipose tissue and causes a metabolic syndrome-like picture in mice. This is in further support of the dehydrogenase hypothesis that states that the metabolic syndrome is caused in part by Cushings disease of fat: normal circulating cortisol levels, but increased adipose tissue glucocorticoid levels [56]. Interestingly in this study, glucocorticoid levels were found to be elevated in the portal circulation, suggesting downstream effects of glucocorticoids, particularly in the liver, could be playing a role in the some of the observed metabolic derangements. At least one other study has suggested a link between features of the metabolic syndrome, specifically insulin resistance, glucocorticoids and the liver. Bernal-Mizrachi et al. demonstrated that activation of PPAR-alpha in the setting of dexamethasone treatment, both in vivo in a mouse model and in vitro in human hepatocytes, resulted in hyperinsulinemia, hypertension and induction of gluconeogenic gene expression. These studies would in turn explain the coupling of increased blood pressure with insulin resistance. Adipose tissue could directly regulate blood pressure by secretion of adipokines such as angiotensinogen. An indirect effect mediated via downstream portal flow of newly generated glucocorticoids might also be a possibility. For a more complete discussion, please see the Chapter dedicated to this subject.

Glucocorticoids and Development and Cardiovascular Risk

Glucocorticoids are essential for fetal health and development and are tightly regulated during the embryonic period. Fetal glucocorticoid levels rise abruptly toward the end of gestation in mammals in preparation for post-natal life [57]. It has been shown that pre-natal glucocorticoid treatment can elicit growth-inhibitory [58] or stimulatory [59] effects in cardiomyocytes, though direct in vivo effects on the heart are unknown. Recent work using a mouse model with global absence of glucocorticoid signaling has shown that fetal heart function is severely impaired under these conditions [60]. The authors of this study proceed to show that deficiency of glucocorticoid signaling in the cardiomyocytes and vascular smooth muscle *only* produce nearly the identical phenotype demonstrating that many aspects of heart maturation are highly dependent on glucocorticoid signaling within these tissues. In addition there are robust data showing that post-natal vessel reactivity and even onset of hypertension may be determined by the pre-natal steroid milieu. Roghair et al. have nicely demonstrated that administration of dexamethasone to pregnant ewes results in increased coronary artery vasoconstriction to acetylcholine and attenuated vasodilatation to adenosine in newborn lambs [61]. In addition lambs from steroid-treated mothers had decreased endothelial nitric oxide synthesis compared to controls. In a follow-up study, coronary arteries from dexamethasone-exposed lambs demonstrated greater vascular smooth muscle cell proliferation and greater production of reactive oxygen species compared to controls and suggest a link between abnormal intrauterine environment and future risk for coronary artery disease [62].

While excess prenatal glucocorticoid has been linked to adverse cardiovascular outcomes in adulthood, prenatal glucocorticoid deficiency is no less worrisome. Preterm birth increases the risk of cardiovascular disease [63], occurring before the surge of cortisol in late gestation. Cardiovascular complications in pre-term infants include persistence of the foramen ovale and the ductus arteriosus, delay in the decrease of pulmonary vascular resistance and delay in the increase in cardiac output necessary to support post-natal life [64]. Many of these complications can be improved by exogenous steroids. While it is difficult to dissect the long-term effects of prematurity vs. the effects of glucocorticoid in humans, animal models are useful in this regard to evaluate the effect of 'programming' [65, 66].

Non-genomic Effects of Glucocorticoids in the Cardiovascular System

It has long been known that glucocorticoids exert many of their effects on the cardiovascular system via genomic effects, i.e. rapid (within hours) nuclear translocation of cytoplasmic glucocorticoid receptor and binding to glucocorticoid response elements (GREs) in the promoter region of target genes. However, genomic effects cannot explain all the effects of glucocorticoids on the cardiovascular system. Non-genomic effects, that is responses that occur on the order of several minutes, in the absence of transcription may also be playing a role. The importance of non-genomic effects in the cardiovascular system is further obscured by the fact that endogenous cortisol rises with stress, which is itself a risk factor for cardiovascular disease.

In general, non-genomic effects may be separated into (i) those with GR and another cytoplasmic signaling molecule and (ii) protein-protein interactions. The rationale for these protein interactions lies in purported alteration in biochemical properties in adjacent membranes via insertion of glucocorticoid in plasma membranes or secondary to altered lipophilicity or polarity [67]. Components of the GR complex released during GR-ligand binding, including heat shock proteins (HSPs), have been shown to influence cell signaling. For example, Src, which is released from this complex has been shown to initiate a signaling cascade in non-cardiovascular systems [68]. GR has also been reported to activate the PI3K-Akt pathway [69] which in turn can activate eNOS and has been shown to reduce myocardial infarct size in a mouse model [70]. Additional non-genomic effects mediated via stimulation of cytoplasmic signaling include dexamethasone-induced increase in coronary lipoprotein lipase by an AMP-activated protein kinase and p38 activated protein kinase in cardiomyocytes resulting in decreased cardiac glucose oxidation [71, 72], stimulation of the phosphoinositide system in vascular smooth muscle cells by cortisol [73], and increased inotropic effect by possible direct potentiation of calcium channels by hydrocortisone in cardiomyocytes [74].

The non-genomic effects of protein-protein interactions are most notably illustrated by the anti-inflammatory effects of dexamethasone in endothelial cells by direct NF-κB-GR interaction [75]. There is also some evidence that corticosteroid-binding globulins (CBG), responsible for transporting glucocorticoids in the blood, may exert a non-genomic role in the cardiovascular system as it has been shown that elevated CBG levels could result in increased blood pressure [76]. It has also been shown that bovine serum albumin-bound glucocorticoids can cause non-genomic effects at the cell surface [77] begging the question of whether glucocorticoid bound

CBG influences functioning of glucocorticoids in many cell types. Currently work is ongoing to identify synthetic glucocorticoids with nongenomic mechanisms of action and fewer systemic side effects [78, 79].

Glucocorticoids as Therapy for Cardiovascular Disease

While the developmental milieu vis-a-vis glucocorticoids is often underappreciated or overlooked, use of these agents as a therapy for adult-onset cardiovascular disease is the subject of a great deal of investigation, both in animal models and in clinical trials in humans. There remains ambiguity as to whether glucocorticoids, and at what doses, are helpful or harmful in cardiovascular disease.

Rodent models have generally been employed to investigate serious morbidities such as stroke or myocardial infarction. It is well known that elevated glucocorticoid levels are a risk factor for cardiovascular disease and, in particular, stroke [80]. Balkaya et al. demonstrated in mouse model that chronic stress, induced by various restraint manipulations resulted in increased blood pressure, reduced endothelial nitric oxide synthase levels and impaired endothelium dependent vasorelaxation in aortic rings and substantial increases in ischemic lesion size when subjected to middle cerebral artery occlusion [81]. Interestingly all of these detrimental effects, which are also risk factors for stroke, were completely reversed by treatment with the glucocorticoid antagonist RU486. Similarly, in an isolated perfused rat heart model that is used to mimic heart failure, cortisol, as well as aldosterone, resulted in increased infarct area and apoptosis [82]. Interestingly perfusion with the synthetic steroid dexamethasone as well as the antagonist RU486 also resulted in worsened outcomes, though there is some evidence for RU486 to be a partial glucocorticoid agonist [83]. The phenotype was improved by treatment with spironolactone which physically occupies the mineralocorticoid receptor as well as by a free-radical scavenger, suggesting that oxidative stress plays a role in the cardiac damage in this model. At least one other rat MI model has shown adverse outcomes with administration of exogenous steroid in the form of methylprednisolone, particularly when given at the late (>2 days post MI) time point, as assayed by matrix metalloproteinase activity, inflammation and caliber of scar formation [84].

In contrast, there are several studies in rodents which demonstrate a beneficial effect of glucocorticoid administration on cardiovascular disease outcomes. For example, Xu et al. have shown that provision of dexamethasone prior to left coronary artery occlusion surgery can induce myocardial transcription of the gene Bcl-xl which protects the heart from ischemia reperfusion injury by preventing mitochondrial release of cytochrome C [85]. These authors also showed that, in vitro, dexamethasone induced Bcl-xL expression in cultured cardiomyocytes. All therapeutic effects of dexamethasone were blocked by the glucocorticoid antagonist, RU486. Interestingly, the dose of dexamethasone that was used in vivo was quite high at 20 mg/kg. Steroids have also been investigated in combination with hyperoxia, which has been shown to provide some benefit in ischemia-reperfusion injury, to determine if the preconditioning effects of hyperoxia and steroids are additive. Kaljusto et al. did indeed show that combined pretreatment of male rats with hyperoxia (95 % O_2) and a modest dose of dexamethasone (3 mg/kg) before 30 min of ischemia followed by 120 min of reperfusion did result in more attenuation of infarct size and enhanced endothelium-independent vessel relaxation compared to groups without dexamethasone pretreatment [86].

As opposed to existing in such black and white terms, there is probably a set of conditions under which steroids will be helpful and another set under which they will be harmful for any given disease. Nakamura et al. explored this using a mouse model of viral myocarditis [87]. In a series of experiments the authors showed that modest dose dexamethasone (0.75 mg/kg) administration before or directly following cocksackie virus infection resulted in improved survival and attenuated fibrosis and systolic dysfunction. However, if the dexamethasone was given at a late time point (>7 days after induction of myocarditis) then these beneficial effects were lost and it resulted in decreased mouse survival and increased cardiac tumor necrosis factor-α. As the beneficial effects were shown to be completely reversed by a selective COX-2 inhibitor, the authors speculated that it was the action of dexamethasone on cardiomyocytes directly and not its anti-inflammatory effects which conferred the beneficial effects at the earlier time point.

While rodent models seem to indicate a somewhat more negative view of using steroids in the treatment of cardiovascular disease, the opposite is generally true for human trials. One cardiovascular intervention which lends itself with relative ease to outcomes testing with regard to steroid effect is drug-eluting stent placement after myocardial infarction. Coronary stent implantation has superior outcomes compared to fibrinolysis or balloon angioplasty [88–90], but is limited by re-stenosis with repeat intervention rates of about 10 % at 6 months post-procedure [89]. Coronary restenosis after stenting is a highly inflammatory process driven by the secretion of cytokines and adhesion molecules in response to endothelial damage [91]. Dexamethasone-eluting stents have been used with good effect and in several trials have been associated with low rates of target lesion revascularization, reinfarction and late stent thrombosis [91, 92].

Oral steroid therapy in combination with bare metal stents has also been studied in patients who had received percutaneous coronary interventions after myocardial infarction. Interestingly there appears to be a prednisone dose response effect in this setting. Known as the IMPRESS (Immunosuppressive Therapy for the Prevention of Restenosis After Coronary Stent Implantation) trials, there were both high dose (HD) and low dose (LD) protocols tested [93, 94]. HD treatment consisted of a total of 45 days of oral prednisone after myocardial infarction with a dose of 1 mg/kg/day prescribed in the first 10 days, 0.5 mg/kg/day for days 11–30 and 0.25 mg/kg/day for days 31–45. LD therapy was considered to be prednisone 1 mg/kg/day for 5 days, 0.5 mg/kg/day for days 5–15 and 0.25 mg/kg/day for days 16–30 [95]. In these studies the LD prednisone regimen was ineffective in preventing major adverse cardiac events and had a restenosis rate of ~5× greater than that observed in the HD group. The authors speculated that the LD protocol, with its abbreviated duration and lower total steroid dose was insufficient to enact the systemic suppression of inflammation that was observed in the HD group. These results argue that the systemic effect of steroids may, in some cases, be necessary for beneficial cardiovascular outcomes and perhaps superior to tissue-specific steroid effects under some conditions.

Conclusions

In summary, despite an enormous body of basic science research as well as a robust clinical experience, the role of glucocorticoids in the regulation of the cardiovascular system remains poorly understood. In vitro studies and work in animal models have generally highlighted the effects or responses of glucocorticoids in a single cell type. In reality, clinical studies in humans are, by necessity much more heterogeneous and cannot as reliably examine cell-specific effects. Though pre-clinical and bench research has and will continue to direct and inform human clinical trials, there is much that still needs to be clarified regarding the role of the glucocorticoids in the development and function of the cardiovascular system.

References

1. Mantero F, Boscaro M. Glucocorticoid-dependent hypertension. J Steroid Biochem Mol Biol. 1992;43:409–13.
2. Baid S, Nieman LK. Glucocorticoid excess and hypertension. Curr Hypertens Rep. 2004; 6:493–9.
3. Etxabe J, Vazquez JA. Morbidity and mortality in Cushing's disease: an epidemiological approach. Clin Endocrinol (Oxf). 1994;40:479–84.
4. Whitworth JA, Mangos GJ, Kelly JJ. Cushing, cortisol, and cardiovascular disease. Hypertension. 2000;36:912–6.
5. Sabharwal P, Fishel RS, Breslow MJ. Adrenal insufficiency—an unusual cause of shock in postoperative patients. Endocr Pract. 1998;4:387–90.
6. De Wachter E, Vanbesien J, De Schutter I, Malfroot A, De Schepper J. Rapidly developing Cushing syndrome in a 4-year-old patient during combined treatment with itraconazole and inhaled budesonide. Eur J Pediatr. 2003;162:488–9.
7. Bertagna X, Bertagna C, Laudat MH, Husson JM, Girard F, et al. Pituitary-adrenal response to the antiglucocorticoid action of RU 486 in Cushing's syndrome. J Clin Endocrinol Metab. 1986;63:639–43.
8. Nieman LK, Chrousos GP, Kellner C, Spitz IM, Nisula BC, et al. Successful treatment of Cushing's syndrome with the glucocorticoid antagonist RU 486. J Clin Endocrinol Metab. 1985;61:536–40.
9. Sartor O, Cutler Jr GB. Mifepristone: treatment of Cushing's syndrome. Clin Obstet Gynecol. 1996;39:506–10.
10. Kalimi M. Role of antiglucocorticoid RU 486 on dexamethasone-induced hypertension in rats. Am J Physiol. 1989;256:E682–5.
11. Grunfeld JP, Eloy L, Moura AM, Ganeval D, Ramos-Frendo B, et al. Effects of antiglucocorticoids on glucocorticoid hypertension in the rat. Hypertension. 1985;7:292–9.
12. Mangos GJ, Whitworth JA, Williamson PM, Kelly JJ. Glucocorticoids and the kidney. Nephrology (Carlton). 2003;8:267–73.
13. Williamson PM, Kelly JJ, Whitworth JA. Dose-response relationships and mineralocorticoid activity in cortisol-induced hypertension in humans. J Hypertens Suppl. 1996;14:S37–41.

14. Montrella-Waybill M, Clore JN, Schoolwerth AC, Watlington CO. Evidence that high dose cortisol-induced Na+ retention in man is not mediated by the mineralocorticoid receptor. J Clin Endocrinol Metab. 1991;72:1060–6.
15. Campen TJ, Vaughn DA, Fanestil DD. Mineralo- and glucocorticoid effects on renal excretion of electrolytes. Pflugers Arch. 1983;399:93–101.
16. Funder JW, Pearce PT, Myles K, Roy LP. Apparent mineralocorticoid excess, pseudohypoaldosteronism, and urinary electrolyte excretion: toward a redefinition of mineralocorticoid action. FASEB J. 1990;4:3234–8.
17. Muller OG, Parnova RG, Centeno G, Rossier BC, Firsov D, et al. Mineralocorticoid effects in the kidney: correlation between alphaENaC, GILZ, and Sgk-1 mRNA expression and urinary excretion of Na+ and K+. J Am Soc Nephrol. 2003;14:1107–15.
18. Stewart PM, Corrie JE, Shackleton CH, Edwards CR. Syndrome of apparent mineralocorticoid excess. A defect in the cortisol-cortisone shuttle. J Clin Invest. 1988;82:340–9.
19. Todd-Turla KM, Schnermann J, Fejes-Toth G, Naray-Fejes-Toth A, Smart A, et al. Distribution of mineralocorticoid and glucocorticoid receptor mRNA along the nephron. Am J Physiol. 1993;264:F781–91.
20. Yan K, Kudo A, Hirano H, Watanabe T, Tasaka T, et al. Subcellular localization of glucocorticoid receptor protein in the human kidney glomerulus. Kidney Int. 1999;56:65–73.
21. Baylis C, Handa RK, Sorkin M. Glucocorticoids and control of glomerular filtration rate. Semin Nephrol. 1990;10:320–9.
22. Welbourne TC. Glucocorticoid control of ammoniagenesis in the proximal tubule. Semin Nephrol. 1990;10:339–49.
23. Rodriguez HJ, Sinha SK, Starling J, Klahr S. Regulation of renal Na+-K+-ATPase in the rat by adrenal steroids. Am J Physiol. 1981;241:F186–95.
24. Freiberg JM, Kinsella J, Sacktor B. Glucocorticoids increase the Na+-H+ exchange and decrease the Na+ gradient-dependent phosphate-uptake systems in renal brush border membrane vesicles. Proc Natl Acad Sci U S A. 1982;79:4932–6.
25. Kinsella J, Cujdik T, Sacktor B. Na+-H+ exchange activity in renal brush border membrane vesicles in response to metabolic acidosis: the role of glucocorticoids. Proc Natl Acad Sci U S A. 1984;81:630–4.
26. Frick A, Durasin I. Proximal tubular reabsorption of inorganic phosphate in adrenalectomized rats. Pflugers Arch. 1980;385:189–92.
27. Welch WJ, Ott CE, Guthrie Jr GP, Kotchen TA. Renin secretion and loop of Henle chloride reabsorption in the adrenalectomized rat. Am J Physiol. 1985;249:F596–602.
28. Doucet A, Hus-Citharel A, Morel F. In vitro stimulation of Na-K-ATPase in rat thick ascending limb by dexamethasone. Am J Physiol. 1986;251:F851–7.
29. Stanton BA. Regulation by adrenal corticosteroids of sodium and potassium transport in loop of Henle and distal tubule of rat kidney. J Clin Invest. 1986;78:1612–20.
30. Naray-Fejes-Toth A, Fejes-Toth G. Glucocorticoid receptors mediate mineralocorticoid-like effects in cultured collecting duct cells. Am J Physiol. 1990;259:F672–8.
31. Naray-Fejes-Toth A, Snyder PM, Fejes-Toth G. The kidney-specific WNK1 isoform is induced by aldosterone and stimulates epithelial sodium channel-mediated Na+ transport. Proc Natl Acad Sci U S A. 2004;101:17434–9.
32. Li C, Li Y, Liu H, Sun Z, Lu J, et al. Glucocorticoid repression of human with-no-lysine (K) kinase-4 gene expression is mediated by the negative response elements in the promoter. J Mol Endocrinol. 2008;40:3–12.
33. Yang CL, Angell J, Mitchell R, Ellison DH. WNK kinases regulate thiazide-sensitive Na-Cl cotransport. J Clin Invest. 2003;111:1039–45.
34. Wilson FH, Disse-Nicodeme S, Choate KA, Ishikawa K, Nelson-Williams C, et al. Human hypertension caused by mutations in WNK kinases. Science. 2001;293:1107–12.
35. Kahle KT, Wilson FH, Leng Q, Lalioti MD, O'Connell AD, et al. WNK4 regulates the balance between renal NaCl reabsorption and K+ secretion. Nat Genet. 2003;35:372–6.
36. Goodwin JE, Zhang J, Velazquez H, Geller DS. The glucocorticoid receptor in the distal nephron is not necessary for the development or maintenance of dexamethasone-induced hypertension. Biochem Biophys Res Commun. 2010;394:266–71.

37. Provencher PH, Saltis J, Funder JW. Glucocorticoids but not mineralocorticoids modulate endothelin-1 and angiotensin II binding in SHR vascular smooth muscle cells. J Steroid Biochem Mol Biol. 1995;52:219–25.
38. Kornel L, Nelson WA, Manisundaram B, Chigurupati R, Hayashi T. Mechanism of the effects of glucocorticoids and mineralocorticoids on vascular smooth muscle contractility. Steroids. 1993;58:580–7.
39. Tsugita M, Iwasaki Y, Nishiyama M, Taguchi T, Shinahara M, et al. Differential regulation of 11beta-hydroxysteroid dehydrogenase type-1 and -2 gene transcription by proinflammatory cytokines in vascular smooth muscle cells. Life Sci. 2008;83:426–32.
40. Wallerath T, Witte K, Schafer SC, Schwarz PM, Prellwitz W, et al. Down-regulation of the expression of endothelial NO synthase is likely to contribute to glucocorticoid-mediated hypertension. Proc Natl Acad Sci U S A. 1999;96:13357–62.
41. Ray KP, Searle N. Glucocorticoid inhibition of cytokine-induced E-selectin promoter activation. Biochem Soc Trans. 1997;25:189S.
42. Yang S, Zhang L. Glucocorticoids and vascular reactivity. Curr Vasc Pharmacol. 2004;2:1–12.
43. Sato A, Suzuki H, Nakazato Y, Shibata H, Inagami T, et al. Increased expression of vascular angiotensin II type 1A receptor gene in glucocorticoid-induced hypertension. J Hypertens. 1994;12:511–6.
44. Kornel L, Prancan AV, Kanamarlapudi N, Hynes J, Kuzianik E. Study on the mechanisms of glucocorticoid-induced hypertension: glucocorticoids increase transmembrane Ca2+ influx in vascular smooth muscle in vivo. Endocr Res. 1995;21:203–10.
45. Molnar GA, Lindschau C, Dubrovska G, Mertens PR, Kirsch T, et al. Glucocorticoid-related signaling effects in vascular smooth muscle cells. Hypertension. 2008;51:1372–8.
46. Ong SL, Zhang Y, Sutton M, Whitworth JA. Hemodynamics of dexamethasone-induced hypertension in the rat. Hypertens Res. 2009;32:889–94.
47. Goodwin JE, Zhang J, Geller DS. A critical role for vascular smooth muscle in acute glucocorticoid-induced hypertension. J Am Soc Nephrol. 2008;19:1291–9.
48. Imai Y, Abe K, Sasaki S, Minami N, Munakata M, et al. Exogenous glucocorticoid eliminates or reverses circadian blood pressure variations. J Hypertens. 1989;7:113–20.
49. Piovesan A, Panarelli M, Terzolo M, Osella G, Matrella C, et al. 24-hour profiles of blood pressure and heart rate in Cushing's syndrome: relationship between cortisol and cardiovascular rhythmicities. Chronobiol Int. 1990;7:263–5.
50. Fallo F, Fanelli G, Cipolla A, Betterle C, Boscaro M, et al. 24-hour blood pressure profile in Addison's disease. Am J Hypertens. 1994;7:1105–9.
51. Wallerath T, Godecke A, Molojavyi A, Li H, Schrader J, et al. Dexamethasone lacks effect on blood pressure in mice with a disrupted endothelial NO synthase gene. Nitric Oxide. 2004;10:36–41.
52. Mitchell BM, Dorrance AM, Mack EA, Webb RC. Glucocorticoids decrease GTP cyclohydrolase and tetrahydrobiopterin-dependent vasorelaxation through glucocorticoid receptors. J Cardiovasc Pharmacol. 2004;43:8–13.
53. Goodwin JE, Zhang J, Gonzalez D, Albinsson S, Geller DS. Knockout of the vascular endothelial glucocorticoid receptor abrogates dexamethasone-induced hypertension. J Hypertens. 2011;29(7):1347–56.
54. Aras-Lopez R, Xavier FE, Ferrer M, Balfagon G. Dexamethasone decreases neuronal nitric oxide release in mesenteric arteries from hypertensive rats through decreased protein kinase C activation. Clin Sci (Lond). 2009;117:305–12.
55. Yudt MR, Cidlowski JA. The glucocorticoid receptor: coding a diversity of proteins and responses through a single gene. Mol Endocrinol. 2002;16:1719–26.
56. Masuzaki H, Paterson J, Shinyama H, Morton NM, Mullins JJ, et al. A transgenic model of visceral obesity and the metabolic syndrome. Science. 2001;294:2166–70.
57. Fowden AL, Li J, Forhead AJ. Glucocorticoids and the preparation for life after birth: are there long-term consequences of the life insurance? Proc Nutr Soc. 1998;57:113–22.
58. Slotkin TA, Seidler FJ, Kavlock RJ, Gray JA. Fetal dexamethasone exposure accelerates development of renal function: relationship to dose, cell differentiation and growth inhibition. J Dev Physiol. 1992;17:55–61.

59. Torres A, Belser 3rd WW, Umeda PK, Tucker D. Indicators of delayed maturation of rat heart treated prenatally with dexamethasone. Pediatr Res. 1997;42:139–44.
60. Rog-Zielinska EA, Thomson A, Kenyon CJ, Brownstein DG, Moran CM, et al. Glucocorticoid receptor is required for foetal heart maturation. Hum Mol Genet. 2013;22:3269–82.
61. Roghair RD, Segar JL, Sharma RV, Zimmerman MC, Jagadeesha DK, et al. Newborn lamb coronary artery reactivity is programmed by early gestation dexamethasone before the onset of systemic hypertension. Am J Physiol Regul Integr Comp Physiol. 2005;289: R1169–76.
62. Volk KA, Roghair RD, Jung F, Scholz TD, Lamb FS, et al. Coronary endothelial function and vascular smooth muscle proliferation are programmed by early-gestation dexamethasone exposure in sheep. Am J Physiol Regul Integr Comp Physiol. 2010;298:R1607–14.
63. Crump C, Sundquist K, Sundquist J, Winkleby MA. Gestational age at birth and mortality in young adulthood. JAMA. 2011;306:1233–40.
64. Huhta JC. Fetal congestive heart failure. Semin Fetal Neonatal Med. 2005;10:542–52.
65. Rog-Zielinska EA, Richardson RV, Denvir MA, Chapman KE. Glucocorticoids and foetal heart maturation; implications for prematurity and foetal programming. J Mol Endocrinol. 2014;52:R125–35.
66. Vuguin PM. Animal models for small for gestational age and fetal programming of adult disease. Horm Res. 2007;68:113–23.
67. Lee SR, Kim HK, Youm JB, Dizon LA, Song IS, et al. Non-genomic effect of glucocorticoids on cardiovascular system. Pflugers Arch. 2012;464:549–59.
68. Losel R, Wehling M. Nongenomic actions of steroid hormones. Nat Rev Mol Cell Biol. 2003;4:46–56.
69. Kfir-Erenfeld S, Sionov RV, Spokoini R, Cohen O, Yefenof E. Protein kinase networks regulating glucocorticoid-induced apoptosis of hematopoietic cancer cells: fundamental aspects and practical considerations. Leuk Lymphoma. 2010;51:1968–2005.
70. Hafezi-Moghadam A, Simoncini T, Yang Z, Limbourg FP, Plumier JC, et al. Acute cardiovascular protective effects of corticosteroids are mediated by non-transcriptional activation of endothelial nitric oxide synthase. Nat Med. 2002;8:473–9.
71. Kewalramani G, Puthanveetil P, Kim MS, Wang F, Lee V, et al. Acute dexamethasone-induced increase in cardiac lipoprotein lipase requires activation of both Akt and stress kinases. Am J Physiol Endocrinol Metab. 2008;295:E137–47.
72. Puthanveetil P, Wang Y, Wang F, Kim MS, Abrahani A, et al. The increase in cardiac pyruvate dehydrogenase kinase-4 after short-term dexamethasone is controlled by an Akt-p38-forkhead box other factor-1 signaling axis. Endocrinology. 2010;151:2306–18.
73. Steiner A, Locher R, Sachinidis A, Vetter W. Cortisol-stimulated phosphoinositide metabolism in vascular smooth muscle cells: a role for glucocorticoids in blood pressure control? J Hypertens Suppl. 1989;7:S140–1.
74. Yano K, Tsuda Y, Kaji Y, Kanaya S, Fujino T, et al. Effects of hydrocortisone on transmembrane currents in guinea pig ventricular myocytes—possible evidence for positive inotropism. Jpn Circ J. 1994;58:836–43.
75. Brostjan C, Anrather J, Csizmadia V, Stroka D, Soares M, et al. Glucocorticoid-mediated repression of NFkappaB activity in endothelial cells does not involve induction of IkappaBalpha synthesis. J Biol Chem. 1996;271:19612–6.
76. Whitworth JA, Kelly JJ, Brown MA, Williamson PM, Lawson JA. Glucocorticoids and hypertension in man. Clin Exp Hypertens. 1997;19:871–84.
77. Hua SY, Chen YZ. Membrane receptor-mediated electrophysiological effects of glucocorticoid on mammalian neurons. Endocrinology. 1989;124:687–91.
78. Schmidt BM, Gerdes D, Feuring M, Falkenstein E, Christ M, et al. Rapid, nongenomic steroid actions: a new age? Front Neuroendocrinol. 2000;21:57–94.
79. Schoneveld JL, Fritsch-Stork RD, Bijlsma JW. Nongenomic glucocorticoid signaling: new targets for immunosuppressive therapy? Arthritis Rheum. 2011;63:3665–7.
80. DeVries AC, Joh HD, Bernard O, Hattori K, Hurn PD, et al. Social stress exacerbates stroke outcome by suppressing Bcl-2 expression. Proc Natl Acad Sci U S A. 2001;98:11824–8.

81. Balkaya M, Prinz V, Custodis F, Gertz K, Kronenberg G, et al. Stress worsens endothelial function and ischemic stroke via glucocorticoids. Stroke. 2011;42:3258–64.
82. Mihailidou AS, Le Loan TY, Mardini M, Funder JW. Glucocorticoids activate cardiac mineralocorticoid receptors during experimental myocardial infarction. Hypertension. 2009;54:1306–12.
83. Schulz M, Eggert M, Baniahmad A, Dostert A, Heinzel T, et al. RU486-induced glucocorticoid receptor agonism is controlled by the receptor N terminus and by corepressor binding. J Biol Chem. 2002;277:26238–43.
84. Garcia RA, Go KV, Villarreal FJ. Effects of timed administration of doxycycline or methylprednisolone on post-myocardial infarction inflammation and left ventricular remodeling in the rat heart. Mol Cell Biochem. 2007;300:159–69.
85. Xu B, Strom J, Chen QM. Dexamethasone induces transcriptional activation of Bcl-xL gene and inhibits cardiac injury by myocardial ischemia. Eur J Pharmacol. 2011;668:194–200.
86. Kaljusto ML, Stenslokken KO, Mori T, Panchenko A, Frantzen ML, et al. Preconditioning effects of steroids and hyperoxia on cardiac ischemia-reperfusion injury and vascular reactivity. Eur J Cardiothorac Surg. 2008;33:355–63.
87. Nakamura H, Kunitsugu I, Fukuda K, Matsuzaki M, Sano M. Diverse stage-dependent effects of glucocorticoids in a murine model of viral myocarditis. J Cardiol. 2013;61:237–42.
88. Keeley EC, Boura JA, Grines CL. Primary angioplasty versus intravenous thrombolytic therapy for acute myocardial infarction: a quantitative review of 23 randomised trials. Lancet. 2003;361:13–20.
89. Stone GW, Grines CL, Cox DA, Garcia E, Tcheng JE, et al. Comparison of angioplasty with stenting, with or without abciximab, in acute myocardial infarction. N Engl J Med. 2002;346:957–66.
90. Nordmann AJ, Hengstler P, Harr T, Young J, Bucher HC. Clinical outcomes of primary stenting versus balloon angioplasty in patients with myocardial infarction: a meta-analysis of randomized controlled trials. Am J Med. 2004;116:253–62.
91. Jimenez-Valero S, Santos B, Pajin F, Canton T, Lazaro E, et al. Clinical outcomes of dexamethasone-eluting stent implantation in ST-elevation acute myocardial infarction. Catheter Cardiovasc Interv. 2007;70:492–7.
92. Liu X, Huang Y, Hanet C, Vandormael M, Legrand V, et al. Study of antirestenosis with the BiodivYsio dexamethasone-eluting stent (STRIDE): a first-in-human multicenter pilot trial. Catheter Cardiovasc Interv. 2003;60:172–8; discussion 179.
93. Ribichini F, Tomai F, Ferrero V, Versaci F, Boccuzzi G, et al. Immunosuppressive oral prednisone after percutaneous interventions in patients with multi-vessel coronary artery disease. The IMPRESS-2/MVD study. EuroIntervention. 2005;1:173–80.
94. Versaci F, Gaspardone A, Tomai F, Ribichini F, Russo P, et al. Immunosuppressive therapy for the prevention of restenosis after coronary artery stent implantation (IMPRESS Study). J Am Coll Cardiol. 2002;40:1935–42.
95. Ferrero V, Ribichini F, Rognoni A, Marino P, Brunelleschi S, et al. Comparison of efficacy and safety of lower-dose to higher-dose oral prednisone after percutaneous coronary interventions (the IMPRESS-LD study). Am J Cardiol. 2007;99:1082–6.

Chapter 14
Glucocorticoids and Cancer

Miles A. Pufall

Abstract Unlike other steroid hormone receptors, the glucocorticoid receptor (GR) is not considered an oncogene. In breast cancer, the estrogen receptor (ER) drives cell growth, proliferation, and metastasis, and the androgen receptor (AR) plays a similar role in prostate cancer. Accordingly, treatment of these diseases has focused on blocking steroid hormone receptor function. In contrast, glucocorticoids (GCs) work through GR to arrest growth and induce apoptosis in lymphoid tissue. Glucocorticoids are amazingly effective in this role, and have been deployed as the cornerstone of lymphoid cancer treatment for decades. Unfortunately, not all patients respond to GCs and dosage is restricted by immediate and long term side effects. In this chapter we review the treatment protocols that employ glucocorticoids as a curative agent, elaborate on what is known about their mechanism of action in these cancers, and also summarize the palliative uses of glucocorticoids for other cancers.

Keywords Leukemia • Apoptosis • Lymphoma • Chemotherapy • Palliative

Glucocorticoids in Cancer

Unlike other highly related nuclear hormone receptors, the glucocorticoid receptor (GR) is not considered an oncogene. In breast cancer, the estrogen receptor (ER) drives cell growth, proliferation and metastasis. The androgen receptor (AR) plays a similar role in prostate cancer. Accordingly, treatment of these diseases has focused on blocking estrogen or testosterone production, or directly blocking steroid binding to their respective receptors. In contrast, glucocorticoids (GCs) work through GR to perform a variety of functions, including arresting growth or inducing apoptosis in lymphocytes. Glucocorticoids are so effective in this role that they are the cornerstone of treatment for all lymphatic cancers, though often

M.A. Pufall, M.S., Ph.D. (✉)
Department of Biochemistry, Carver College of Medicine, Holden Comprehensive Cancer Center, 51 Newton Road, Bowen Science Building, Room 4-430, Iowa City, IA 52242, USA
e-mail: miles-pufall@uiowa.edu

J.-C. Wang, C. Harris (eds.), *Glucocorticoid Signaling*, Advances in Experimental Medicine and Biology 872, DOI 10.1007/978-1-4939-2895-8_14

hampered by a panoply of off-target consequences. In this chapter, we will review the treatment protocols that employ glucocorticoids as a curative agent, elaborate on what is known about the mechanism of how they function in those cancers, and also summarize the palliative uses of glucocorticoids in treatment of a variety of cancers, and the implications of that use.

Lymphoid Cancers

Glucocorticoids in Childhood Leukemia

Much of what is known about glucocorticoids as a chemotherapeutic, and indeed many other cytotoxic agents, was learned from treatment of childhood leukemia. The predominant role of glucocorticoids in cancer is in the treatment of lymphoid malignancies. Building on an observation that there is an inverse relationship between the size of the adrenal cortex and thymus [1], Dougherty and White demonstrated that administration of ACTH reduced the size of most lymphoid tissue (excluding the spleen) [2]. Around the same time, cells from a tumor of unknown origin, later described to be "lymphosarcoma," were injected into mice and were found to form tumors. These tumors did not spontaneously regress, but instead shrunk when exposed to Compound E, otherwise known as cortisone [3]. A similar effect was observed in rats [4]. The increased catabolism observed at the time was thought to be responsible for the consumption of the tumor. These findings were rapidly moved to the clinic where is was reported in 1949 that either ACTH or cortisone acetate dramatically reduced the size of lymphoid tumors or leukemias, but not other carcinomas [5]. Emboldened by these findings, much larger studies were undertaken to explore the effect of cortisone and ACTH on hematological disorders. Patients were recruited with lymphoid malignancies, including CLL, lymphoma, Hodgkin's, acute lymphoblastic leukemia, and multiple myeloma (plasma cell myeloma), as well as the myeloid derived AML and CML. Beneficial effects were observed specifically in lymphoid, but not myeloid disease, ranging from symptomatic relief (multiple myeloma) to complete, but *temporary* remission in childhood ALL [6]. Despite the effectiveness of GCs in reducing lymphoid disease, as single agents they did not produce durable remissions, much less a cures.

Combination Therapy: Glucocorticoids Have a Central Role

Arround the same time, in the early 1950s, other agents were being tested for their anti-tumorigenic properties. Folic acid had been shown to increase the growth of lymphoid tumors, likely by upregulating amino acid and purine biosynthetic pathways. Much as with GCs blocking these pathways with anti-folates, such as the early aminopterin and later methotrexate, produced, a dramatic, but temporary remission in lymphoid malignancies. A dark, but serendipitous, observation brought another agent to the fore. Exposure to mustard gases in World War I was shown to deplete bone marrow and lymph nodes. The alkylating properties of these gasses were developed into drugs, such as cyclophosphamide, that proved to be potent anti-tumor agents in rodent models and later in human trials. Inhibition of nucleotide synthesis was effective as well, with 6-mercaptopurine showing promise both in acute leukemias and other cancers [7]. Interestingly, although each of these classes of drugs was shown to reduce or even clear the cancerous disease, none of the remissions proved durable.

Three treatment breakthroughs came in the early 1960s. First, vinca alkaloids (later vincristine), isolated from plants, were found to disrupt microtubules and have potent anti-tumorogenic properties [7]. Next, it was discovered that leukemic blasts are unable to make their own asparagine, and thus needed to absorb it from their medium to survive. To exploit this, a component of guinea pig serum, l-asparaginase, was isolated that converts free asparagine to aspartic acid, effectively starving the cells [8]. Perhaps most importantly, in the early 1960s intrepid cancer physicians took the radical step of administering agents in combination. The first successful protocol, called VAMP (vincristine, amethopterin, 6-mercaptopurine, and prednisone), improved. The 5 year survival of children with leukemia from 25 % with single agents to over 60 % [7].

For childhood leukemia, these basic components have evolved over the years to today's treatment, which is administered in three phases: remission induction, intensification (consolidation), and maintenance. Glucocorticoids can be administered during all three phases, but are used most intensely during remission induction, with the goal of eliminating greater than 99 % of the disease tissue (minimum residual disease, MRD) [9]. During this phase a glucocorticoids are administered with vincristine and asparaginase and are 98 % effective in inducing remission in childhood B-cell leukemias. In intensification, mercaptopurine is combined with polyethylene glycol conjugated asparaginase (pegasparaginase), and methotrexate. Intensification may also include cyclophosphamide or cytarabine (araC) or cycles of reinduction, using the same agents described above. GCs may again be administered during maintenance therapy, though at a lower dose, and less frequently [9]. These advances in treatment have led to cure rates approaching 90 % in children with both B and T-cell acute lymphoblastic leukemia, making it one of the most treatable cancers. In addition, the success of these trials have informed the treatment of other lymphoid cancers, discussed below.

GC Response Predicts Treatment Response

Though only effective as a curative agent in combination, GCs are central to the effectiveness of treatment. This has been best elucidated in European studies from the Berlin-Frankfurt-Muenster (BFM) group, who have developed treatment protocols in parallel with those in the US. This consortium showed that the initial response of infants and children with leukemia to prednisone alone was the best predictor of eventual outcome to full treatment [10, 11]. Importantly, other groups showed that patient response to prednisone, and overall response, could also be predicted by treatment of leukemic blasts *ex vivo* with GCs [12, 13]. This suggested that the function of GCs in inducing leukemic blast cell death was not necessarily dependent on the environment, but the cell autonomous program initiated by the drug.

The importance of GCs in treatment of leukemias is perhaps best highlighted by comparisons of patient response to dexamethasone and prednisone. Dex and pred are both derivatives of cortisol, with dex differing by addition of a fluorine at the 9α and a methyl at C16. These two differences make dex more specific for GR, with little to no MR activity, and about 10–16× more potent according to established indices [14]. In clinical trials, substituting dex for pred in high risk ALL patients improves outcome by over 10 % (81–94 % overall survival) (COG AALL0232), despite each inducing indistinguishable MRD after induction. This indicates that response to GCs in not only predictive of eventual outcome, but is a major determinant of outcome.

Side and Late Effects

Unfortunately, despite the clear benefits of using high-dose potent GCs in disease treatment, the side effects and potential late effects limit what can be administered. Although dex is much more effective than pred in children, it is only well-tolerated in children under ten. For children over ten, and adults, dex is significantly more likely to cause avascular necrosis (AVN), psychiatric issues, muscle wasting, and mortality [14, 15]. The late effects, or effects that arise years after cessation of treatment, are also a concern. The muscle wasting, osteoporosis and metabolic effects of GCs can persist after treatment ends, and the eventual neuro-psychiatric effects are of concern. More recently it has been shown that pulsing dex during maintenance can produce the same outcomes with more acceptable side effects. Nonetheless, physicians are currently challenged with the choice of which GC to use, what dose, and for how long. A more complete understanding of how GCs function in leukemic blasts and other tissues will help inform these choices.

Mechanism of Action

The biological function of GCs in hematopoietic cells and why they induce cell death of lymphoid cells is not clear. However, seminal work performed by John Ashwell at the NIH indicates that glucocorticoids serve as a negative signal in lymphoid development. By knocking out GR in developing thymocytes he showed that the mice were immunocompromised due to a reduction in T cell repertoire. He further showed that intact glucocorticoid signaling was important for proper T cell *selection*, perhaps pointing to a role for endogenous GCs in developing lymphocytes [16]. More recently, experiments in mice indicate that GCs may have a similar role in B cells [17].

Although non-genomic effects of GR have been proposed, cell death appears to require GC-induced gene regulation. General blocks of transcription or translation by actinomycin D and cycloheximide, respectively, block GC-induced cell death

[18]. In addition, a mutation that weakens DNA-induced GR dimerization and blunts activation also impairs GC-induced cell death. Further, the effect of GCs appears to proceed in two steps; an initial growth arrest that lasts about 24 h, and subsequent cell death. Continuous administration of GCs is required through arrest to induce cell death, which takes 2–3 days more [19]. The most frequently reported mechanism of cell death is apoptosis, though cases of necrosis [20] and necroptosis [21] have also been reported.

The GR-regulated genes that are required to induce apoptosis have not been well defined, but studies on both patient samples and cell lines have lent some insight. The clearest role is in driving apoptosis genes. Activation of GR has been shown to tip the balance of the BH3 domain containing Bcl2 family of apoptotic factors towards apoptosis through activation of pro-apoptotic *BIM* (BCL2L11), and down regulation of the anti-apoptotic *BCL2* [22, 23]. Less directly, GCs consistently upregulate thioredoxin interacting protein (*TXNIP*), which results in accumulation of reactive oxygen species and apoptosis [24]. Part of the initial growth arrest likely involves regulation of cell-survival genes, such as repression of pro-growth c-*MYC* and Hexokinase II, and cell cycle genes [19]. Although blocking these pathways does not necessarily block GC induced apoptosis, they may contribute eventually to cell death [25]. There is some evidence that by inhibiting NFκB and growth signaling pathways (e.g. ERK, MEK) GR can arrest cells and induce cell death, though a clear direct mechanism has not been elaborated [24, 26]. In addition, there is now also evidence that GR suppresses miRNA expression, and that repression of miR-17-92 correlates with apoptosis. Thus, although GCs can alter apoptotic and cell survival pathways it is not clear how prevalent either mechanism is in inducing cell death, or whether these pathways are directly regulated by GR.

Glucocorticoid Resistance

Misregulation of apoptotic factors have been implicated in resistance. For example, resistant ALL patients exhibit higher expression levels of the anti-apoptotic genes *BCL2* and *MCL1* [25, 27, 28], which likely blunt the apoptotic signal affected by GCs. Overexpression of *MCL1* has been linked to activation of mTor signaling, and can be alleviated by inhibition of this pathway with rapamycin. On the other side of the coin, activation of Akt stimulates the mTor pathway (as well as perhaps inhibiting GR directly, see below) and makes cells resistant [28]. It has also been observed that a failure of GCs to activate the pro-apoptotic *BIM* contributes to resistance [29]. Thus, in cells where the balance of Bcl2 family members cannot be sufficiently biased toward pro-apoptosis, GC-induced cell death is likely to be impaired.

In addition to apoptosis, pathway analysis of resistant samples compared to sensitive ones has implicated repression of cell cycle genes [30], and increased carbohydrate consumption through overexpression of associated genes, including carbonic anhydrase 4 (CA4), glucose transporter 3 (GLUT3/SLC2A3), and glyceraldehyde-3-phosphate dehydrogenase (GAPDH) [27, 31, 32].

More general changes in transcription have been also been linked to GC resistance. For example, ALLs that have translocations in the *MLL* gene are more likely to not respond to treatment [33]. *MLL* encodes a histone methyl transferase that methylates lysine 4 of histone H3 and is a mark of active enhancers including response elements and active genes [34, 35]. Translocations that impair the methyltransferase activity of *MLL* show widespread changes in the chromatin state that are thought to reprogram, and generally downregulate gene expression [36]. In addition, mutational analysis of diagnostic and relapsed patients shows a significant enrichment for transcription factor mutations over other gene sets, again implicating transcriptional defects [37].

More recently, alterations in GR regulation have been implicated in resistance. Phosphorylation of S134 of GR was shown in to be associated with GC resistance in a T-ALL line (CEM). In a heterologous cell system, AKT phosphorylation of GR at this site appeared to impair translocation, providing a mechanism of resistance [38]. This finding, however, is at odds with another study showing that phosphorylation of S134 is able to translocate to the nucleus, but alters gene expression [39]. In other studies, a failure to increase GR levels through a positive feedback loop has been shown to impair GC induced apoptosis, though how this feedback is disrupted has not been established [40, 41]. Lastly, not all GR isoforms have the ability to induce apoptosis, suggesting that mechanisms that regulate isoform selection may play a role [42].

Adult Acute Lymphoblastic Leukemia

The treatment for adult ALL is modeled on the treatments that are so successful in children. Unfortunately, response rates in adults are significantly worse, with an 80–90 % initial response rate, but only 25–50 % disease free survival after 5 years. Why the response of adults is so much worse is not clear, but has been attributed to two factors. The first is that significantly more mutations accumulate in adult ALL blasts, though few have been directly attributable to treatment response. The second is that adults are not able to tolerate the treatment regimens as well as children. For example, with similar dosing of vincristine, daunorubicin, and dexamethasone during induction, treatment related death is almost 10 % in adults compared to ~1 % in children. To avoid some of these side effects, dex is given in pulses, rather than continuously during induction [43]. Because response rates are worse, bone marrow transplants are often the best route to a durable remission.

Intensification (consolidation) and maintenance are also similar to the children's protocol. Typical consolidation includes methotrexate, cytarabine, cyclophosphamide, and asparaginase. Clinical trials using hyper CVAD (cyclophosphamide, vincristine, doxorubicin, and dex) have shown somewhat better responses [44]. Maintenance therapy consists of 6-mecaptopurine and methotrexate with monthly pulses of vincristine and prednisone.

Treatment of T-ALLs are similar [43], except for cases with mutation in the NOTCH pathway. Notch is a cell surface receptor, whose intracellular domain is liberated by cleavage with γ-secretase upon ligand binding. This domain translocates to the nucleus and acts as a coactivator of transcription. Activating mutations in NOTCH are found in ~50 % of T cell ALLs and correlate with GC resistance. Administration of γ-secretase inhibitors reverses this resistance, but causes gut toxicity. Fortunately, GCs protect against gut toxicity [45], allowing inhibitors to be used in clinical trials (NCT01088763).

Multiple Myeloma

Multiple Myeloma is the clonal expansion of plasma cells, the mature B cells that emerge from germinal centers. The resulting cells both overpopulate the bone marrow and secrete excessive immunoglobulin, resulting in impaired immune function, renal disease, and bone lesions. At this point, multiple myeloma is not curable, but can be managed with chemotherapy. Like ALL, chemotherapeutic regimens have dramatically improved prognosis, from months in 1950s to 7 years or more for standard risk patients today [46].

The treatment for multiple myeloma has, until recently, involved alkylating agents and glucocorticoids almost exclusively. Like ALL, mustard gases and their derivatives, such as malphalan, were initially used for treatment of the disease. They also induced remission, but at a lower rate (about 1/3 response), and also quickly relapsed. Prednisone also exhibited initial effectiveness, with an average complete response rate of 44 %. In 1969 these two agents were combined to produce a much more robust response rate of 60 %. Unlike ALL, however, these treatments did not fully cure the disease, but improved survival. In the early 1970s, multiple alkylating agents (carmustine, cyclophosphamide, and melphalan) were combined with GCs and vincristine to increase initial response rates, though survival was not improved. Subsequently, dex was combined with doxorubicin and vincristine under the VAD protocol, which was used for years as the main treatment [46]. While these formulations were being used in the clinic, researchers were exploring the use of thalidomide, the teratogenic treatment for nausea and morning sickness used in the early 1960s. In 1990s, thalidomide was shown to have anti-angiogenic properties, including specific action on multiple myeloma. Combination of thalidomide and its derivatives with GCs and cyclophosphamide proved effective in treatment of relapsed multiple myeloma and was installed as the main treatment for most patients [47].

Proteasome inhibitors have also emerged as having activity in myeloma and synergy with GCs [48]. Combination therapy of bortezomib with GCs began in 2003, with other proteasome inhibitors being developed since then that have proven effective both in initial and relapse treatment. These include carfilzomib, which target different proteolytic activities within the proteasome.

Mechanism

Although less is known about how GCs induce cell death in multiple myeloma, there are some clear parallels with their mechanism of action in ALL. First, they modulate the expression of Bcl2 family members, tipping the balance to apoptotis. Second, GCs also inhibit proliferation by suppressing c-MYC. Lastly, it has been shown that GCs also affect the redox balance of multiple myeloma cells, which makes them more susceptible to cell death. GC activation of the transcription factor GilZ has also been implicated in apoptosis. GilZ is regulated by GR in all known tissues, but, in contrast to other tissues, induces apoptosis in multiple myeloma [49].

Resistance to GCs show similarities, but also some differences. Like ALL, activation of Akt attenuates the cytotoxic effects of GCs. However, in multiple myeloma the disease microenvironment shows a clear effect. Secretion of IL6 by either the disease itself or the supporting tissue severely impairs the response of multiple myeloma to GCs and treatment in general, and appears to work though NFκB [50]. Like ALL, unfortunately, not enough is known about the GC-induced program of cell death to account for how resistance arises in most cases.

Hodgkin's Disease

Hodgkin's disease occurs within lymph nodes as the clonal expansion of mature B cells. The disease is characterized by starting in one node then spreading systemically. First described in 1832, it was initially treated with radiation. Response was poor, and the treatment was stopped in 1920s. Many of the agents that worked in ALL were tried for Hodgkin's disease, including alkylating agents such as chlorambucil and cyclophosphamide, vincristine, and glucocorticoids. ACTH and cortisone used as single agents were found to induce remarkable, but temporary remission. It was not until 1967 that an effective combination therapy was formulated. The MOPP protocol included a mustard alkylator, vincristine (oncovin), procarbazine (another alkylator), and prednisone and achieved a 50 % cure rate. When combined with involved field radiation, MOPP produced even better results, with response rates as high as 70 % [43]. The MOPP protocol was unfortunately associated with a number of late effects, including nerve damage, infertility, and secondary malignancies such as acute leukemia. A better tolerated alternative called ABVD (Adriamycin, bleomycin, vinblastine, and dacarbazine) was developed in 1970s, and by 1990s had supplanted MOPP as more effective an better tolerated. Over the last few years, a new protocol, once again involving GCs, was developed called BEACOPP (bleomycin, etoposide, Adriamycin, cyclophosphamide, vincristine, procarbazine, and prednisone), and has demonstrated better initial response. This response comes at a cost. First, the BEACOPP treatment has a higher mortality rate, a higher secondary malignancy rate, and sterility risk. Second, the rate of recovery for those who fail to respond or relapse, called the "salvage rate" is lower. When taken into account, the rate of initial response plus the salvage rate for ABVD and BEACOPP are not significantly different [51]. The relative benefits of BEACOPP vs. ABVD are the subject of debate as of the writing of this chapter, though it is clear that ABVD is better tolerated by most patients.

Chronic Lymphocytic Leukemia

Chronic lymphocytic leukemia differs from ALL in that it is a disease of later stage B cells that accumulates over time, rather than as a result of hyperproliferation. CLL cells are refractory to apoptosis, and accumulate in the blood stream, lymph nodes, and bone marrow. The disease becomes pathological when CLL cells crowd out the production of other blood cells. In the early 1950s, the effect of ACTH or GCs on CLL was tested along with several other lymphoid malignancies [52]. When tested as a monotherapy, only ~11 % of patients had even a partial response [53]. Later, when used as part of combination therapies such as CHOP (cyclophosphamide, doxorubicin, vincristine, and prednisone), GCs were found to have no effect on eventual outcome while still causing side effects, and were not included in treatment regimens. When high dose prednisone was tested in patients in 1990s, a better initial response was observed, but the response was not durable [53]. The most severe side effects in these studies were opportunistic infections, and have limited the usefulness of GCs in CLL.

More recently monoclonal antibodies (mAbs) have become an essential agent in treatment of CLL. The most common therapy for CLL is FCR which consists of: an alkylating agent, such as cyclophosphamide; flutarabine, a nucleoside analog; and rituximab, a mAb directed against the B cell specific CD20 cell surface marker. This combination therapy has a very high initial response rate (some reports as high as 90 %) with an overall response rate over 50 % [43]. Based on the success of rituximab, other mAbs have been developed, including Ofatumumab (also against CD20), and alemtuzumab (against CD52). Recently, high dose GCs have been combined with mAbs in clinical trials for CLLs refractory to standard therapy. Although overall response rates have been over 50 %, the median progression free survival is less than a year [53].

Use of GCs in CLL is still being considered because of their role not in cell death, but in lymphocyte redistribution. In the late 1940s and early 1950s, patients with CLL treated with ACTH or cortisone experienced a reduction in nodal or splenic tumor masses. Surprisingly, this was accompanied by an increase in the circulating leukocytes, called leukocytosis. It was thought that, in addition to probably modest cell death, GCs induce a redistribution of leukocytes to the blood stream that is reversed upon removal of GCs. This behavior mirrors the normal, circadian redistribution of B cells. Under non-pathological conditions, when GCs are low, B and T cell circulation is high, but when cortisol spikes in the morning, cells home to tissue locations. Why CLL cells would leave tumors is not clear, but GC induced expression of CXCR4 causes B and T cells to enter the bloodstream, and eventually migrate to environments, such as the bone marrow or lymph nodes, that express CXCR12 [54]. This window of time when lymphocytes are circulating provides an opportunity for other agents to attack.

Non-Hodgkin's Lymphomas

Non-Hodgkin's Lymphoma (NHL) is the most common hematological malignancy diagnosed in the US. It represents a collection of over 30 subtypes of lymphoid malignancies that are distinct, for the most part, from leukemias in that they are not circulating. The subtypes are distinguished by their lineage, developmental stage, or location. Despite this heterogeneity, most NHLs are treated with a similar protocol that involves GCs [55]. Follicular, Mantle cell, diffuse large B cell lymphoma (DLBCL), and T cell lymphomas are treated with the CHOP protocol, which comprises cyclophosphamide, doxorubicin (hydroxyduanomycin), vincristine (Oncovin), and prednisone. Outcome can be improved in most B cell lymphomas with the addition of the mAb rituximab targeted against the B-cell specific antigen CD20 (R-CHOP). For more advanced case of DLBCL, ACVBP (doxycycline, cyclophosphamide, vincristine, bleomycin, and prednisone) or R-CHOP with etoposide can be used. Higher grade, or more aggressive NHLs are can be treated with hyperCVAD (cyclophosphamide, vincristine, doxorubicin, and dexamethasone), which is the same combination as CHOP, but with the more potent dexamethasone [43]. As observed in childhood ALL, the upgrade to Dex is clearly more effective, but harbors the risk of more side effects and late effects.

Solid Tumors

Prostate

Prostate cancer afflicts about one in five men in the US, making it their second most common cancer. The growth and proliferation of prostate cancer is driven by androgens, which work through the androgen receptor exclusively. Accordingly, an important part of treatment for prostate cancer is to block production of androgens, typically through castration. Though this is effective in blocking production of testosterone and inducing cancer regression, the cancer often returns in 2–3 years [56]. The relapsed tumor is able to grow in the apparent absence of testosterone, and is termed either castration resistant prostate cancer (CRPC) or Hormone Refractory Prostate Cancer (HRPC). Small molecule inhibitors that block testosterone binding to the androgen receptor have been developed, and can once again induce regression [57]. However, once resistance to these anti-androgens develops, the treatment options for CRPC are significantly less effective. Glucocorticoids are not used in initial therapy, but the glucocorticoid receptor has two opposing functions in anti-androgen refractory CRPC.

Prostate cancer is diagnosed and staged based on biopsies of the prostate gland and invasion to nearby tissues. Concomitant measurement of Prostate Specific Antigen (PSA), a gene driven specifically by AR in the prostate, serves as a marker for activity of AR in the prostate and a potential indicator of PC growth.

Although the accuracy of PSA as an indicator of prostate cancer progression or aggressiveness is controversial, it is nonetheless a widely accepted metric of AR activity in the prostate [43]. After successful initial therapy, when the prostate is either removed or regressed by androgen deprivation therapy followed by radiation therapy, PSA levels drop precipitously (normal, though it can vary widely, is ~4 ng/mL in the blood) [58]. During subsequent monitoring, a PSA level doubling time of 15 months or more is not associated with poor outcome, whereas a doubling time of <3 months is an indicator of recurrent disease. Radiographic evidence is more definitive. In these recurrent patients, AR activity is still observed despite low levels (<50 ng/dL) of circulating testosterone. This behavior is indicative of androgen independent activity of AR, hypersensitivity of AR to even low levels of testosterone, or an AR independent mechanism [58].

The former of these two possibilities, androgen independence and hypersensitivity to androgens, often result from overexpression of AR and are difficult to distinguish. One common therapy for such patients is glucocorticoids either alone or combined with other chemotherapeutics, such as paclitaxel or mitoxantrone. The effectiveness of this therapy varies widely, with 20–79 % of patients exhibiting suppressed PSA levels [56].

How patients derive any benefit from glucocorticoids is not clear. The administration of GCs provides a negative feedback on the pituitary gland, suppressing production of adrenal testosterone [59]. Though the adrenal gland is thought to be a minor source of androgens, a decrease may nonetheless provide relief if the patient is androgen hypersensitive. Quite separate from this mechanism, GR may act as a tumor suppressor in PC itself. GR is highly expressed in normal prostate, but often suppressed in PC. Studies in cell line models of PC show that GCs can suppress pro-growth or tumorogenic pathways such as IL-6, NFκB, and MAP and ERK kinases, as well as induce growth arrest through upregulation of TGFβ, p21 and p27 [60, 61]. However, for as many examples of GR regulated genes that potentially block PC proliferation, there are examples of the opposite, such as downregulation of p53 or upregulation of anti-apoptotic S100P. A definitive mechanism for how GCs prevent PC growth and progression awaits further study in primary tissue.

For locally advanced PC, the androgen deprivation therapy in combination with competitive AR blockers and radiation therapy are often used. Goserelin, a gonadotropin releasing hormone (GnRH) agonist, blocks androgen production by interrupting the endogenous feedback loop in the pituitary gland [57]. Androgen analogs, such as bicalutamide and flutamide, then inhibit AR function by blocking the testosterone-binding domain without activating the receptor. Despite the success of this therapy, resistance can emerge, and for those patients second-generation AR competitive antagonists have been developed, including enzaludamide [58]. Two modes of resistance to these second generation inhibitors have been described: mutation of AR and surprisingly upregulation of GR. High GR levels have been observed in PC bone metastases of enzaludamide resistant patients. In a preclinical model of PC in which GR is overexpressed, it was shown that GR could substitute for AR by regulating some of the same genes, including SGK1, STK39, and the PSA gene (KLK3). In all, about 80 % of GR regulated genes overlapped with those regulated by AR. Further, it was shown that GCs could induce growth of these cells in the presence of AR antagonists, and that effect could be blocked by GR antagonists [62]. This recent work suggests that in cells conditioned with anti-androgens, the highly homologous AR and GR can complement each other. The cellular factors that allow this complementation have not been identified, but this dependence on GR function for PC growth suggests that combination therapy with anti-glucocorticoids, such as RU-486, may be useful in resistant PC.

These two examples highlight the potential dangers of using GC therapy in hormone dependent cancers in which the mechanism is not well understood. GCs have a clear though modest, effect on CRPC, with a >20 % response rate. However, GCs can also be AR-like in hormone resistant PCs. Since the cellular determinants of GC action in these two relapsed PCs have not been determined, administering GCs may carry substantial risk.

Kaposi Sarcoma

Kaposi's Sarcoma is a virally induced cancer that is best known for being activated in patients with HIV/AIDS. Although mostly disfiguring, KS can have cause serious problems, including lymphedema, gastrointestinal blockage, and in rare cases, death. First-line treatment includes ABV (doxorubicin, bleomycin, and vincristine) among several formulations. These treatments manage the disease, but are not a cure.

A retrospective study concluded that the Hodgkin's formulation of EVAD (etoposide, vincristine, doxorubicin and dexamethasone) (EVAD) was an effective treatment for advanced or relapsed Kaposi's Sarcoma, though no mechanism was proposed [63].

Cancers Where Glucocorticoids Are Not Used as a Curative Agent

Glucocorticoids are often administered to help patients tolerate treatment, rather than as a chemotherapeutic that targets the cancer itself. Reflective of the biology of glucocorticoids described elsewhere in this book, their palliative effects are diverse. In some chemotherapeutic regimens, for example those that include cisplatin, GCs are first-line antiemetics (see below). For others, such as folate inhibitors, they are used to blunt hypersensitivity, which can result in severe skin rashes (see Table 1.1). Glucocorticoids are used for their anti-inflammatory properties to relieve bone pain other discomfort that may arise from metastatic disease and CNS compression due to metastatic disease. Though effective for these purposes, the use of glucocorticoids in patients with cancer caries some risk of protecting the tumor against chemotherapeutic agents, or even increasing proliferation rates.

Table 1.1 Palliative uses for glucocorticoids

Tumor type	Chemotherapy regimen	Glucocorticoid	Purpose	Reference
Metastatic bladder cancer	Pemetrexed (Altima), folic acid, vitamin B12 (vitamins)	Dexamethasone (4 mg)	Prevent skin rashes	[64]
Metastatic lung cancer	Pemetrexed (Altima) and cisplatin or bevacizumab (Avastin) and vitamins	Dexamethasone (4 mg)	Prevent skin rashes	[65–67]
Metastatic lung cancer	Paclitaxel, carboplatin premedicate with: diphenhydramine cimetidine, Ranitidine or Zantac	Dexamethsone (20, 8 mg if no hypersensitivity)	Blunt hypersensitivity	[68]
Mesoltheliomа	Cisplatin or carboplatin and pemetrexed	Dexamethasone (4 mg)	Prevent skin rashes	[69, 70]
Kaposi's sarcoma	Paclitaxel premedicate with: diphenhydramine, cimetidine	Dexamethsone (20, 8 mg if no hypersensitivity)	Blunt hypersensitivity	[71]
Breast cancer	Ixabepilone premedicate with: diphenhydramine, Ranitidine	Dexamethasone (20 mg)	Allergic reaction	[72–74]

Use of GCs as Antiemetics

Chemotherapeutics, radiation, and surgery, all of which are important tools in the fight against cancer, are often poorly tolerated by patients. In addition to their side effects, they can cause severe nausea and vomiting that result in weakness, dehydration, and an unwillingness of patients to continue with therapy. The relative emetic risk of chemotherapeutic agents have been categorized and published by American Society of Clinical Oncology [75]. The categories range from high likelihood (>90 %), with cisplatin being at the top of this range, to minimal likelihood (<10 %), which includes agents such as vincristine and rituximab. The virtual universal reaction of patients to cisplatin, which began being used in 1978, prompted the search for effective anti-emetic agents [76, 77].

There are a number of agents that are used to ameliorate these effects, including: dopaminergic blockers (e.g. metoclopramide); serotonin type 3 (5-HT$_3$) receptor inhibitors antagonists; NK1 receptor inhibitors (aprepitant); and low dose glucocorticoids such as dexamethasone and methylprednisolone. The ASCO guidelines recommend how to administer these classes of drugs. For moderately emetogenic treatments, a combination of dex and 5-HT is recommended. For highly emetogenic agents, such as cisplatin, combinations of GCs, 5-HT, and NK1 inhibitors are recommended. Although the mechanism for dopaminergic blockers, 5-HT, and NK1 inhibitors are well established, how GCs work is not well understood [78].

Some mechanisms for how GCs work have been hypothesized. First, physiological levels of GCs appear to be required for general well-being. Low levels of GC in and of themselves have been linked to nausea and vomiting [79]. Second, the anti-inflammatory actions may be sufficient on some cases. Cyclooxegenase inhibitors or ibuprofen, both of which suppress inflammation, can ameliorate the effects of both radiation and some chemotherapeutic agents [80]. Third, GCs have been shown to reduce 5-HT production, perhaps effectively blocking serotonin receptors in the vagal nerve complexes that transmit the vomiting response [81]. Lastly, GCs inhibit production of prostaglandins and substance P, both of which have been implicated in the vomiting response [80]. Other mechanisms have been suggested as well, such as reducing pain and direct effects on brain centers, but further research needs to be done to uncover which GC effects are most beneficial. In addition GCs are used as appetite stimulants in patients with cancer cachexia [82].

The Future of GCs in Cancer Treatment

GCs are still a critical component of chemotherapy for hematopoietic malignancies. As described above, they are very effective in treatment of lymphoid malignancies, including leukemia, lymphomas, and multiple myeloma, and much work is being done to enhance their effect and overcome resistance. However, the use of GCs as chemotherapeutic agents is considerably limited by their side effects, most prominently osteonecrosis [14]. A good deal of effort has been invested in the development of selective GR modulators (SGRMs, also sometime referred to as SEGRMs)—compounds that work through GR to enhance the beneficial effect and minimize or eliminate the side effects [83]. The prevailing model has been that GCs exert their beneficial effects by repression of genes with side effects resulting from gene activation [84]. Accordingly, the search for SGRMs has been focused on development of what are known as dissociating compounds, or those that selectively only allow GR to repress, but not activate genes. As more has been learned about GR gene regulation in a variety of tissues and conditions, this model has proven too simplistic (though it should be noted that the general trend holds). BIM and BCL2, positive and negative regulators of apoptosis, respectively, are good examples of how this model fails. To induce efficient apoptosis in leukemic blasts, GCs *activate* BIM, but *repress* BCL2 [22, 85]. Thus in this tissue, both activation and repression are beneficial. Further, both BIM and BCL2 appear to be similarly regulated in bone, contributing to osteonecrosis [86]. Therefor, for leukemia, a SGRM that represses but doesn't activate would be a less effective chemotherapeutic because BIM would not be activated. Further, even if a dissociating SGRM was developed to preserve regulation of BIM and BCL2, it would still have severe osteonecrotic side effects.

The future use of SGRMs for treatment of hematopoietic malignancies thus requires a model of GC function with tissue and gene-specific resolution. One method is to first identify genes that are regulated by GR specifically in lymphoid cells that contribute to cell death, but do not perform similar function in bone. Recently, KLF13 was identified as a GR regulated gene that helps coordinate B and

T cell development and GC-induced cell death [85]. As this gene does not yet appear to have a function in bone, development of compounds that separate allow KLF13 regulation, but not BIM or BCL2, might allow GC-induced apoptosis in leukemic blasts but not bone. There are other strategies currently under investigation to develop activity and tissue specific GC function. One is to develop a deeper of understanding of not just which genes are regulated, but *how* they are regulated in different tissues to identify alternative targets that enhance specific GR functions. One key may be in understanding which GR cofactors are used in each tissue. For examples, the GR cofactors NCOA1, 2, and 3, are differentially expressed in tissues [87]. If one, such as NCOA2, is the primary GR cofactor in bone, but not B cells, then it could be targeted to block bone-cell death but still allow GC-induced apoptosis in leukemic blasts. Lastly, it may be possible to develop GCs whose chemical makeup or delivery method partition uptake specifically to lymphoid cells over bone. In this way, the drug would provide selective modulation of GR function at the tissue level, but not at the level of gene regulation. Each of these strategies are the subject of current research efforts.

References

1. Kendall EC. Hormones. Annu Rev Biochem. 1941;10:285–336.
2. Dougherty TF, White A. Effect of pituitary adrenotropic hormone on lymphoid tissue. Exp Biol Med. 1943;53(2):132–3. doi:10.3181/00379727-53-14219P.
3. Heilman FR, Kendall EC. The influence of 11-dehydr0-17-hydroxycorticosterone (compound E) on the growth of a malignant tumor in the Mouse1. Endocrinology. 1944;34(6):416–20.
4. Murphy JB, Sturm E. The effect of adrenal cortical and pituitary adrenotropic hormones on transplanted leukemia in rats. Science. 1944;99(2572):303. doi:10.1126/science.99.2572.303.
5. Pearson OH, Eliel LP, Rawson RW, Dobriner K, Rhoads CP. Acth- and cortisone-induced regression of lymphoid tumors in man. A preliminary report. Cancer. 1949;2(6):943–5. doi:10.1002/1097-0142(194911)2:6<943::AID-CNCR2820020602>3.0.CO;2-P.
6. Pearson OH, Eliel LP. Use of pituitary adrenocorticotropic hormone (ACTH) and cortisone in lymphomas and leukemias. J Am Med Assoc. 1950;144(16):1349–53.
7. DeVita VT, Chu E. A history of cancer chemotherapy. Cancer Res. 2008;68(21):8643–53. doi:10.1158/0008-5472.CAN-07-6611.
8. Broome JD. Evidence that the L-asparaginase of guinea pig serum is responsible for its antilymphoma effects. J Exp Med. 1963;118:121–48.
9. Pui C-H, Evans WE. Treatment of acute lymphoblastic leukemia. N Engl J Med. 2006;354(2):166–78. doi:10.1056/NEJMra052603.
10. Dördelmann M, Reiter A, Borkhardt A, et al. Prednisone response is the strongest predictor of treatment outcome in infant acute lymphoblastic leukemia. Blood. 1999;94(4):1209–17.
11. Lonnerholm G, Thorn I, Sundstrom C, et al. In vitro cellular drug sensitivity at diagnosis is correlated to minimal residual disease at end of induction therapy in childhood acute lymphoblastic leukemia. Leuk Res. 2009;33(1):46–53. doi:10.1016/j.leukres.2008.06.012.
12. Hongo T, Yajima S, Sakurai M, Horikoshi Y, Hanada R. In vitro drug sensitivity testing can predict induction failure and early relapse of childhood acute lymphoblastic leukemia. Blood. 1997;89(8):2959–65.
13. Kaspers GJ, Veerman AJ, Pieters R, et al. In vitro cellular drug resistance and prognosis in newly diagnosed childhood acute lymphoblastic leukemia. Blood. 1997;90(7):2723–9.

14. Inaba H, Pui C-H. Glucocorticoid use in acute lymphoblastic leukaemia. Lancet Oncol. 2010;11(11):1096–106. doi:10.1016/S1470-2045(10)70114-5.
15. Teuffel O, Kuster SP, Hunger SP, et al. Dexamethasone versus prednisone for induction therapy in childhood acute lymphoblastic leukemia: a systematic review and meta-analysis. Leukemia. 2011;25(8):1232–8. doi:10.1038/leu.2011.84.
16. Mittelstadt PR, Monteiro JP, Ashwell JD. Thymocyte responsiveness to endogenous glucocorticoids is required for immunological fitness. J Clin Invest. 2012;122(7):2384–94. doi:10.1172/JCI63067.
17. Gruver-Yates AL, Quinn MA, Cidlowski JA. Analysis of glucocorticoid receptors and their apoptotic response to dexamethasone in male murine B cells during development. Endocrinology. 2014;155(2):463–74. doi:10.1210/en.2013-1473.
18. Smith LK, Cidlowski JA. Glucocorticoid-induced apoptosis of healthy and malignant lymphocytes. Prog Brain Res. 2010;182:1–30. doi:10.1016/S0079-6123(10)82001-1.
19. Ploner C, Schmidt S, Presul E, et al. Glucocorticoid-induced apoptosis and glucocorticoid resistance in acute lymphoblastic leukemia. J Steroid Biochem Mol Biol. 2005;93(2–5):153–60. doi:10.1016/j.jsbmb.2004.12.017.
20. Lambrou GI, Vlahopoulos S, Papathanasiou C, et al. Prednisolone exerts late mitogenic and biphasic effects on resistant acute lymphoblastic leukemia cells: relation to early gene expression. Leuk Res. 2009;33(12):1684–95. doi:10.1016/j.leukres.2009.04.018.
21. Bonapace L, Bornhauser BC, Schmitz M, et al. Induction of autophagy-dependent necroptosis is required for childhood acute lymphoblastic leukemia cells to overcome glucocorticoid resistance. J Clin Invest. 2010;120(4):1310–23. doi:10.1172/JCI39987DS1.
22. Ploner C, Rainer J, Niederegger H, et al. The BCL2 rheostat in glucocorticoid-induced apoptosis of acute lymphoblastic leukemia. Leukemia. 2007;22(2):370–7. doi:10.1038/sj.leu.2405039.
23. Schmidt S, Rainer J, Ploner C, Presul E, Riml S, Kofler R. Glucocorticoid-induced apoptosis and glucocorticoid resistance: molecular mechanisms and clinical relevance. Cell Death Differ. 2004;11:S45–55. doi:10.1038/sj.cdd.4401456.
24. Tissing WJE, den Boer ML, Meijerink JPP, et al. Genomewide identification of prednisolone-responsive genes in acute lymphoblastic leukemia cells. Blood. 2007;109(9):3929–35. doi:10.1182/blood-2006-11-056366.
25. Sionov RV, Spokoini R, Kfir-Erenfeld S, Cohen O, Yefenof E. Chapter 6 mechanisms regulating the susceptibility of hematopoietic malignancies to glucocorticoid-induced apoptosis. Adv Cancer Res. 2008;101:127–248. doi:10.1016/S0065-230X(08)00406-5.
26. Schmidt S. Identification of glucocorticoid-response genes in children with acute lymphoblastic leukemia. Blood. 2006;107(5):2061–9. doi:10.1182/blood-2005-07-2853.
27. Holleman A, Cheok MH, den Boer ML, et al. Gene-expression patterns in drug-resistant acute lymphoblastic leukemia cells and response to treatment. N Engl J Med. 2004;351(6):533–42. doi:10.1056/NEJMoa033513.
28. Wei G, Twomey D, Lamb J, et al. Gene expression-based chemical genomics identifies rapamycin as a modulator of MCL1 and glucocorticoid resistance. Cancer Cell. 2006; 10(4):331–42. doi:10.1016/j.ccr.2006.09.006.
29. Bachmann PS. Dexamethasone resistance in B-cell precursor childhood acute lymphoblastic leukemia occurs downstream of ligand-induced nuclear translocation of the glucocorticoid receptor. Blood. 2005;105(6):2519–26. doi:10.1182/blood-2004-05-2023.
30. Cario G, Fetz A, Bretscher C, et al. Initial leukemic gene expression profiles of patients with poor in vivo prednisone response are similar to those of blasts persisting under prednisone treatment in childhood acute lymphoblastic leukemia. Ann Hematol. 2008;87(9):709–16. doi:10.1007/s00277-008-0504-x.
31. Hulleman E, Kazemier KM, Holleman A, et al. Inhibition of glycolysis modulates prednisolone resistance in acute lymphoblastic leukemia cells. Blood. 2009;113(9):2014–21. doi:10.1182/blood-2008-05-157842.
32. Buentke E, Nordström A, Lin H, et al. Glucocorticoid-induced cell death is mediated through reduced glucose metabolism in lymphoid leukemia cells. Blood Cancer J. 2011;1(7):e31–9. doi:10.1038/bcj.2011.27.

33. Chen CS, Sorensen PH, Domer PH, et al. Molecular rearrangements on chromosome 11q23 predominate in infant acute lymphoblastic leukemia and are associated with specific biologic variables and poor outcome. Blood. 1993;81(9):2386–93.
34. Briggs SD, Bryk M, Strahl BD, et al. Histone H3 lysine 4 methylation is mediated by Set1 and required for cell growth and rDNA silencing in Saccharomyces cerevisiae. Genes Dev. 2001;15(24):3286–95. doi:10.1101/gad.940201.
35. Milne TA, Briggs SD, Brock HW, et al. MLL targets SET domain methyltransferase activity to Hox gene promoters. Mol Cell. 2002;10(5):1107–17.
36. Krivtsov AV, Armstrong SA. MLL translocations, histone modifications and leukaemia stem-cell development. Nat Rev Cancer. 2007;7(11):823–33. doi:10.1038/nrc2253.
37. Mullighan CG, Phillips LA, Su X, et al. Genomic analysis of the clonal origins of relapsed acute lymphoblastic leukemia. Science. 2008;322(5906):1377–80. doi:10.1126/science.1164266.
38. Piovan E, Yu J, Tosello V, et al. Direct reversal of glucocorticoid resistance by AKT inhibition in acute lymphoblastic leukemia. Cancer Cell. 2013;24(6):766–76. doi:10.1016/j.ccr.2013.10.022.
39. Galliher-Beckley AJ, Williams JG, Cidlowski JA. Ligand-independent phosphorylation of the glucocorticoid receptor integrates cellular stress pathways with nuclear receptor signaling. Mol Cell Biol. 2011;31(23):4663–75. doi:10.1128/MCB.05866-11.
40. Gruber G, Carlet M, Türtscher E, et al. Levels of glucocorticoid receptor and its ligand determine sensitivity and kinetics of glucocorticoid-induced leukemia apoptosis. Leukemia. 2009; 23(4):820–3. doi:10.1038/leu.2008.360.
41. Schmidt S, Irving JAE, Minto L, et al. Glucocorticoid resistance in two key models of acute lymphoblastic leukemia occurs at the level of the glucocorticoid receptor. FASEB J. 2006;20(14):2600–2. doi:10.1096/fj.06-6214fje.
42. Wu I, Shin SC, Cao Y, et al. Selective glucocorticoid receptor translational isoforms reveal glucocorticoid-induced apoptotic transcriptomes. Cell Death Dis. 2013;4, e453. doi:10.1038/cddis.2012.193.
43. DeVita VT, Lawrence TS, Rosenberg SA. DeVita, Hellman, and Rosenberg's cancer: principles & practice of oncology, vol. 2. Philadelphia, PA: Lippincott Williams & Wilkins; 2008.
44. Thomas DA. Outcome with the hyper-CVAD regimens in lymphoblastic lymphoma. Blood. 2004;104(6):1624–30. doi:10.1182/blood-2003-12-4428.
45. Real PJ, Tosello V, Palomero T, et al. γ-secretase inhibitors reverse glucocorticoid resistance in T cell acute lymphoblastic leukemia. Nat Med. 2008;15(1):50–8. doi:10.1038/nm.1900.
46. Rajkumar SV. Multiple myeloma. Curr Probl Cancer. 2009;33(1):7–64. doi:10.1016/j.currproblcancer.2009.01.001.
47. Kyle RA, Rajkumar SV. Multiple myeloma. Blood. 2008;111(6):2962–72. doi:10.1182/blood-2007-10-078022.
48. Munshi NC, Anderson KC. New strategies in the treatment of multiple myeloma. Clin Cancer Res. 2013;19(13):3337–44. doi:10.1158/1078-0432.CCR-12-1881.
49. Grugan KD, Ma C, Singhal S, Krett NL, Rosen ST. Dual regulation of glucocorticoid-induced leucine zipper (GILZ) by the glucocorticoid receptor and the PI3-kinase/AKT pathways in multiple myeloma. J Steroid Biochem Mol Biol. 2008;110(3–5):244–54. doi:10.1016/j.jsbmb.2007.11.003.
50. Raab MS, Podar K, Breitkreutz I, Richardson PG, Anderson KC. Multiple myeloma. Lancet. 2009;374(9686):324–39. doi:10.1016/S0140-6736(09)60221-X.
51. Viviani S, Zinzani PL, Rambaldi A. ABVD versus BEACOPP for Hodgkin's lymphoma when high-dose salvage is planned. N Engl J Med. 2011;365(3):203–12. doi:10.1056/NEJMoa1100340.
52. Rosenthal MC, Saunders RH, Schwartz LI, Zannos L, Perez Santiago E, Dameshek W. The use of adrenocorticotropic hormone and cortisone in the treatment of leukemia and leukosarcoma. Blood. 1951;6(9):804–23.
53. Smolej L. The role of high-dose corticosteroids in the treatment of chronic lymphocytic leukemia. Expert Opin Investig Drugs. 2012;21(7):1009–17. doi:10.1517/13543784.2012.690393.
54. Burger JA, Montserrat E. Coming full circle: 70 years of chronic lymphocytic leukemia cell redistribution, from glucocorticoids to inhibitors of B-cell receptor signaling. Blood. 2013;121(9):1501–9. doi:10.1182/blood-2012-08-452607.

55. Connors JM. Non-Hodgkin lymphoma: the clinician's perspective—a view from the receiving end. Mod Pathol. 2013;26(s1):S111–8. doi:10.1038/modpathol.2012.184.
56. Kassi E, Moutsatsou P. Glucocorticoid receptor signaling and prostate cancer. Cancer Lett. 2011;302(1):1–10. doi:10.1016/j.canlet.2010.10.020.
57. Agarwal N, Di Lorenzo G, Sonpavde G, Bellmunt J. New agents for prostate cancer. Ann Oncol. 2014;25(9):1700–9. doi:10.1093/annonc/mdu038.
58. Cookson MS, Roth BJ, Dahm P, et al. Castration-resistant prostate cancer: AUA guideline. J Urol. 2013;190(2):429–38. doi:10.1016/j.juro.2013.05.005.
59. Montgomery B, Mostaghel E, Cheng H, Drechsler J. Glucocorticoids and prostate cancer treatment: friend or foe? Asian J Androl. 2014;16(3):354–8. doi:10.4103/1008-682X.125392.
60. Yemelyanov A, Czwornog J, Chebotaev D, et al. Tumor suppressor activity of glucocorticoid receptor in the prostate. Oncogene. 2006;26(13):1885–96. doi:10.1038/sj.onc.1209991.
61. Cavarretta IT, Neuwirt H, Untergasser G, et al. The antiapoptotic effect of IL-6 autocrine loop in a cellular model of advanced prostate cancer is mediated by Mcl-1. Oncogene. 2006;26(20):2822–32. doi:10.1038/sj.onc.1210097.
62. Arora VK, Schenkein E, Murali R, et al. Glucocorticoid receptor confers resistance to antiandrogens by bypassing androgen receptor blockade. Cell. 2013;155(6):1309–22. doi:10.1016/j.cell.2013.11.012.
63. Zhong DT, Shi CM, Chen Q, Huang JZ, Liang JG, Lin D. Etoposide, vincristine, doxorubicin and dexamethasone (EVAD) combination chemotherapy as second-line treatment for advanced AIDS-related Kaposi's sarcoma. J Cancer Res Clin Oncol. 2011;138(3):425–30. doi:10.1007/s00432-011-1109-7.
64. Sweeney CJ. Phase II, study of pemetrexed for second-line treatment of transitional cell cancer of the urothelium. J Clin Oncol. 2006;24(21):3451–7. doi:10.1200/JCO.2005.03.6699.
65. Scagliotti GV, Parikh P, von Pawel J, et al. Phase III study comparing cisplatin plus gemcitabine with cisplatin plus pemetrexed in chemotherapy-naive patients with advanced-stage non-small-cell lung cancer. J Clin Oncol. 2008;26(21):3543–51. doi:10.1200/JCO.2007.15.0375.
66. Hanna N. Randomized phase III trial of pemetrexed versus docetaxel in patients with non-small-cell lung cancer previously treated with chemotherapy. J Clin Oncol. 2004;22(9):1589–97. doi:10.1200/JCO.2004.08.163.
67. Herbst RS, O'Neill VJ, Fehrenbacher L, et al. Phase II study of efficacy and safety of bevacizumab in combination with chemotherapy or erlotinib compared with chemotherapy alone for treatment of recurrent or refractory non small-cell lung cancer. J Clin Oncol. 2007;25(30):4743–50. doi:10.1200/JCO.2007.12.3026.
68. Belani CP, Ramalingam S, Perry MC, et al. Randomized, phase III study of weekly paclitaxel in combination with carboplatin versus standard every-3-weeks administration of carboplatin and paclitaxel for patients with previously untreated advanced non-small-cell lung cancer. J Clin Oncol. 2008;26(3):468–73. doi:10.1200/JCO.2007.13.1912.
69. Vogelzang NJ. Phase III, study of pemetrexed in combination with cisplatin versus cisplatin alone in patients with malignant pleural mesothelioma. J Clin Oncol. 2003;21(14):2636–44. doi:10.1200/JCO.2003.11.136.
70. Ceresoli GL, Phase II. Study of pemetrexed plus carboplatin in malignant pleural mesothelioma. J Clin Oncol. 2006;24(9):1443–8. doi:10.1200/JCO.2005.04.3190.
71. Gill PS, Tulpule A, Espina BM, et al. Paclitaxel is safe and effective in the treatment of advanced AIDS-related Kaposi's sarcoma. J Clin Oncol. 1999;17(6):1876–83.
72. Perez EA, Lerzo G, Pivot X, et al. Efficacy and safety of ixabepilone (BMS-247550) in a phase II study of patients with advanced breast cancer resistant to an anthracycline, a taxane, and capecitabine. J Clin Oncol. 2007;25(23):3407–14. doi:10.1200/JCO.2006.09.3849.
73. Roche H, Yelle L, Cognetti F, et al. Phase II clinical trial of ixabepilone (BMS-247550), an epothilone B analog, as first-line therapy in patients with metastatic breast cancer previously treated with anthracycline chemotherapy. J Clin Oncol. 2007;25(23):3415–20. doi:10.1200/JCO.2006.09.7535.

74. Thomas E, Tabernero J, Fornier M, et al. Phase II clinical trial of ixabepilone (BMS-247550), an epothilone B analog, in patients with taxane-resistant metastatic breast cancer. J Clin Oncol. 2007;25(23):3399–406. doi:10.1200/JCO.2006.08.9102.
75. American Society of Clinical Oncology, Kris MG, Hesketh PJ, et al. American Society of Clinical Oncology guideline for antiemetics in oncology: update 2006. J Clin Oncol. 2006;24(18):2932–47. doi:10.1200/JCO.2006.06.9591.
76. Aapro MS, Alberts DS. Dexamethasone as an antiemetic in patients treated with cisplatin. N Engl J Med. 1981;305(9):520.
77. Rich WM, Abdulhayoglu G, DiSaia PJ. Methylprednisolone as an antiemetic during cancer chemotherapy—a pilot study. Gynecol Oncol. 1980;9(2):193–8.
78. Chu C-C, Hsing C-H, Shieh J-P, Chien C-C, Ho C-M, Wang J-J. The cellular mechanisms of the antiemetic action of dexamethasone and related glucocorticoids against vomiting. Eur J Pharmacol. 2014;722(C):48–54. doi:10.1016/j.ejphar.2013.10.008.
79. Hursti TJ, Fredrikson M, Steineck G, Börjeson S, Fürst CJ, Peterson C. Endogenous cortisol exerts antiemetic effect similar to that of exogenous corticosteroids. Br J Cancer. 1993;68(1):112–4.
80. Rhen T, Cidlowski JA. Antiinflammatory action of glucocorticoids—new mechanisms for old drugs. N Engl J Med. 2005;353(16):1711–23. doi:10.1056/NEJMra050541.
81. Mantovani G, Maccio A, Massa E, Lai P, Esu S. Cisplatin induces serotonin release from human peripheral blood mononuclear cells of cancer patients and methylprednisolone inhibits this effect. Oncol Rep. 1997;4(5):1051–3.
82. Inui A. Cancer anorexia-cachexia syndrome: current issues in research and management. CA Cancer J Clin. 2002;52(2):72–91.
83. Quax RAM, Peeters RP, Feelders RA. Selective glucocorticoid receptor modulators: future of glucocorticoid immunosuppressive therapy? Endocrinology. 2011;152(8):2927–9. doi:10.1210/en.2011-1258.
84. Clark AR, Belvisi MG. Maps and legends: the quest for dissociated ligands of the glucocorticoid receptor. Pharmacol Ther. 2012;134(1):54–67. doi:10.1016/j.pharmthera.2011.12.004.
85. Jing D, Bhadri VA, Beck D, et al. Opposing regulation of BIM and BCL2 controls glucocorticoid-induced apoptosis of pediatric acute lymphoblastic leukemia cells. Blood. 2014;125(2):273–83. doi:10.1182/blood-2014-05-576470.
86. Moutsatsou P, Kassi E, Papavassiliou AG. Glucocorticoid receptor signaling in bone cells. Trends Mol Med. 2012;18(6):348–59. doi:10.1016/j.molmed.2012.04.005.
87. McKenna NJ, O'Malley BW. Combinatorial control of gene expression by nuclear receptors and coregulators. Cell. 2002;108(4):465–74.

Part IV
Miscellaneous Topics

Chapter 15
Animal Models of Altered Glucocorticoid Signaling

Charles Harris

Abstract In this chapter we will review genetically engineered mice with alterations in glucocorticoid signaling. Most of the mice involve direct alterations to the glucocorticoid receptor locus, but we will touch briefly on other relevant models including 11-β-HSD transgenics which alter tissue levels of ligand as well as mice with glucocorticoid excess. Of course, the number of mice with mutations in genes such as GR targets and transcriptional coregulators is beyond the scope of this chapter.

Keywords Altered glucocorticoid signaling • Animal models • 11-β-HSD transgenics • Hypomorphs • GR-null mice • Tissue specific KO • Tissue-specific transgenics

Version 1.0: Hypomorphs

The scientific community was quite eager to know the function of the glucocorticoid receptor in vivo. Transgenic mice were introduced in 1981 [1], but gene targeting via homologous recombination was not introduced until 1990 [2] and did not become commonplace until the mid-1990s. GR was cloned in 1985 [3–5]. Therefore, in order to probe the function of GR in vivo in the early 1990s, multiple groups created transgenic animals with transgenes encoding antisense for GR coding sequence, thereby creating in essence a "tissue specific knockdown" governed by the promoter driving the anti-sense construct. The first paper expressed a rat antisense GR sequence under the control of a human neurofilament gene promoter. These mice were obese, had decreased but detectable glucocorticoid binding and GR mRNA in the brain, mildly increased serum corticosterone and very increased ACTH levels [6]. Another group used a similar approach to knockdown GR with a transgene encoding 3′ UTR rat GR antisense driven by lck, an early T-cell specific promoter. These mice had small

C. Harris (✉)
Division of Endocrinology, Metabolism and Lipid Research, Department of Internal Medicine, Washington University School of Medicine, 660 S Euclid Avenue, Campus Box 8127, St. Louis, MO 63110, USA
e-mail: caharris@dom.wustl.edu

J.-C. Wang, C. Harris (eds.), *Glucocorticoid Signaling*, Advances in Experimental Medicine and Biology 872, DOI 10.1007/978-1-4939-2895-8_15

thymi and had decreased number of thymocytes due to decreased differentiation of mature thymocytes as well as increased apoptosis [7].

Version 2.0: Whole Body Knockouts

GR-null mice have been generated separately from several different groups [8–10]. GR-null mice are born at the expected Mendelian frequency, but die within hours of birth due to respiratory difficulties and defects in lung maturation. This finding is concordant with clinical observations showing infants born prematurely suffer from similar defects in lung maturity if mothers are not given glucocorticoids shortly before birth. In addition, in GR-null mice, circulating ACTH and corticosterone levels are high due to loss of negative feedback. The elevated corticosterone does not activate GR in these animals (there is none), but would be predicted to activate MR and cause a syndrome of apparent mineralocorticoid excess. This phenomenon has been described in persons with resistance to glucocorticoid hormone [11]. GR-null mice also have cortical adrenal hyperplasia, presumably from increased ACTH levels stimulating intact ACTH receptors (MC2R) in the adrenal cortex. The role of GR in the adrenal chromaffin lineage has been subject to reinterpretation over the years. GR-null mice were originally reported to be missing adrenal medullae [8], but more careful studies showed the existence of adrenal chromaffin cells in normal numbers with the more specific phenotype of conversion of adrenergic to noradrenergic cells due to loss of GR-mediated activation of the enzyme PNMT (Phenylethanolamine *N*-methyltransferase), that converts noradrenaline to adrenaline). More recently, a study of tissue-specific loss of GR in noradrenergic cells (using the cre-lox system and a dopamine beta hydroxylase driven cre line) revealed loss of chromaffin cells as adults [12]. The synthesis of these findings is that GR is not required for the formation of adrenal chromaffin cells but does help maintain these cells into adulthood. An attempt to rescue neonatal mortality due to the lung phenotype by crossing GR-null mice to transgenic mice expressing GR under the SP-C promoter was unsuccessful [13]. GR-null mice have a defect in epidermal differentiation and skin barrier function [14]. Because neonatal mortality has precluded analysis of adult GR-null animals, several investigators have alternatively studied GR-null cells in vitro, in transplant experiments, GR heterozygous mice in vivo [10] and later tissue specific null mice. GR-null erythroid progenitors have decreased erythropoiesis in vitro [15]. In an elegant set of experiments it was demonstrated that E18.5 GR-null mice have a disorganized endocrine pancreas. GR-null pancreas appeared normal at E15.5 and the disorganization that occurred between E15 and E18.5 could be rescued by transplanting the pancreas under the kidney capsule of a GR expressing mouse. The conclusion of these experiments was the disorganization of the GR-null pancreas was an indirect effect [16].

Version 3.0: Tissue Specific KO and Tissue-Specific Transgenics

The advent of cre lox technology allows specific deletion of gene products from tissues or cell types of interest. A thorough review of the subject can be found here [17]. Briefly, a critical region of the gene of interest is modified to be flanked by 34 base pair loxP sites, which are introduced by homologous recombination in such a manner so as not to disrupt gene function. So called "floxed" (flanking loxP mice) are then crossed to transgenic mice expressing the bacteriophage cre recombinase driven by the promoter of interest. The cre recombinase is expressed only in the tissue of interest where it excises the DNA between the two loxP sites, resulting in a null allele.

The pleiotropic actions of glucocorticoids have made it difficult to draw direct links between specific gene activations with physiological observations. For example, glucocorticoids are known to cause osteoporosis, but this could be due to direct actions on bone, both osteoblasts and osteoclasts, but also indirectly through reductions in sex hormones, reductions in Growth Hormone, along with actions on the intestinal absorption of calcium. Another example is the known sarcopenic effects of glucocorticoids which could be explained by direct actions on muscle, but indirectly through suppression of the GH-IGF1 axis as well. Tissue specific deletion of GR has allowed a dissection of these pleiotropic actions to determine the contribution of cell-autonomous effects on respective phenotypes. Tissue-specific deletion of GR has confirmed effects on bone (osteoblast) and muscle are largely cell autonomous (see below). This is of clinical importance as novel therapeutics with tissue-specific delivery of GR modulators becomes a possible avenue in the future [18–20]. Glucocorticoid Receptor has been deleted in many different tissues. We do not have sufficient room to include all of these mice, but rather describe a handful and direct the reader to MGI for a complete listing. With respect to GR several different floxed mice have been generated with loxP sites flanking different exons of GR genomic locus [21–25]. Recently a comparison between exon 2 floxed mice and exon 3 floxed mice was used using the Sim1-cre mice with expression in the hypothalamus. Exon 3 mice gave an interesting expected phenotype of Cushing's syndrome (due to loss of GR-mediated negative feedback on CRH in the hypothalamus, but exon 2 flanked mice crossed to the same cre transgenic mouse did not display the phenotype. This may be due to inefficient excision of exon 2 in exon 2 floxed mice bearing certain cre transgenes. Alternatively, there may be truncated GR after exon 2 excision that can retain some transcriptional activity [26]. These observations do not derail previous results obtained with exon 2 floxed mice, as some cre lines do result in complete excision and loss of function when exon 2 is excised. However, given the potential for residual function with exon 2, we anticipate greater use of exon 3 floxed mice for future studies.

Lung

Since GR-null mice died due to defects in lung maturation, it was of great interest to know which cell type was responsible for the phenotype. Unfortunately there has not been consensus among investigators in this area. One group found that mesenchymal expression of GR governed the phenotype since mesenchymal-specific deletion of GR using col1-cre or dermo1-cre mice recapitulated the neonatal mortality phenotype [27, 28] but deletion using epithelial SPC-cre did not have a phenotype whereas another group observed decreased viability with an epithelial specific GR KO mouse that was generated by crossing GR floxed mice to mice carrying SPC-rtTA and tetO-Cre transgenes [29]. So, there is disagreement as to the necessity of GR in these cell types. As mentioned above, attempts to rescue the GR-null mouse by crossing to a transgenic expressing GR under the control of surfactant protein C regulatory elements were unsuccessful. It would be interesting to attempt a similar rescue with a mesenchymal targeted GR transgene. However, it is not clear whether a GR-null mouse rescued from the lung phenotype would survive given defects in other cell types or gain of function phenotype from excessive action on the mineralocorticoid receptor (syndrome of apparent mineralocorticoid excess).

Skin

The skin phenotype of global GR-null mice was described above. This phenotype is recapitulated in keratinocyte specific GR KO mice generated using cre lox technology and keratin14-cre mice [30]. These epidermal GR KO mice also have increased susceptibility to chemical induced skin tumors [31]. Overexpression of GR in transgenic mice driven by the keratin5 promoter results in abnormal skin including hypoplasia and impaired hyperplastic response to topical TPA [32].

Cardiovascular System

GR action in the distal nephron is not critical for glucocorticoid mediated hypertension [33] as mice with GR deleted in the distal nephron (Ksp-cre) had a hypertensive response to dexamethasone similar to WT mice. In contrast, mice with endothelial specific GR KO mice were created by using tie2-cre mice. Endothelial GR KO mice are resistant to dexamethasone mediated hypertension and have increased sensitivity to hemodynamic shock associated with sepsis as mimicked by systemic LPS injections [34]. This effect is thought to be mediated by loss of negative regulation of iNOS in maintaining vascular tone. A conditional induction of a GR transgene in the heart resulted in arrhythmia [35]. Deletion of GR in the heart (α-MHC-cre) results in death at age 6 months from heart failure, indicating a previously unknown role for GR signaling in the heart [22]. These findings have potential clinical implications for the development of GR antagonists in metabolic disease. In addition global GR-null embryos were found to have decreased cardiac function [36].

Muscle

Muscle specific deletion of GR was accomplished using muscle creatine kinase-cre mice. Muscle specific GR KO mice were completely protected from dexamethasone mediated muscle atrophy [37] indicating this is a cell autonomous process. Induction of the atrogins MURF1 and MAFbx1 was abrogated in muscle specific GR KO mice treated with glucocorticoid. This was in contrast to the lack of effect seen in muscle-specific GRKO mice on muscle atrophy and atrogin gene activation seen with sciatic nerve transection. There was a partial effect on both atrogin gene induction and muscle atrophy seen with nutritional deprivation suggesting that glucocorticoids combine with other signals in that paradigm. A separate group has shown muscle GR is required for muscle atrophy following either chemotherapy or endotoxin [38, 39]. These reports did not indicate the contribution of muscle GR to whole body changes in glucose homeostasis, but may be the subject of future studies.

Gastrointestinal System

GR is dispensable for the development of the intestine as GR-null intestines are morphologically normal [40]. However, the perinatal death of GR-null mice precluded examination of GR in adult intestine. Enterocyte specific GR KO mice were created using villin-cre. Intestine-specific GR KO mice had loss of dexamethasone-stimulated intestinal glucose absorption that was thought to be mediated by transcriptional activation of genes encoding glucose transporters [41]. Recently, it was shown dexamethasone caused gastroparesis in wild type mice. This effect was not mediated by enterocytes as it was preserved in enterocyte specific GR KO animals (villin-cre). Rather, the authors believed it is due to loss of NO signaling in the stomach [42].

Immune System

Because GR-null animals die at birth one must use either tissue-specific knockouts or alternatively one can isolate fetal liver from GR-null animals and perform bone marrow transplants or perform bone marrow transplants from a tissue-specific KO to narrow down the cell type. Such bone marrow transplants were performed to show that macrophage GR was dispensable for atherogenesis in LDLR-null mice, but macrophage GR contributed to vascular calcification [43]. Glucocorticoids are effective in multiple sclerosis and appear to act on T-cells in mouse models of autoimmune encephalitis [44] as evidenced by T-cell specific (lck) and macrophage specific KO (LysM-cre) mice. In a contact hypersensitivity model myeloid cells seemed to be the predominant mediator of glucocorticoid effects as myeloid specific GR KO mice (LysM-cre) were resistant to the anti-inflammatory effects of dexamethasone whereas T-cell (lck-Cre) and keratinocyte (K14cre-ER following tamoxifen injection) were unaffected [45]. T-cells were identified as the critical site of action for the therapeutic effects of glucocorticoid in an antigen-induced model of rheumatoid arthritis [46].

Bone

One group found osteoblast specific deletion of GR (Runx2-cre) rendered mice resistant to glucocorticoid-induced osteoporosis whereas deletion of GR in osteoclasts (LysM-cre) had no effect [47]. This same group did not see any defects in bone development as assessed by GR-null mice analyzed at E18.5 and also reported complete sensitivity of GR dim mice (see below) to glucocorticoid induced osteoporosis, implicating GR monomers in this process. Another group has implicated an anabolic function of glucocorticoids in bone development from the phenotype of transgenic mice expressing the glucocorticoid inactivating 11-β-HSD2 in osteoblasts [48]. Another group showed that action of GR in osteoclasts was the dominant determinant on the bone phenotype using a similar LysM-cre breeding strategy to create osteoclast deficiency of GR [49]. From these studies it is clear GR action on the osteoblast is critical for glucocorticoid induced osteoporosis, but there may be a role in the osteoclast as well. The actions of GR mediating osteoporosis do not require functional dimers. This has implications for the possible development of selective GR modulators as treatments for osteoporosis or to be used in conjunction with another glucocorticoid to attenuate glucocorticoid induced osteoporosis.

Tissues Involved in Metabolism

Deletion of GR in hepatocytes using albumin-cre resulted in neonatal mortality with incomplete penetrance [50]. These mice displayed fasting hypoglycemia and amelioration of streptozotocin induced diabetes [51]. The growth impairment in these mice phenocopied liver-specific deletion of Stat5 using the same cre line. The authors went on to show via microarrays that GR and Stat5 regulate similar sets of genes in the liver [52]. Deletion of GR in the liver with a different albumin-cre displayed no such viability defects. In addition, the latter mice had normal response to hepatectomy mediated hepatic regeneration [53]. The discrepant data with respect to viability is likely due to the timing of the two liver-specific cre activations, with early deletion being responsible for mortality and runting. The albumin-cre lines used in these studies are similar, but made by different investigators. The early expressing albumin-cre that resulted in mortality and a growth phenotype contained additional alpha fetoprotein enhancer elements which may explain its earlier expression. Alternatively the timing of expression could have been affected by the genomic integration site. Liver-specific GR KO mice have also demonstrated a role for GR in bile acid homeostasis [54]. Specific deletion of GR in pancreatic beta cells using RIP-cre had no effect on alpha or beta cell mass whereas deletion of GR in the precursor, using Pdx1-cre resulted in a doubling of beta cell mass [55]. Furthermore, as noted above GR-null embryos have disorganized pancreata.

CNS and HPA Axis

Glucocorticoids are known to cause negative feedback on hypothalamic CRH and pituitary ACTH secretion and the phenotypes of tissue-specific GR-null mice are informative as to the relative contribution of these sites. Mice with neuron specific deletion of GR (nestin-Cre) displayed Cushing's syndrome due to loss of negative feedback at the hypothalamic CRH producing neurons. These mice also had reduced anxiety [25]. The phenotype of this mouse is similar to the CRH-Tg mouse (see below) as well as the mice with deletion of GR in the hypothalamus using Sim1-cre and to the phenotype of mice with the (-120) mutation in the CRH promoter (see below), all resulting in loss of negative feedback at the level of the hypothalamus. Mice with loss of GR in the ACTH producing cells of the pituitary (POMC-cre) have elevated corticosterone levels in the first week of life, but normal peak and nadir corticosterone levels as adults [56]. Pituitary GR KO mice did have impaired suppression of corticosterone in a dexamethasone suppression test. In contrast deletion of GR in both the pituitary and hypothalamus (calcium/calmodulin-dependent protein kinase II alpha-cre) results in death in the first week of life from an advanced form of Cushing's syndrome [57]. Therefore, loss of pituitary negative feedback has modest effects, loss of the hypothalamic site results in Cushing's syndrome but is compatible with life, but complete loss of negative feedback is not. Given the role of stress on psychiatric conditions a number of GR cell type specific KO mice have been analyzed in behavioral paradigms. GR in dopamine responding neurons (dopamine receptor 1-cre) was found to be a critical mediator of social aggression whereas GR in dopaminergic neurons (dopamine active transporter, DAT-cre) was dispensable. Mice subject to social aggression show characteristic behaviors which were abolished in mice lacking GR in dopaminoceptive neurons [58]. Mice with deletion of GR in dopaminoceptive neurons (D1R-cre) self-administered less cocaine than wildtype mice, but no effect was seen when GR was deleted from dopaminergic neurons (DAT-cre) [59]. GR has been found to be a mediator of MPTP-induced Parkinsonism in microglia. Microglial GR KO mice (LysM-cre) had more death of dopaminergic neurons following MPTP-insult, whereas the loss was not effected by deletion of GR from dopaminergic neurons (DAT-cre) [60]. Microglia from microglia GR KO mice also had increased activation, consistent with a role for glucocorticoids in regulating microglial activation status as an anti-inflammatory.

Other Tissues

Deletion of GR in the mammary epithelium (whey acidic protein-cre) had no effect on lactation, but there was a modest defect in glandular proliferation during pregnancy [61]. Deletion of GR in sertoli cells (using anti-Mullerian hormone-cre) resulted in decreased sertoli cell number, circulating FSH and secondarily decreased LH and abnormal Leydig cell morphology, but these mice were fertile [62]. As mentioned above, deletion of GR in noradrenergic cells (using dopamine beta hydroxylase cre) resulted in decreased numbers of adrenal chromaffin cells in adults.

Version 4.0: Qualitative Differences (Point Mutants)

The most well studied GR knockin mouse is the dim mouse (MGI:1931329). Site directed mutagenesis studies identified this mutation in vitro as inhibiting transactivation while leaving repression intact [63]. Alanine to threonine mutation was generated by homologous recombination. Unlike GR-null animals, GR dim mice are viable [64]. However, we have observed most GRdim/dim mice on a C57/BL6 background die at birth [65]. Interestingly, viability is normal on a mixed (BALBC and C57) background (C.H., unpublished observation). This could be due to genetic modifiers affecting GR signaling. In addition, although the dim allele was originally described as being a loss of function allele for all transactivation targets [64], more detailed analysis in the age of genomics has uncovered that some targets are still activated by the dim allele and in addition the dim allele has some gain of function properties [66, 67]. In other words, a subset of genes is activated in a ligand dependent manner by the dim allele that is not normally activated by wildtype GR. These so called "rogue" genes have been identified in vitro [66, 68] as well as in vivo [69 and our unpublished results]. One potential explanation for gain of function properties of the dim allele is gene activation by an increased concentration of GR monomers in the absence of dimer formation. Therefore, though the dim mouse is a powerful tool to dissect whether biological effects are mediated by dimers experiments using them must be interpreted with this in mind.

GR dim mice had blunted response to anemia by phenylhydrazine and also did not increase red blood cell counts in response to hypoxia [15]. As opposed to wildtype mice, GR dim mice did not potentiate hippocampal calcium currents or enhance serotonin response in the hippocampus when treated with glucocorticoids [70]. Water maze experiments demonstrated that GR dim mice have impairments in spatial memory [71]. GR dimerization is required for inhibition of contact hypersensitivity by glucocorticoids [45]. The defects in skin observed in GR-null mice were not seen in GR dim mice suggesting that GR dimers are dispensable for skin development [14]. GR dim mice undergo a muscle atrophy similar to WT mice when treated with glucocorticoids despite attenuated induction of MuRF1, indicating that muscle atrophy is a dimer independent process [72]. Glucocorticoids induce osteoporosis in GR dim mice indicating this is a dimer independent process [47, 65]. GR dim mice were resistant to the therapeutic effects of glucocorticoid in an antigen induced arthritis model of rheumatoid arthritis [46] indicating GR dimers are essential for this process. GR dim mice did not increase intestinal glucose uptake in response to glucocorticoids indicating this process is dependent on GR dimers [41]. GR dim mice were not immune to glucocorticoid-mediated changes in triglyceride metabolism in adipose, indicating this is a GR dimer independent process [65].

Another knockin mouse (MGI strain 3842978), of interest is a mutation that made GR more sensitive to ligand by mutation M610L in the LBD. These mice did not have a phenotype of glucocorticoid excess because their HPA axis was reset such that the circulating levels of glucocorticoids were lower than WT [73]. Muglia generated a knockin for a GFP-GR fusion protein allowing visualization of GR in vivo (MGI:3577992) [74].

Version 5.0: The Future of Mice with Altered GR Signaling

The advent of CRISPRs has made gene targeting more accessible given reduced cost, time and labor involved. First generation CRISPRs experiments were able to introduce double-strand breaks repaired in a stochastic manner involving non-homologous end-joining resulting in deletions and insertions [75]. Newer approaches allowed coinjection of a large single stranded donor oligonucleotide or double stranded DNA donor plasmid to induce targeted mutations via homologous recombination. Further advances have been made rapidly allowing the introduction of loxP sites to create floxed mice for conditional null alleles [76]. We predict this technology will lead to the creation of even more mice genetically modified in GR signaling. What mice will we see in the near future? Additional knock-in mice with point mutations leading to altered AF1 and AF2 function [77] would be insightful. In addition, point mutations for GR residues shown to be post-translationally modified would be informative as to the role of these post-translational modifications in vivo. In addition, there are a number of additional polymorphisms in GR that have been linked to human disease states including metabolic syndrome and depression [78] which could be informative in the function of GR in vivo. Finally, the ease of using CRISPRs to generate targeted germ line mutations could be the solution to a problem that has existed in the field of transcriptional regulation. We have access to whole genome transcriptional data and chromatin-immunoprecipitation data for regulation by glucocorticoids. It has been difficult to demonstrate that a given binding site is directly responsible for a given gene regulation event. Often this correlation is made by way of parsimony: Ockham's razor argues that the closest GR binding region to a gene regulated by GR is responsible for that gene's regulation. However, it is clear that there are a number of transcriptional events regulated from distal DNA elements. Therefore, gene targeting of a GRE would enable one to determine what genes are governed by the GRE. While it is not feasible to do this on a wide scale, and it is more feasible to do this in cell lines, we would not be surprised if we saw some GRE targeted CRISPRs in the near future. A handful of such mice might reveal rules to the logic of transcriptional regulation previously undecipherable. Candidates for such approaches would be GR primary target genes that are critical in a glucocorticoid-mediated phenotype with a well-defined GRE.

Non-GR Mutants Relevant to Glucocorticoid Signaling

We will not have space to describe all of the genetically altered mice related to glucocorticoid action, but a few notable mutations are described here. Given the regulation of glucocorticoid synthesis by the hypothalamic–pituitary–adrenal axis, it was of great interest to know the phenotype of CRH-null mice. CRH-null mice have reduced but detectable corticosterone levels. CRH-null animals born to heterozygous mothers survive, but CRH-null mice born to CRH-null mothers die at birth due to defects in lung maturation, similar to GR-null animals. If CRH-null mothers are given corticosterone in the drinking water, their CRH-null pups will survive. The peptide ACTH is encoded by the POMC gene which gives rise to several other peptides. Therefore, the phenotype of the POMC-null mice can not be attributed only to defects in ACTH. Mice lacking 7B2, a cofactor for proconvertase 2 (a protease that processes several endocrine protein hormones) display Cushing's disease, due to hypersecretion of ACTH [79]. Recently, a Cushingoid mouse was identified as part of an ENU mutagenesis screen [80]. The mutation was mapped to the CRH promoter (120 basepairs upstream of the transcriptional start site) in a caudal-type homeobox response element (CDXARE). The mutation leads to increased CRH expression and is recapitulated with a luciferase reporter. The phenotype of this mouse is similar to the CRH-Tg mouse as well as the mice with deletion of GR in tissues governing negative feedback in the hypothalamus including Sim1-cre and Nestin-Cre. Vale's group created a Cushingoid mouse by placing the CRH gene under a constitutive metallothionein promoter [81]. Interestingly, the investigators found that the transgene was only expressed in the hypothalamus and CRH levels could not be detected in the circulation. Yet, these mice had increased ACTH levels. The investigators postulated that a cryptic regulatory site within the coding region was responsible for this tissue specific transgene expression. While their peak corticosterone levels are not significantly different from WTs, their nadir levels are much higher, abrogating the circadian rhythmicity of corticosterone production similar to patients with endogenous glucocorticoid excess. CRH-Tg mice have increased adiposity, decreased bone density, decreased skin thickness and skin collagen production. Interestingly, CRH-Tg mice have large increases in adipose triglyceride futile cycling with increased rate of both lipolysis and reesterification [82]. CRH-Tg mice display age-dependent malabsorption that is due to a "transdifferentiation" of exocrine pancreas to express hepatocyte markers [83].

Dehydrogenase Mutants

To test the hypothesis that increased glucocorticoid signaling in adipose tissue plays a role in the etiology of metabolic syndrome Flier's group created an aP2-11-β-HSD1 transgenic mouse. Selective expression of this GC activating enzyme in adipose should raise tissue corticosterone levels in adipose. These mice displayed features of metabolic syndrome while on a chow diet [84]. Similarly, the converse was shown, i.e. transgenic expression of 11-β-HSD2, the GC deactivating enzyme, protected mice from metabolic abnormalities when fed a high-fat diet [85]. A similar approach was used to modulate glucocorticoid levels in the bone [48].

GR Targets

A full description of genetic modifications to primary GR causative target genes is beyond the scope of this chapter, but we would like to highlight a few key observations. Given that glucocorticoid receptor alters transcription of hundreds to thousands of genes in most cell types it is amazing that elimination of a single GR target gene can affect the phenotype associated with glucocorticoid exposure. Furthermore, many glucocorticoid targets have functions distinct from their regulation by glucocorticoids. Glucocorticoid-mediated muscle atrophy occurs via transcriptional activation of atrogins such as MuRF1 and MAFbx (atrogin1). Deletion of MURF1 attenuates glucocorticoid-mediated muscle atrophy [86]. Similarly, deletion of the GR target Hes1 attenuates metabolic effects of glucocorticoids in the liver [87].

Summary and Looking Ahead

The last 25 years have seen a proliferation of our knowledge about glucocorticoid signaling. Essential to this has been the ability to probe the role of GR in vivo in mice using first anti-sense transgenics, the traditional knockouts, and then tissue specific knockouts. The tissue-specific GR KO mice have allowed a better understanding of which effects of glucocorticoids are cell autonomous and which are indirect. For example, muscle atrophy which could have been due to effects directly on muscle as well as indirectly through effects on liver (IGF1 secretion), the HPA axis (reduced sex steroids) appears to be largely cell autonomous. Similarly, a priori, effects on bone could be direct or indirect through reductions in other pituitary axes including GH-IGF1, hypothalamic–pituitary–gonadal, intestinal absorption of calcium, etc. Here, too effects seem to be direct on bone, and while the osteoblast appears to be critical, the osteoclast may play a role. Furthermore, some phenotypes are agreed upon by multiple investigators, such as the muscle specific KO mouse which has been made by three different investigators. Areas of discrepant data include the critical cell type in the lung responsible for the lethality of GR-null mice. The apparent discrepancy in phenotype of liver specific GR-null mice can be explained by the timing of activation of the cre lines used by the different investigators. Given the recent breakthrough in gene editing with CRISPRs, it is likely many more GR mutant mice will be generated and contribute further to our understanding of how GR signals. In addition CRISPRs may enable investigators that use other experimental models such as rat to also generate additional animals with gene modifications.

After detailing the multiple effects of the glucocorticoid receptor in the sections above, it is quite remarkable that the GR-null mouse makes it to birth and is not embryonic lethal. In other words, although GR has been implicated in cell fate decisions, GR-null mice do not have an obvious loss of an essential cell type. The exceptions are the maintenance of adrenal chromaffin cells as adults and maintenance of some pancreatic cell types as mentioned above. Similarly, glucocorticoid excess alters some cell fate decisions with the best evidence in vivo being hepatic conversion of exocrine pancreas in CRH-Tg mice. Although the focus regarding neonatal mortality of GR-null mice has been on the lung, it is possible than even if this were rescued, GR-null mice might die due to effects on other tissues, such as skin or the heart. We predict the answers to these questions will come in the years ahead.

Additional Resources

In addition to this chapter, we suggest the reader use the following websites to assemble the most up to date information. Mouse genome informatics (http://www.informatics.jax.org/). This is a comprehensive listing of mouse genes with links to individual mice (i.e. separate listings for different strains created by different labs including knockout projects). There is a separate listing for the GR gene (NR3C1) as well as a listing for each mutant mouse. Each mutant mouse line is assigned a unique MGI number which is occasionally referenced in this chapter. There are links to references, as well as to many other data bases, including IMSR (international mouse strain resource, http://www.findmice.org/), a database listing where various mouse strains are held internationally.

Nuclear Receptor Signaling Atlas (NURSA, http://www.nursa.org/) contains valuable information on nuclear receptors and coregulators including links to MGI pages.

References

1. Brinster RL, et al. Somatic expression of herpes thymidine kinase in mice following injection of a fusion gene into eggs. Cell. 1981;27:223–31.
2. Thomas KR, Capecchi MR. Targeted disruption of the murine int-1 proto-oncogene resulting in severe abnormalities in midbrain and cerebellar development. Nature. 1990;346:847–50. doi:10.1038/346847a0.
3. Miesfeld R, et al. Characterization of a steroid hormone receptor gene and mRNA in wild-type and mutant cells. Nature. 1984;312:779–81.
4. Weinberger C, et al. Identification of human glucocorticoid receptor complementary DNA clones by epitope selection. Science. 1985;228:740–2.
5. Govindan MV, Devic M, Green S, Gronemeyer H, Chambon P. Cloning of the human glucocorticoid receptor cDNA. Nucleic Acids Res. 1985;13:8293–304.
6. Pepin MC, Pothier F, Barden N. Impaired type II glucocorticoid-receptor function in mice bearing antisense RNA transgene. Nature. 1992;355:725–8. doi:10.1038/355725a0.
7. King LB, et al. A targeted glucocorticoid receptor antisense transgene increases thymocyte apoptosis and alters thymocyte development. Immunity. 1995;3:647–56.

8. Cole TJ, et al. Targeted disruption of the glucocorticoid receptor gene blocks adrenergic chromaffin cell development and severely retards lung maturation. Genes Dev. 1995;9:1608–21.
9. Brewer JA, Kanagawa O, Sleckman BP, Muglia LJ. Thymocyte apoptosis induced by T cell activation is mediated by glucocorticoids in vivo. J Immunol. 2002;169:1837–43.
10. Michailidou Z, et al. Glucocorticoid receptor haploinsufficiency causes hypertension and attenuates hypothalamic-pituitary-adrenal axis and blood pressure adaptions to high-fat diet. FASEB J. 2008;22:3896–907. doi:10.1096/fj.08-111914.
11. Chrousos GP, et al. Primary cortisol resistance in man. A glucocorticoid receptor-mediated disease. J Clin Invest. 1982;69:1261–9.
12. Parlato R, et al. Conditional inactivation of glucocorticoid receptor gene in dopamine-beta-hydroxylase cells impairs chromaffin cell survival. Endocrinology. 2009;150:1775–81. doi:10.1210/en.2008-1107.
13. Gagnon S, et al. Transgenic glucocorticoid receptor expression driven by the SP-C promoter reduces neonatal lung cellularity and midkine expression in GRhypo mice. Biol Neonate. 2006;90:46–57. doi:10.1159/000091844.
14. Bayo P, et al. Glucocorticoid receptor is required for skin barrier competence. Endocrinology. 2008;149:1377–88. doi:10.1210/en.2007-0814.
15. Bauer A, et al. The glucocorticoid receptor is required for stress erythropoiesis. Genes Dev. 1999;13:2996–3002.
16. Gesina E, et al. Glucocorticoid signalling affects pancreatic development through both direct and indirect effects. Diabetologia. 2006;49:2939–47. doi:10.1007/s00125-006-0449-3.
17. Bouabe H, Okkenhaug K. Gene targeting in mice: a review. Methods Mol Biol. 2013;1064:315–36. doi:10.1007/978-1-62703-601-6_23.
18. Shah K, et al. Discovery of liver selective non-steroidal glucocorticoid receptor antagonist as novel antidiabetic agents. Bioorg Med Chem Lett. 2012;22:5857–62. doi:10.1016/j.bmcl.2012.07.078.
19. Zinker B, et al. Liver-selective glucocorticoid receptor antagonism decreases glucose production and increases glucose disposal, ameliorating insulin resistance. Metabolism. 2007;56:380–7. doi:10.1016/j.metabol.2006.10.021.
20. von Geldern TW, et al. Liver-selective glucocorticoid antagonists: a novel treatment for type 2 diabetes. J Med Chem. 2004;47:4213–30. doi:10.1021/jm0400045.
21. Jeanneteau FD, et al. BDNF and glucocorticoids regulate corticotrophin-releasing hormone (CRH) homeostasis in the hypothalamus. Proc Natl Acad Sci U S A. 2012;109:1305–10. doi:10.1073/pnas.1114122109.
22. Oakley RH, et al. Essential role of stress hormone signaling in cardiomyocytes for the prevention of heart disease. Proc Natl Acad Sci U S A. 2013;110:17035–40. doi:10.1073/pnas.1302546110.
23. Mittelstadt PR, Monteiro JP, Ashwell JD. Thymocyte responsiveness to endogenous glucocorticoids is required for immunological fitness. J Clin Invest. 2012;122:2384–94. doi:10.1172/JCI63067.
24. Brewer JA, et al. T-cell glucocorticoid receptor is required to suppress COX-2-mediated lethal immune activation. Nat Med. 2003;9:1318–22. doi:10.1038/nm895.
25. Tronche F, et al. Disruption of the glucocorticoid receptor gene in the nervous system results in reduced anxiety. Nat Genet. 1999;23:99–103. doi:10.1038/12703.
26. Mittelstadt PR, Ashwell JD. Disruption of glucocorticoid receptor exon 2 yields a ligand-responsive C-terminal fragment that regulates gene expression. Mol Endocrinol. 2003;17:1534–42. doi:10.1210/me.2002-0429.
27. Habermehl D, et al. Glucocorticoid activity during lung maturation is essential in mesenchymal and less in alveolar epithelial cells. Mol Endocrinol. 2011;25:1280–8. doi:10.1210/me.2009-0380.
28. Li A, et al. Deletion of mesenchymal glucocorticoid receptor attenuates embryonic lung development and abdominal wall closure. PLoS One. 2013;8:e63578. doi:10.1371/journal.pone.0063578.

29. Manwani N, et al. Reduced viability of mice with lung epithelial-specific knockout of glucocorticoid receptor. Am J Respir Cell Mol Biol. 2010;43:599–606. doi:10.1165/rcmb.2009-0263OC.
30. Sevilla LM, Latorre V, Sanchis A, Perez P. Epidermal inactivation of the glucocorticoid receptor triggers skin barrier defects and cutaneous inflammation. J Invest Dermatol. 2013;133:361–70. doi:10.1038/jid.2012.281.
31. Latorre V, Sevilla LM, Sanchis A, Perez P. Selective ablation of glucocorticoid receptor in mouse keratinocytes increases susceptibility to skin tumorigenesis. J Invest Dermatol. 2013;133:2771–9. doi:10.1038/jid.2013.255.
32. Perez P, et al. Altered skin development and impaired proliferative and inflammatory responses in transgenic mice overexpressing the glucocorticoid receptor. FASEB J. 2001;15:2030–2. doi:10.1096/fj.00-0772fje.
33. Goodwin JE, Zhang J, Velazquez H, Geller DS. The glucocorticoid receptor in the distal nephron is not necessary for the development or maintenance of dexamethasone-induced hypertension. Biochem Biophys Res Commun. 2010;394:266–71. doi:10.1016/j.bbrc.2010.02.123.
34. Goodwin JE, Zhang J, Gonzalez D, Albinsson S, Geller DS. Knockout of the vascular endothelial glucocorticoid receptor abrogates dexamethasone-induced hypertension. J Hypertens. 2011;29:1347–56. doi:10.1097/HJH.0b013e328347da54.
35. Sainte-Marie Y, et al. Conditional glucocorticoid receptor expression in the heart induces atrioventricular block. FASEB J. 2007;21:3133–41. doi:10.1096/fj.07-8357com.
36. Rog-Zielinska EA, et al. Glucocorticoid receptor is required for foetal heart maturation. Hum Mol Genet. 2013;22:3269–82. doi:10.1093/hmg/ddt182.
37. Watson ML, et al. A cell-autonomous role for the glucocorticoid receptor in skeletal muscle atrophy induced by systemic glucocorticoid exposure. Am J Physiol Endocrinol Metab. 2012;302:E1210–20. doi:10.1152/ajpendo.00512.2011.
38. Braun TP, et al. Muscle atrophy in response to cytotoxic chemotherapy is dependent on intact glucocorticoid signaling in skeletal muscle. PLoS One. 2014;9:e106489. doi:10.1371/journal.pone.0106489.
39. Braun TP, et al. Cancer- and endotoxin-induced cachexia require intact glucocorticoid signaling in skeletal muscle. FASEB J. 2013;27:3572–82. doi:10.1096/fj.13-230375.
40. Gartner H, Graul MC, Oesterreicher TJ, Finegold MJ, Henning SJ. Development of the fetal intestine in mice lacking the glucocorticoid receptor (GR). J Cell Physiol. 2003;194:80–7. doi:10.1002/jcp.10189.
41. Reichardt SD, et al. Glucocorticoids enhance intestinal glucose uptake via the dimerized glucocorticoid receptor in enterocytes. Endocrinology. 2012;153:1783–94. doi:10.1210/en.2011-1747.
42. Reichardt SD, et al. Glucocorticoids induce gastroparesis in mice through depletion of l-arginine. Endocrinology. 2014;155:3899–908. doi:10.1210/en.2014-1246.
43. Preusch MR, et al. Critical role of macrophages in glucocorticoid driven vascular calcification in a mouse-model of atherosclerosis. Arterioscler Thromb Vasc Biol. 2008;28:2158–64. doi:10.1161/ATVBAHA.108.174128.
44. Wust S, et al. Peripheral T cells are the therapeutic targets of glucocorticoids in experimental autoimmune encephalomyelitis. J Immunol. 2008;180:8434–43.
45. Tuckermann JP, et al. Macrophages and neutrophils are the targets for immune suppression by glucocorticoids in contact allergy. J Clin Invest. 2007;117:1381–90. doi:10.1172/JCI28034.
46. Baschant U, et al. Glucocorticoid therapy of antigen-induced arthritis depends on the dimerized glucocorticoid receptor in T cells. Proc Natl Acad Sci U S A. 2011;108:19317–22. doi:10.1073/pnas.1105857108.
47. Rauch A, et al. Glucocorticoids suppress bone formation by attenuating osteoblast differentiation via the monomeric glucocorticoid receptor. Cell Metab. 2010;11:517–31. doi:10.1016/j.cmet.2010.05.005.
48. Sher LB, et al. Transgenic expression of 11beta-hydroxysteroid dehydrogenase type 2 in osteoblasts reveals an anabolic role for endogenous glucocorticoids in bone. Endocrinology. 2004;145:922–9. doi:10.1210/en.2003-0655.

49. Kim HJ, et al. Glucocorticoids suppress bone formation via the osteoclast. J Clin Invest. 2006;116:2152–60. doi:10.1172/JCI28084.
50. Tronche F, et al. Glucocorticoid receptor function in hepatocytes is essential to promote postnatal body growth. Genes Dev. 2004;18:492–7. doi:10.1101/gad.284704.
51. Opherk C, et al. Inactivation of the glucocorticoid receptor in hepatocytes leads to fasting hypoglycemia and ameliorates hyperglycemia in streptozotocin-induced diabetes mellitus. Mol Endocrinol. 2004;18:1346–53. doi:10.1210/me.2003-0283.
52. Engblom D, et al. Direct glucocorticoid receptor-Stat5 interaction in hepatocytes controls body size and maturation-related gene expression. Genes Dev. 2007;21:1157–62. doi:10.1101/gad.426007.
53. Shteyer E, Liao Y, Muglia LJ, Hruz PW, Rudnick DA. Disruption of hepatic adipogenesis is associated with impaired liver regeneration in mice. Hepatology. 2004;40:1322–32. doi:10.1002/hep.20462.
54. Rose AJ, et al. Molecular control of systemic bile acid homeostasis by the liver glucocorticoid receptor. Cell Metab. 2011;14:123–30. doi:10.1016/j.cmet.2011.04.010.
55. Gesina E, et al. Dissecting the role of glucocorticoids on pancreas development. Diabetes. 2004;53:2322–9.
56. Schmidt MV, et al. Postnatal glucocorticoid excess due to pituitary glucocorticoid receptor deficiency: differential short- and long-term consequences. Endocrinology. 2009;150:2709–16. doi:10.1210/en.2008-1211.
57. Erdmann G, Schutz G, Berger S. Loss of glucocorticoid receptor function in the pituitary results in early postnatal lethality. Endocrinology. 2008;149:3446–51. doi:10.1210/en.2007-1786.
58. Barik J, et al. Chronic stress triggers social aversion via glucocorticoid receptor in dopaminoceptive neurons. Science. 2013;339:332–5. doi:10.1126/science.1226767.
59. Ambroggi F, et al. Stress and addiction: glucocorticoid receptor in dopaminoceptive neurons facilitates cocaine seeking. Nat Neurosci. 2009;12:247–9. doi:10.1038/nn.2282.
60. Ros-Bernal F, et al. Microglial glucocorticoid receptors play a pivotal role in regulating dopaminergic neurodegeneration in parkinsonism. Proc Natl Acad Sci U S A. 2011;108:6632–7. doi:10.1073/pnas.1017820108.
61. Wintermantel TM, Bock D, Fleig V, Greiner EF, Schutz G. The epithelial glucocorticoid receptor is required for the normal timing of cell proliferation during mammary lobuloalveolar development but is dispensable for milk production. Mol Endocrinol. 2005;19:340–9. doi:10.1210/me.2004-0068.
62. Hazra R, et al. In vivo actions of the Sertoli cell glucocorticoid receptor. Endocrinology. 2014;155:1120–30. doi:10.1210/en.2013-1940.
63. Heck S, et al. A distinct modulating domain in glucocorticoid receptor monomers in the repression of activity of the transcription factor AP-1. EMBO J. 1994;13:4087–95.
64. Reichardt HM, et al. DNA binding of the glucocorticoid receptor is not essential for survival. Cell. 1998;93:531–41.
65. Roohk DJ, et al. Dexamethasone-mediated changes in adipose triacylglycerol metabolism are exaggerated, not diminished, in the absence of a functional GR dimerization domain. Endocrinology. 2013;154:1528–39. doi:10.1210/en.2011-1047.
66. Rogatsky I, et al. Target-specific utilization of transcriptional regulatory surfaces by the glucocorticoid receptor. Proc Natl Acad Sci U S A. 2003;100:13845–50. doi:10.1073/pnas.2336092100.
67. Schiller BJ, Chodankar R, Watson LC, Stallcup MR, Yamamoto KR. Glucocorticoid receptor binds half sites as a monomer and regulates specific target genes. Genome Biol. 2014;15:418. doi:10.1186/s13059-014-0418-y.
68. Jewell CM, Scoltock AB, Hamel BL, Yudt MR, Cidlowski JA. Complex human glucocorticoid receptor dim mutations define glucocorticoid induced apoptotic resistance in bone cells. Mol Endocrinol. 2012;26:244–56. doi:10.1210/me.2011-1116.
69. Frijters R, et al. Prednisolone-induced differential gene expression in mouse liver carrying wild type or a dimerization-defective glucocorticoid receptor. BMC Genomics. 2010;11:359. doi:10.1186/1471-2164-11-359.

70. Karst H, et al. Corticosteroid actions in hippocampus require DNA binding of glucocorticoid receptor homodimers. Nat Neurosci. 2000;3:977–8. doi:10.1038/79910.
71. Oitzl MS, Reichardt HM, Joels M, de Kloet ER. Point mutation in the mouse glucocorticoid receptor preventing DNA binding impairs spatial memory. Proc Natl Acad Sci U S A. 2001;98:12790–5. doi:10.1073/pnas.231313998.
72. Waddell DS, et al. The glucocorticoid receptor and FOXO1 synergistically activate the skeletal muscle atrophy-associated MuRF1 gene. Am J Physiol Endocrinol Metab. 2008;295:E785–97. doi:10.1152/ajpendo.00646.2007.
73. Zhang J, et al. Characterization of a novel gain of function glucocorticoid receptor knock-in mouse. J Biol Chem. 2009;284:6249–59. doi:10.1074/jbc.M807997200.
74. Brewer JA, Sleckman BP, Swat W, Muglia LJ. Green fluorescent protein-glucocorticoid receptor knockin mice reveal dynamic receptor modulation during thymocyte development. J Immunol. 2002;169:1309–18.
75. Wang H, et al. One-step generation of mice carrying mutations in multiple genes by CRISPR/Cas-mediated genome engineering. Cell. 2013;153:910–8. doi:10.1016/j.cell.2013.04.025.
76. Yang H, et al. One-step generation of mice carrying reporter and conditional alleles by CRISPR/Cas-mediated genome engineering. Cell. 2013;154:1370–9. doi:10.1016/j.cell.2013.08.022.
77. Iniguez-Lluhi JA, Lou DY, Yamamoto KR. Three amino acid substitutions selectively disrupt the activation but not the repression function of the glucocorticoid receptor N terminus. J Biol Chem. 1997;272:4149–56.
78. Koper JW, van Rossum EF, van den Akker EL. Glucocorticoid receptor polymorphisms and haplotypes and their expression in health and disease. Steroids. 2014;92:62–73. doi:10.1016/j.steroids.2014.07.015.
79. Sarac MS, Zieske AW, Lindberg I. The lethal form of Cushing's in 7B2 null mice is caused by multiple metabolic and hormonal abnormalities. Endocrinology. 2002;143:2324–32. doi:10.1210/endo.143.6.8808.
80. Bentley L, et al. An N-ethyl-N-nitrosourea induced corticotropin-releasing hormone promoter mutation provides a mouse model for endogenous glucocorticoid excess. Endocrinology. 2014;155:908–22. doi:10.1210/en.2013-1247.
81. Stenzel-Poore MP, Cameron VA, Vaughan J, Sawchenko PE, Vale W. Development of Cushing's syndrome in corticotropin-releasing factor transgenic mice. Endocrinology. 1992;130:3378–86. doi:10.1210/endo.130.6.1597149.
82. Harris C, et al. Large increases in adipose triacylglycerol flux in Cushingoid CRH-Tg mice are explained by futile cycling. Am J Physiol Endocrinol Metab. 2013;304:E282–93. doi:10.1152/ajpendo.00154.2012.
83. Wallace K, Flecknell PA, Burt AD, Wright MC. Disrupted pancreatic exocrine differentiation and malabsorption in response to chronic elevated systemic glucocorticoid. Am J Pathol. 2010;177:1225–32. doi:10.2353/ajpath.2010.100107.
84. Masuzaki H, et al. A transgenic model of visceral obesity and the metabolic syndrome. Science. 2001;294:2166–70. doi:10.1126/science.1066285.
85. Kershaw EE, et al. Adipocyte-specific glucocorticoid inactivation protects against diet-induced obesity. Diabetes. 2005;54:1023–31.
86. Baehr LM, Furlow JD, Bodine SC. Muscle sparing in muscle RING finger 1 null mice: response to synthetic glucocorticoids. J Physiol. 2011;589:4759–76. doi:10.1113/jphysiol.2011.212845.
87. Lemke U, et al. The glucocorticoid receptor controls hepatic dyslipidemia through Hes1. Cell Metab. 2008;8:212–23. doi:10.1016/j.cmet.2008.08.001.

Chapter 16
The Dehydrogenase Hypothesis

Conor Woods and Jeremy W. Tomlinson

Abstract Circulating glucocorticoid (GC) levels are controlled by the Hypothalamo–Pituitary–Adrenal (HPA) axis, but within tissues, GC availability is controlled by the isoforms of 11β (Beta)-Hydroxysteroid Dehydrogenase 11β (Beta)-HSD that interconvert inactive cortisone and active cortisol. Two isoforms have been identified; in key metabolic target tissues (including liver and adipose), expression of 11β (Beta)-HSD1 predominates that *in vivo* converts cortisone to cortisol and thus amplifies local GC action. In contrast, in mineralocorticoid target tissues 11β (Beta)-HSD2 is the isoform that is most abundantly expressed. This inactivates cortisol to cortisone and offers protection for the mineralocorticoid receptor form occupation and activation by cortisol. Dysregulated 11β (Beta)-HSD1 activity has been implicated in many metabolic diseases such as obesity and diabetes and inhibition of 11β (Beta)-HSD1 represents a promising therapeutic target. Mutations within the gene encoding 11β (Beta)-HSD2 cause the Syndrome of Apparent Mineralocorticoid Excess and decreases in activity are linked to hypertension as well as impairment in placental function and neonatal growth. We will discuss the molecular biology and enzymology of 11β (Beta)-HSD and its role in normal physiology and discuss altered 11β (Beta)-HSD activity in pathological states and the potential for therapeutic targeting.

Keywords 11β (Beta)-HSD • Endoplasmic reticulum (ER) • Cortisol • Cortisone • H6PDH • NADPH

C. Woods, M.B.B.Ch., M.R.C.P.
Department of Diabetes and Endocrinology, St Vincent's University Hospital, Dublin, Leisnter, Ireland

J.W. Tomlinson (✉)
Oxford Centre for Diabetes, Endocrinology & Metabolism, University of Oxford, Radcliffe Department of Medicine, Churchill Hospital, Oxford, UK
e-mail: jeremy.tomlinson@ocdem.ox.ac.uk

J.-C. Wang, C. Harris (eds.), *Glucocorticoid Signaling*, Advances in Experimental Medicine and Biology 872, DOI 10.1007/978-1-4939-2895-8_16

Introduction

Glucocorticoids (GC) (cortisol in man and corticosterone in rodents) are crucial in regulating many important physiological functions including glucose and amino acid metabolism, inflammation, immunity and general health and well-being [1]. GCs bind to the glucocorticoid receptor (GR), a member of the nuclear receptor family and alter gene transcription up-regulating anti-inflammatory protein synthesis and reducing expression of pro-inflammatory cytokines.

Since their discovery in the 1940s by Kendall and Hench, GCs have become a mainstay of therapy in many clinical conditions including rheumatoid arthritis, asthma and as a cornerstone in anti-rejection medication regimes in organ transplant recipients.

Whereas circulating GC levels are regulated by the Hypothalamo–pituitary–adrenal (HPA) axis, tissue cortisol metabolism is under the control of the enzymes of the 11β (Beta)-Hydroxysteroid Dehydrogenase (11β (Beta)-HSD) system. Two isozymes exist, namely 11β (Beta)-HSD1 and 11β (Beta)-HSD2. In recent decades, attention has focussed on tissue cortisol metabolism and the pre-receptor regulation of GCs within tissues such as fat, muscle and liver [2–6]. 11β (Beta)-HSD1 is a bi-directional enzyme, which *in vivo* acts primarily as an oxo-reductase, converting inactive cortisone (and inactive 11-dehydrocorticosterone in rodents) to active cortisol. 11β (Beta)-HSD1 is located in a number of key metabolic tissues including liver, skeletal muscle, gonads and adipose tissue [6, 7] and has key metabolic influence acting as an amplification step in tissue cortisol exposure. Tissue exposure to GC therefore is not only reliant on circulating levels of cortisol but can also be altered by pre receptor regulation via 11β (Beta)-HSD1. Dysregulated 11β (Beta)-HSD1 activity has recently been implicated in many metabolic and inflammatory diseases and is a promising target to alter tissue glucocorticoid exposure. Selective 11β (Beta)-HSD1 inhibitors exist and have been trialled in early phase 2 studies primarily in type 2 diabetes [8].

11β (Beta)-HSD2 is expressed in aldosterone selective tissues, mainly in the distal nephrons, colonic epithelium, salivary and sweat glands (mineralocorticoid receptor (MR) expressing tissues) and in the foetus and placenta during gestation. It acts uniquely as a dehydrogenase deactivating cortisol to inactive cortisone, protecting tissues from excess glucocorticoid exposure [6, 9]. 11β (Beta)-HSD2 deficiency or inhibition causes hypertension secondary to apparent mineralocorticoid excess whereby cortisol activates the MR for which it shares the same affinity as aldosterone. A similar syndrome manifests in people ingesting large amounts of European licorice which contains an inhibitor of this enzyme. 11β (Beta)-HSD2 is important in gestation with significant activity in both placenta and fetal brain development by preventing early exposure to GCs.

Hypothalamic Pituitary Adrenal Axis and Cortisol Production

GCs, synthesized from cholesterol precursors, are produced primarily by the adrenal cortex (zona fasiculata) under the influence of the HPA axis in a classical circadian rhythm, regulated by negative feedback.

Healthy adults secrete 10–15 mg s cortisol/day [10]. The majority of cortisol is protein bound to cortisol binding globulin (CBG) although some is albumin bound. Estimates suggest that only 5 % of circulating cortisol is "free" and biologically active [11, 12]. The half-life of free cortisol is only a few minutes whereas protein bound cortisol has a much longer half life of between 70 and 120 min [11, 13, 14]. Cortisone has similar unbound concentrations to that of cortisol but has a lower affinity for both CBG and GR [12, 15]. Importantly the biological availability of GCs represents a balance between synthesis/secretion and metabolism/clearance.

Cortisol Metabolism and Clearance

There are multiple pathways involved in cortisol and cortisone metabolism. These include A-ring reduction to from tetrahydrocortisol and it's 5α (alpha)-isomer allo-tetrahydrocortisol, hydroxylation to 6β (Beta) hydroxycortisol and reduction of the 20-oxo group giving cortols. Among the various distinct pathways, perhaps the most critical of these is the inter-conversion of cortisol and cortisone mediated by the 11β (Beta)-HSD isozymes. 11β (Beta)-HSD interconverts cortisol and cortisone by altering the hydroxyl group at C11 (see Fig. 16.1).

Tissue Cortisol Metabolism: 11β (Beta) HSD1

11β (Beta) HSD1: Discovery

Glucocorticoids were discovered in the 1940s and 1950s and were heralded as a potentially curative treatment for many ailments [16]. Kendall et al. published the discovery of what they believed to be a treatment that could reverse rheumatoid arthritis in the 1950s [17]. What they referred to as Compound E; was actually cortisone, a

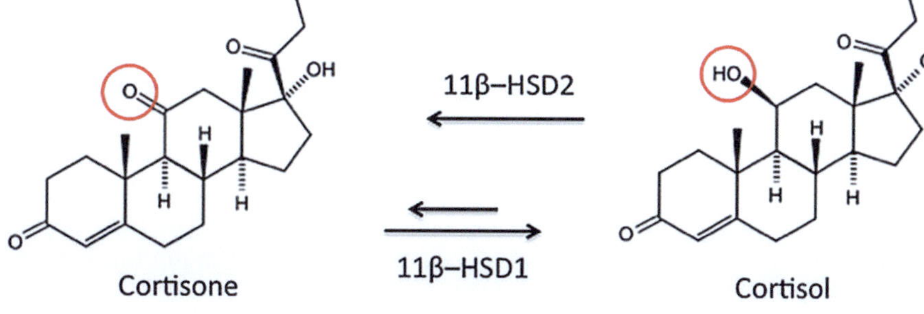

Fig. 16.1 Isoforms of 11β (Beta)-Hydroxysteroid dehydrogenase as pre-receptor regulators of glucocorticoid availability

precursor of cortisol (Compound F). We now know that the cortisone administered, was converted to cortisol by 11β (Beta)-HSD1. The hydroxyl group at C11 is crucially important for cortisol to be effective. 11β (Beta)-HSD1 reduces inactive cortisone to active cortisol by adding a hydroxyl group at C11. Thus, unbeknownst to the authors they had actually made two discoveries—the first that the glucocorticoid cortisol has dramatic therapeutic effect on arthritis and second that the enzyme 11β (Beta)-HSD1 has the ability to convert inactive cortisone to active steroid.

Subsequently to Kendall's paper, cortisol, and not cortisone, was established as the active steroid. It also became apparent that inter-conversion between cortisone and cortisol is possible [18, 19]. This conversion is made possible by the enzyme 11β (Beta)-HSD1, a member of the dehydrogenase/reductase superfamily. Short-chain Dehydrogenase/Reducatse (SDR) enzymes are NADP(H) dependent enzymes, that are involved primarily in breaking down hormones and chemical messengers [20]. There are more than 3,000 members in the SDR family [7].

Early studies identified 11β (Beta)-HSD activity in different tissues such as placenta, kidney and liver. However, 11β (Beta)-HSD did not have the same activity between different tissues. The "set point" of the enzyme varied from tissue to tissue with dehydrogenase activity (cortisol conversion to cortisone) predominating in the placenta and kidney and reductive activity (cortisone conversion to cortisol) predominating in liver [7]. The answer lay in the fact that there are two distinct isozymes of 11β (Beta)-HSD namely type 1 being predominantly reductive and type 2 acting solely as a dehydrogenase. In addition 11β (Beta)-HSD type 1 is a bi-directional capable of acting in either direction but dependent on co-factors such as NADPH. *In vivo* in intact tissues the 11β (Beta)-HSD1 enzyme acts as an oxo-reductase converting cortisone to cortisol. Purified 11β (Beta)-HSD1 enzyme behaves principally as a dehydrogenase, oxidizing cortisol to cortisone.

Human 11β (Beta)-HSD1

11β (Beta)-HSD1 was initially purified and cloned from rodents in the 1980s by Carl Monder's group [21, 22]. In humans the gene for 11β (Beta)-HSD1 (HSD11B1) is located on chromosome 1, is 30 kb in length and has six exons and five introns. It comprises of 292 amino acids and shares 77 % homology with rat amino acid sequence [23]. Human 11β (Beta)-HSD1, cloned in 2002 [24], exits as a dimer [25] and is bound to the endoplasmic reticulum (ER) with its catalytic domain within the ER lumen [26] (see Fig. 16.2). The catalytic directionality of the enzyme is based on the position of 11β (Beta)-HSD1 within the ER lumen where it co-localises with Hexose 6 Phosphate Dehydrogenase (H6PD). H6PD generates the reduced co-substrate NADPH. Thus the ratio of NADPH/NADP confers directionality to 11β (Beta)-HSD1 [27, 28]. Purified 11β (Beta)-HSD1, in the absence of H6PD, behaves principally as a dehydrogenase, oxidizing cortisol to cortisone. The enzyme 11β (Beta)-HSD1, has Michaelis Menten (K_m) constants of 1.83 ± 0.06 μm for corticosterone and 17.3 ± 2.2406 μm for cortisol [7].

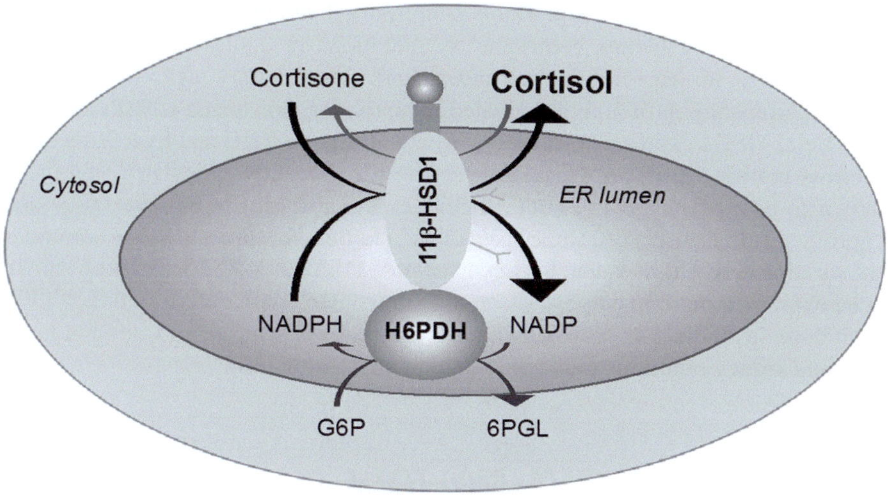

Fig. 16.2 Localization and co-factor dependency of 11β (Beta)-HSD1 within the endoplasmic reticulum

11β (Beta)-HSD1 Ontogeny

The ontogeny of 11β (Beta)-HSD1 has been studied predominantly in rodents and sheep animal models. The expression of mammalian 11β (Beta)-HSD1 is predominantly post-natal. In general 11β (Beta)-HSD1 is detectable in many tissues but there is a lack of activity in early gestation with reductase activity only becoming apparent after delivery and rising steadily throughout infancy [7, 29]. The ontogeny of 11β (Beta)-HSD1 in humans is not well documented with few published studies. Cortisone therapy is not useful for treating congenital adrenal hyperplasia in early infancy most likely due to absent or significantly reduced liver 11β (Beta)-HSD1 activity [30]. Both reductase and dehydrogenase activity has been demonstrated in fetal lung tissue [31, 32]. 11β (Beta)-HSD1 activity remains unchanged throughout childhood in both boys and girls [33]. At puberty there is a reduction in 11β (Beta)-HSD1 activity in women [33] and this appears to remain. In adults, there is a well described dimorphism in cortisol metabolism between men and women with an apparent reduction in 11β (Beta)-HSD1 activity in women [34, 35]. However, not all studies have shown this dimorphism [36].

Regulation of 11β (Beta)-HSD1 Expression

11β (Beta)-HSD1 expression increases steadily until 1 year of age at which time peak levels are reached. 11β (Beta)-HSD1 is expressed in many tissues including liver, adipose tissue, gonads, GI tract, kidney, eye, anterior pituitary, leukocytes and bone [7]. 11β (Beta)-HSD1 expression is highest in liver, brain gonads and adipose tissues [37]. Many factors contribute to alterations in tissue expression of 11β (Beta)-HSD1. In general glucocorticoids, pro-inflammatory cytokines (TNFα, IL-1β) peroxisome proliferator-activated receptor γ agonists and CEBPs increase 11β (Beta)-HSD1 expression whereas Growth Hormone (GH) and liver X receptor agonists inhibit 11β (Beta)-HSD1 expression [7]. Recently salicylates have been shown to down regulate 11β (Beta)-HSD1 expression [38] in adipose tissue and improve insulin sensitivity. The effects of sex steroids, insulin and other hormones are variable across tissues and between species. 11β (Beta)-HSD1 has been shown to be expressed more in adipose tissue in lean women compared to lean men [39]. High dose oestradiol has been shown to repress 11β (Beta)-HSD1 expression in rat liver and kidney but testosterone did not alter expression [40].

Tissue-Specific Role of 11β (Beta)-HSD1

Adipose Tissue

11β (Beta)-HSD1 is highly expressed in human adipose tissue [3, 41]. Broadly speaking, studies have shown similar expression levels of 11β (Beta)-HSD1 between visceral and subcutaneous compartments [39, 42, 43]. However, H6PD and GR have been shown to have different levels of expression between adipose tissue depots [43].

As in other tissues, 11β (Beta)-HSD1 acts primarily as a reductase generating active GCs. Its expression and activity are induced by GCs and inflammatory cytokines such as interleukin 1 and TNFα [7, 44–47]. 11β (Beta)-HSD1 expression and activity increases upon adipocyte differentiation [48] and inhibition of 11β (Beta)-HSD1 blocks cortisone-induced adipocyte differentiation [49]. 11β (Beta)-HSD1 activity within adipose tissue depots is controversial with some studies demonstrating increased activity in omental tissue compared to subcutaneous adipose tissue.

Dysregulation of 11β (Beta)-HSD1 activity has been postulated to be critical in the pathogenesis of metabolic complications associated with obesity. In rodent models increased 11β (Beta)-HSD1 activity has been demonstrated in visceral adipose tissue of obese rodents compared to wild type [50, 51] although expression in adipose tissue was reduced following high fat feeding in mice [52]. Short term high fat diet reduced the activity of 11β (Beta)-HSD1 in both subcutaneous and visceral adipose tissue in Wistar rats [53]. Genetically modified rodent models have contributed significantly to our understanding of the role of 11β (Beta)-HSD1 in adipose tissue. 11β (Beta)-HSD1 knockout mice are resistant to the metabolic side effects of a high fat diet compared to wild type [54]. Conversely over expression of 11β (Beta)-HSD1 in adipose tissue leads to weight gain and metabolic complications compared to wild type [55] even on a chow diet.

Clinical studies have yielded varied and sometimes conflicting results. In human adipose tissue, the majority of studies have examined subcutaneous adipose tissue. 11β (Beta)-HSD1 activity in subcutaneous adipose tissue has been shown to be increased in obese patients compared to non-obese controls [56]. Both 11β (Beta)-HSD1 activity and expression were shown to correlate positively with BMI by Lindsay et al. in 2003 [57]. Whilst the majority of studies show a positive relationship of 11β (Beta)-HSD1 expression and activity with BMI, this is not the case for all studies. Data with regards to the omental depot are more sparse, but some studies have identified increased expression in women in association with increased omental adiposity [58]. The impact of diet induced obesity on 11β (Beta)-HSD1 activity and expression is not clear. Obese Zucker rats have increased visceral adipose tissue 11β (Beta)-HSD1 expression compared to lean rats [51] however Wistar rats, fed a short term high fat diet show reduced 11β (Beta)-HSD1 activity [53]. Diet induced obese mice also have reduced expression of 11β (Beta)-HSD1 [59]. In these diet induced obese mice 11β (Beta)-HSD1 expression is increased by NFκ (Kappa) B and reduced by HIF-1. H6PD is also important in the directionality of 11β (Beta)-HSD1 and some studies suggest a role in visceral adipose tissue [60]. Pro inflammatory cytokines and glucocorticoids (increased in obesity) are also known to induce 11β (Beta)-HSD1 expression in adipose tissue [7]. In vitro studies highlight the role of transcription factors (CEBPs) for controlling 11β (Beta)-HSD1 expression in adipocytes [61, 62].

Liver

GCs have key metabolic effects on the liver, augmenting insulin stimulated lipogenesis [63] and reducing lipolysis in hepatic tissue [64]. Indirectly, they increase hepatic lipid accumulation by inducing surrounding adipose tissue lipolysis thus increasing the delivery of free fatty acids to the liver via the portal circulation [65].

11β (Beta)-HSD1 expression and activity has been extensively studied in hepatic tissue. In animal studies global 11β (Beta)-HSD1 knock out mice are protected from diet induced hepatic steatosis when fed a high fat diet [54, 55]. Transgenic mice with hepatic 11β (Beta)-HSD1 overexpression, develop hypertension, hepatic steatosis and dyslipidemia but interestingly do not develop steatohepatitis and minimal insulin resistance [66]. Other factors including HPA axis activation and adipose tissue 11β (Beta)-HSD1 activity may play a role in the development of hepatosteatitis. Liver specific HSD1-null mice have minimal phenotype (ref) suggesting visceral adipose HSD1 is the key site of action and excess adipose-generated GCs act on the liver via visceral portal drainage.

11β (Beta)-HSD1 is highly expressed in human liver [23], predominantly in a centripetal pattern histologically with maximum expression around the central vein [67]. Hepatic 11β (Beta)-HSD1 appears to act exclusively as an oxo-reductase generating active GC [68]. In animal rodent models over expression of 11β (Beta)-HSD1 in visceral adipose tissue was associated with Non Alcoholic Fatty Liver Disease (NAFLD) [69]. In humans, some studies have demonstrated increased expression of 11β (Beta)-HSD1 in liver tissue in obese persons with metabolic syndrome [70] however other studies have not confirmed these findings [71, 72]. Liver 11β (Beta)-HSD1 activity, as measured by urinary steroid metabolite analysis is reduced in obesity compared to non-obese controls [73]. Liver 11β (Beta)-HSD1 activity (using serum cortisol generation form oral cortisone) is also reduced in obesity compared to non-obese controls [73]. These reductions in both liver 11β (Beta)-HSD1 activity and expression contrast with increased activity and expression in adipose tissue in obesity. Ahmed et al. have described a switch in the expression of 11β (Beta)-HSD1 across the spectrum of liver disease from lower expression of 11β (Beta)-HSD1 in steatosis to higher levels of 11β (Beta)-HSD1 expression associated with steatohepatitis [74]. The authors concluded these changes might reflect a response to a more inflammatory phenotype.

Skeletal Muscle

GCs play an important role in protein metabolism. In Cushing's syndrome the effect of excess GCs is clearly seen with myopathy and muscle atrophy. 11β (Beta)-HSD1 is expressed in skeletal muscle [75, 76], whereas 11β (Beta)-HSD2 is not. 11β (Beta)-HSD1 activity has been demonstrated in skeletal muscle [77] and it may have a role in insulin sensitivity and metabolic disease. Increased expression [78] and activity [79] have been demonstrated in rats and humans with type 2 DM.

The exact role of 11β (Beta)-HSD1 in sarcopaenia remains unexplained. It is thought that increased 11β (Beta)-HSD1 activity may play a role in allowing increased muscle cortisol exposure. Increased 11β (Beta)-HSD1 expression in skeletal muscle is associated with reduced muscle strength in older adults compared to younger controls [80].

Skin

Data on tissue cortisol metabolism within skin is only recently becoming an area of interest and investigation. Skin has been shown to be an active site of cortisol production and metabolism [81, 82]. Excess skin exposure to GC's cause skin changes similar to the natural aging process. These include reduced elasticity, reduced collagen and fibroblast numbers, thinning of dermis and epidermis and a general reduction in the repair capacity of skin [83]. Increased exposure to GC's has been postulated as a factor in age related changes, inflammatory and auto-immunity changes seen in skin [84]. It has been postulated that skin changes seen over time are in part as a result of 11β (Beta)-HSD1 activity [82].

Both 11β (Beta)-HSD1 and 2 are expressed in skin [81, 82, 85]. 11β (Beta)-HSD2 is expressed in association with the mineralocorticoid receptor on sweat glands however its role (if any) within the dermis and epidermis is debated. In wound healing, 11β (Beta)-HSD2 expression has been shown to be induced 48 h after tissue injury with subsequent return to basal levels at 96 h [81]. This has been postulated to be a mechanism to reduce local cortisol excess following inflammation.

11β (Beta)-HSD1 is widely expressed in human and mouse dermis and epidermis [82, 85, 86]. Upon differentiation of keratinocytes 11β (Beta)-HSD1 expression increases [85], somewhat akin to the changes seen with pre adipocyte differentiation [87]. Interestingly despite reducing levels of expression of 11β (Beta)-HSD1 in elderly subjects, a paradoxical rise in 11β (Beta)-HSD1 activity is seen with increasing age in both humans and mice [82]. This gives credence to the concept of age related skin atrophic changes being in part due to increased cortisol exposure secondary to increased 11β (Beta)-HSD1 activity.

11β (Beta)-HSD1 has been shown to have a pivotal role in skin repair following injury [88] and tissue remodeling [85]. In mice 11β (Beta)-HSD1 is contributory to impaired wound healing. Blocking 11β (Beta)-HSD1 improved wound healing in mice and prevented age induced skin changes [89]. These data suggest that 11β (Beta)-HSD1 generated local cortisol is critically important in wound healing and in aging skin changes. Inhibitors of 11β (Beta)-HSD1 inhibitors (topical or oral) may therefore have therapeutic potential.

Cardiovascular System

Both isozymes are expressed in blood vessel walls [90] and heart [91], however oxo reductase directionality (11β (Beta)-HSD1) predominates in vascular smooth muscle [88]. There is evidence linking 11β (Beta)-HSD1 activity with atherosclerosis. Mediastinal adipose tissue 11β (Beta)-HSD1 has been linked with coronary atherosclerosis [92]. The same authors demonstrated increased 11β (Beta)-HSD1 expression in aortas of obese patients with the metabolic syndrome [93]. 11β (Beta)-HSD1 inhibition in apoE knockout mice achieved significant reduction atherosclerotic load suggesting a role in plaque formation [94]. Carbenoxolone treatment has been shown to reduce atherosclerosis in mice [95]. 11β (Beta)-HSD1 knock out mice show improved angiogenesis to infarcted regions possibly through reduced GC regeneration locally [96].

Central Nervous System (CNS)

GCs are required for normal brain development and normal brain function. Excess GCs are associated with alterations in mood, memory and brain function [97, 98]. Both 11β (Beta)-HSD1 and 2 are expressed in the brain [99–102], but 11β (Beta)-HSD2 is expressed at lower levels. 11β (Beta)-HSD1 acts principally as an oxo-reductase in brain tissue [103] but interestingly H6PD does not always co-localise with 11β (Beta)-HSD1 in the CNS [101]. This lack of universal co-localisation suggests other enzymes providing the necessary co factors for 11β (Beta)-HSD1. Elevated 11β (Beta)-HSD1 is seen with ageing and is associated with cognitive decline. Altered 11β (Beta)-HSD1 activity in the CNS is associated with changes in appetite, affective behavior and circadian rhythm [104].

A putative role for 11β (Beta)-HSD1 in cognitive impairment is postulated along with investigations looking at the link between 11β (Beta)-HSD1 and human eye disease including thyroid eye disease [105] and glaucoma. 11β (Beta)-HSD1 inhibitors have been shown to improve cognitive function in elderly persons with T2 DM [106].

Inflammation and Immunity

Glucocorticoids in pharmacological doses are immunosuppressive and produce powerful anti-inflammatory effects [107]. They achieve this by altering gene transcription and altering pro and anti-inflammatory mediators including cytokines and signaling pathways. 11β (Beta)–HSD1 is believed to play a key role in local inflammation and immune response to stimuli and allergens [108].

11β (Beta)-HSD1 is expressed on numerous immune cell types in humans including mast cells, mononuclear cells and macrophages, lymphocytes, B cells and dendritic cells [109–112]. 11β (Beta)-HSD1 is not expressed in non-stimulated monocytes but 11β (Beta)-HSD1 expression is increased upon differentiation of monocytes cells into macrophages with interleukin 4 and 13 [110]. A further increase in 11β (Beta)–HSD1 expression is seen when monocytes differentiate into dendritic cells under the influence of interleukin-4 and granulocyte-macrophage colony-stimulating factor. Other pro-inflammatory cytokines do not increase 11β (Beta)-HSD1. In macrophages, 11β (Beta)-HSD1 activity increases with exposure to lipopolysaccharide [110]. 11β (Beta)-HSD1 expression has also been seen in murine lymphocytes and B cells with increases seen when CD4-positive lymphocytes polarize into Th1 or Th2 subsets [111].

Separate to expression, the functional role of 11β (Beta)-HSD1 in immunity remains debated as it is difficult to separate out the specific subsets of immune cells and to be confident of the role (if any) of 11β (Beta)-HSD1. Our understanding of 11β (Beta)-HSD1 expression and activity on human immune cell actions and inflammation is incomplete and warrants further investigation.

There is evidence both in animal models and in human studies that demonstrate that 11β (Beta)-HSD1 is implicated in immune response to infection and inflammation. In 11β (Beta)-HSD1 knock out mice defective macrophage phagocytosis of apoptotic neutrophils is seen in peritonitis [113]. KO mice have increased susceptibility to endotoxaemia compared to wild type [114]. Of note, a study in mice demonstrated that in various models of joint and lung inflammation 11β (Beta)-HSD1 knock out mice developed inflammation earlier, substantially more inflammation and slower resolution of inflammation compared to wild type controls [115]. This study suggests that 11β (Beta)-HSD1 plays a key role as an anti-inflammatory regulator and raises concerns for drug therapy inhibiting 11β (Beta)-HSD1.

In humans (and rats) 11β (Beta)-HSD1 expression is elevated in colitis [116, 117] with a concomitant reduction in 11β (Beta)-HSD2 expression. Acute exacerbations of inflammatory bowel disease are associated with elevated 11β (Beta)-HSD1 expression [118] but interestingly 11β (Beta)-HSD1 activity was elevated in patients who were in remission, suggesting that high local glucocorticoid levels are important in limiting inflammation. In a rat model, chemically induced inflammation increased 11β (Beta)-HSD1 expression both in colonic tissue and lymphoid tissue [119].

Recently, 11β (Beta)-HSD1 has been shown to provide tonic inhibition of mast cell deactivation. 11β (Beta)-HSD1 knock out mice had increased mast cell numbers and a lower threshold for deactivation suggesting that reduced 11β (Beta)-HSD1 activity increases allergy and anaphylaxis [109].

Bone and Joint

11β (Beta)-HSD1 is expressed in human bone [120], predominantly in osteoblasts [121].

In rheumatoid arthritis, both 11β (Beta)-HSD1 and 11β (Beta)-HSD2 isozymes are expressed in synovial tissue with conflicting reports as to which is the dominant isoform. Global 11β (Beta)-HSD1 activity is increased in rheumatoid arthritis (as measured by urinary corticosteroid metabolites) compared to non-arthritic controls [122]. Enzyme activity correlates with synovial inflammation severity [123]. In animal models 11β (Beta)-HSD1 knock out mice develop more severe arthritis and earlier compared to wild type controls [115]and mice treated with 11β (Beta)-HSD1 inhibitors (Carbenoxolone) developed worse arthritis [124].

Cortisone Reductase Deficiency (CRD) and Apparent Cortisone Reductase Deficiency (ACRD)

Genetic defects in both HSDB1 and H6PD encoding genes demonstrate the effect of alterations in tissue 11β (Beta)-HSD1 on clinical phenotype. Both Cortisone Reductase Deficiency (CRD) from HSDB1 gene defects and apparent Cortisone Reductase Deficiency (ACRD) from H6PD gene defects show the impact of reduction in tissue 11β (Beta)-HSD1 activity with low urine cortisol, significantly elevated cortisone with subsequent compensatory increased HPA activity leading to hyperandrogenism, early adrenarche and PCOS in women [7, 125].

CRD was first described in the 1980s. The majority of cases are female and present with clinical and biochemical hyperandrogenism, with males presenting with precocious puberty [7, 37]. CRD has been described as the "human 11β (Beta)-HSD1 knockout" [7]. The condition is ameliorable to dexamethasone therapy. It shares some clinical and biochemical features and should not be confused with non-classical congenital adrenal hyperplasia.

11β (Beta)-HSD1 Inhibition in Clinical Studies

Due to evidence that has implicated 11β (Beta)-HSD1 in the pathogenesis of disease states including obesity, diabetes and the metabolic syndrome, it represents an exciting therapeutic target to limit local GC availability [2, 126]. Many inhibitors of the 11β (Beta)-HSD enzymes have been described. These include naturally occurring inhibitor compounds such as liquorice derived glycyrrhetinic acid [127], flavanone/hydroxyl flavanones [128], bile acids [129], progesterone metabolites [130] and even coffee [131]. Most of these naturally occurring compounds inhibit both 11β (Beta)-HSD1 and 11β (Beta)-HSD2, with subsequent hypertension and hypokalaemia limiting their possible benefits.

Carbenoxolone, a non-selective inhibitor derived from glycyrrhetinic acid, was the first drug to show benefit in human studies [4, 132]. In healthy volunteers it improved insulin sensitivity and reduced glucose production rates via a reduction in glycogenolysis but not gluconeogenesis in patients with type 2 diabetes [132]. Carbenoxolone has also been shown to reduce local cortisol availability in subcutaneous adipose tissue and inhibits glucocorticoid induced lipolysis [4]. These early small "proof of principal" studies demonstrated that 11β (Beta)-HSD1 in metabolic disease could be targeted for drug manipulation and importantly that tissue specific effect could be demonstrated, despite a relative lack of specificity of carbenexolone on 11β (Beta)-HSD1. Several pharmaceutical companies have developed potent selective 11β (Beta)-HSD1 inhibitors. Indeed 11β (Beta)-HSD inhibition has become a significant area of investment for many companies [133] and an extensive array of chemical compounds have been patented and reviewed extensively elsewhere [134]. In general, they are highly selective for 11β (Beta)-HSD1 over 11β (Beta)-HSD2 and have high potency for inhibition.

A small number of clinical trials have been published with relatively short durations of intervention (all no more than 12 weeks) and have demonstrated improvements in biomarkers including cholesterol profiles, weight, glycaemic control and blood pressure [8, 135, 136]. The first outcome study to be published investigated the addition of an 11β (Beta)-HSD1 inhibitor (INCN13739) to metformin in patients with type 2 Diabetes. The drug was well tolerated and the 12 week study demonstrated improvements in weight, glycaemic control and lipid profiles in those people that received the inhibitor [135]. Of note, in this trial there was a compensatory increase in the HPA axis with a dose dependent increase in ACTH and subsequent increase in certain androgens. In women there was a small and significant increase in testosterone levels but biologically active testosterone was felt to be unchanged as sex hormone binding globulin also increased. Whilst all blood results remained in the normal reference range these alterations and small elevations in androgens from HPA activation will remain a clinical concern in the long term, especially for women.

In a further study using a different compound (MK0916) again in the setting of type 2 diabetes and metabolic syndrome, modest improvements in glycaemic control with no reduction in fasting glucose levels were seen [8]. Again however there was mild activation of HPA with elevations in adrenal androgen secretion. Overall, the long-term side effects of 11β (Beta)-HSD1 inhibition remain unknown and while the benefits may outweigh any side effects the possible disruption of HPA axis will remain a concern alongside the magnitude of the clinical response.

Tissue Cortisone Metabolism: 11β (Beta)-HSD2

11β (Beta)-HSD2 Discovery

In 1993, Seckl et al. isolated an enzyme with exclusive 11β (Beta)-HSD dehydrogenase activity from both human placenta and rat kidney [137]. In 1994, Krozoski et al. isolated human 11β (Beta)-HSD from human kidney [138], identical to the dehydrogenase enzyme found in placenta. This second enzyme was found to be distinct from

11β (Beta)-HSD1 and was called 11β (Beta)-HSD2 and is also a member of the SDR family [139]. Both 11β (Beta)-HSD1 and 11β (Beta)-HSD2 isozymes are members of the Short-Chain Dehydrogenase/Reductase (SDR) superfamily of enzymes, however each isozyme has a distinct gene sequence with exons found on different chromosomes. There is little similarity or overlap in sequence between isozymes (18 % identity), except for similar co-factor binding regions at the NH_2-terminal [6].

The human 11β (Beta)-HSD2 gene is located on chromosome 16, has 5 exons and is only 6 kbs in length [140]. Human 11β (Beta)-HSD2 measures 405 amino acids in length and has a molecular mass of 44 kDa [138]. It is also anchored to the ER and loses its dehydrogenase activity once removed from tissue membranes [141]. 11β (Beta)-HSD2 universally acts as a dehydrogenase across species [6] and has a Km for cortisol of 50–60 nM and 10–13 nM for cortisone [6, 142]. Mutations in 11β (Beta)-HSD2 leading to apparent mineralocorticoid excess (AME) and hypertension have been extensively reported [143–145].

11β (Beta)-HSD2 Ontogeny

11β (Beta)-HSD2 has an important role in fetal development with intra uterine "programming" affecting subsequent adult physiology. 11β (Beta)-HSD2 plays an important role in gestation in humans and mammals protecting tissues against GC exposure prematurely. 11β (Beta)-HSD2 is expressed and active in placenta [36, 51] and steadily rises throughout gestation and declines 2 weeks prior to labour [29, 137, 146]. Placenta 11β (Beta)-HSD2 is localized to the syncytiotrophoblast where it has been described as a barrier to maternal corticosteroid which is considerably more concentrated [6]. GCs play a critical role in the development of fetal organs, in particular, towards the end of pregnancy, in lung tissue. Excess GC exposure in utero is associated with physiological and metabolic complications [147–149]. There is mounting evidence that 11β (Beta)-HSD2 plays a key protective role in normal development of the fetus and in particular brain development [148]. Altered or disrupted 11β (Beta)-HSD2 activity with subsequent excess intra-uterine exposure to glucocorticoid has a "programming" effect on the fetus leading to low birth weight and lifelong physiological consequences such as increased cardiovascular, metabolic and psychiatric complications [150]. This role in the development of the fetus and its role in subsequent lifelong physiology has led some to consider the degree of prenatal GC exposure as a potential prognostic biomarker [6].

Regulation 11β (Beta)-HSD2 Expression

Unlike 11β (Beta)-HSD1, there is considerable data published on the epigenetic influence on 11β (Beta)-HSD2 activity in humans and in rodent models.

The 11β (Beta)-HSD2 gene is susceptible to epigenetic influence, with methylation of the promoter region of particular interest. Increased methylation of this region has been inversely associated with 11β (Beta)-HSD2 expression and influences the development of hypertension, intrauterine growth, birth weight and neurobehavioural movement [151, 152] in rats. Intrauterine growth retardation has been associated with increased methylation of 11β (Beta)-HSD2 gene promoter with subsequent repression of 11β (Beta)-HSD2 in adult kidneys [153].

Factors that increase 11β (Beta)-HSD1 expression tend to reduce 11β (Beta)-HSD2 such as TNFα [154]. Oestrogen increases 11β (Beta)-HSD2 expression [40, 155]. Vasopressin has been shown to stimulate 11β (Beta)-HSD2 [156]. Glucocorticoids down-regulate 11β (Beta)-HSD2 in foetal placenta but not foetal kidney [157]. Dexamethasone up-regulates 11β (Beta)-HSD2 in lung cells [158]. Hypoxia has been shown to reduce 11β (Beta)-HSD2 expression [159]. In colonic epithelium, aldosterone increases 11β (Beta)-HSD2 expression [160].

Tissue-Specific Role of 11β (Beta)-HSD2

CNS

In the adult brain, 11β (Beta)-HSD2 is expressed in a select few regions [161]. Before birth, 11β (Beta)-HSD2 is expressed in several additional brain regions, including the thalamus and cerebellum, where this enzyme protects proliferating granule cells from the growth-limiting effects of glucocorticosteroids [162, 163]. In adults, however, mRNA and protein are no longer detectable in these regions. Instead, 11β (Beta)-HSD2 expression is found in just a few small sites in adult mice and rats [161, 164] most prominently in a group of neurons found inside the nucleus of the solitary tract (NTS). NTS neurons with 11β (Beta)-HSD2 expression are the only cells in the brain shown to express both this enzyme and the mineralocorticoid receptor (MR); these "HSD2 neurons" are activated along with salt appetite after dietary sodium deprivation or volume depletion, and may trigger salt appetite. The only other brain sites in which 11β (Beta)-HSD2 expression (mRNA and protein) was identified consistently in adult animals are the subcommissural organ (a circumventricular organ comprised of modified ependymal cells which do not express MR) and a small subdivision of the ventromedial hypothalamic nucleus (neurons that lack MR immunolabeling). Information regarding 11β (Beta)-HSD2 expression remains incomplete for non-neuronal tissues such as the meninges, choroid plexus, ventricular ependyma, and cerebral vasculature.

Cardiovascular System

11β (Beta)-HSD2 is expressed in vascular endothelium [90]. 11β (Beta)-HSD2 knockout mice develop endothelial dysfunction [165]. Lack of 11β (Beta)-HSD2 and MR activation is implicated in generation of severe atherosclerosis in mouse models [166].

Kidney

11β (Beta)-HSD2 is perhaps best known for its role in the kidney where it protects the mineralocorticoid receptor from excess exposure to GC. 11β (Beta)-HSD2 is widely expressed in distal nephrons [141]. Although the inherent enzyme ability of 11β (Beta)-HSD2 to clear cortisol (converting it to cortisone) should not be enough, given concentrations and binding affinities, in reality it protects the mineralocorticoid receptor from GC exposure [6, 167]. Lack of 11β (Beta)-HSD2 in kidney (AME discussed below), leads to hypertension and other sequelae. 11β (Beta)-HSD2 activity, measured by urinary metabolite ratios, reduces with age suggesting a role in age related hypertension [168]. Some studies show a role of 11β (Beta)-HSD2 in hypertension although not all studies are in agreement [169]. In kidney disease reduced 11β (Beta)-HSD2 activity has been shown in persons with hypertension [170, 171].

Colon

11β (Beta)-HSD2 is expressed in colonic epithelium [172]. Expression is increased by aldosterone in rats [160]. In Inflammatory bowel disease 11β (Beta)-HSD2 expression is down regulated in both humans and rats [116]. This is accompanied by an increase in 11β (Beta)-HSD1 expression and so is presumed to be an attempt to locally control GC exposure to inflamed tissue. Zhang et al. showed that inhibiting 11β (Beta)-HSD2 reduces colon carcinogenesis by inhibiting COX 2 pathways. The reduction in 11β (Beta)-HSD2 blocked colorectal adenocarcinoma angiogenesis and metastasis [173].

Salivary Gland and Skin

As mineralocorticoid target tissues, both skin and salivary glands express 11β (Beta)-HSD2. In the skin expression is mainly restricted to sweat glands [174] 11β (Beta)-HSD2 is expressed in both parotid and sub-mandibular glands [174, 175]. Measuring salivary cortisone has been postulated a potential biomarker of serum free cortisol [176]. In addition, reduced activity of 11β (Beta)-HSD2 in sweat glands has also been linked with essential hypertension [177].

Pituitary

In normal anterior pituitary tissue 11β (Beta)-HSD2 mRNA expression is seen, but immunofluorescence reveals absent 11β (Beta)-HSD2 isozyme. Interestingly, ACTH secreting tumours induce 11β (Beta)-HSD2 expression and may in part explain the re-setting of GC feedback control seen in Cushing's disease [99].

Apparent Mineralocorticoid Excess (AME)

AME is a rare clinical disease that presents in early childhood with hypertension, sodium retention, potassium loss, suppressed renin activity and a metabolic alkalosis [178]. This condition was hallmarked by an increased ratio of urinary cortisol to cortisone metabolites as a result of a lack of conversion of cortisol to cortisone. There was also a low level of circulating of mineralocorticoid despite having evidence of apparent excess with metabolic derangement. Usually fatal in childhood adults with AME were discovered and successfully treated with dexamethasone. Physiological replacement with cortisol caused a return of symptoms and signs [179]. Stewart et al. in 1996 demonstrated that AME was caused by a defect in the 11β (Beta)-HSD2 gene [180] (similar to the effect seen with liquorice ingestion [181]). These and other observations led to the understanding that 11β (Beta)-HSD2 in the kidney protected the mineralocorticoid receptor from glucocorticoid binding. Under normal circumstances 11β (Beta)-HSD2 deactivates cortisol to cortisone and thus protects the MR receptor. In AME, with loss of 11β (Beta)-HSD2 activity unopposed cortisol binds to MR with subsequent clinical mineralocorticoid excess.

The Future of the Dehydrogenase Hypothesis

There is strong evidence that dysregulated 11β (Beta)-HSD1 activity is involved in many pathological processes including obesity and type 2 diabetes. We have also begun to see early clinical trials that demonstrate clinical benefit in 11β (Beta)-HSD1 inhibition. However, whether or not these compounds eventually enter the market with a licence to treat diabetes and metabolic disease remains to be seen. However, they may have utility in other conditions including glaucoma, idiopathic intracranial hypertension and low bone mineral density and clinical trials in these areas are ongoing. In addition, recent work based on observations in a patient with Cushing's disease [182] has suggested that 11β (Beta)-HSD1 may have a role in regulating the phenotype of circulating active GC excess [183]. This not only raises the possibility for the use of 11β (Beta)-HSD1 inhibitors in the treatment of Cushing's syndrome, but also that basal activity may predict the susceptibility to the adverse effects of exogenous GCs and that 11β (Beta)-HSD1 inhibitors could ameliorate the adverse effects, without compromise to the desired actions, of therapeutically indicated GCs.

11β (Beta)-HSD2 has an established role in the regulation of blood pressure and this is highlighted in patients with AME. However, evidence points to an additional role in placental development and function that may have implications for neonatal growth, either directly or through programming. Inhibiting 11β (Beta)-HSD2 action has been shown to stop colorectal adenocarcinoma spread and development and this warrants further investigation [173]. Importantly though, whilst much attention in recent years has focussed on 11β (Beta)-HSD1 activity and therapeutic inhibition, we must not forget that these two enzymes are tightly associated not least of all because the activity of 11β (Beta)-HSD1 is entirely dependent upon substrate availability (cortisone) that is generated by 11β (Beta)-HSD2. It is entirely plausible that cortisone availability may represent a rate-limiting step regulating 11β (Beta)-HSD1 activity through substrate availability and this needs to be further explored [184].

Inhibition of 11β (Beta)-HSD1 and specifically within tissues is a promising target as a potential disease prevention/modifying pathway. There is strong evidence to date that dysregulated 11β (Beta)-HSD1 activity is involved in many disease states including obesity and diabetes among many. We have also begun to see early human clinical trials that demonstrate clinical benefit with 11β (Beta)-HSD1 inhibition. However, as outlined above both the small number, and the short nature of studies to date have not yielded strong robust evidence to use 11β (Beta)-HSD1 inhibitors. Limited benefits and possible side effects including HPA activation will likely impede and slow the process of these inhibitor compounds from entering phase 3 trials and into clinical use.

One area of possible clinical use, which has not been looked at yet, is to help antagonise the effects of excess glucocorticoid side effects. Published data demonstrate a key role of 11β (Beta)-HSD1 in contributing to GC side effects in Cushing's [182] and in bone metabolism [184]. 11β (Beta)-HSD1 global knock out mice treated with excess GC are protected against GC side effects when compared to wild type controls [183]. This suggests that in conditions with excess circulating GC's, it is the re-activation of cortisone to cortisol within tissues by 11β (Beta)-HSD1, rather than simple cortisol delivery to tissue from the circulation that is the crucial step determining Cushingoid side effects. Data on the role of 11β (Beta)-HSD1 activity in humans with excess GC's is lacking. GC's are widely prescribed with estimates of 1–2 % [185] of the population taking prescribed steroids for various inflammatory conditions. Despite their efficacy, up to 70 % of patients experience an adverse systemic side-effect profile [186]. Inhibiting 11β (Beta)-HSD1 activity may play a beneficial role in preventing glucocorticoid side effects therefore making 11β (Beta)-HSD1 an exciting therapeutic target for patients with Cushing's syndrome.

Lastly, as mentioned previously there is evidence in the literature that 11β (Beta)-HSD1 activity increases with age and is associated with tissue damage and dysfunction. Therefore using tissue specific inhibitors or 11β (Beta)-HSD1 may be a target in conditions such as sarcopaenia and osteoporosis.

Conclusions

Tissue-specific GC metabolism is under the control of the 11β (Beta)-HSD dehydrogenase/reductase enzymes. 11β (Beta)-HSD2 exists to primarily protect mineralocorticoid receptor from excess cortisol exposure in tissues such as kidney. 11β (Beta)-HSD1 exists in many tissues and amplifies tissue exposure to cortisol by activating cortisol from inert cortisone. Both isoforms have a potent ability to manipulate clinical phenotype entirely independent of circulating GC levels. The complexity of this system and the intricate and finely tuned control that is able to exert at a tissue-specific level to govern ligand access to corticosteroid receptors has highlighted a fundamental shift in our approach. It adds weight to the argument that simple measurement of circulating steroid hormone levels provides an over-simplistic and perhaps misleading view of GC action.

References

1. Munck A, Náray-Fejes-Tóth A. The ups and downs of glucocorticoid physiology. Permissive and suppressive effects revisited. Mol Cell Endocrinol. 1992;90(1):C1–4. http://www.ncbi.nlm.nih.gov/pubmed/1301388. Accessed 31 Dec 2013.
2. Anagnostis P, Katsiki N, Adamidou F, et al. 11beta-Hydroxysteroid dehydrogenase type 1 inhibitors: novel agents for the treatment of metabolic syndrome and obesity-related disorders? Metabolism. 2013;62(1):21–33. doi:10.1016/j.metabol.2012.05.002.
3. Bujalska IL, Kumar SSP. Does central obesity reflect "Cushing's disease of the omentum"? Lancet. 1997;349(9060):1210–3.
4. Tomlinson JW, Sherlock M, Hughes B, et al. Inhibition of 11beta-hydroxysteroid dehydrogenase type 1 activity in vivo limits glucocorticoid exposure to human adipose tissue and decreases lipolysis. J Clin Endocrinol Metab. 2007;92(3):857–64. doi:10.1210/jc.2006-2325.
5. Tomlinson JW, Moore JS, Clark PMS, Holder G, Shakespeare L, Stewart PM. Weight loss increases 11beta-hydroxysteroid dehydrogenase type 1 expression in human adipose tissue. J Clin Endocrinol Metab. 2004;89(6):2711–6. doi:10.1210/jc.2003-031376.
6. Chapman K, Holmes M, Seckl J. 11β-Hydroxysteroid dehydrogenases: intracellular gatekeepers of tissue glucocorticoid action. Physiol Rev. 2013;93(3):1139–206. doi:10.1152/physrev.00020.2012.
7. Tomlinson JW, Walker EA, Bujalska IJ, et al. 11Beta-hydroxysteroid dehydrogenase type 1: a tissue-specific regulator of glucocorticoid response. Endocr Rev. 2004;25(5):831–66. doi:10.1210/er.2003-0031.
8. Feig PU, Shah S, Hermanowski-Vosatka A, et al. Effects of an 11β-hydroxysteroid dehydrogenase type 1 inhibitor, MK-0916, in patients with type 2 diabetes mellitus and metabolic syndrome. Diabetes Obes Metab. 2011;13(6):498–504. doi:10.1111/j.1463-1326.2011.01375.x.
9. Edwards CR, Stewart PM, Burt D, et al. Localisation of 11 beta-hydroxysteroid dehydrogenase—tissue specific protector of the mineralocorticoid receptor. Lancet. 1988;2(8618):986–9. http://www.ncbi.nlm.nih.gov/pubmed/2902493. Accessed 23 Feb 2014.
10. Esteban NV, Loughlin T, Yergey AL, et al. Daily cortisol production rate in man determined by stable isotope dilution/mass spectrometry. J Clin Endocrinol Metab. 1991;72(1):39–45. doi:10.1210/jcem-72-1-39.
11. Keenan DM, Roelfsema F, Veldhuis JD. Endogenous ACTH concentration-dependent drive of pulsatile cortisol secretion in the human. Am J Physiol Endocrinol Metab. 2004;287(4):E652–61. doi:10.1152/ajpendo.00167.2004.
12. Siiteri PK, Murai JT, Hammond GL, Nisker JA, Raymoure WJ, Kuhn RW. The serum transport of steroid hormones. Recent Prog Horm Res. 1982;38:457–510. http://www.ncbi.nlm.nih.gov/pubmed/6750727. Accessed 30 Dec 2013.
13. Dorin RI, Qiao Z, Qualls CR, Urban FK. Estimation of maximal cortisol secretion rate in healthy humans. J Clin Endocrinol Metab. 2012;97(4):1285–93. doi:10.1210/jc.2011-2227.
14. Toothaker RD, Welling PG. Effect of dose size on the pharmacokinetics of intravenous hydrocortisone during endogenous hydrocortisone suppression. J Pharmacokinet Biopharm. 1982;10(2):147–56. http://www.ncbi.nlm.nih.gov/pubmed/7120045. Accessed 9 Feb 2014.

15. Meulenberg PM, Hofman JA. Differences between concentrations of salivary cortisol and cortisone and of free cortisol and cortisone in plasma during pregnancy and postpartum. Clin Chem. 1990;36(1):70–5. http://www.ncbi.nlm.nih.gov/pubmed/2297937. Accessed 7 Mar 2014.
16. Kendall EC. Cortisone: memoirs of a hormone hunter. New York: Charles Scribner's Sons; 1971.
17. Hench PS, Kendall EC. The effect of a hormone of the adrenal cortex (17-hydroxy-11-dehydrocorticosterone; compound E) and of pituitary adrenocorticotropic hormone on rheumatoid arthritis. Proc Staff Meet Mayo Clin. 1949;24(8):181–97. http://www.ncbi.nlm.nih.gov/pubmed/18118071.
18. Burton RB, Keutmann EH, Waterhouse C, Schuler EA. The conversion of cortisone acetate to other alphaketolic steroids. J Clin Endocrinol Metab. 1953;13(1):48–63. doi:10.1210/jcem-13-1-48.
19. Amelung D, Hubener HJ, Roka L, Meyerheim G. Conversion of cortisone to compound F. J Clin Endocrinol Metab. 1953;13(9):1125–6. doi:10.1210/jcem-13-9-1125.
20. Kavanagh KL, Jörnvall H, Persson B, Oppermann U. Medium- and short-chain dehydrogenase/reductase gene and protein families: the SDR superfamily: functional and structural diversity within a family of metabolic and regulatory enzymes. Cell Mol Life Sci. 2008;65(24):3895–906. doi:10.1007/s00018-008-8588-y.
21. Agarwal AK, Monder C, Eckstein B, White PC. Cloning and expression of rat cDNA encoding corticosteroid 11 beta-dehydrogenase. J Biol Chem. 1989;264(32):18939–43. http://www.ncbi.nlm.nih.gov/pubmed/2808402.
22. Lakshmi V, Monder C. Purification and characterization of the corticosteroid 11 beta-dehydrogenase component of the rat liver 11 beta-hydroxysteroid dehydrogenase complex. Endocrinology. 1988;123(5):2390–8. doi:10.1210/endo-123-5-2390.
23. Tannin GM, Agarwal AK, Monder C, New MI, White PC. The human gene for 11 beta-hydroxysteroid dehydrogenase. Structure, tissue distribution, and chromosomal localization. J Biol Chem. 1991;266(25):16653–8.
24. Nobel CSI, Dunås F, Abrahmsén LB. Purification of full-length recombinant human and rat type 1 11beta-hydroxysteroid dehydrogenases with retained oxidoreductase activities. Protein Expr Purif. 2002;26(3):349–56. http://www.ncbi.nlm.nih.gov/pubmed/12460758. Accessed 5 Jan 2014.
25. Maser E, Völker B, Friebertshäuser J. 11 Beta-hydroxysteroid dehydrogenase type 1 from human liver: dimerization and enzyme cooperativity support its postulated role as glucocorticoid reductase. Biochemistry. 2002;41(7):2459–65. http://www.ncbi.nlm.nih.gov/pubmed/11841241. Accessed 31 Dec 2013.
26. Odermatt A. The N-terminal anchor sequences of 11beta-hydroxysteroid dehydrogenases determine their orientation in the endoplasmic reticulum membrane. J Biol Chem. 1999;274(40):28762–70. doi:10.1074/jbc.274.40.28762.
27. Bujalska IJ, Draper N, Michailidou Z, et al. Hexose-6-phosphate dehydrogenase confers oxo-reductase activity upon 11 beta-hydroxysteroid dehydrogenase type 1. J Mol Endocrinol. 2005;34(3):675–84. doi:10.1677/jme.1.01718.
28. Draper N, Walker EA, Bujalska IJ, et al. Mutations in the genes encoding 11beta-hydroxysteroid dehydrogenase type 1 and hexose-6-phosphate dehydrogenase interact to cause cortisone reductase deficiency. Nat Genet. 2003;34(4):434–9. doi:10.1038/ng1214.
29. Murphy VE, Clifton VL. Alterations in human placental 11beta-hydroxysteroid dehydrogenase type 1 and 2 with gestational age and labour. Placenta. 2003;24(7):739–44. http://www.ncbi.nlm.nih.gov/pubmed/12852864. Accessed 12 Feb 2014.
30. Jinno K, Sakura N, Nomura S, Fujitaka M, Ueda K, Kihara M. Failure of cortisone acetate therapy in 21-hydroxylase deficiency in early infancy. Pediatr Int. 2001;43(5):478–82. http://www.ncbi.nlm.nih.gov/pubmed/11737708. Accessed 22 Jan 2014.
31. Murphy BE. Cortisol production and inactivation by the human lung during gestation and infancy. J Clin Endocrinol Metab. 1978;47(2):243–8. doi:10.1210/jcem-47-2-243.

32. Abramovitz M, Branchaud CL, Murphy BE. Cortisol-cortisone interconversion in human fetal lung: contrasting results using explant and monolayer cultures suggest that 11 beta-hydroxysteroid dehydrogenase (EC 1.1.1.146) comprises two enzymes. J Clin Endocrinol Metab. 1982;54(3):563–8. doi:10.1210/jcem-54-3-563.
33. Dimitriou T, Maser-Gluth C, Remer T. Adrenocortical activity in healthy children is associated with fat mass. Am J Clin Nutr. 2003;77(3):731–6. http://www.ncbi.nlm.nih.gov/pubmed/12600869.
34. Toogood AA, Taylor NF, Shalet SM, Monson JP. Sexual dimorphism of cortisol metabolism is maintained in elderly subjects and is not oestrogen dependent. Clin Endocrinol (Oxf). 2000;52(1):61–6. http://www.ncbi.nlm.nih.gov/pubmed/10651754. Accessed 22 Jan 2014.
35. Vierhapper H, Heinze G, Nowotny P. Sex-specific difference in the interconversion of cortisol and cortisone in men and women. Obesity (Silver Spring). 2007;15(4):820–4. doi:10.1038/oby.2007.592.
36. Finken MJ, Andrews RC, Andrew R, Walker BR. Cortisol metabolism in healthy young adults: sexual dimorphism in activities of A-ring reductases, but not 11beta-hydroxysteroid dehydrogenases. J Clin Endocrinol Metab. 1999;84(9):3316–21. doi:10.1210/jcem.84.9.6009.
37. Gathercole LL, Lavery GG, Morgan SA, et al. 11β-hydroxysteroid dehydrogenase 1: translational and therapeutic aspects. Endocr Rev. 2013;34(4):525–55. doi:10.1210/er.2012-1050.
38. Nixon M, Wake DJ, Livingstone DE, et al. Salicylate downregulates 11β-HSD1 expression in adipose tissue in obese mice and in humans, mediating insulin sensitization. Diabetes. 2012;61(4):790–6. doi:10.2337/db11-0931.
39. Paulsen SK, Pedersen SB, Fisker S, Richelsen B. 11Beta-HSD type 1 expression in human adipose tissue: impact of gender, obesity, and fat localization. Obesity (Silver Spring). 2007;15(8):1954–60. doi:10.1038/oby.2007.233.
40. Gomez-Sanchez EP, Ganjam V, Chen YJ, et al. Regulation of 11 beta-hydroxysteroid dehydrogenase enzymes in the rat kidney by estradiol. Am J Physiol Endocrinol Metab. 2003;285(2):E272–9. doi:10.1152/ajpendo.00409.2002.
41. Quirk SJ, Slattery JA, Funder JW. Epithelial and adipose cells isolated from mammary glands of pregnant and lactating rats differ in 11 beta-hydroxysteroid dehydrogenase activity. J Steroid Biochem Mol Biol. 1990;37(4):529–34. http://www.ncbi.nlm.nih.gov/pubmed/2278836. Accessed 31 Dec 2013.
42. Goedecke JH, Wake DJ, Levitt NS, et al. Glucocorticoid metabolism within superficial subcutaneous rather than visceral adipose tissue is associated with features of the metabolic syndrome in South African women. Clin Endocrinol (Oxf). 2006;65(1):81–7. doi:10.1111/j.1365-2265.2006.02552.x.
43. Veilleux A, Laberge PY, Morency J, Noël S, Luu-The V, Tchernof A. Expression of genes related to glucocorticoid action in human subcutaneous and omental adipose tissue. J Steroid Biochem Mol Biol. 2010;122(1–3):28–34. doi:10.1016/j.jsbmb.2010.02.024.
44. Tomlinson JW, Moore J, Cooper MS, et al. Regulation of expression of 11β-hydroxysteroid dehydrogenase type 1 in adipose tissue: tissue-specific induction by cytokines. Endocrinology. 2001;142(5):1982–9.
45. Friedberg M, Zoumakis E, Hiroi N, Bader T, Chrousos GP, Hochberg Z. Modulation of 11 beta-hydroxysteroid dehydrogenase type 1 in mature human subcutaneous adipocytes by hypothalamic messengers. J Clin Endocrinol Metab. 2003;88(1):385–93. doi:10.1210/jc.2002-020510.
46. Handoko K, Yang K, Strutt B, Khalil W, Killinger D. Insulin attenuates the stimulatory effects of tumor necrosis factor alpha on 11beta-hydroxysteroid dehydrogenase 1 in human adipose stromal cells. J Steroid Biochem Mol Biol. 2000;72(3–4):163–8. http://www.ncbi.nlm.nih.gov/pubmed/10775808. Accessed 31 Dec 2013.
47. Esteves CL, Kelly V, Breton A, et al. Proinflammatory cytokine induction of 11β-hydroxysteroid dehydrogenase type 1 (11β-HSD1) in human adipocytes is mediated by MEK, C/EBPβ, and NF-κB/RelA. J Clin Endocrinol Metab. 2014;99(1):E160–8. doi:10.1210/jc.2013-1708.
48. Bujalska IJ, Walker EA, Hewison M, Stewart PM. A switch in dehydrogenase to reductase activity of 11 beta-hydroxysteroid dehydrogenase type 1 upon differentiation of human

omental adipose stromal cells. J Clin Endocrinol Metab. 2002;87(3):1205–10. doi:10.1210/jcem.87.3.8301.

49. Bujalska IJ, Gathercole LL, Tomlinson JW, et al. A novel selective 11beta-hydroxysteroid dehydrogenase type 1 inhibitor prevents human adipogenesis. J Endocrinol. 2008; 197(2):297–307. doi:10.1677/JOE-08-0050.
50. Livingstone DE, Kenyon CJ, Walker BR. Mechanisms of dysregulation of 11 beta-hydroxysteroid dehydrogenase type 1 in obese Zucker rats. J Endocrinol. 2000;167(3):533–9. http://www.ncbi.nlm.nih.gov/pubmed/11115781. Accessed 7 Mar 2014.
51. Livingstone DE, Jones GC, Smith K, et al. Understanding the role of glucocorticoids in obesity: tissue-specific alterations of corticosterone metabolism in obese Zucker rats. Endocrinology. 2000;141(2):560–3. doi:10.1210/endo.141.2.7297.
52. Morton NM, Ramage L, Seckl JR. Down-regulation of adipose 11beta-hydroxysteroid dehydrogenase type 1 by high-fat feeding in mice: a potential adaptive mechanism counteracting metabolic disease. Endocrinology. 2004;145(6):2707–12. doi:10.1210/en.2003-1674.
53. Drake AJ, Livingstone DEW, Andrew R, Seckl JR, Morton NM, Walker BR. Reduced adipose glucocorticoid reactivation and increased hepatic glucocorticoid clearance as an early adaptation to high-fat feeding in Wistar rats. Endocrinology. 2005;146(2):913–9. doi:10.1210/en.2004-1063.
54. Morton NM, Holmes MC, Fiévet C, et al. Improved lipid and lipoprotein profile, hepatic insulin sensitivity, and glucose tolerance in 11beta-hydroxysteroid dehydrogenase type 1 null mice. J Biol Chem. 2001;276(44):41293–300. doi:10.1074/jbc.M103676200.
55. Masuzaki H, Paterson J, Shinyama H, et al. A transgenic model of visceral obesity and the metabolic syndrome. Science. 2001;294(5549):2166–70. doi:10.1126/science.1066285.
56. Rask E. Tissue-specific dysregulation of cortisol metabolism in human obesity. J Clin Endocrinol Metab. 2001;86(3):1418–21. doi:10.1210/jc.86.3.1418.
57. Lindsay RS. Subcutaneous adipose 11-hydroxysteroid dehydrogenase type 1 activity and messenger ribonucleic acid levels are associated with adiposity and insulinemia in Pima Indians and Caucasians. J Clin Endocrinol Metab. 2003;88(6):2738–44. doi:10.1210/jc.2002-030017.
58. Veilleux A, Rhéaume C, Daris M, Luu-The V, Tchernof A. Omental adipose tissue type 1 11 beta-hydroxysteroid dehydrogenase oxoreductase activity, body fat distribution, and metabolic alterations in women. J Clin Endocrinol Metab. 2009;94(9):3550–7. doi:10.1210/jc.2008-2011.
59. Lee JH, Gao Z, Ye J. Regulation of 11β-HSD1 expression during adipose tissue expansion by hypoxia through different activities of NF-κB and HIF-1α. Am J Physiol Endocrinol Metab. 2013;304(10):E1035–41. doi:10.1152/ajpendo.00029.2013.
60. McCormick KL, Wang X, Mick GJ. Evidence that the 11 beta-hydroxysteroid dehydrogenase (11 beta-HSD1) is regulated by pentose pathway flux. Studies in rat adipocytes and microsomes. J Biol Chem. 2006;281(1):341–7. doi:10.1074/jbc.M506026200.
61. Gout J, Tirard J, Thévenon C, Riou J-P, Bégeot M, Naville D. CCAAT/enhancer-binding proteins (C/EBPs) regulate the basal and cAMP-induced transcription of the human 11beta-hydroxysteroid dehydrogenase encoding gene in adipose cells. Biochimie. 2006;88(9):1115–24. doi:10.1016/j.biochi.2006.05.020.
62. Esteves CL, Kelly V, Bégay V, et al. Regulation of adipocyte 11β-hydroxysteroid dehydrogenase type 1 (11β-HSD1) by CCAAT/enhancer-binding protein (C/EBP) β isoforms LIP and LAP. PLoS One. 2012;7(5):e37953. doi:10.1371/journal.pone.0037953.
63. Amatruda JM, Danahy SA, Chang CL. The effects of glucocorticoids on insulin-stimulated lipogenesis in primary cultures of rat hepatocytes. Biochem J. 1983;212(1):135–41. http://www.pubmedcentral.nih.gov/articlerender.fcgi?artid=1152020&tool=pmcentrez&rendertype=abstract.
64. Dolinsky VW, Douglas DN, Lehner R, Vance DE. Regulation of the enzymes of hepatic microsomal triacylglycerol lipolysis and re-esterification by the glucocorticoid dexamethasone. Biochem J. 2004;378(Pt 3):967–74. doi:10.1042/BJ20031320.

65. Baxter JD, Forsham PH. Tissue effects of glucocorticoids. Am J Med. 1972;53(5):573–89. http://www.ncbi.nlm.nih.gov/pubmed/4342884. Accessed 22 Jan 2014.
66. Paterson JM, Morton NM, Fievet C, et al. Metabolic syndrome without obesity: hepatic overexpression of 11beta-hydroxysteroid dehydrogenase type 1 in transgenic mice. Proc Natl Acad Sci U S A. 2004;101(18):7088–93. doi:10.1073/pnas.0305524101.
67. Ricketts ML, Verhaeg JM, Bujalska I, Howie AJ, Rainey WE, Stewart PM. Immunohistochemical localization of type 1 11beta-hydroxysteroid dehydrogenase in human tissues. J Clin Endocrinol Metab. 1998;83(4):1325–35. doi:10.1210/jcem.83.4.4706.
68. Jamieson PM, Walker BR, Chapman KE, Andrew R, Rossiter S, Seckl JR. 11 beta-hydroxysteroid dehydrogenase type 1 is a predominant 11 beta-reductase in the intact perfused rat liver. J Endocrinol. 2000;165(3):685–92. http://www.ncbi.nlm.nih.gov/pubmed/10828853. Accessed 7 Mar 2014.
69. Candia R, Riquelme A, Baudrand R, et al. Overexpression of 11β-hydroxysteroid dehydrogenase type 1 in visceral adipose tissue and portal hypercortisolism in non-alcoholic fatty liver disease. Liver Int. 2012;32(3):392–9. doi:10.1111/j.1478-3231.2011.02685.x.
70. Baudrand R, Carvajal CA, Riquelme A, et al. Overexpression of 11beta-hydroxysteroid dehydrogenase type 1 in hepatic and visceral adipose tissue is associated with metabolic disorders in morbidly obese patients. Obes Surg. 2010;20(1):77–83. doi:10.1007/s11695-009-9937-0.
71. Konopelska S, Kienitz T, Hughes B, et al. Hepatic 11beta-HSD1 mRNA expression in fatty liver and nonalcoholic steatohepatitis. Clin Endocrinol (Oxf). 2009;70(4):554–60. doi:10.1111/j.1365-2265.2008.03358.x.
72. Torrecilla E, Fernández-Vázquez G, Vicent D, et al. Liver upregulation of genes involved in cortisol production and action is associated with metabolic syndrome in morbidly obese patients. Obes Surg. 2012;22(3):478–86. doi:10.1007/s11695-011-0524-9.
73. Valsamakis G, Anwar A, Tomlinson JW, et al. 11Beta-hydroxysteroid dehydrogenase type 1 activity in lean and obese males with type 2 diabetes mellitus. J Clin Endocrinol Metab. 2004;89(9):4755–61. doi:10.1210/jc.2003-032240.
74. Ahmed A, Rabbitt E, Brady T, et al. A switch in hepatic cortisol metabolism across the spectrum of non alcoholic fatty liver disease. PLoS One. 2012;7(2):e29531. doi:10.1371/journal.pone.0029531.
75. Whorwood CB, Donovan SJ, Flanagan D, Phillips DIW, Byrne CD. Increased glucocorticoid receptor expression in human skeletal muscle cells may contribute to the pathogenesis of the metabolic syndrome. Diabetes. 2002;51(4):1066–75. http://www.ncbi.nlm.nih.gov/pubmed/11916927.
76. Whorwood CB, Donovan SJ, Wood PJ, Phillips DI. Regulation of glucocorticoid receptor alpha and beta isoforms and type I 11β-hydroxysteroid dehydrogenase expression in human skeletal muscle cells: a key role in the pathogenesis of insulin resistance? J Clin Endocrinol Metab. 2001;86(5):2296–308.
77. Morgan SA, Sherlock M, Gathercole LL, et al. 11beta-hydroxysteroid dehydrogenase type 1 regulates glucocorticoid-induced insulin resistance in skeletal muscle. Diabetes. 2009;58(11):2506–15. doi:10.2337/db09-0525.
78. Abdallah BM, Beck-Nielsen H, Gaster M. Increased expression of 11beta-hydroxysteroid dehydrogenase type 1 in type 2 diabetic myotubes. Eur J Clin Invest. 2005;35(10):627–34. doi:10.1111/j.1365-2362.2005.01552.x.
79. Zhang M, Lv X-Y, Li J, Xu Z-G, Chen L. Alteration of 11beta-hydroxysteroid dehydrogenase type 1 in skeletal muscle in a rat model of type 2 diabetes. Mol Cell Biochem. 2009;324(1–2):147–55. doi:10.1007/s11010-008-9993-0.
80. Kilgour AHM, Gallagher IJ, Maclullich AMJ, et al. Increased skeletal muscle 11βHSD1 mRNA is associated with lower muscle strength in ageing. PLoS One. 2013;8(12):e84057. doi:10.1371/journal.pone.0084057.
81. Vukelic S, Stojadinovic O, Pastar I, et al. Cortisol synthesis in epidermis is induced by IL-1 and tissue injury. J Biol Chem. 2011;286(12):10265–75. doi:10.1074/jbc.M110.188268.

82. Tiganescu A, Walker EA, Hardy RS, Mayes AE, Stewart PM. Localization, age- and site-dependent expression, and regulation of 11β-hydroxysteroid dehydrogenase type 1 in skin. J Invest Dermatol. 2011;131(1):30–6. doi:10.1038/jid.2010.257.
83. Fisher GJ, Varani J, Voorhees JJ. Looking older: fibroblast collapse and therapeutic implications. Arch Dermatol. 2010;144(5):666–72. doi:10.1001/archderm.144.5.666.Looking.
84. Slominski A, Zbytek B, Nikolakis G, et al. Steroidogenesis in the skin: implications for local immune functions. J Steroid Biochem Mol Biol. 2013;137:107–23. doi:10.1016/j.jsbmb.2013.02.006.
85. Terao M, Murota H, Kimura A, et al. 11β-Hydroxysteroid dehydrogenase-1 is a novel regulator of skin homeostasis and a candidate target for promoting tissue repair. PLoS One. 2011;6(9):e25039. doi:10.1371/journal.pone.0025039.
86. Cirillo N, Prime SS. Keratinocytes synthesize and activate cortisol. J Cell Biochem. 2011;112(6):1499–505. doi:10.1002/jcb.23081.
87. Napolitano A, Voice MW, Edwards CR, Seckl JR, Chapman KE. 11Beta-hydroxysteroid dehydrogenase 1 in adipocytes: expression is differentiation-dependent and hormonally regulated. J Steroid Biochem Mol Biol. 1998;64(5–6):251–60. http://www.ncbi.nlm.nih.gov/pubmed/9618026. Accessed 3 Feb 2014.
88. Tiganescu A, Hupe M, Uchida Y, Mauro T, Elias PM, Holleran W. Increased glucocorticoid activation during mouse skin wound healing. J Endocrinol. 2014;221(1):51–61. doi:10.1530/JOE-13-0420.
89. Tiganescu A, Tahrani AA, Morgan SA, et al. 11β-Hydroxysteroid dehydrogenase blockade prevents age-induced skin structure and function defects. J Clin Invest. 2013;123(7):3051–60. doi:10.1172/JCI64162.
90. Brem AS, Bina RB, King TC, Morris DJ. Localization of 2 11beta-OH steroid dehydrogenase isoforms in aortic endothelial cells. Hypertension. 1998;31(1 Pt 2):459–62. http://www.ncbi.nlm.nih.gov/pubmed/9453345. Accessed 14 Feb 2014.
91. Walker BR, Yau JL, Brett LP, et al. 11 beta-hydroxysteroid dehydrogenase in vascular smooth muscle and heart: implications for cardiovascular responses to glucocorticoids. Endocrinology. 1991;129(6):3305–12. doi:10.1210/endo-129-6-3305.
92. Atalar F, Gormez S, Caynak B, et al. The role of mediastinal adipose tissue 11β-hydroxysteroid dehydrogenase type 1 and glucocorticoid expression in the development of coronary atherosclerosis in obese patients with ischemic heart disease. Cardiovasc Diabetol. 2012;11:115. doi:10.1186/1475-2840-11-115.
93. Atalar F, Vural B, Ciftci C, et al. 11β-hydroxysteroid dehydrogenase type 1 gene expression is increased in ascending aorta tissue of metabolic syndrome patients with coronary artery disease. Genet Mol Res. 2012;11(3):3122–32. doi:10.4238/2012.August.31.10.
94. Hermanowski-Vosatka A, Balkovec JM, Cheng K, et al. 11beta-HSD1 inhibition ameliorates metabolic syndrome and prevents progression of atherosclerosis in mice. J Exp Med. 2005;202(4):517–27. doi:10.1084/jem.20050119.
95. Nuotio-Antar AM, Hachey DL, Hasty AH. Carbenoxolone treatment attenuates symptoms of metabolic syndrome and atherogenesis in obese, hyperlipidemic mice. Am J Physiol Endocrinol Metab. 2007;293(6):E1517–28. doi:10.1152/ajpendo.00522.2007.
96. Small GR, Hadoke PWF, Sharif I, et al. Preventing local regeneration of glucocorticoids by 11beta-hydroxysteroid dehydrogenase type 1 enhances angiogenesis. Proc Natl Acad Sci U S A. 2005;102(34):12165–70. doi:10.1073/pnas.0500641102.
97. Brown ES. Effects of glucocorticoids on mood, memory, and the hippocampus. Treatment and preventive therapy. Ann N Y Acad Sci. 2009;1179:41–55. doi:10.1111/j.1749-6632.2009.04981.x.
98. Swaab DF, Bao A-M, Lucassen PJ. The stress system in the human brain in depression and neurodegeneration. Ageing Res Rev. 2005;4(2):141–94. doi:10.1016/j.arr.2005.03.003.
99. Korbonits M, Bujalska I, Shimojo M, et al. Expression of 11 beta-hydroxysteroid dehydrogenase isoenzymes in the human pituitary: induction of the type 2 enzyme in corticotropinomas and other pituitary tumors. J Clin Endocrinol Metab. 2001;86(6):2728–33. doi:10.1210/jcem.86.6.7563.

100. Moisan MP, Seckl JR, Edwards CR. 11 beta-hydroxysteroid dehydrogenase bioactivity and messenger RNA expression in rat forebrain: localization in hypothalamus, hippocampus, and cortex. Endocrinology. 1990;127(3):1450–5. doi:10.1210/endo-127-3-1450.
101. Gomez-Sanchez EP, Romero DG, de Rodriguez AF, Warden MP, Krozowski Z, Gomez-Sanchez CE. Hexose-6-phosphate dehydrogenase and 11beta-hydroxysteroid dehydrogenase-1 tissue distribution in the rat. Endocrinology. 2008;149(2):525–33. doi:10.1210/en.2007-0328.
102. Lakshmi V, Sakai RR, McEwen BS, Monder C. Regional distribution of 11 beta-hydroxysteroid dehydrogenase in rat brain. Endocrinology. 1991;128(4):1741–8. doi:10.1210/endo-128-4-1741.
103. Rajan V, Edwards RW, Seckl JR. 11 beta-hydroxysteroid dehydrogenase in cultured cells reactivates inert 11-dehydrocorticosterone, potentiating neurotoxicity hippocampal. J Neurosci. 1996;76(1):65–70.
104. Wyrwoll CS, Holmes MC, Seckl JR. 11β-hydroxysteroid dehydrogenases and the brain: from zero to hero, a decade of progress. Front Neuroendocrinol. 2011;32(3):265–86. doi:10.1016/j.yfrne.2010.12.001.
105. Tomlinson JW, Durrani OM, Bujalska IJ, et al. The role of 11beta-hydroxysteroid dehydrogenase 1 in adipogenesis in thyroid-associated ophthalmopathy. J Clin Endocrinol Metab. 2010;95(1):398–406. doi:10.1210/jc.2009-0873.
106. Sandeep TC, Yau JLW, MacLullich AMJ, et al. 11Beta-hydroxysteroid dehydrogenase inhibition improves cognitive function in healthy elderly men and type 2 diabetics. Proc Natl Acad Sci U S A. 2004;101(17):6734–9. doi:10.1073/pnas.0306996101.
107. Rhen T, Cidlowski JA. Antiinflammatory action of glucocorticoids—new mechanisms for old drugs. N Engl J Med. 2005;353(16):1711–23. doi:10.1056/NEJMra050541.
108. Chapman KE, Coutinho AE, Gray M, Gilmour JS, Savill JS, Seckl JR. The role and regulation of 11beta-hydroxysteroid dehydrogenase type 1 in the inflammatory response. Mol Cell Endocrinol. 2009;301(1-2):123–31. doi:10.1016/j.mce.2008.09.031.
109. Coutinho AE, Brown JK, Yang F, et al. Mast cells express 11β-hydroxysteroid dehydrogenase type 1: a role in restraining mast cell degranulation. PLoS One. 2013;8(1):e54640. doi:10.1371/journal.pone.0054640.
110. Thieringer R, Le Grand CB, Carbin L, et al. 11 Beta-hydroxysteroid dehydrogenase type 1 is induced in human monocytes upon differentiation to macrophages. J Immunol. 2001;167(1):30–5. http://www.ncbi.nlm.nih.gov/pubmed/11418628.
111. Zhang TY, Ding X, Daynes RA, Alerts E. Regulation of glucocorticoid activities 1. J Immunol. 2005;174(2):879–89.
112. Freeman L, Hewison M, Hughes SV, et al. Expression of 11β-hydroxysteroid dehydrogenase type 1 permits regulation of glucocorticoid bioavailability by human dendritic cells. Blood. 2005;106(6):2042–9. doi:10.1182/blood-2005-01-0186.
113. Gilmour JS, Coutinho AE, Cailhier J, et al. Local amplification of glucocorticoids by 11β-hydroxysteroid dehydrogenase type 1 promotes macrophage phagocytosis of apoptotic leukocytes. J Immunol. 2006;176(12):7605–11.
114. Zhang TY, Daynes RA. Macrophages from 11beta-hydroxysteroid dehydrogenase type 1-deficient mice exhibit an increased sensitivity to lipopolysaccharide stimulation due to TGF-beta-mediated up-regulation of SHIP1 expression. J Immunol. 2007;179(9):6325–35.
115. Coutinho AE, Gray M, Brownstein DG, et al. 11β-Hydroxysteroid dehydrogenase type 1, but not type 2, deficiency worsens acute inflammation and experimental arthritis in mice. Endocrinology. 2012;153(1):234–40. doi:10.1210/en.2011-1398.
116. Zbánková S, Bryndová J, Leden P, Kment M, Svec A, Pácha J. 11beta-hydroxysteroid dehydrogenase 1 and 2 expression in colon from patients with ulcerative colitis. J Gastroenterol Hepatol. 2007;22(7):1019–23. doi:10.1111/j.1440-1746.2006.04529.x.
117. Stegk JP, Ebert B, Martin H-J, Maser E. Expression profiles of human 11beta-hydroxysteroid dehydrogenases type 1 and type 2 in inflammatory bowel diseases. Mol Cell Endocrinol. 2009;301(1–2):104–8. doi:10.1016/j.mce.2008.10.030.

118. Cooper MS, Kriel H, Sayers A, et al. Can 11β-hydroxysteroid dehydrogenase activity predict the sensitivity of bone to therapeutic glucocorticoids in inflammatory bowel disease? Calcif Tissue Int. 2011;89(3):246–51. doi:10.1007/s00223-011-9512-2.
119. Ergang P, Vytáčková K, Svec J, Bryndová J, Mikšík I, Pácha J. Upregulation of 11β-hydroxysteroid dehydrogenase 1 in lymphoid organs during inflammation in the rat. J Steroid Biochem Mol Biol. 2011;126(1–2):19–25. doi:10.1016/j.jsbmb.2011.04.002.
120. Cooper MS, Walker EA, Bland R, Fraser WD, Hewison M, Stewart PM. Expression and functional consequences of 11beta-hydroxysteroid dehydrogenase activity in human bone. Bone. 2000;27(3):375–81. http://www.ncbi.nlm.nih.gov/pubmed/10962348. Accessed 14 Feb 2014.
121. Bellows CG, Ciaccia A, Heersche JN. Osteoprogenitor cells in cell populations derived from mouse and rat calvaria differ in their response to corticosterone, cortisol, and cortisone. Bone. 1998;23(2):119–25. http://www.ncbi.nlm.nih.gov/pubmed/9701470. Accessed 14 Feb 2014.
122. Hardy R, Rabbitt EH, Filer A, et al. Local and systemic glucocorticoid metabolism in inflammatory arthritis. Ann Rheum Dis. 2008;67(9):1204–10. doi:10.1136/ard.2008.090662.
123. Schmidt M, Weidler C, Naumann H, Anders S, Schölmerich J, Straub RH. Reduced capacity for the reactivation of glucocorticoids in rheumatoid arthritis synovial cells: possible role of the sympathetic nervous system? Arthritis Rheum. 2005;52(6):1711–20. doi:10.1002/art.21091.
124. Ergang P, Leden P, Vagnerová K, et al. Local metabolism of glucocorticoids and its role in rat adjuvant arthritis. Mol Cell Endocrinol. 2010;323(2):155–60. doi:10.1016/j.mce.2010.03.003.
125. Lavery GG, Walker EA, Tiganescu A, et al. Steroid biomarkers and genetic studies reveal inactivating mutations in hexose-6-phosphate dehydrogenase in patients with cortisone reductase deficiency. J Clin Endocrinol Metab. 2008;93(10):3827–32. doi:10.1210/jc.2008-0743.
126. Morgan SA, Tomlinson JW. 11beta-hydroxysteroid dehydrogenase type 1 inhibitors for the treatment of type 2 diabetes. Expert Opin Investig Drugs. 2010;19(9):1067–76. doi:10.1517/13543784.2010.504713.
127. Monder C, Lakshmi V. Evidence for kinetically distinct forms of corticosteroid 11 beta-dehydrogenase in rat liver microsomes. J Steroid Biochem. 1989;32(1A):77–83. http://www.ncbi.nlm.nih.gov/pubmed/2913404. Accessed 23 Jan 2014.
128. Schweizer RAS, Atanasov AG, Frey BM, Odermatt A. A rapid screening assay for inhibitors of 11beta-hydroxysteroid dehydrogenases (11beta-HSD): flavanone selectively inhibits 11beta-HSD1 reductase activity. Mol Cell Endocrinol. 2003;212(1–2):41–9. http://www.ncbi.nlm.nih.gov/pubmed/14654249. Accessed 23 Jan 2014.
129. Diederich S, Grossmann C, Hanke B, et al. In the search for specific inhibitors of human 11beta-hydroxysteroid-dehydrogenases (11beta-HSDs): chenodeoxycholic acid selectively inhibits 11beta-HSD-I. Eur J Endocrinol. 2000;142(2):200–7. http://www.ncbi.nlm.nih.gov/pubmed/10664531.
130. Latif SA, Pardo HA, Hardy MP, Morris DJ. Endogenous selective inhibitors of 11beta-hydroxysteroid dehydrogenase isoforms 1 and 2 of adrenal origin. Mol Cell Endocrinol. 2005;243(1–2):43–50. doi:10.1016/j.mce.2005.08.006.
131. Atanasov AG, Dzyakanchuk AA, Schweizer RAS, Nashev LG, Maurer EM, Odermatt A. Coffee inhibits the reactivation of glucocorticoids by 11beta-hydroxysteroid dehydrogenase type 1: a glucocorticoid connection in the anti-diabetic action of coffee? FEBS Lett. 2006;580(17):4081–5. doi:10.1016/j.febslet.2006.06.046.
132. Andrews RC, Rooyackers O, Walker BR. Effects of the 11 beta-hydroxysteroid dehydrogenase inhibitor carbenoxolone on insulin sensitivity in men with type 2 diabetes. J Clin Endocrinol Metab. 2003;88(1):285–91. doi:10.1210/jc.2002-021194.
133. Reuters T. The changing role of chemistry in drug discovery. 2011. http://thomsonreuters.com/content/dam/openweb/documents/pdf/pharma-life-sciences/report/international-year-of-chemistry-report-drug-discovery.pdf.
134. Boyle CD, Kowalski TJ. 11beta-hydroxysteroid dehydrogenase type 1 inhibitors: a review of recent patents. Expert Opin Ther Pat. 2009;19(6):801–25. doi: 10.1517/13543770902967658.

135. Rosenstock J, Banarer S, Fonseca VA, et al. The 11-beta-hydroxysteroid dehydrogenase type 1 inhibitor INCB13739 improves hyperglycemia in patients with type 2 diabetes inadequately controlled by metformin monotherapy. Diabetes Care. 2010;33(7):1516–22. doi:10.2337/dc09-2315.
136. Shah S, Hermanowski-Vosatka A, Gibson K, et al. Efficacy and safety of the selective 11β-HSD-1 inhibitors MK-0736 and MK-0916 in overweight and obese patients with hypertension. J Am Soc Hypertens. 2011;5(3):166–76. doi:10.1016/j.jash.2011.01.009.
137. Brown RW, Chapman KE, Edwards CR, Seckl JR. Human placental 11 beta-hydroxysteroid dehydrogenase: evidence for and partial purification of a distinct NAD-dependent isoform. Endocrinology. 1993;132(6):2614–21. doi:10.1210/endo.132.6.8504762.
138. Albiston AL, Obeyesekere VR, Smith RE, Krozowski ZS. Cloning and tissue distribution of the human 11 beta-hydroxysteroid dehydrogenase type 2 enzyme. Mol Cell Endocrinol. 1994;105(2):R11–7. http://www.ncbi.nlm.nih.gov/pubmed/7859916. Accessed 5 Feb 2014.
139. Persson B, Kallberg Y, Bray JE, et al. The SDR (short-chain dehydrogenase/reductase and related enzymes) nomenclature initiative. Chem Biol Interact. 2009;178(1–3):94–8. doi:10.1016/j.cbi.2008.10.040.
140. Agarwal AK, Rogerson FM, Mune T, White PC. Gene structure and chromosomal localization of the human HSD11K gene encoding the kidney (type 2) isozyme of 11 beta-hydroxysteroid dehydrogenase. Genomics. 1995;29(1):195–9. doi:10.1006/geno.1995.1231.
141. Brown RW, Chapman KE, Kotelevtsev Y, et al. Cloning and production of antisera to human placental 11β-hydroxysteroid dehydrogenase type 2. Biochem J. 1996;313:1007–17.
142. Stewart PM, Murry BA, Mason JI. Human kidney 11 beta-hydroxysteroid dehydrogenase is a high affinity nicotinamide adenine dinucleotide-dependent enzyme and differs from the cloned type I isoform. J Clin Endocrinol Metab. 1994;79(2):480–4. doi:10.1210/jcem.79.2.8045966.
143. Dave-Sharma S, Wilson RC, Harbison MD, et al. Examination of genotype and phenotype relationships in 14 patients with apparent mineralocorticoid excess. J Clin Endocrinol Metab. 1998;83(7):2244–54. doi:10.1210/jcem.83.7.4986.
144. Wilson RC, Krozowski ZS, Li K, et al. A mutation in the HSD11B2 gene in a family with apparent mineralocorticoid excess. J Clin Endocrinol Metab. 1995;80(7):2263–6. doi:10.1210/jcem.80.7.7608290.
145. Mune T, Rogerson FM, Nikkilä H, Agarwal AK, White PC. Human hypertension caused by mutations in the kidney isozyme of 11 beta-hydroxysteroid dehydrogenase. Nat Genet. 1995;10(4):394–9. doi:10.1038/ng0895-394.
146. Osinski PA. Steroid 11beta-ol dehydrogenase in human placenta. Nature. 1960;187:777. http://www.ncbi.nlm.nih.gov/pubmed/14429221. Accessed 31 Dec 2013.
147. Waffarn F, Davis EP. Effects of antenatal corticosteroids on the hypothalamic-pituitary-adrenocortical axis of the fetus and newborn: experimental findings and clinical considerations. Am J Obstet Gynecol. 2012;207(6):446–54. doi:10.1016/j.ajog.2012.06.012.
148. Wyrwoll CS, Holmes MC. Prenatal excess glucocorticoid exposure and adult affective disorders: a role for serotonergic and catecholamine pathways. Neuroendocrinology. 2012;95(1):47–55. doi:10.1159/000331345.
149. Huang WL, Beazley LD, Quinlivan JA, Evans SF, Newnham JP, Dunlop SA. Effect of corticosteroids on brain growth in fetal sheep. Obstet Gynecol. 1999;94(2):213–8. http://www.ncbi.nlm.nih.gov/pubmed/10432130. Accessed 14 Feb 2014.
150. Seckl JR, Holmes MC. Mechanisms of disease: glucocorticoids, their placental metabolism and fetal "programming" of adult pathophysiology. Nat Clin Pract Endocrinol Metab. 2007;3(6):479–88. doi:10.1038/ncpendmet0515.
151. Alikhani-Koopaei R, Fouladkou F, Frey FJ, Frey BM. Epigenetic regulation of 11 beta-hydroxysteroid dehydrogenase type 2 expression. J Clin Invest. 2004;114(8):1146–57. doi:10.1172/JCI21647.
152. Marsit CJ, Maccani MA, Padbury JF, Lester BM. Placental 11-beta hydroxysteroid dehydrogenase methylation is associated with newborn growth and a measure of neurobehavioral outcome. PLoS One. 2012;7(3):e33794. doi:10.1371/journal.pone.0033794.

153. Baserga M, Kaur R, Hale MA, et al. Fetal growth restriction alters transcription factor binding and epigenetic mechanisms of renal 11β-hydroxysteroid dehydrogenase type 2 in a sex-specific manner. Am J Physiol Regul Integr Comp Physiol. 2010;299(1):334–42. doi:10.1152/ajpregu.00122.2010.
154. Kostadinova RM, Nawrocki AR, Frey FJ, Frey BM. Tumor necrosis factor alpha and phorbol 12-myristate-13-acetate down-regulate human 11beta-hydroxysteroid dehydrogenase type 2 through p50/p50 NF-kappaB homodimers and Egr-1. FASEB J. 2005;19(6):650–2. doi:10.1096/fj.04-2820fje.
155. Low SC, Assaad SN, Rajan V, Chapman KE, Edwards CR, Seckl JR. Regulation of 11 beta-hydroxysteroid dehydrogenase by sex steroids in vivo: further evidence for the existence of a second dehydrogenase in rat kidney. J Endocrinol. 1993;139(1):27–35. http://www.ncbi.nlm.nih.gov/pubmed/8254291. Accessed 13 Feb 2014.
156. Rubis B, Krozowski Z, Trzeciak WH. Arginine vasopressin stimulates 11beta-hydroxysteroid dehydrogenase type 2 expression in the mineralocorticosteroid target cells. Mol Cell Endocrinol. 2006;256(1–2):17–22. doi:10.1016/j.mce.2006.04.032.
157. Clarke KA, Ward JW, Forhead AJ, Giussani DA, Fowden AL. Regulation of 11 beta-hydroxysteroid dehydrogenase type 2 activity in ovine placenta by fetal cortisol. J Endocrinol. 2002;172(3):527–34. http://www.ncbi.nlm.nih.gov/pubmed/11874701. Accessed 13 Feb 2014.
158. Suzuki S, Koyama K, Darnel A, et al. Dexamethasone upregulates 11beta-hydroxysteroid dehydrogenase type 2 in BEAS-2B cells. Am J Respir Crit Care Med. 2003;167(9):1244–9. doi:10.1164/rccm.200210-1139OC.
159. Heiniger CD, Kostadinova RM, Rochat MK, et al. Hypoxia causes down-regulation of 11 beta-hydroxysteroid dehydrogenase type 2 by induction of Egr-1. FASEB J. 2003;17(8):917–9. doi:10.1096/fj.02-0582fje.
160. Fukushima K, Funayama Y, Yonezawa H, et al. Aldosterone enhances 11beta-hydroxysteroid dehydrogenase type 2 expression in colonic epithelial cells in vivo. Scand J Gastroenterol. 2005;40(7):850–7. doi:10.1080/00365520510015700.
161. Geerling JC, Loewy AD. Aldosterone in the brain. Am J Physiol Renal Physiol. 2009;297(3):F559–76. doi:10.1152/ajprenal.90399.2008.
162. Holmes MC, Sangra M, French KL, et al. 11beta-hydroxysteroid dehydrogenase type 2 protects the neonatal cerebellum from deleterious effects of glucocorticoids. Neuroscience. 2006;137(3):865–73. doi:10.1016/j.neuroscience.2005.09.037.
163. Noguchi KK, Lau K, Smith DJ, Swiney BS, Farber NB. Glucocorticoid receptor stimulation and the regulation of neonatal cerebellar neural progenitor cell apoptosis. Neurobiol Dis. 2011;43(2):356–63. doi:10.1016/j.nbd.2011.04.004.
164. Roland BL, Li KX, Funder JW. Hybridization histochemical localization of 11 beta-hydroxysteroid dehydrogenase type 2 in rat brain. Endocrinology. 1995;136(10):4697–700. doi:10.1210/endo.136.10.7664691.
165. Christy C, Hadoke PWF, Paterson JM, Mullins JJ, Seckl JR, Walker BR. 11beta-hydroxysteroid dehydrogenase type 2 in mouse aorta: localization and influence on response to glucocorticoids. Hypertension. 2003;42(4):580–7. doi:10.1161/01.HYP.0000088855.06598.5B.
166. Deuchar GA, McLean D, Hadoke PWF, et al. 11β-hydroxysteroid dehydrogenase type 2 deficiency accelerates atherogenesis and causes proinflammatory changes in the endothelium in apoe-/- mice. Endocrinology. 2011;152(1):236–46. doi:10.1210/en.2010-0925.
167. Leckie C, Chapman KE, Edwards CR, Seckl JR. LLC-PK1 cells model 11 beta-hydroxysteroid dehydrogenase type 2 regulation of glucocorticoid access to renal mineralocorticoid receptors. Endocrinology. 1995;136(12):5561–9. doi:10.1210/endo.136.12.7588309.
168. Henschkowski J, Stuck AE, Frey BM, et al. Age-dependent decrease in 11beta-hydroxysteroid dehydrogenase type 2 (11beta-HSD2) activity in hypertensive patients. Am J Hypertens. 2008;21(6):644–9. doi:10.1038/ajh.2008.152.
169. Ferrari P. The role of 11β-hydroxysteroid dehydrogenase type 2 in human hypertension. Biochim Biophys Acta. 2010;1802(12):1178–87. doi:10.1016/j.bbadis.2009.10.017.
170. Mongia A, Vecker R, George M, et al. Role of 11βHSD type 2 enzyme activity in essential hypertension and children with chronic kidney disease (CKD). J Clin Endocrinol Metab. 2012;97(10):3622–9. doi:10.1210/jc.2012-1411.

171. Watson B, Bergman SM, Myracle A, Callen DF, Acton RT, Warnock DG. Genetic association of 11 beta-hydroxysteroid dehydrogenase type 2 (HSD11B2) flanking microsatellites with essential hypertension in blacks. Hypertension. 1996;28(3):478–82. http://www.ncbi.nlm.nih.gov/pubmed/8794836. Accessed 2 Mar 2014.
172. Whorwood CB, Ricketts ML, Stewart PM. Epithelial cell localization of type 2 11 beta-hydroxysteroid dehydrogenase in rat and human colon. Endocrinology. 1994;135(6):2533–41. doi:10.1210/endo.135.6.7988441.
173. Zhang M-Z, Xu J, Yao B, et al. Inhibition of 11beta-hydroxysteroid dehydrogenase type II selectively blocks the tumor COX-2 pathway and suppresses colon carcinogenesis in mice and humans. J Clin Invest. 2009;119(4):876–85. doi:10.1172/JCI37398.
174. Smith RE, Maguire JA, Stein-Oakley AN, et al. Localization of 11 beta-hydroxysteroid dehydrogenase type II in human epithelial tissues. J Clin Endocrinol Metab. 1996;81(9):3244–8. doi:10.1210/jcem.81.9.8784076.
175. Shimojo M, Ricketts ML, Petrelli MD, et al. Immunodetection of 11 beta-hydroxysteroid dehydrogenase type 2 in human mineralocorticoid target tissues: evidence for nuclear localization. Endocrinology. 1997;138(3):1305–11. doi:10.1210/endo.138.3.4994.
176. Perogamvros I, Keevil BG, Ray DW, Trainer PJ. Salivary cortisone is a potential biomarker for serum free cortisol. J Clin Endocrinol Metab. 2010;95(11):4951–8. doi:10.1210/jc.2010-1215.
177. Bocchi B, Kenouch S, Lamarre-Cliche M, et al. Impaired 11-beta hydroxysteroid dehydrogenase type 2 activity in sweat gland ducts in human essential hypertension. Hypertension. 2004;43(4):803–8. doi:10.1161/01.HYP.0000121362.64182.ad.
178. Ulick S, Levine LS, Gunczler P, et al. A syndrome of apparent mineralocorticoid excess associated with defects in the peripheral metabolism of cortisol. J Clin Endocrinol Metab. 1979;49(5):757–64. doi:10.1210/jcem-49-5-757.
179. Stewart PM, Corrie JE, Shackleton CH, Edwards CR. Syndrome of apparent mineralocorticoid excess. A defect in the cortisol-cortisone shuttle. J Clin Invest. 1988;82(1):340–9. doi:10.1172/JCI113592.
180. Stewart PM, Krozowski ZS, Gupta A, et al. Hypertension in the syndrome of apparent mineralocorticoid excess due to mutation of the 11 beta-hydroxysteroid dehydrogenase type 2 gene. Lancet. 1996;347(8994):88–91. http://www.ncbi.nlm.nih.gov/pubmed/8538347. Accessed 12 Feb 2014.
181. Stewart PM, Wallace AM, Valentino R, Burt D, Shackleton CH, Edwards CR. Mineralocorticoid activity of liquorice: 11-beta-hydroxysteroid dehydrogenase deficiency comes of age. Lancet. 1987;2(8563):821–4. http://www.ncbi.nlm.nih.gov/pubmed/2889032. Accessed 12 Feb 2014.
182. Tomlinson JW, Draper N, Mackie J, et al. Absence of Cushingoid phenotype in a patient with Cushing's disease due to defective cortisone to cortisol conversion. J Clin Endocrinol Metab. 2002;87(1):57–62. doi:10.1210/jcem.87.1.8189.
183. Morgan SA, McCabe EL, Gathercole LL, Hassan-Smith ZK, Larner DP, Bujalska IJ, Stewart PM, Tomlinson JW, Lavery GG. 11β-HSD1 is the major regulator of the tissue-specific effects of circulating glucocorticoid excess. Proc Natl Acad Sci USA. 2014;111(24):E2482-91. doi: 10..1073/pnas.1323681111
184. Cooper MS, Syddall HE, Fall CHD, et al. Circulating cortisone levels are associated with biochemical markers of bone formation and lumbar spine BMD: the Hertfordshire Cohort Study. Clin Endocrinol (Oxf). 2005;62(6):692–7. doi:10.1111/j.1365-2265.2005.02281.x.
185. Overman RA, Yeh J-Y, Deal CL. Prevalence of oral glucocorticoid usage in the United States: a general population perspective. Arthritis Care Res (Hoboken). 2013;65(2):294–8. doi:10.1002/acr.21796.
186. Fardet L, Flahault A, Kettaneh A, et al. Corticosteroid-induced clinical adverse events: frequency, risk factors and patient's opinion. Br J Dermatol. 2007;157(1):142–8. doi:10.1111/j.1365-2133.2007.07950.x.

Chapter 17
Conclusions and Future Directions

Jen-Chywan Wang and Charles Harris

We hope this book has provided a robust introduction to glucocorticoid action. We have provided a historical perspective on discoveries related to glucocorticoid action: from the discovery of the hormones, their use as therapeutics for inflammatory disease, biochemical identification of receptors and finally the cloning of the glucocorticoid receptor. This has led to identification of not only the primary amino acid sequence, but structural information gleaned from both X-ray crystallography as well as nuclear magnetic resonance approaches. We hope the reader comes away with a better understanding of the many actions of glucocorticoids on various tissues and cell types. Indeed, one of the unanswered questions in our field is how one hormone and receptor can have such varied transcriptional responses in different cell types. Surely, some of this will be explained by epigenetic modifications that occur in different cell types. Recent work has also made it clear that GR does not act alone to modulate transcription as numerous transcriptional coregulators have been described to modulate GR's transcriptional program. Additional combinatorial complexity comes about from composite response elements between GR and other transcription factors. Subtle changes in the glucocorticoid response element as well as levels of GR (both the major alpha isoform as well as alternatively spliced and beta and gamma isoforms) could also modulate the transcriptional program. Analysis of mice with altered glucocorticoid signaling has reinforced the critical importance of this pathway. Mice lacking GR die at birth due to defects in lung maturation and

J.-C. Wang (✉)
Department of Nutritional Sciences and Toxicology, University of California Berkeley,
119 Morgan Hall, Berkeley, CA, USA
e-mail: walwang@berkeley.edu

C. Harris (✉)
Division of Endocrinology, Metabolism and Lipid Research, Department of Internal
Medicine, Washington University School of Medicine, 660 S Euclid Avenue,
Campus Box 8127, St. Louis, MO, USA
e-mail: caharris@dom.wustl.edu

J.-C. Wang, C. Harris (eds.), *Glucocorticoid Signaling*, Advances in Experimental
Medicine and Biology 872, DOI 10.1007/978-1-4939-2895-8_17

it is a well described phenomenon that premature infants can have their lung maturation accelerated by administering glucocorticoids when pregnant mothers are threatening preterm labor. The treatment of these women with glucocorticoids has greatly reduced infant mortality due to prematurity. Thus it is clear, that understanding glucocorticoid action has great clinical benefits. In addition to treating patient with adrenal insufficiency (Addison's disease) glucocorticoids are used to treat a wide variety of common inflammatory diseases such as asthma, inflammatory bowel disease and rheumatoid arthritis. However, treatment of inflammatory disease with glucocorticoids has powerful adverse effects including osteoporosis, insulin, resistance, hypertension and behavioral problems. Therefore, the identification of selective GR modulators that retain anti-inflammatory properties with less adverse effects has great promise for safer steroids in the twenty-first century. Hope from this model comes from prior successes developing selective modulators of the estrogen receptor for the treatment of osteoporosis. Key to the development of selective GR modulators will be a better understanding of how these effects are mediated. It is clear that early models linking GR transactivation to adverse effects and repression to immunosuppression are overly simplistic. We are hopeful, that new models will replace them so that such selective molecules will be possible in the future. One barrier to a more complete understanding of glucocorticoid action is linking GR binding sites identified in whole genome approaches (such as Chip-seq) to the genes that are regulated by these binding sites. Early hopes that a binding site would regulate the gene nearest to it do not appear to always be the case. Techniques such as HiC and FaireSeq have shown that binding sites can regulate genes far away due to looping of chromatin. Therefore, understanding which genes are regulated by which binding sites is currently a hurdle. New techniques such as CRISPRs have facilitated genetic modification so that in the future it might be possible to mutate binding sites individually to determine the exact transcriptional consequences. We are confident the coming years will be an exciting time for research on glucocorticoid action.

Index

J.-C. Wang, C. Harris (eds.), *Glucocorticoid Signaling*, Advances in Experimental Medicine and Biology 872, DOI 10.1007/978-1-4939-2895-8

MIX
Papier aus verantwortungsvollen Quellen
Paper from responsible sources
FSC® C105338

If you have any concerns about our products,
you can contact us on
ProductSafety@springernature.com

In case Publisher is established outside the EU,
the EU authorized representative is:
Springer Nature Customer Service Center GmbH
Europaplatz 3, 69115 Heidelberg, Germany

Printed by Libri Plureos GmbH
in Hamburg, Germany